KU-535-334

Contents

Introduction

This book has been designed with you in mind. It has a dual purpose: to support your learning, and to be a useful reference book long after you have gained your qualification.

The book has been written to support the NVQ Level 3 Diploma in Plumbing and Heating. This qualification has been developed by SummitSkills in consultation with employers, the main awarding organisations and training providers. While awarding organisations will design their own assessment content, they will all follow the same unit structure and assessment strategy.

This book is designed to support:

- 6189 City and Guilds Level 3 NVQ Diploma in Plumbing and Heating, and
- 600/1252/3 EAL Level 3 NVQ Diploma in Domestic Plumbing and Heating.

These qualifications are approved on the Qualifications and Credit Framework (QCF), a government framework that regulates all vocational qualifications to make sure they are structured and titled consistently, and are quality-assured.

Who will take the qualification

The new Diploma qualification is aimed both at new entrants to the profession, such as apprentices or adults changing career, and at members of the existing workforce who are looking to improve their skills. It is intended to train and assess candidates so that they can be recognised as occupationally competent by the plumbing industry.

Learners will:

- gain the skills to work as a competent plumber at this high level
- achieve a qualification recognised by the Joint Industry Board (JIB) for professional grading to the industry
- complete an essential part of the SummitSkills Advanced Apprenticeship.

The Level 3 Diploma in Plumbing and Heating comprises six core units, all of which are mandatory. To gain the Diploma, you will need to complete these six core units plus your choice of one of four routes: gas, oil, solid fuels or environmental technologies.

The qualification is also aligned to the relevant registration bodies. Once you have completed all the mandatory units plus your chosen route, you can be registered with the registration bodies relevant to that route, without the need for further assessment. These bodies are:

- Water regulations – DEFRA (WRAS)
- Unvented hot water – Building Regulations/Standards

- Energy efficiency – Building Regulations/Standards (Part L1 of the Building Regulations in England and Wales)
- Gas – gas registration provider
- Oil – OFTEC
- Solid fuel – HETAS
- Electrical – limited scope Part P electrics
- Emerging technologies – MTC proposals (Competent Persons Schemes).

Your centre will be able to give you details of how to register once you have completed the qualification.

About this book

This book has been prepared by expert JTL trainers, who have many years of experience of training learners and delivering plumbing qualifications.

Each unit of the book relates to a particular unit of the Diploma and provides all the information needed to attain the required knowledge and understanding of that area. The content of each unit will underpin the various topics that your awarding organisation will assess you on.

Each unit has knowledge tests throughout, as well as a set of multiple-choice questions at the end, so that you can measure your knowledge and understanding as you go along.

Our intention is that this book will also be a useful reference tool for you in your professional life once you become a practicing plumber.

Reference feature

As you go through the book, you will find boxes in the margin that give references, such as 'WSR Schedule 2, paras 17–24'. These references indicate which standards and legislation are relevant to the subject you are reading about. If you wish to, you can then find out more about what that particular standard or piece of legislation says.

WSR Schedule 2, paras 17–24

The electrical content covered by Unit 7

Before the introduction of this Level 3 Diploma, the electrical unit, which was taken at Level 2, included an appreciation of electrical circuits, earthing circuits, identification of components, and the safe isolation of appliances. Even if you had completed it, you would not have been considered competent to install and test electrical systems and components, as you would not have undertaken your limited scope Part P of the Building Regulations.

Because of the introduction of more controls within the plumbing industry, employers wanted training to be modified to include the testing and maintenance of controls, so that limited scope Part P could be aligned to the new Level 3 Diploma.

The electrical content in this book covers the information you will need as a plumber to work within the Competent Person's Scheme as a limited scope Part P plumber. The competencies only cover electrically operated mechanical services components and controls up to 230V single-phase supply. Having the qualification will not mean that you are considered competent to test a complete electrical installation.

In places, material is cross-referenced between Unit 7 and other units, to avoid repetition. This means that, at various stages in your course, you will need to read Unit 7 in conjunction with other units. The cross-referencing is clearly marked in the text.

Specific installations – note

It is important to note that this book is intended for training purposes, rather than as a guide to specific installations. Always refer to the relevant British Standards or manufacturer's data when designing a plumbing installation.

Features of this book

This book has been fully illustrated with artworks and photographs. These will help to give you more information about a concept or a procedure, as well as helping you to follow a step-by-step procedure or identify a particular tool or material. Within the index, key terms are emboldened – this is so you can find the definitions easily.

This book also contains a number of different features to help your learning and development.

Key term

These are essential technical words. They are picked out in **bold** in the text and then defined in the margin.

Remember

This highlights key facts or concepts, sometimes from earlier in the text, to remind you of important things you will need to think about.

Did you know?

This feature gives you interesting facts about the plumbing trade.

Safety tip

This feature gives you guidance for working safely within the profession.

Professional Practice

This feature gives you a chance to read about and debate a real-life work scenario or problem. Why has the situation occurred? What would you do?

Activity

These are short activities and research opportunities, designed to help you gain further information about, and understanding of, a topic area.

Progress check

These are a series of short questions, usually appearing at the end of each learning outcome, which gives you the opportunity to check and revise your knowledge. Answers to the questions are supplied on the Training Resource disk.

Preparation for assessment

This feature provides guidance for preparing for the practical assessment. It will give you advice on using the theory you have learnt about in a practical way.

Check your knowledge

This is a series of multiple-choice questions at the end of each unit, referencing to Pearson Level 2 resource. Answers to the questions are supplied on the Training Resource disk.

References are given in the text to Level 2 NVQ Diploma Plumbing (978-0-435031-31-2) also published by Pearson.

Acknowledgments

Pearson and JTL would like to thank Barry Spick and Shaun O'Malley for their hard work and dedication in writing Units 1–6 and reviewing all the material for Unit 7. They would also like to thank Dave Allan for writing Unit 7.

Pearson would like to thank Sarah Christopher for her detailed development work on the manuscript and Stephen Blair, Keith Powell and Jim Dawson for their thorough review of the material.

Barry Spick would like to thank Hull College for giving him the opportunity to take part in writing this book. He would also like to thank his colleagues for their invaluable help and encouragement when finding information for the book.

Shaun O'Malley would like to thank his wife and family (Sue, Emma, Laura, Rebecca) for their help and patience. He would like to thank Kendal College for their assistance and help.

The publisher would like to thank the following for their kind permission to reproduce their photographs:

(Key: b-bottom; c-centre; l-left; r-right; t-top)

Alamy Images: Andrew Lincott 97, Andrew Twort 254tl, Art Directors & TRIP 312, Ashley Cooper 163, Backyard Productions 275, Corbis Bridge 170, Frank Irwin 274, GlowImages RM 251cr, Jean Schweitzer 278, LianeM 322cr, Macana 447l, Marshall ikonography 254cl, MediaColour's 322br, Peter jordan 223, Philip Traill 90; **Construction Photography:** David Potter 322cl, Gavin Wright 108b; **Corbis:** 96; **Dave Allen:** 447br; **Getty Images:** Flickr / Stephen G. de. Polo 159, Photolibrary / Ed Lalla 13; **Graham Hare:** 442; **Imagestate Media:** John Foxx Collection 16; **John Guest Ltd:** 39; **Masterfile UK Ltd:** Andrew Douglas 341; **Megger Limited:** 482t; **Pearson Education Ltd:** Clark Wiseman, Studio 8 23, Gareth Boden 15, 178b, 179tr, 179cl, 181,

182, 183, 273, 357l, 357r, 358, 361, 362, 397, 398t, 398b, 399t, 409, 413, 446bl, 446br, 449tl, 449tc, 449tr, Lord and Leverett 27, Jules Selmes 354; **Robert Harding World Imagery:** age fotostock 417; **Shutterstock.com:** Hamsterman 251br, Justin Kral 100, Kris Schmidt 95, LoopAll 410, Olivier le Queinec 47, Plumdesign 108t, ra3rn 399c, Skyline 213; **SuperStock:** Fancy Collection 1, imagebroker.net 269; **Wavin Limited:** 323tl, 323tc, 323tr, 323bc, 323br, 328, 332, 324 (all), 325 (all)

Cover images: *Front:* **Pearson Education Ltd:** Clark Wiseman, Studio 8

All other images © Pearson Education Ltd / Naki Photography

Every effort has been made to trace the copyright holders and we apologise in advance for any unintentional omissions. We would be pleased to insert the appropriate acknowledgement in any subsequent edition of this publication.

The following materials have been reproduced with kind permission from the following organisations: p39: from *The Push-Fit Solution for Plumbing and Heating Systems*, p.7, John Guest Speedfit Limited © copyright 2011; p54: from The Building Regulations 2000 (2010 Edition) Conservation of Fuel and Power - Approved Document L1A, Stationery Office, © Crown Copyright. p71: from *DOE Fundamentals Handbook Mechanical Science*, Volume 2 of 2, U.S. Department of Energy p.48, DOE-HDBK-1018/2-93, January 1993, source: U.S. Department of Energy, Washington, D.C. 20585; p106: from Hot and Cold Water Supply, 3rd edition, BSI, British Standards Institution (Robert H. Garrett, 2008), Figure 8.11, p.256. Reproduced with permission of Blackwell Publishing Ltd; p113: 'Surrey & York Flange Range', http://www.watermillshowers.co.uk, copyright © Grundfos Watermill Limited; p227, p228 and p239: from The Building Regulations 2000 (2002 Edition) Drainage and Waste Disposal - Approved Document H, Stationery Office, © Crown Copyright; p249: "Initial Planning/Suitability of your floor" and "Installing the shower drain", www.diywetroom.com, copyright © DIY Wetroom.com; p250 "Exploded view of trap", www.dallmer.de. Reproduced by permission of Dallmer Limited; p305: "Symptoms indicating that a system needs a power flush", and p306: "Survey sheet" adapted from Power flushing survey and check list, www.kamco.co.uk, copyright © Kamco Limited. Reproduced with permission; p313: from "Installation & Servicing Instructions, Baxi System 100 HE Plus. Wall Mounted Powered Flue Condensing Gas Fired Central Heating Boiler", p.46. Comp No 5110481 - Iss 3 - 03/05, copyright © Baxi Heating UK Ltd, 2008. www.baxi.co.uk; p320, 321, p324 and p325: from "Underfloor Heating The Hidden Advantage. An Introduction to Solutions from OSMA Underfloor Heating", pp.4,8,10,11 http://cms.esi.info/Media/documents/Wavin_underfloorheat_ML.pdf, copyright © Wavin Limited; p329 and p330: from "Additional guidance for underfloor heating systems on how to comply with 2010 Building Regulations Part L", *UHMA*, p.4 17/02/2011. Reproduced by permission of Beam and UHMA; p337: from "Thermoboard Composite Manifold Troubleshooting", www.plumbed-in.com, copyright © Thermoboard, Wavin Limited; p362: 'Combi boiler with room stat fitted', p363: 'Combi wiring with programmable room thermostat' and 'Combi wiring with wireless room thermostat and programmer', and p378: 'Pump overrun' adapted from Danfoss Randall Central heating Control Seminar, copyright © Danfoss Randall, http://danfoss-randall.co.uk; p374: 'Multi-zone control valve system' adapted from 'Multi-zone control valve system', copyright © Honeywell Control Systems Ltd; p375: 'Multi boiler installation' and 'Wiring installation for multi boiler installation', Remeha/Broag Boilers. Reproduced by permission of Sans Frontiere Marketing.

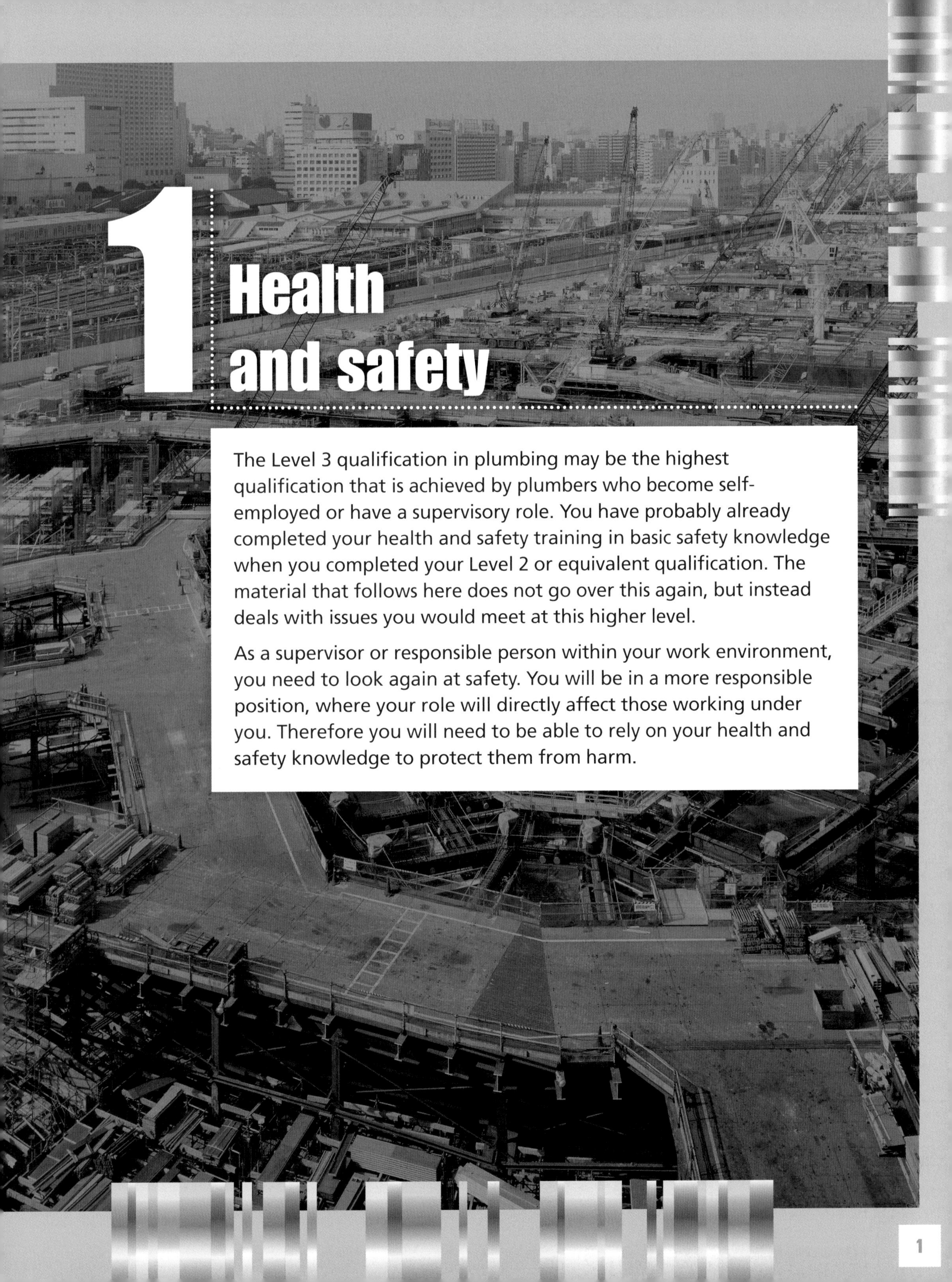

1 Health and safety

The Level 3 qualification in plumbing may be the highest qualification that is achieved by plumbers who become self-employed or have a supervisory role. You have probably already completed your health and safety training in basic safety knowledge when you completed your Level 2 or equivalent qualification. The material that follows here does not go over this again, but instead deals with issues you would meet at this higher level.

As a supervisor or responsible person within your work environment, you need to look again at safety. You will be in a more responsible position, where your role will directly affect those working under you. Therefore you will need to be able to rely on your health and safety knowledge to protect them from harm.

Did you know?

Under the Health and Safety at Work etc. Act 1974, it is a requirement to provide health and safety training to all staff.

The focus on health and safety in recent years, especially in the construction industry, has led to many improvements, but the number of accidents at work is still high (see Table 1.1). Data is available from the HSE website, via the Index of data tables.

Type	Description	2006/7	2007/8	2008/9	2009/10	2010/11p
Part 1 (Notifiable in relation to any place of work)						
1	Failure, collapse or overturning of lifting machinery, excavator, pile driving frame or mobile powered access platform	1023	1128	1046	853	820
2	The failure of any closed vessel including boiler or of any associated pipework, in which the internal pressure was above or below atmospheric pressure	119	79	97	93	124
3	The failure of any freight container in any of its load-bearing parts while it is being raised, lowered or suspended	12	6	10	4	4
4	Plant or equipment either comes into contact with overhead electric line in which the voltage exceeds 200 volts or causes an electrical discharge	124	130	120	79	77
5	Electrical short circuit which results in the stoppage of the plant for more than 24 hours	196	224	227	195	195
6	Unintentional ignition or explosion of explosives	123	106	65	94	255
7	The release or escape of a biological agent likely to cause human infection or illness	348	351	363	338	382
8	The malfunction of radiation generators	18	3	21	18	7
9	Failure of breathing apparatus in service	93	92	93	176	112
10	Failure of any lifting or life-support equipment during a diving operation which puts a diver at risk	31	31	30	33	34
11	Complete or partial collapse of scaffold over 5 m high	75	60	39	24	26
12	Any unintended collision of a train with any other train or vehicle (other than one recorded in part 4 of this table) which caused, or might have caused, the death of or major injury to any person	–	2	2	4	9
13	Incidents in relation to a well (other than a well sunk for the purposes of the abstraction of water)	40	56	73	28	36
14	Incidents in respect of a pipeline or pipeline works	348	344	256	221	185
15	Failure of fairground equipment in use or under test	16	14	14	10	10
16	Overturning or serious damage to a tank while conveying by road prescribed dangerous substances, or the uncontrolled release or fire involving the substance being conveyed	4	12	15	13	7
17	Uncontrolled release or escape of a dangerous substance, or a fire involving the dangerous substance, when being conveyed by road in a vehicle	6	7	9	6	7

Type	Description	2006/7	2007/8	2008/9	2009/10	2010/11p
Part 1 (Notifiable in relation to any place of work)						
18	Collapse or partial collapse of any building or structure under construction involving over 5 tonnes of materials or any floor or wall of a building used as a place of work	138	149	111	69	71
19	An explosion or fire occurring in any plant or premises which results in the stoppage of that plant for more than 24 hours	245	257	245	209	209
20	The sudden, uncontrolled release of flammable substances	303	258	180	204	203
21	The accidental release or escape of any substance in a quantity sufficient to cause the death, major injury or any other damage to the health of any person	706	705	735	738	701
Part 1 (Notifiable in relation to any place of work)		3968	4014	3751	3409	3474
Part 2 (Notifiable in relation to mines)		71	56	85	48	71
Part 3 (Notifiable in relation to quarries)		67	65	57	52	67
Part 4 (Notifiable in relation to railways)		5170	4803	3609	3578	3522
Part 5 (Notifiable in relation to offshore workplaces)		350	414	363	339	332
Total – reported dangerous occurences		9626	9352	7865	7426	7466

Dangerous occurrences reported to all enforcing authorities 2006/7 – 2010/11p

Notes: p = provisional

From 2008 a different system was adopted for the recording of RIDDOR incidents notifiable in relation to railways, which are reported to the Office of Rail Regulation (ORR). For more information see Chapter 10 of the following ORR publication: National Rail Trends.

Table 1.1: RIDDOR data on accidents at work

Anyone who employs others or is self-employed has a duty of care and needs to be aware of the legislation and procedures around health and safety.

Employer's liability

To make sure that claims are kept to a minimum, insurance companies are strict in ensuring that companies comply with health and safety legislation. An employer has to have employer's liability insurance under the Employers' Liability (Compulsory Insurance) Act 1969. This sets out the duty of employers to ensure and maintain insurance against any injury to their employees, up to a total of £2 000 000 per claim per employee.

If a reportable accident occurs, the Health and Safety Inspectors have specific powers to:

- enter premises, undertake investigations, take photos and ask questions
- serve an improvement notice on a person where there is a breach of regulations
- serve a prohibition notice on persons controlling activities

Activity

Find out how much it costs to take out liability insurance for your firm or a local business.

Safety tip

Make sure the accident book at work is always up to date. You will be preventing the accidents of the future.

Key terms

Hazards – things with a potential to cause harm.

RIDDOR – Reporting of Injuries, Diseases and Dangerous Occurrences Regulations 1995.

- seize and destroy articles and substances that may cause personal injury
- give information to safety representatives regarding **hazards** and notices, and so on.

Reporting incidents

The seven online **RIDDOR** reporting forms are:

- F2508 Report of an Injury
- F2508 Report of a Dangerous Occurrence
- F2508A Report of a Case of Disease
- OIR9B Report of an Injury Offshore
- OIR9B Report of a Dangerous Occurrence Offshore
- F2508G1 Report of a Flammable Gas Incident
- F2508G2 Report of a Dangerous Gas Fitting.

Fatal and major injuries and incidents can still be reported to HSE's Incident Contact Centre by phone. If a fatal injury occurs on site, a prohibition notice may have to be served.

Professional Practice

An HSE inspector is called to a site where a fatal accident has occurred. He notices that the fatal fall from height was caused by improper installation of scaffolding. Which notice does the inspector fill in: a prohibition notice or an improvement notice?

The main thrust however needs to be on accident prevention rather than prosecution. To this end any supervisor, employer or self-employed person needs to consider other pieces of legislation apart from the Health and Safety at Work etc. Act 1974. Here are some of the most important areas.

Construction (Design and Management) Regulations 2007

The Construction (Design and Management) Regulations 2007 (CDM 2007) came into force in Great Britain on 6 April 2007. The CDM 2007 Regulations are divided into five parts.

- Part 1 deals with the application of the regulations and definitions.
- Part 2 covers general duties that apply to construction projects.
- Part 3 contains additional duties that apply to notifiable construction projects: those lasting more than 30 days or involving more than 500 person-days of construction work.
- Part 4 contains practical requirements that apply to construction sites.
- Part 5 contains the transitional arrangements and revocations.

The Construction (Design and Management) Regulations 2007 (CDM) can help you to:

- improve health and safety in your industry
- have the right people for the right job at the right time to manage the risks on site
- focus on effective planning and managing the risk – not on the paperwork.

Everyone controlling site work has health and safety responsibilities. Checking that working conditions are healthy and safe before work begins, and ensuring that the proposed work is not going to put others at risk, require planning and organisation. This applies whatever the size of the site.

CDM 2007 places legal duties on virtually everyone involved in construction work. Those with legal duties, commonly known as 'dutyholders', are:

- clients
- CDM co-ordinators
- designers
- principal contractors
- contractors
- workers.

The client on a domestic project has no legal requirements to appoint a CDM co-ordinator (see page 8 of *Level 2 NVQ Diploma Plumbing*). However, other duties do apply to your domestic client and, if you are designing the building services in their property, you are responsible for informing the client of their duties. This involves all parts except Regulation 3, which they are exempt from. Those with duties under the CDM Regulations 2007 must satisfy themselves that the businesses they employ or appoint are competent. This means making reasonable checks that the organisation or individual is competent to do the relevant work and can also allocate adequate resources to it. The people taken on to do the work must also be sure that they are competent to carry out the required tasks before agreeing to take on the work. If you are asked to take on an installation of a solar panel and have not received training on it, you should turn down the work or book yourself on the next course *before* accepting the work (see Tables 1.2 to 1.4 for examples of acceptable levels of individual competence).

Remember

Part 3 of the CDM Regulations contains additional duties that only apply to notifiable construction projects.

Activity

What is public liability? Do some research to find out.

Individual competence

Tables 1.2 to 1.4 give details of what the HSE expects from workers at different levels in terms of their individual competence. This information is taken from the HSC's (Health and Safety Commission) *Managing health and safety in construction*, which is available to download free on the HSE website.

Safety tip

As a supervisor, as well as encouraging people to learn new skills, you can help them spot the things they cannot yet do – and should not do, if they are to stay safe.

Trainee	Description	Example of attainment
Risk control knowledge	Adequate knowledge of tasks to be undertaken; understands what is expected and when to ask for help; understands role and importance of supervisor; can identify key risks of activities; knows how to react to basic risks; knows main health hazards and why PPE is important	CITB-CS Health and Safety Test (CSCS) or CCNSG Certificate or equivalent recognised passport training S/NVQ Level 1
Experience and ability	From no experience; has physical capability to carry out duties; minimum standard of language skills; can identify deteriorating conditions that may lead to increased risk; is aware of personal responsibility for self and others; is aware of what constitutes a good attitude	Attends site induction; attends mandatory in-house training; works safely to agreed standard under supervision; demonstrates safe behaviour and wears appropriate PPE at all times

Table 1.2: Individual competence for a trainee

Site worker	Description	Example of attainment
Risk control knowledge	As for trainee, plus: knows standards of health and safety required for site operations; can identify all foreseeable risks arising from their work activity and knows what actions to take to control these risks; can apply existing knowledge to new circumstances	As for trainee, plus: S/NVQ Level 2 or 3
Experience and ability	As for trainee, plus: consistently works to agreed standards of health and safety; quickly identifies defects and unacceptable risks; demonstrates good attitude and example at work; capable of working safely with minimal supervision	As for trainee, plus: commensurate with Level 2 achievement; plays full role in site consultation; demonstrates ability to report unsafe conditions to supervisor; demonstrates motivation to learn

Table 1.3: Individual competence for a site worker

Supervisor	Description	Example of attainment
Risk control knowledge	As for site worker plus: knows how to lead in identifying remedial actions to mitigate risk in all foreseeable circumstances; understands implications of own decisions on others; knows when to ask for specialist help	As for site worker plus: S/NVQ Level 3. Knowledge of supervision equivalent to CITB-CS 2-day supervisors' course, NEBOSH certificate, etc.
Experience and ability	Able to identify causes of problems and to deploy resources to solve problems on own initiative; demonstrates leadership skills, appropriate communication strategies; can read plans, think through problems and is flexible to adapt to changing circumstances	3–5 years' experience of this operation; trained and qualified to a level where able to describe risks of the range of work activities they are responsible for; capable of identifying remote risks, and anticipating problems of change

Table 1.4: Individual competence for a supervisor

From Table 1.4 you can see that, as a supervisor, you need to have achieved a Level 3 qualification, through your training centre/college and the experience you have gained at work. It is essential that you realise that those under your supervision may not be as experienced as you, especially when it comes to assessing risks, so it is part of your duties to keep them out of harm's way.

Activity

Find out what qualifications your employer has in health and safety.

Duties of different roles

Knowing who does what – and who is responsible for what – is vital for your team to work efficiently and safely. The roles of designer and contractor are especially important for you as a Level 3 plumber, because you may be taking on these duties yourself.

Designer duties

Imagine that you have accepted a job that requires you to design a new bathroom for the client. Your role at the start of the project is that of designer and because of this you have specific responsibilities under CDM.

The first thing that designers need to do is eliminate hazards from their designs so far as is reasonably practicable, taking account of other design considerations. Examples would be to design plant that will need regular maintenance at ground level, so there is no need for work at height, or providing permanent safe access for work at height. Eliminating hazards removes the associated risk; it is the best option, and should always be the first choice.

This may not be as easy as it sounds. There are many factors to consider, and things can go wrong, so you have to think of safety as an aspect of the project in its entirety, from start to finish.

Remember

When selecting the best system for the client, you also need also to look at how safe it is to install in that particular building.

Professional Practice

A client has asked you to design a bathroom and wishes to have a wood-burning stove added to an existing cistern-fed indirect hot water system, which in turn is heated by a gas boiler. Make a list of safety factors that you will need to design into the job.

Remember

Accidents don't just happen – they are caused.

Contractor duties

Once the job is designed and agreed **you** become the contractor and you now have specific duties to perform. Contractors and those actually doing the building services work, are the most at risk of injury and ill health. They have a key role to play, in planning and managing the work to ensure that risks are adequately controlled.

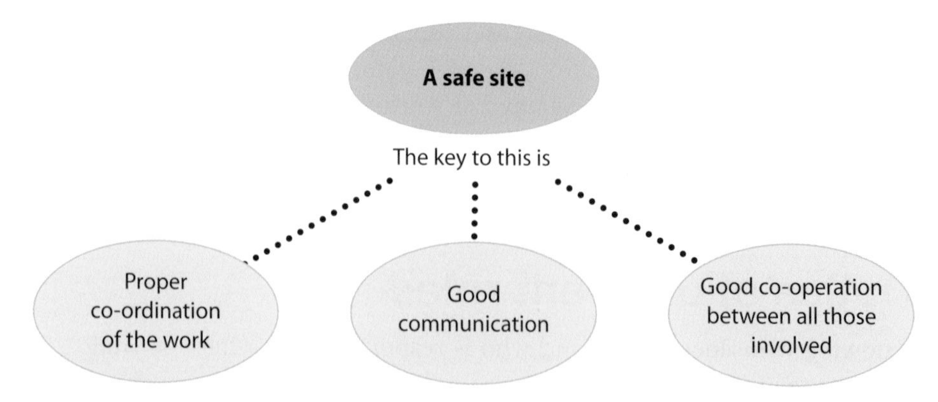

Figure 1.1: Key factors in making a site safe

All contractors and the self-employed nominated by the client have a part to play in ensuring that the site is a safe place to work.

For all projects, contractors must:

- check clients are aware of their duties
- satisfy themselves that they and anyone they employ are competent and adequately resourced
- plan, manage and monitor their own work to make sure that workers under their control are safe from the start of their work on site
- ensure that any contractor who they appoint or engage to work on the project is informed of the minimum amount of time which will be allowed for them to plan and prepare before starting work on site
- provide workers under their control (whether employed or self-employed) with any necessary information, including about relevant aspects of other contractors' work, and site induction (where not provided by a principal contractor) which they need to work safely, to report problems or to respond appropriately in an emergency
- ensure that any design work they do complies with CDM Regulation 11
- comply with any requirements listed in Schedule 2 and Part 4 of the CDM Regulations.

(From The Construction (Design and Management) Regulations 2007)

CDM Schedule 2 and Part 4

To do this the contractor should make a written plan. One of the key points is when other trades will be needed on site, such as an electrician or plasterer, and how are you going to ensure they are aware of hazards, such as lifted floorboards. How will you be aware of the risks in the first place, and what will the level of risk be?

Professional Practice

Shaun, a supervisor for a large plumbing company, has been sent to a new refurbishment job to survey the house. The new boiler installation is to be put into the cellar of the house, and the pipework is to run through to the kitchen above. He notices that access to the cellar is through a narrow stairway in poor repair, the cellar has inadequate lighting and the walls are damp.

1 What are Shaun's duties under the CDM Regulations?

2 What immediate actions should he take?

Risk assessment

Under the Management of Health and Safety at Work Regulations 1999 (MHSWR 1999), an employer has 'a duty to undertake suitable and sufficient assessment of risk to employees and others while undertaking his business'.

You will have covered risk assessment at Level 2 (see *Level 2 NVQ Diploma Plumbing*). However, at Level 3, assessing the risk to all the operatives on the job involves a lot more responsibility. The process of risk assessment and control requires careful thought and should follow the procedure shown in Table 1.5.

Classify work activities
Classify hazards
Identify risk control
Estimate risk
Determine the tolerability of risk
Prepare risk control plan to improve risk control
Review adequacy of action plan – confirm whether risks are now tolerable or not
Ensure risk assessment and controls are effective and up to date

Table 1.5: Risk assessment stages (taken from BS 8800FIG E1 page 41)

MHSWR 1999

Did you know?

'Tolerable risk' means that the risk has been reduced to the lowest level that is reasonably practicable.

A lot of accidents are due to human mistakes: it is estimated that 80 per cent of accidents are due to people's actions or omissions. The HSE divides human failures into two groups: errors or violations.

There are a number of factors that the self-employed or managers have to consider with human error:

- inadequate information: such as not aware of how equipment is used or a way of dealing safely with a situation
- lack of understanding: or in other words lack of communication so people make assumptions that can lead to errors
- inadequate design: a system of work that does not take into consideration human error is in itself a cause of accidents

- lapses of attention: due to the complexity of the task highly skilled people can make mistakes because they are concentrating on the specific task in front of them and are not aware of their surroundings
- mistaken actions: doing the wrong thing when thinking it is the correct thing to do
- misperceptions: often in a stressful situation competing information causes the person to develop tunnel vision so they continue to do a task wrongly until it leads to an accident
- mistaken priorities: when an individual has misunderstood safety as a priority within an organisation to be set aside to do a task as quickly as possible, e.g. working from a ladder rather than erecting scaffolding because of time constraints
- wilfulness: not following safety procedures because they see them as slowing them down and not being able to get on with the job, etc.

To eliminate the above risks a supervisor or self-employed worker will need to consider them and factor them into any risk assessments and procedures at the design stage.

For more on risk assessment, see Unit 2, pages 30–35.

Professional Practice

Jay, a self-employed plumber, asks his apprentice Paul to fix a pipe at high level and to use the long stepladders that are in the van. Paul sees a shorter set by the wall close to where the pipework needs to be installed. He uses the shorter step as it saves time, but he falls off when reaching too far and breaks his leg.

1. List what could be done to have prevented this accident.
2. What reports will Jay need to fill in?
3. Will Jay be liable under his liability insurance?

Remember

Stress can affect anyone – including your employer.

Stress at work

It has taken many years for stress at work to become a recognised health and safety problem. Factors recognised as contributing to occupational stress are:

- repetitive or monotonous work
- uncertainty
- unclear objectives
- interpersonal conflict
- inflexible and over-demanding schedules.

Factors outside work will also contribute to the stress levels of staff. Stress is not always bad for a person, but being overstressed can lead to illness and loss of work time, which can become an employment issue. The risks associated with stress also have to be considered when planning work activities, to help prevent accidents in the workplace.

The ability of a good employer to reduce stress for their employees and themselves will increase productivity and reduce risk in the workplace – but it is not an easy task. This has just been a taster of the requirements that you will need to successfully manage in the workplace. To find out more, there are many good books available on the subject of management and health and safety.

Check your knowledge

1. What legislation obliges employers to have employer's liability insurance?
   - **a** Employers' Liability (Compulsory Insurance) Act 1969
   - **b** Health and Safety at Work etc Act 1974
   - **c** Employers' Responsibility Act 2005
   - **d** Race Relations Act

2. When did The Construction (Design and Management) Regulations 2007 (CDM 2007) come into force in Great Britain?
   - **a** 1 April 1961
   - **b** 6 April 2007
   - **c** 21 October 2007
   - **d** 1 October 2007

3. The CDM 2007 Regulations are divided into five parts. Which one does *not* include domestic sites?
   - **a** Part 1
   - **b** Part 2
   - **c** Part 3
   - **d** Part 9

4. Contractors have a key role to play in planning and managing the work to ensure that risks are:
   - **a** adequately controlled
   - **b** put in place
   - **c** installed
   - **d** moved to another part of the site.

5. For all projects, what do the CDM Regulations say that contractors must do?
   - **a** Check clients are aware of their duties
   - **b** Check that clients own the property
   - **c** Check that the client is a nice person
   - **d** Check that the client has a minimum Level 3 qualification

6. Under MHSWR, what must employers undertake suitable and sufficient assessment of?
   - **a** Pay for employees and others while undertaking their business
   - **b** Risk to employees and others while undertaking their business
   - **c** Transport of employees to and from the site
   - **d** New methods of installation practice

7. What is the estimated percentage of accidents that is due to human error?
   - **a** 20%
   - **b** 50%
   - **c** 90%
   - **d** 80%

8. Which of the following is not a factor that the self-employed or managers have to consider with regard to human error?
   - **a** Lack of understanding
   - **b** Pay
   - **c** Wilfulness
   - **d** Mistaken actions

9. Which of these factors is recognised as contributing to occupational stress?
   - **a** Too much money
   - **b** A large pension
   - **c** Early retirement
   - **d** Interpersonal conflict

2 Understand how to organise resources within building services engineering

Before reading this unit, it would be advisable to review Level 2 Unit 2 *Understand how to communicate with others within building services engineering*. As a Level 3 qualified plumber, you will at some point in your career have to organise your workforce, especially if you own the company. You will need to be able to communicate with a range of different management professionals on the site, such as the architect and clerk of works, and to deal with customers from domestic contracts.

This unit covers the following learning outcomes:

- know the responsibilities of relevant people in the building services industry
- know how to oversee building services work
- know how to produce risk assessments and method statements for the building services industry
- know how to plan work programmes for work tasks in the building services industry.

Professional Practice

Lucas has been given the responsibility of organising personnel, plant and materials for a single dwelling contract. During this unit the Professional Practice features will give you tasks to complete as if you had the same responsibility as Lucas.

1. Know the responsibilities of relevant people in the building services industry

Types of client

Key terms

Private customer – a customer who arranges and pays for work themselves.

Contracting customer – a customer who employs a specialist business to carry out work.

Internal customer – a customer from within the same company.

As an employee of a business, the first step is to recognise who your customers are. You will encounter **private**, **contracting** and **internal customers**.

A **private customer** is where your company will have been invited to do a contract by the customer, either directly or through a customer representative, such as a managing agent. Usually the customer will not have any technical knowledge of the work to be carried out, and will put their trust in you and your company. However, the customer will have certain expectations of the work. Trying to meet their expectations before they need to state them is the start point for good customer care.

A **contracting customer** occurs when your company does contract work for organisations such as property developers, housing associations or local authorities. Do not assume that the customer is different for this type of work – your customer is the organisation. That organisation will have various staff representatives who take the lead in running the contract, such as a site agent or clerk of works. These people can be thought of as your frontline customers. They too will have expectations of what they need from you and your company, which may not be so different from those of a private householder. However, the customer's representative may have an in-depth technical knowledge of the service you are providing.

Finally, an **internal customer** is found when you work for a plumbing company that is part of a larger building services or construction company. These would be known as internal customers. In this situation, it may be easy to forget customer care issues because the customer feels very distant from the work. Your customer is the parent company representative, who, in these days of competitive contracting, can go outside your company for services if customer care is lacking.

Communication

A great deal of plumbing work involves communication – the passing and receiving of information between one or more people. Poor communication can lead to dispute and disagreement on a contract, so your ability to communicate effectively with customers and fellow workers is important.

Communication and the passing of information takes place in a number of ways:

- by email
- in writing/by post
- by fax
- verbally
- by word of mouth, via a second party
- by phone
- visually.

Each of these methods of communication has advantages and disadvantages, as you can see from Table 2.1.

Method of communication	Advantages	Disadvantages
Email To: customer@waterlane.co.uk Cc: Subject: Drawing for estimate Attachments: Plumbing Drawing 2.jpg A Leak Plumbing and heating Dear Sir Find attached the proposed plumbing layout drawing for the installation contract at No.1 Water Lane, Waterford. Could you please reply at yor earliest opportunity if you accept the proposed layout at the above address. Regards A.Leek	• Widely used thanks to the internet and new phone technology	• Only works if all parties are online
Writing/by post	• Will be able to send large items and drawings for working on site	• Will not arrive until the next day, and then only if posted first class
Fax	• Good for sending drawing details, but limited to A4 size	• Only if the other party has a fax machine, which are not so common these days
Verbally face to face	• Can go into greater detail if there are any concerns with the contract • Will be able to take the other party to show the problem with the contract • Other party can put their concerns within the contract	• Can lead to disagreements with other party

Method of communication	Advantages	Disadvantages
Verbally via a second party	• Passing instructions on for work	• Second party may not pass on correct information leading to disagreements or incorrect installations
Phone	• Possibly the fastest when on site • Good for chasing up material deliveries • Most people now have a mobile phone	• May be in an area where there is no signal • Battery can go flat • Could cause accident if operative is not paying attention to work area
Visually	• Is graphic • Allows you to interpret and use body language	• Some operatives may not understand what is being passed on

Table 2.1: Advantages and disadvantages of different methods of communication

Communicating with clients

Communicating well with your clients will be key to the success of your work, and potentially of your business. However skilled you are, and whatever qualification level you have reached, your abilities can only be put to the best use if you can understand what your clients want, and explain how best they can achieve it.

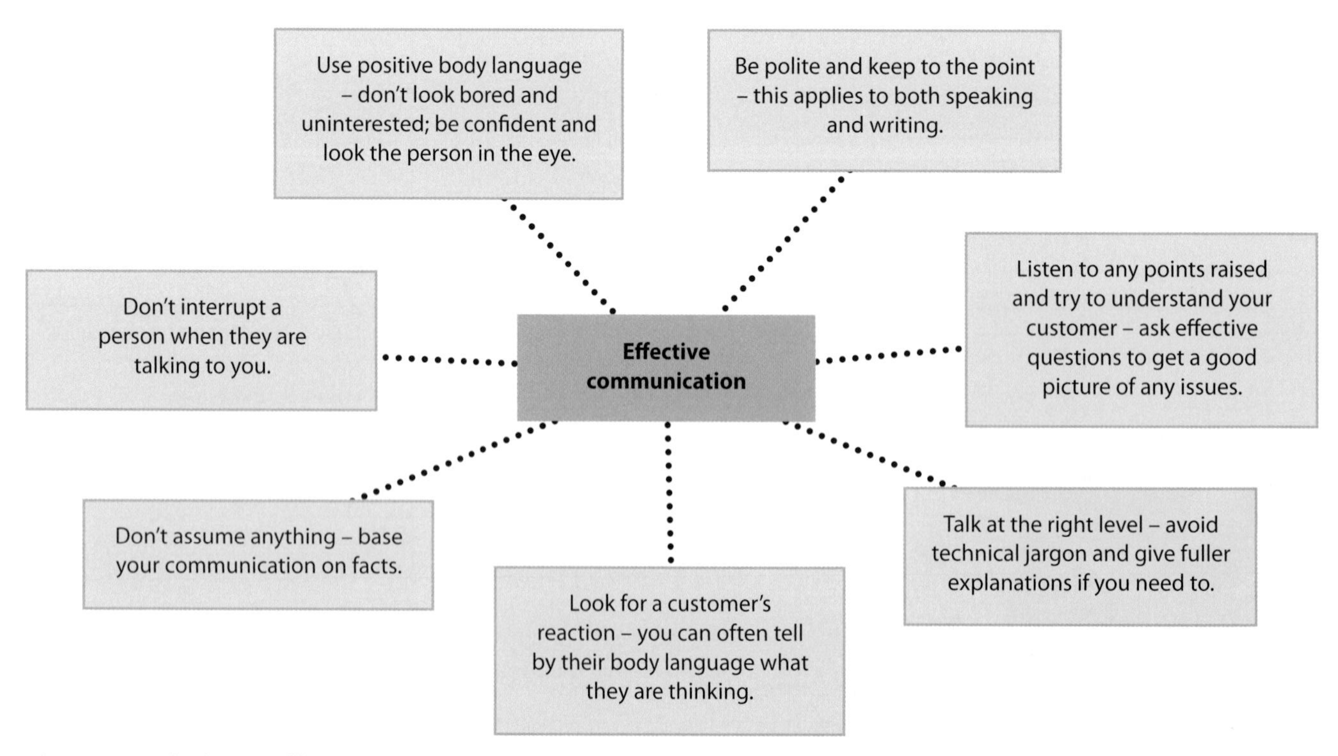

Use positive body language – don't look bored and uninterested; be confident and look the person in the eye.
Be polite and keep to the point – this applies to both speaking and writing.
Don't interrupt a person when they are talking to you.
Effective communication
Listen to any points raised and try to understand your customer – ask effective questions to get a good picture of any issues.
Don't assume anything – base your communication on facts.
Look for a customer's reaction – you can often tell by their body language what they are thinking.
Talk at the right level – avoid technical jargon and give fuller explanations if you need to.

Figure 2.1: The key to effective communication

The best way to determine what a client needs is to discuss the work with them. You will have catalogues for the client to look at, and can listen carefully to what the customer wants in terms of specifications, quality and the location of appliances and components.

You should consider any special needs, such as for people with disabilities, older people, or households with young children.

Figure 2.1 shows you some ways to ensure your communications with clients are as effective as possible.

Remember

A small error in communication can result in a big mistake – in terms of time, costs and reputation.

Communicating with the site management team

At some point on a construction site, whether you are the owner of a small company or you are running a contract for a larger company, you will have to communicate with different members of the site management team. The way you communicate with the team and with your own personnel is essential to the smooth running of the contract.

Generally, the best form of communication with the site management team is written, as this provides clear and permanent records that can be tracked throughout the contract. It is essential that you keep all evidence of discussions, decisions and agreements in the form of relevant emails, letters and variation orders, and that you log your telephone conversations, detailing when they took place, who they were with and what was discussed.

Here we look at some of the key roles in any site management team and how you can best handle your communications with each of them.

Architect

The architect's main role is to plan and design buildings. The range of their work varies widely and can include the design and procurement (buying) of new buildings, alteration and refurbishment of existing buildings and conservation work. The architect's work includes:

- meeting and negotiating with clients
- creating design solutions
- preparing detailed drawings and specifications
- obtaining planning permission and preparing legal documents
- choosing building materials
- planning and sometimes managing the building process
- liaising with the construction team
- inspecting work on site
- advising the client on their choice of contractor.

On larger contracts it would be unusual for you to communicate directly with the architect. The architect has a representative on large sites who is usually the clerk of works, but can also be known as the project manager. It is essential that you get written communication when the clerk of works requests any works to be done, as instructed by the architect. Ensure that you save all communication notices, whether emails, faxes or letters.

On a small contract such as an extension to a dwelling, it would still be unusual for you to communicate directly with the architect; you would normally have to use a third party such as the building contractor who would have direct contact with the architect. You should make sure that any verbal conversations or requests passed onto the third party for extra works or alterations are followed up with written confirmation from the architect.

Remember

Do not begin any extra works until you receive a written confirmation.

Quantity surveyor

The quantity surveyor can also be referred to as a cost planner or cost engineer. The term quantity surveyor derives from their role in quantifying the various resources that it takes to construct a given project.

The quantity surveyor:

- advises on and monitors the costs of a project
- allocates work to specialist subcontractors cost-effectively
- manages costs
- negotiates with the client's quantity surveyor on payments and final account
- arranges payments to subcontractors
- conducts feasibility studies to estimate materials, time and labour costs
- advises on a range of legal and contractual issues.

As the quantity surveyor has written the bill of quantities, you would need to inform them if any items have been missed from it or seemed to be incorrect – something that may not have been picked up during the **tendering** stage.

To be paid by the quantity surveyor for **interim payments**, you would have to submit an **invoice** for works completed since the last interim payment. For any works other than the works that are in the bill of quantities, the quantity surveyor would expect a written **variation order** from the architect confirming this. The variation order works in a similar way to an invoice, where everything is itemised, including labour costs for any work that is undertaken.

All your communications with the quantity surveyor need to be clear and detailed. Miscommunication about quantities and materials can be very costly.

Key terms

Tendering – a process where suppliers make offers for a contract.

Interim payment – a part-payment for the total contract, to a timing set during the contractual agreement stage.

Invoice – a written record of goods or services provided and the amount charged for them, sent to a customer or employer as a request for payment.

Variation order – written confirmation of any change or addition to the bill of quantities.

Quotation – an accurate working out of the cost, made as an offer of a fixed price for the contract, which cannot be changed once the customer/client has accepted it.

Buyer and estimator

The buyer can also be known as the procurement officer. The buyer:

- identifies suppliers of materials
- obtains **quotations** from suppliers
- purchases all the construction materials needed for a job
- negotiates on prices and delivery costs
- resolves quality or delivery problems
- liaises with other members of the construction team.

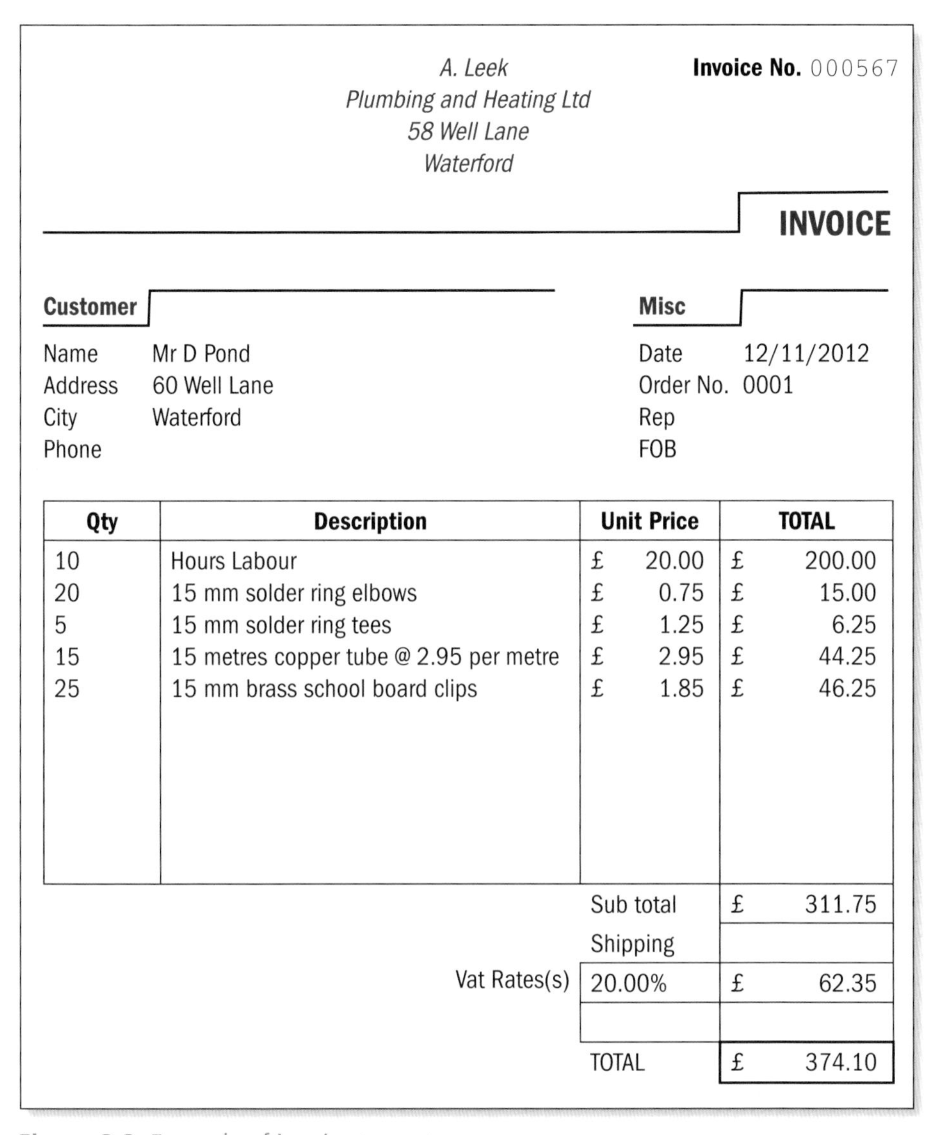

A. Leek
Plumbing and Heating Ltd
58 Well Lane
Waterford

Invoice No. 000567

INVOICE

Customer

Name Mr D Pond
Address 60 Well Lane
City Waterford
Phone

Misc

Date 12/11/2012
Order No. 0001
Rep
FOB

Qty	Description	Unit Price	TOTAL
10	Hours Labour	£ 20.00	£ 200.00
20	15 mm solder ring elbows	£ 0.75	£ 15.00
5	15 mm solder ring tees	£ 1.25	£ 6.25
15	15 metres copper tube @ 2.95 per metre	£ 2.95	£ 44.25
25	15 mm brass school board clips	£ 1.85	£ 46.25
		Sub total	£ 311.75
		Shipping	
	Vat Rates(s)	20.00%	£ 62.35
		TOTAL	£ 374.10

Figure 2.2: Example of invoice to customer

Estimator

The estimator:

- works out **estimates** for how much a project will cost including plant, materials and labour
- identifies the most cost-effective construction methods
- establishes costs for labour, plant, equipment and materials
- calculates cashflows and margins
- seeks clarification on contract issues affecting costs.

If you are to plan your workforce's activities successfully, good communication with the buyer and estimator is essential. The buyer is purchasing the materials that you are going to use, so will need a clear, written order for what you will need. The buyer is also responsible for delivery issues, so will need to keep you informed about delivery dates and any delays as they arise.

Key term

Estimate – an educated guess at the approximate cost of the contract, which is not binding.

Professional Practice

Lucas found an unforeseen problem when he went for a site visit to survey the dwelling. He discovered that he would be unable to install the pipework as drawn on the installation plan. What should Lucas do to overcome this problem? Should he:

- carry on regardless of the cost
- contact the site manager
- go direct to the architect?

Give reasons for your answer.

Surveyor

A building surveyor is a professional trained in understanding and interpreting building law. They are authorised to assess building plans with a view to ensuring they are compliant with the Building Regulations. In addition to having recognised qualifications, a building surveyor must be registered.

Building surveyors are involved in the maintenance, alteration, repair, refurbishment and restoration of existing buildings. A building surveyor's work includes:

- organising and carrying out structural surveys
- ensuring that buildings are safe
- ensuring that buildings are energy efficient
- legal work, including negotiating with local authorities
- preparing plans and specifications
- advising people on building matters such as conservation and insulation.

They will resolve any queries on the Building Regulations and allied legislation presented by staff or other persons on the contract.

Your communication with the surveyor could be regarding requirements of the Building Regulations for the installation of heating and hot water systems. You may wish to clarify some aspects of the Building Regulations or associated legislation before the contract starts.

Project manager/clerk of works

The project manager or clerk of works is the site representative of the architect, so they will want to make sure that you have installed your appliances/equipment as specified by the architect.

The project manager or clerk of works will keep records of:

- daily weather conditions
- plant and materials that have been delivered or removed from site
- any stoppages, including industrial disputes (strikes)
- official visitors to the site, such as the Building Control Officer
- the number of personnel on site, on a daily basis

- drawings received from the architect
- any variation orders passed on to contractors
- the progress of works undertaken for the week.

The project manager also prepares **snagging lists** of defects needing remedial action before sign-off at the end of the contract.

Your dealings with the project manager or clerk of works may involve trying to come to some agreement on the scope of works when undertaking the snagging. Using Figure 2.3 as an example, you can see that some of the snags that have been itemised may be the project manager's own personal opinion. You will have to prove that you have installed equipment and pipework to the contract specification and to British Standards, such as BS 8000: Part 15 for hot and cold water supplies. As one of the items in the snagging list is about clipping, you would have to prove that you have clipped to correct distances as specified in BS 8000: Part 15.

Key term

Snagging list – a list identifying small defects with the work that has been installed.

BS 8000: Part 15

Plumbing snagging list		
Plot/room	**Snag**	**Completed**
Kitchen	Sink plug missing	
	No handle on cold washing machine valve	
Cloakroom	Insufficient clips on hot and cold pipework	
	Cold water tap drips, staining wash basin	
	Flux and solder not cleaned off pipework	
Bedroom 1	Thermostatic valve loose	
Bedroom 2	Radiator loose	

Figure 2.3: Plumbing snagging list

Structural engineer

Structural engineers are involved in the structural design of buildings and structures such as bridges and viaducts. The primary role of the structural engineer is to ensure that these structures function safely. They can also be involved in the assessment of existing structures, perhaps for insurance claims, to advise on repair work or to analyse the viability of alterations and adaptations.

Communication with the structural engineer concerns safe installation of pipework, equipment and systems within the fabric of the building, such as:

- drilling through floor slabs for service pipework where holes have not been left out in the concrete when cast
- checking if the structure is capable of taking the weight of equipment.

If the structure were not strong enough to take the weight of equipment, it would be in the structural engineer's contract to come up with a solution to the problem. This could also involve the architect, building services engineer, contracts manager and construction manager.

Building services engineer

Building services engineers are responsible for designing, installing and maintaining water, heating, lighting, electrical, gas, communications and other mechanical services, such as lifts and escalators, in domestic, public, commercial and industrial buildings. A building services engineer's work includes:

- designing the services, mostly using computer-aided design packages
- planning, installing, maintaining and repairing services
- making detailed calculations and drawings.

Most building services engineers work for manufacturers, large construction companies, engineering consultants, architects' practices or local authorities. Their role often involves working with other professionals as part of a team on the design of buildings: for example, with architects, structural engineers and contractors. When working for a consultant, this job is mainly office-based at the design stage. Once construction starts, there will be site visits to liaise with the contractors installing the services. When working for a contractor, the building services engineer may oversee the job and even manage the workforce and is therefore likely to be site-based. When working for a services supplier, the role will require being involved in design, manufacture and installation and may involve spending a lot of time travelling between the office and various sites.

You would communicate with this person:

- where the installation as drawn on the installation plan is not possible; the building services engineer would have to work out a viable alternative and may need your input to do so
- where an appliance specified in the contract specification cannot be sourced; the building services engineer would have to agree to an alternative, ensuring that the specification is the same as for the original appliance that was specified.

Contracts manager

The contracts manager:

- is responsible for running several contracts
- works closely with the construction management team
- acts as the link between the other sections of the business and the Managing Director or Chief Executive
- makes sure the job is running to cost and programme.

The contracts manager is not permanently on site during the contract, so communication must be through a third party, such as the construction manager or when the contracts manager attends site meetings.

Construction manager

The construction manager, also known as site manager, site agent or building manager, has the overall responsibility for the running of the contract (or for a section of a large project). The construction manager:

- develops a strategy for the project
- plans ahead to solve problems before they happen
- makes sure site and construction processes are carried out safely
- communicates with clients to report progress and seek further information
- motivates the workforce.

As the plumber in charge, you would communicate with the construction manager about progress, normally through site meetings, as well as reporting any delays to delivery of materials, which could stop the contract progressing. You would also need to talk to the construction manager to make sure the contract was adequately staffed to progress as per the **work programme**.

Key term

Work programme – the timetable for a construction project.

Typical site responsibilities for craft operatives

As your career progresses you will need to understand your level of responsibility. Most companies have different ways of dealing with this, so it is important that you find out what your company requires at each level.

Apprentice plumber

- Works directly with a qualified member of the staff.
- Should be given the necessary level of work instruction and supervision to undertake the work that is set for them.
- Not usually directly involved in communication with customers and co-contractors.
- Should pass all problems and work issues to supervisor.

You will need to understand your level of responsibility

Level 2 qualified plumber

- Does not usually take full responsibility for the contract.
- Normally works under minimal supervision.
- Responds to queries from customers and co-workers.
- Should forward requests for additional work to supervisor for action.

Level 3 qualified plumber

- Usually takes full responsibility for the contract including dealing with queries from customers and co-contractors.
- Responsible for dealing with requests for additional work, confirmation of the work and pricing.
- Deals with complaints from customers.

Professional Practice

Lucas has been informed by the customer that the operative on the contract has not carried out the installation work to the promised standard. How should Lucas deal with this complaint? Should he:

- ignore the complaint
- visit the contract to see for himself
- send a third party to check the installation
- send the operative back to correct it without a visit
- visit the job and then send a different operative?

Activity

Look at the responsibilities you have in your current role. How confident do you feel? What sorts of skills do you need to develop to improve your performance as a supervisor?

Supervising individuals

Apprentices are under the supervision of a plumber qualified to Level 2 and also the Level 3 plumber. The plumber will have the responsibility of guiding the apprentice in all aspects of plumbing work. They will have to ensure that the apprentice is working safely and, if working on a construction site, abiding by the site rules.

The apprentice may be attached to one plumber for the duration of the contract and that plumber would be responsible for the well-being of the apprentice. They would also be responsible for the on-site training, which should always be done by the company.

Job responsibilities when supervising staff

The job supervisor, who would be a Level 3 qualified plumber, would monitor the on-site personnel, ensuring that progress is as per the work programme. If the work progress falls behind, this could result in more labour being required to complete the contract so impacting on profits. The job supervisor is also responsible for understanding the competence levels of the people they are supervising, making sure that they are only given tasks that are within their abilities. Responsibility for health and safety also comes under the supervisor's role, so as a supervisor you would need to ensure those working under you had the correct health and safety training, and that this was regularly refreshed and kept fully up to date.

It would be the supervisor's responsibility to report back to the senior management any problems that have arisen on site, such as:

- safety being compromised (for example, working off a ladder with no one footing the bottom)
- trade disputes
- discipline (for example, the operative being continually late for work or arriving at the contract site late)
- any dispute between client and contractor.

Where a dispute is about quality, it would be best practice to get a third party involved who is independent of both client and contractor, to establish the quality of the installation. This can often settle the dispute, especially with the private customer.

PROGRESS CHECK

1. What is the role of an architect?
2. What documentation is normally used to request payments?
3. What is an interim payment?
4. Who would produce a snagging list?
5. Who would an apprentice work with?

2. Know how to oversee building services work

Dealing with variations to works

On some contracts, even though you have made contingency plans for the unexpected, there may be times when you have to deal with variations to works. This section deals with the correct procedures to follow when that occurs.

Variation orders

A variation order is a contractually binding document (usually an A4 sheet in triplicate) that allows an architect or official company representative to make changes to the design, quality or quantity of the building and/or its components. Variation orders are usually associated with larger contracts, and are issued for:

- additions, omissions or substitution of any work
- alterations to the kind or standard of materials
- changes to the work programme
- restrictions imposed by the client, such as:
  - access to the site or parts of the site
  - limitations of working space
  - limitations of working hours
- a specific order of work.

The variation order will clearly state exactly what will change from the original specification, and will form the basis for claims for any additional costs or time. In practice, the variation order is usually issued by the architect's representative on site, the clerk of works.

On smaller contracts there will not be an architect, so you may have to deal with issues like this yourself, although it is usual practice for you to pass them on to your employer to deal with. The main point is that you should not commit to doing any additional work without confirmation from the customer, which must be in writing. In the event that your employer or supervisor is not able to deal with the matter, you should get confirmation of the work requested, preferably on company letterhead, including the customer's signature.

Architect's Instruction

Issued by: Ivor Kingston Associates — Job Reference: IK/AL/001

Address: Kingston Road

Employer: Mr A Waterside

Address: 60 Well Lane — **Variation Order No:** 001

Contractor: A.Leak Plumbing and Heating — **Issue date**: 11th September 2012

Address: 58 Well Lane

Works: New dwelling — Sheet 1 of 1

Situated at: 60 Well Lane

Under the terms of the above mentioned contract, I/we issue the following instructions

	Office use: Approximate costs	
	£ omit	£ add
1. Re-route cold water supply pipework services	0.00	200.00
2. Re-route hot water distribution pipework services	0.00	200.00
1) Approximate totals	0.00	400.00
2) Signed: *Ivor Kingston*		

Figure 2.4: Example of a variation order

On most contracts large or small there could be a variation to the work that you will be involved with. The causes for these variations can be due to:

- an obstruction in the work environment that has not been identified on a drawing
- the client wishing to change the position of appliances and/or equipment.

How should the problem with the client be approached?

Professional Practice

The client for Lucas's job has requested that the architect change the layout of a first-floor construction. This will result in Lucas having to change pipework runs significantly.

1 What paperwork would Lucas need to request, and from whom?
2 What different changes might Lucas need to allow for?

Key term

Daywork rate – a rate of pay per hour of work agreed before the start of the contract.

Did you know?

Written records can be used as proof of an agreement between two or more parties.

Communication is the key to this. When communicating with the client make sure that you have got your facts correct and the reason why you would not be able to proceed as specified. You can make your initial communication with the client verbally and make sure that you follow up with a written explanation.

You will need to come to some agreement about the extra time and materials that would be involved in making the alteration.

Labour costs would also have to be taken into consideration; on larger contracts **daywork rates** are agreed before the contract begins.

You may have to give the client an estimate for the work to be undertaken.

The level of responsibility that you have may well be different from business to business. For example, in a company specialising in service and maintenance you may be expected to give a price for the contract and collect the money before you leave.

It is common for a customer to require a reasonable estimate of the cost before work begins. Your company should have procedures in place for you to get the price of any parts. They will also have a standard cost of labour calculation.

Many companies make use of a Personal Digital Assistant (PDA). These can hold pre-set pricing information for various types of task, including:

- labour cost, usually at a quarter-hour rate
- material cost
- set rates for a given task, such as changing a float valve.

Using a PDA to get costing information

PROGRESS CHECK

1. Who would usually issue a variation order?
2. What is a PDA?
3. If any alterations are undertaken without a variation order, what could happen?

Professional Practice

Lucas needs to invoice the client for the extra works that have been completed. Using computer software, put together an invoice for the work that has been completed for the example variation order shown in Figure 2.4.

Monitoring progress against the work programme

During the length of the contract, whether it is a small domestic contract or a larger multi-dwelling contract, you will have to supervise the work's progress. The factors that you will need to monitor are safety, cost-effectiveness and quality.

Safety

According to the Health and Safety Executive, during 2010/11 in the construction industry there were 50 fatal injuries, 2250 major injuries, 4784 reported over-three-day injuries and two fatal injuries to members of the public. Even though the general trend of injuries has gone down, they are still a major contributor to sickness from work in our industry.

Your first responsibility is to ensure the safety of personnel, customers and the general public by making sure that all site and company safety policies are being adhered to. This could be done by:

- unannounced visits to site
- **toolbox talks**
- having method statements in place
- having the supervisor keep a safety log
- minor occurrences being logged
- collecting photographic evidence.

Key term

Toolbox talk – a short presentation to the workforce on a single aspect of health and safety.

Cost-effectiveness

You will need to monitor your costs against the bill of quantities and the price that has been given to the client. Factors you will need to look at include:

- how many personnel you have on the contract
- whether you are purchasing materials from suppliers who give a good discount
- whether you can take equipment off hire.

Quality

Throughout all work procedures the quality of the product being installed should be to an acceptable level. You will be monitoring quality against cost-effectiveness. It is essential that the quality of the product does not drop, even when costs are going up.

Quality can be checked by:

- spot checks
- giving the operative a snagging list of work to do if the work is not up to standard.

Dealing with deficiencies in work performance

Monitoring progress in these areas is only really of use if you are going to do something about any deficiencies you find. As a Level 3 plumber, you will need to be able to take action to put problems right and learn how to improve performance.

Safety

Operatives need to be constantly reminded that they should always put safety first and foremost when working. This principle can be enforced by a weekly or daily toolbox talk about the company's or site's safety policies. The toolbox talk has been widely accepted as a practical way to raise workers' awareness of specific problems on site, to encourage consultation and to help remind them that health and safety is an important part of the working day.

Timekeeping

Operatives need to arrive and finish at the contract at the agreed working day times, as set by the company. Lateness and leaving the contract early can lengthen the contract, and will incur extra costs for the company; as the labour costs to a contract go up, so profits will come down. It would be the responsibility of the site operative to tackle this problem with the person concerned.

Cost-effectiveness

Operatives are expected to give a fair day's work for a fair day's pay, but some operatives do not fulfil this. If this is due to a lack of experience of the type of work undertaken, these operatives would need training to upskill them, and costs for the company would be incurred. However, an operative may just be someone who gets to the job late, who continually stops and leaves the work area, whose standard of work is not up to company and the contract standards, who leaves early and has longer than allowed breaks.

Quality

It is your responsibility to ensure that the quality of installations is to the highest standard. If during the contract your customer or the clerk of works is complaining about the standard of work, you will need to address this problem with the operative responsible. Some operatives may not be capable of working to the high standard that is expected from some companies, but others may need greater supervision or further instruction and training within the company in order to attain the level required of them.

Professional Practice

Lucas has been supervising operatives and has identified that several of them are not wearing the required PPE when drilling into brickwork. What should Lucas do to ensure that health and safety procedures are being used by everyone under his charge?

3. Know how to produce risk assessments and method statements for the building services industry

Key terms

Hazard – anything with the potential to cause harm, such as working at height on scaffolding.

Risk – the likelihood that a hazard will cause a specified harm to someone or something: for example, if there are no guard rails on the scaffolding, it is likely that a construction worker will fall.

As a Level 3 plumber, you should be able to:

- describe a **hazard**
- understand what a **risk** is
- assess the levels of risk for a particular work situation.

It will be your responsibility as the plumber or supervisor in charge to compile risk assessments for the contract that you are undertaking, as part of a process known as risk management.

Managing and assessing risk

Risk management is a process that involves assessing the risks that arise in your workplace, putting sensible health and safety measures in place to control them and then making sure they work in practice. These measures need to be 'reasonably practicable'. This means that you have to take action to control the health and safety risks in your workplace except where the cost of doing so (in terms of time and effort as well as money) is 'grossly disproportionate' to the reduction in the risk.

For example, on a job where you need to run pipework inside a building, you might decide to erect an independent scaffold, which would incur a great cost. A more reasonable (and more adaptable) option could be to use a mobile scaffold or scissor lift.

You can work this out for yourself, or simply apply accepted good practice.

What is a risk assessment?

A risk assessment is nothing more than a careful examination of what, in your work, could cause harm to people, so that you can weigh up whether you have taken enough precautions or should do more to prevent harm.

Levels of risk

In the context of risk assessment, you may come across the abbreviations ALARP and SFAIRP. ALARP stands for 'as low as reasonably practicable', while SFAIRP stands for 'so far as is reasonably practicable'. ALARP is the term used by risk practitioners while SFAIRP is the term most often used in the Health and Safety at Work etc. Act and in other health and safety regulations.

You might use ALARP when undertaking a risk assessment for guttering. You would need to reduce the level of risk as far as possible: working at height is always dangerous, no matter which method of access you use. In this case you would be right to use a mobile scaffold to reduce the risk, due to the cost of having a permanent scaffold erected.

You can use the tables on page 33 to determine the level of risk for a particular task.

Hazards presented by work situations

When compiling your risk assessment you will need to be able to identify the hazards that you may come across on a particular contract. These hazards could include:

- working at height, which is the most common cause of construction accidents
- soldering, including high-temperature brazing using oxyacetylene bottles
- using solvents, ensuring that there is ventilation available where the installation is being put in
- using specialist tools, such as a power threading machine, press-fit tool and hydraulic bender.

Activity

Find as many hazards associated with plumbing as you can think of. Which of these present the greatest risk?

Carrying out a risk assessment for a task

During your time working as a plumber, at some point you have probably found yourself thinking: 'That's risky.' As a Level 3 plumber, where you are the supervisor on the contract, you may be required to identify those risks and then you would be expected to write a risk assessment.

When you did your Level 2 qualification, you needed to understand about risk assessments and why you have them. Now as a Level 3 plumber, it is you who will be responsible for writing the risk assessment.

Activity

How many situations have you come across that would be classed as 'risky'? Make a list of these, giving as much detail as you can.

To write a proper risk assessment, you need to follow the Health and Safety Executive's 'Five Steps to Risk Assessment', which are now the accepted standard.

HSE's Five Steps to Risk Assessment

Step 1: Identify the hazards

Walk around the site and look to see what hazards there are. You will need to work out how people could be harmed.

Step 2: Decide who might be harmed and how

The best way of managing the risk is by being clear about who might be harmed and by which hazard.

Step 3: Evaluate the risks and decide on precautions

Now that you have spotted the risks, it is up to you to decide what to do about them. The law requires you to do everything 'reasonably practicable' to protect people from harm. You will need to consider these options:

- Can I get rid of the hazard altogether?
- If not, how can I control the risks so that harm is unlikely?

When you are controlling the risks, you need to apply the following principles:

- Try a less risky option.
- Prevent access to the hazard (for example, by guarding).
- Organise the work to reduce exposure to the hazard.
- Issue personal protective equipment (such as clothing, footwear or eye protection).
- Provide welfare facilities (such as appropriate first aid and washing facilities).

Step 4: Record your findings and implement them

When writing down your results, keep things simple: for example, 'Fume from lead welding: local exhaust ventilation used and regularly checked'.

Your risk assessment may not be perfect but it must be suitable and sufficient. You will need to show that:

- a proper check was made
- you asked all those who might be affected
- you dealt with all the significant hazards, taking into account the number of people who could be involved
- the precautions are reasonable
- the remaining risk is low
- you involved your staff or their representatives in the process.

Activity

Take one of the 'risky' work situations you thought of earlier and think about how you might apply the HSE's Five Steps to it.

Step 5: Review your risk assessment and update if necessary

Risk calculation formula

Risk factors are calculated using a simple formula:

Likelihood × Consequence = Risk

The outcome of the likelihood of an accident occurring and the maximum consequences should it happen will reveal a risk factor of between 1 and 25.

You can use Tables 2.2 and 2.3 to assess the risk for any tasks undertaken.

Likelihood of accident occurring	
Likelihood	**Scale value**
No likelihood	0
Very unlikely	1
Unlikely	2
Likely	3
Very likely	4
Certainty	5

Maximum consequences of an accident	
Injury or loss	**Scale value**
No injury or loss	0
Treated by first aid	1
Up to 3 days off work	2
Over 3 days off work	3
Specified major injury	4
Fatality	5

- A figure between 1 and 7 = minor risk, can be disregarded but closely monitored
- A figure between 8 and 15 = significant risk, requires immediate control measures
- A figure between 16 and 25 = critical risk, activity must cease until risk is reduced

Table 2.2 and 2.3: Likelihood and consequence ratings for risks

How to present a risk assessment

Figure 2.5 shows an example of a risk assessment for working with specialist tools. It is based on the model that has been developed by the British Plumbing Employers Council (BPEC).

The factors that are in the risk assessment are discussed in more detail below.

Risk exposure

Risk exposure describes the individuals or groups of people that may be affected by the work activity or process. Control measures must take account of all those people.

Safeguards hardware

Safeguards hardware describes the in-built safety features of work equipment. For example, on powered machines this would include machine guards or trip switches. In this particular example safeguards hardware is nil.

TASK
MANUAL HANDLING OF LOADS
SPECIALIST TOOLS

APPLICATION OF EQUIPMENT	**APPLICATION OF SUBSTANCES**
PIPE-BENDING MACHINES, STILSONS, ROPES, LEAD DRESSERS, BENDING SPRINGS, BLOCK AND TACKLE, SPANNERS, ETC.	N/A

ASSOCIATED HAZARD
RISK OF MUSCLE STRAINS
RISK OF SPRAINS
RISK OF MUSCULO-SKELETAL INJURY

LIKELIHOOD	**CONSEQUENCE**	**RISK FACTOR**
3	3	3 x 3 = 9

RISK EXPOSURE	**SAFEGUARDS**
EMPLOYEES	NIL

CONTROL MEASURES
1 SPECIFIC TRAINING AND INSTRUCTION TO EMPLOYEES – KINETIC LIFTING
2 INDIVIDUAL ASSESSMENT TO BE PERFORMED FOR ALL TASKS
3 WORKPLACE INSPECTIONS CONDUCTED AT 3-MONTHLY INTERVALS
4 RANDOM SAFETY INSPECTIONS
5 SUITABLE AND SUFFICIENT PERSONAL PROTECTIVE EQUIPMENT
6 MEDICAL SCREENING FOR STAFF AT RISK

Figure 2.5: An example risk assessment

Control measures

These describe the additional safeguards that underpin your arrangements. Where these are identified they must be followed through, and a record kept of any outcomes, etc.

The list below shows some of the most common risks you will face during your plumbing career.

Professional Practice

Lucas's employer has informed him that the contract requires a risk assessment for working at height on a mobile scaffold. Consider the requirements and write the risk assessment for this task using the BPEC format.

Producing a method statement for areas of work with safety risk

A method statement – sometimes called a 'safe system of work' – is a document detailing how a particular process will be carried out.

When producing a method statement, the first task is to carry out a risk assessment. The method statement should outline the hazards involved and include a step-by-step guide to how to proceed safely, given the hazards and levels of risk involved. The method statement must also detail which control measures have been introduced to ensure the safety of anyone who is affected by the task or process.

Information to be included

A method statement is commonly used to describe how construction or installation works can be carried out safely. Your method statement should include:

- background details of your company
- the site address and an overview of the project
- details of the type of activity the method statement is for.

In the next section of the method statement, you identify the hazards that are associated with that task (there can be as few or as many of these as are required, but your list should be as comprehensive as you can make it).

You should also detail the possible dangers and risks associated with your particular part of the project and the methods of control to be established, to show how the work will be managed safely.

Presentation of a method statement

Figure 2.6 shows a completed method statement for the removal of a cistern and WC.

Background Information	
Company details	A Leak Ltd 58 Pump Lane, Tel 555678 Fax 555670 Email aleak@aservice.co.uk
Site address	Contact name Mrs A User Address 60 Pump Lane Contact No 555789
Activity-Risk	To access and work safely during the replacement of a WC cistern and pan
Implementation and Control of Risk	
Hazardous Task-Risk	**Method of control**
Access roof space	Where access to the roof space is required use a suitably secured stepladder or loft ladder of the correct height for the task
Access working area	Access to work area to be kept clear and free from obstructions. Electric lead lights will illuminate all access and work areas
Replacement of existing cistern and WC	Pipework to cistern will be isolated prior to cutting. The cistern and WC will not be broken under any circumstance. Pipework and drainage to be temporarily capped
Use of blowlamp – hot soldering	All pipe lagging within 500 mm is to be removed. A suitable fire extinguisher will be adjacent to the working area
Removal of cistern and WC	The cistern and WC will be carefully removed from dwelling. Water in WC trap to be emptied before removal from dwelling
Safe deposit of cistern and toilet pan	Cistern and toilet pan to be removed and skipped at company workshops
Site Control	
Inspection of equipment	All equipment such as stepladder, ladders, and blowlamp shall be regularly inspected before commencement of work. Ensure electrical equipment has a current PAT test certificate
Customer awareness	The customer will be notified of all potential dangers throughout the contract. The customer will be notified of any delays to the contract
Protection of customer floor coverings	Ensure that all carpets are protected where access is required with dust sheets

Figure 2.6: Method statement for the removal of a cistern and WC

PROGRESS CHECK

1. How is a risk level calculated?
2. What is the purpose of a method statement?
3. Which process should be followed when putting together a risk assessment?
4. Where would information regarding risk management be found?
5. What is a hazard?

4. Know how to plan work programmes for work tasks in the building services industry

Types of work programme

Different types of job will require different types of work programme. More complex jobs always need more careful planning and documentation. Simpler tasks may be easy to organise, but there may still be benefits from drawing up a short work programme, for you and your customer.

Private installation work

The work programme for this would normally be inside your head, based on an agreed start and finish date with the customer. You will need to gauge whether a more formal, written work programme will be of benefit.

Private service/maintenance work

Work like this would usually be set out on a contract sheet provided by the employer. This would give all the details of the contract along with the date and time to start the work.

New-build installation contract work

On larger contracts involving more dwellings the approach is more scientific, and a contract programme will be provided. This could consist of an overall programme for all site trades, as well as separate programmes for each trade, including plumbing. These sorts of work programme can be produced using project-planning software, which will produce a Gantt chart (similar to a bar chart, but showing timings and interdependencies too).

Figure 2.7 shows an example of a programme for a plumbing installation.

Service or maintenance contract work

On large service or maintenance contract work, you would normally follow a work programme. This is particularly important for preventative maintenance that follows a standard company procedure, which would be set out in a maintenance or service document.

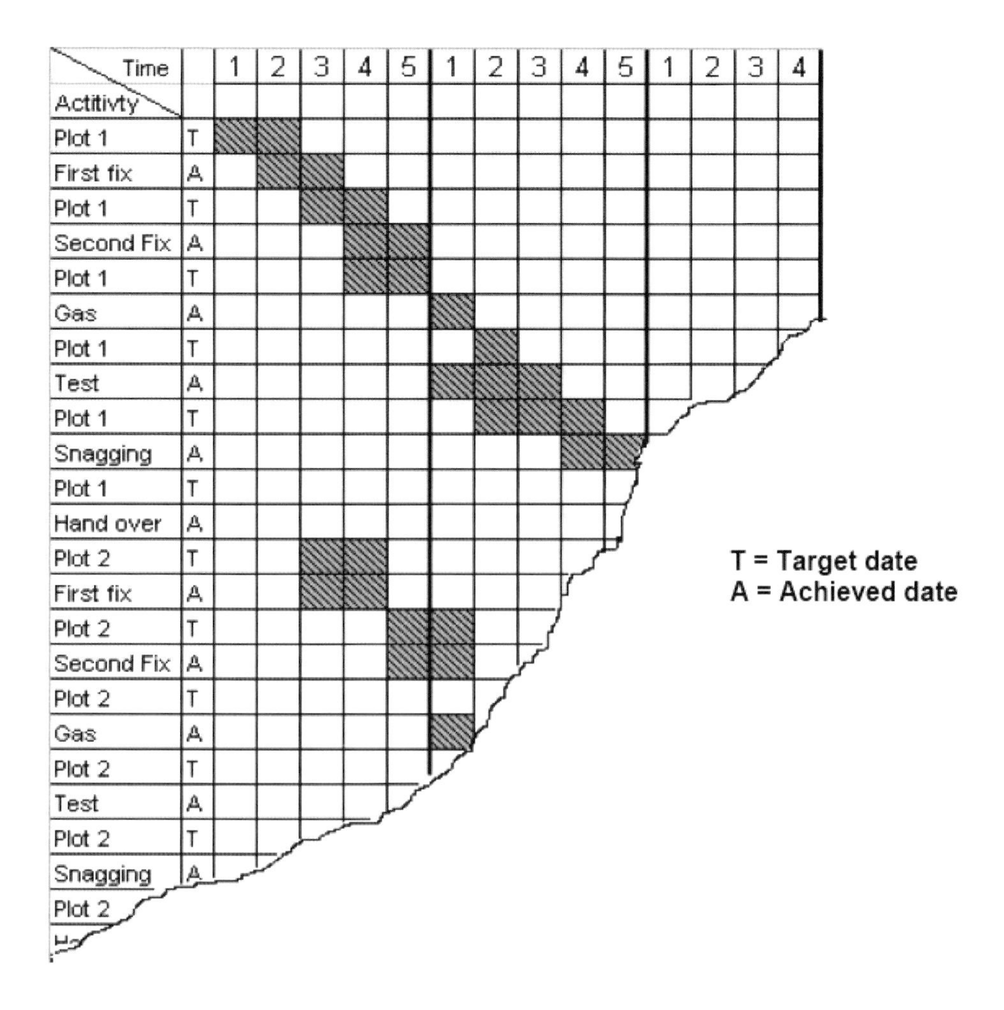

Figure 2.7: Work programme for a plumbing installation

Planning work activities against contract specifications

On larger installation contracts, a bill of quantities is produced to specify how the work will be carried out, as well as the quality and quantity of the materials involved. The bill of quantities is produced by measuring all the quantities, based on a drawing – a process referred to as 'taking off'.

Figure 2.8 is a typical example of an extract from a bill of quantities.

The bill of quantities is produced by a quantity surveyor, usually on behalf of the architect, and its main purpose is to cost a contract in great detail. It will also be used throughout the contract to control costs and provide milestones for contractors' payments. The architect, on behalf of the client, will use the bill of quantities before a contract starts as part of the tender documentation. It is sent out to contractors with the rate and cost columns blank for them to fill in when tendering for the work.

Item No.	Description	Quantity	Unit	Rate	Cost £
A	All sanitaryware supplied by a manufacturer 'Hiline' pedestal-mounted wash basin in white vitreous china to BS 3402, 67 cm x 53 cm	10	Item	55.00	550.00
B	'Starly' bath in cast white acrylic sheet 170 cm x 70 cm, and slip-resistant base	10	Item	250.00	2500.00
C	'Space' close-coupled WC suite with washdown bowl and box flushing rim in vitreous china to BS 3402, 39 cm wide x 79 cm high	10	Item	150.00	1500.00
D	15 mm copper tube grade X half-hard to EN 1057, supplied in 6 m lengths	120	M	1.10	132.00
E	Table copper integral solder ring fittings to EN 1254, 15 mm				
	Tees	20	Item	1.20	24.00
	Elbows	40	Item	0.80	32.00
	Tap connectors	30	Item	1.80	54.00
	Straight connectors	40	Item	0.40	16.00

Figure 2.8: Extract from a typical bill of quantities

Remember

It is essential that manufacturer's instructions are left with the user.

A material schedule contains information similar to the above, but would be used by a plumber on site as a working document to provide details of what materials would be specified for a particular dwelling. The material schedule is unlikely to have any costings in it.

Component and appliance details

Generally speaking, most plumbers use manufacturers' instructions for specific details of components or appliances. This type of information is supplied with the component or appliance in the delivery packaging. If working on an existing appliance on a maintenance contract (particularly boilers), you must have access to the manufacturer's instructions. You should be able to obtain copies of instructions from most manufacturers.

Figure 2.9 shows a typical page from a manufacturer's installation instructions.

Make sure to always use a Pipe Insert when using Speedfit Pipe

This makes the pipe completely round ensuring the best possible seal between the 'O' ring and the outside diameter of the pipe. When using Speedfit Pipe with a compression fitting the Insert gives rigidity to ensure the pipe does not collapse under pressure as the compression olive is tightened.

Figure 2.9: Installation instruction

> **Activity**
>
> If you have Internet access, take a few minutes to explore a few manufacturers' websites. Try typing 'plumbing manufacturers' literature' in the search engine and see what comes up.

Scope, purpose and requirements of the work

The scope, purpose and requirements of the work to be undertaken are usually described using drawings (see Unit 3 pages 118–121). You need to ensure that the installation meets industry standards and the requirements as set out in the contract specification.

Materials used in plumbing installations should be to the relevant EN or BS number. British Standards also make recommendations on design and installation practice: for example, BS 6700. In addition to British Standards, the following legislation places statutory responsibilities on plumbers:

- Water Supply (Water Fittings) Regulations 1999
- Gas Safety (Installation and Use) Regulations 1998
- Electricity Supply Regulations 1998 and Electricity At Work Regulations 1989
- Building Regulations 2000
- Health and Safety at Work etc. Act 1974.

> **Remember**
>
> Many of the requirements laid down in legislation are absolute. If you do not comply with them, severe penalties can follow.

Identifying work responsibilities

It will be the Level 3 qualified plumber who will take overall control of the contract. They will organise the work on a daily basis while following the work programme. The Level 2 qualified plumber will work with minimum supervision.

External factors that affect time frame

Delays to the progress of works can lead to losses for the contracting company. These delays can cause a dispute with the main contractor or client, and on some contracts penalties can be incurred.

Delays can be caused in three ways:

- by the main contractor
- by the employer (taking personnel away from the contract)
- by other events that are out of the control of either the main contractor or contractor. These could be things such as delays with the delivery of materials, unexpected weather and other causes.

If delays happen, the site supervisor, if working on a site, or the customer must be informed. If agreed, the work programme can be amended to take this into consideration.

Activity

What could be the consequences of delays in a project that you are currently involved in? Think what the impact could be on:

- you
- your contractors
- your clients.

Selecting resources against a contract specification

Larger contracts will always have a job specification, which has been prepared by the quantity surveyor. This itemises all the things you will need to do each job, from materials to specialist equipment, so it is a key document for you to be able to understand and use.

Materials

To ensure that you are sourcing the correct materials you will need to consult the job specification for details of the type, the manufacturer and the standard of work expected.

For the amount of a specified appliance you would consult the bill of quantities. The total amount, which your estimator would have priced against, is held in this document. Some bills of quantities are structured in a room-by-room format: for example, Kitchen: 1 Left-hand drainer S/S sink.

There may be times when some materials originally specified when the contract was first assembled by the architect are no longer available, perhaps because they are no longer being produced. When this occurs, other materials to the same standard need to be sourced and approved by the architect and customer or client. This may incur costs and may delay the progress of the installation. A situation like this can be difficult, as the process of materials being accepted can be slow.

Job Specification

Contract: Single dwelling, 60 Well Lane, Waterford

Item	Type	Standard
Copper tube	All copper tube to be Grade X half-hard manufactured to BS EN 1057	• Installed to current water regulations and to BS 8000: Part 15 • Pipework to be tested to BS 6700:2006+A1:2009b *Metallic pipework*
Copper fittings	Soft soldered solder ring fitting to BS EN 1254. Manufacturer: Yorkshire Imperial Fittings	All solder to be lead free to BS EN 29453:1994/ISO 9453:1990, *Soft solder alloys – Chemical compositions and forms*
Fluxes	Water soluble self-cleaning Manufacturer: Everflux	Flux is not to be used excessively. Excess must be cleaned and removed on completion of joint as per the manufacturer's instruction

Figure 2.10: Example of a job specification

Plant

Plant would be such things as scissor lifts, cherry pickers and mobile scaffolds. These all have to be taken into consideration when planning for the contract, and would have to be included at the tendering stage. The most suitable plant needs to be hired when needed for the shortest period possible in order to keep costs down.

Scissor lifts could be required if installing pipework either externally or internally where, if a mobile scaffold was used, it might have to be dismantled and re-assembled to avoid high-level obstructions.

A cherry picker could be used in a place that is difficult to access. Because of the 'reach', the operative would be able to get access easily to such things as chimneys for installing flue liners, and tops of roofs for the installation of leadwork and guttering.

Vehicles

For a small company, getting the right type of transport is essential. You may only be on a contract for two hours and materials that you do not carry in your van stock may need to be sourced from merchants. Larger contracting companies will have at least one vehicle, usually of the mini-bus type, available for use at all times. However, even for larger companies, it is uneconomical to have many vehicles standing unused all day.

Equipment

On some contracts you will need to get in specialist equipment such as threading machines, oxyacetylene brazing equipment and barriers (for example, where you could be working in an excavation to replace some below-ground drainage pipework). This equipment can be hired as and when needed, to keep the costs down.

Material delivery and availability

Remember

Keeping tools and materials secure is important. Where materials and tools are stored on site, ensure that they are locked in a secure lock-up. During breaks in the working day, use a lockable tool vault.

For security reasons and storage problems on site take the example of a multi-dwelling housing estate, noting that the dwellings will be in plot numbers; you can at this point use the work programme to organise delivery of materials.

By arranging with your supplier (many of them will do this), you could have a first-fix kit which would contain all materials required for the first-fix of hot and cold water pipework and central heating. Depending on the construction of the building, it may also include the radiators, soil and waste. It may also include materials for the guttering and rainwater pipes, which need to be installed before the dismantling of the scaffolding.

Activity

What are the advantages of only ordering materials when you need them? What are the disadvantages? Write some notes, thinking about:

- types of materials
- stages of the job
- risks and benefits.

For the second-fix kit the suppliers would deliver, when ordered by the buyer if you are a large company or by your immediate boss if it is a small company, the appliances, boiler, radiators (if they have not already been fitted) and all the necessary fittings and pipework to complete the second fix.

If there is a delay from the supplier this will affect progress of the contract. The site manager (or customer if it is a small contract) would need to be informed.

Allocating work time

On contracts both large and small there are factors that can affect the completion time of the contract. These can be varied but here are a few examples:

- operatives being off sick
- lack of skilled personnel, unable to recruit in the area
- on larger contracts there could be strikes by personnel
- other trades not progressing as expected on the work programme
- dispute with customer (the customer may have had a disagreement with the plumber who is installing).

Where external works are undertaken, the weather could have an impact on progress.

Where changes are necessary you may need to create variations orders, which were covered earlier. Look back now at pages 25–27.

Producing a simple work programme

A work programme helps you keep on course so that you can complete any contract to the correct timings.

Figure 2.11 shows a good example of a simple work programme. The grid shown is just for the plumbing; the site manager would hold a 'master' work programme that included all the trades, including plumbing.

To draw up a work programme for the plumbing tasks, you would need to get key dates or 'milestones' from the master. These dates might be predicted handover dates or payment dates. From your plumbing work programme you should be able to identify your labour requirements for the contract.

Activity

In Figure 2.11, you can see that there is an overlap between the work for Plot 1 and Plot 2. How might this affect time allocation and labour requirements?

Using Figure 2.11 as an example imagine you have been asked to produce a detailed plumbing programme for a small housing development. It will consist of three new-build three-bedroom detached dwellings of traditional construction, and three existing three-bedroom detached dwellings which are to be completely stripped of their existing plumbing systems and refurbished. The master bedroom will have en-suite facilities, and there will be a downstairs cloakroom. Plumbing will also be required for a dishwasher, washing machine and outside tap. Gas appliances include inset living-flame-type fire, boiler and hob.

Draw up the programme, to include:

- the points at which other trades/persons may be needed, including:
    - external groundworks for the mains service
    - joiner for the cold water storage system (cwsc) supports, and lifting floorboards to existing dwellings
    - bricklayer for cutting holes and chases
    - building site manager for progress meetings or organising labour
    - water company approvals/approved installer

	Week commencing	**1/01/2012**					**08/01/2012**					**15/01/2012**					**22/01/12**		
Plot number	**Activity**	1	2	3	4	5	1	2	3	4	5	1	2	3	4	5	1	2	3
Plot 1	Guttering and rainwater pipes																		
	First-fix																		
	Second-fix																		
	Commissioning																		
	Snagging																		
Plot 2	Guttering and rainwater pipes																		
	First-fix																		

Figure 2.11: A simple work programme

- what materials need to be ordered and when (think about having money tied up in materials, and having to store them)
- laying cold water service pipe from external stop tap to dwelling
- first-fix carcassing for hot and cold water, heating and above ground discharge pipework
- first-fix carcassing for gas pipework
- second-fix heating appliances, components and controls
- second-fix sanitary appliances
- second-fix pipework
- testing for water tightness
- testing for gas tightness
- commissioning systems and components
- handing over to the client
- carrying out snagging.

Working around other trades, the contract period is six weeks for all the installation activities, followed by a two-week snagging period. Decide on an appropriate plumbing specification, to give an indication of installation time.

Consider:

- cold water supply: direct or indirect
- type of heating system, including number of radiators: fully pumped, sealed
- type of gas boiler and controls: energy-efficiency considerations
- unvented hot water system or vented
- type and quality of sanitary appliances: will you include a bidet?

Give some thought to the amount of time required to do the work. Will one plumber be able to cope, or will it need more labour?

Check your knowledge

1. How should the plumbing supervisor handle the supplier if two deliveries of sanitaryware have failed to come up to specification?
   a Speak calmly to the supplier and ask them not to do it again
   b Make demands of the supplier using a firm and assertive tone to explain the situation
   c Use a low, calm voice to explain the problem and seek a solution
   d Demand talks with the Managing Director and refuse to take no for an answer

2. Good customer relations for a business can start with:
   a acting promptly to requests for information
   b polite dealings with potential customers
   c deposits before work commences
   d both **a** and **b**.

3. On a large new building project, to whom is the plumbing company ultimately accountable for the quality of customer service?
   a Client
   b Public
   c Architect
   d Building Control Officer

4. What is the responsibility of the clerk of works?
   a To check that the work complies with the specification laid down by the architect
   b Compliance with safety regulations
   c To inspect the work on behalf of the local authority
   d The co-ordination of suppliers and subcontractors

5. At intervals a plumbing subcontractor's representative may visit the site to measure the amount of materials that have been installed. What is the purpose of this exercise?
   a To check the availability of adequate plumbing materials to complete the job
   b To assess the level of theft
   c To create materials orders to suppliers
   d To prepare interim valuations (invoices) for stage payments

6. On the construction of a large office building, any changes or alterations to the plumbing specification will be made by the:
   a Client
   b Clerk of works
   c Building Control Officer
   d Architect.

7. What is one of the principal roles of the quantity surveyor?
   a Measuring all materials used on site
   b Carrying out valuations
   c Organising the subcontractors required for a job
   d All of the above

8. What is another name for a procurement officer in a plumbing or contractor company?
   a Planner
   b Site engineer
   c Buyer
   d Estimator

9. Which is the preferred method for checking and chasing deliveries of materials to a plumbing job?
   a Semaphore
   b Email
   c Phone
   d Fax

10. Which of the following best describes what a work programme does?
   a It identifies when plumbing jobs should start and be completed
   b It is used to help the contractor plan the number of plumbers needed for each task
   c It sets out dates when payments are due for each bit of work completed
   d It gives the plumber an indication of material quantities

Preparation for assessment

Now that you have completed all of your learning outcomes, it is time to prepare for assessment.

Use the following revision strategy for the underpinning knowledge test.

- Do not try to take in all of the information in one go; take your time and do a little each day.
- Reread Unit 2 *Understand How To Organise Resources Within Building Services Engineering*.
- Go over the notes you have taken during the weeks of teaching.
- Research more information about anything which you are not sure of.
- Check any queries with your tutor.
- Do the practice questions in your book.
- Check your answers with your tutor.

There are no practical assessments with this unit; it is a knowledge only unit.

Good luck and we hope you do well.

3 Understand and apply domestic cold water system installation, commissioning, service & maintenance techniques

As a plumber the bulk of the work you will carry out will be covered by Regulations, British Standards and Codes of Practice. These include the Building Regulations, health and safety legislation and many British Standards and Codes of Practice, particularly BS 6700.

In this unit you will focus on the requirements of the Water Supply (Water Fittings) Regulations 1999, which generally cover the requirements of cold and hot water supply installations. You will also look in more detail at pumped supplies to showers and private water supplies from wells and boreholes.

This unit covers the following learning outcomes:

- know the legislation relating to the installation and maintenance of cold water supplied for domestic purposes
- know the types of cold water system layout used in multi-storey dwellings
- know the types of cold water system layout used with single-occupancy dwellings fed by private water supplies
- know the requirements for backflow protection in plumbing systems
- know the uses of specialist components in cold water systems
- know and be able to apply the design techniques for cold water systems
- know and be able to apply the fault diagnosis and rectification procedures for cold water systems and components
- know and be able to apply the commissioning requirements of cold water systems and components.

1. Know the legislation relating to the installation and maintenance of cold water supplied for domestic purposes

A note on references to relevant legislation in this unit

This section looks at the legislation that is relevant to the installation and use of cold and hot water services. It includes a brief overview of the current Regulations, the Regulations that are relevant to your job, a discussion of some important issues arising from the Regulations, and the European perspective.

Where a particular Regulation is being discussed, the relevant name, number and paragraph are listed beside the text. Copies of the Water Industry Act 1991 and the Water Supply (Water Fittings) Regulations 1999 are published on the website of Her Majesty's Stationery Office.

Abbreviations for relevant Regulations and documents are:

- WIA – Water Industry Act
- WSR – Water Supply (Water Fittings) Regulations 1999
- GD – Guidance Document to the WSR
- BS – British Standard.

Background to the legislation

Interpreting the legislation

The control of water supply installations in England and Wales has been completely revised by the introduction of the Water Supply (Water Fittings) Regulations 1999. The Secretary of State for the Department of the Environment, Transport and the Regions (DETR) exercised his powers under the Water Industry Act 1991 to enforce a set of Water Regulations that control the installation and use of water fittings, resulting in the making of the Water Supply (Water Fittings) Regulations 1999. These Regulations apply only in England and Wales, but similar requirements have been made by the Scottish Office and Northern Ireland Office.

Before this, the UK had a long history of Water Byelaws, which were managed and enforced by local water suppliers. Newly introduced byelaws expired after ten years, but they were renewed or updated as necessary. The last renewal of the Water Byelaws was in 1986, before they were finally replaced by the Water Regulations on 1 July 1999.

The Water Regulations are national regulations made by the Department for Environment, Food and Rural Affairs (DEFRA). They apply to all installations in England and Wales that are supplied from a public main by a **water undertaker**. Water undertakers are responsible for the enforcement of the Regulations.

The Regulations have similar aims to the old byelaws, but are applied differently. They have introduced a significant number of changes in the way water fittings have to be installed and used.

The principal legislation governing the creation of the Water Regulations is the Water Industry Act 1991, with sections 73, 74, 75, 84 and 213(2) being particularly relevant. Table 3.1 gives an overview of what these sections discuss.

Key term

Water undertaker – a water supply company (this is the technical term used in the Regulations).

Section	Contents of section
73	Offences of contaminating, wasting and misusing water (legal action)
74	Regulations for preventing contamination, waste, etc. with respect to water fittings
75	Power to prevent damage, taking steps to prevent contamination, waste, etc.
84	Local authority rights of entry, etc.
213(2)	Powers to make Regulations

Table 3.1: Relevant sections of the Water Industry Act 1991, which govern the creation of the Water Supply (Water Fittings) Regulations 1999

Did you know?

The Regulations are a means of preventing waste, undue consumption, misuse, contamination and the erroneous measurement of water.

Water supplied from a water undertaker

WIA Section 74(1)

An extract from Section 74 is reproduced below:

74 Regulations for preventing contamination, waste, etc. and with respect to water fittings

(1) The Secretary of State may by Regulations make such provision as he considers appropriate for any of the following purposes, that is to say –

(a) for securing –

(i) that water in a water main or other pipe of a water undertaker is not contaminated; and

(ii) that its quality and suitability for particular purposes is not prejudiced, by the return of any substance from any premises to that main or pipe;

(b) for securing that water which is in any pipe connected with any such main or other pipe or which has been supplied to any premises by a water undertaker is not contaminated, and that its quality and suitability for particular purposes is not prejudiced, before it is used;

(c) for preventing the waste, undue consumption and misuse of any water at any time after it has left the pipes of a water undertaker for the purpose of being supplied by that undertaker to any premises; and

(d) for securing that water fittings installed and used by persons to whom water is or is to be supplied by a water undertaker are safe and do not cause or contribute to the erroneous measurement of any water or the reverberation of any pipes.

In other words, Section 74 outlines that the Water Regulations have been made to:

- make sure water is not contaminated, and that its quality and suitability for a purpose is not harmed before or after being supplied to a premise
- prevent waste, undue consumption and misuse of water supplied by the undertaker
- make sure that water fittings are safe and do not cause or lead to erroneous measurements or vibration and noise in pipes.

This shows that the Regulations have been written to protect the water supply and to protect users against their own actions.

The Water Supply (Water Fittings) Regulations 1999

The Water Supply (Water Fittings) Regulations 1999 are made up of 14 Regulations. These are divided into three parts and are supported by three Schedules. The Schedules should be treated as part of the Regulations. A brief outline of the Regulations is given in Table 3.2.

Regulation	Content of Regulation
Part I	• Gives the date when the Regulations came into force and some interpretations to help understand the Regulations. • Makes statements as to how the Regulations should be applied.
Schedule 1	• Supports Part I. • Outlines the fluid risk categories that may occur within and downstream of a water supply network. • Needed for the backflow requirements of Schedule 2.
Part II	• Defines what is expected of a person(s) installing water fittings. • Outlines how water fittings should be installed and used to prevent waste or contamination. • Puts conditions on materials and fittings that may be used. • Requires contractors to notify the water suppliers of certain installations, and encourages the introduction of Approved Contractors Schemes.
Schedule 2	• Supports Regulation 4(3) (Requirements for Water Fittings). • Has 31 separate requirements. • Looks at all aspects of water fittings. • Deals with the practical aspects of Part II Regulations.

Regulation	Content of Regulation
Part III	• Deals with the enforcement of the Regulations. • Sets out penalties for breaking the Regulations. • Sets out disputes procedures.
Schedule 3	• Supports Regulation 14. • Lists byelaws of various water undertakers that have been replaced with the Water Industry (Water Fittings) Regulations 1999.

Table 3.2: Summary of the main parts of the Water Supply (Water Fittings) Regulations 1999

Important aspects of Part I

Part I helps with the interpretation of the Regulations by defining terms used in them and explaining what they apply to.

Regulation 1 definitions

Regulation 1 makes some important definitions:

- **Approved contractor:**
  - a person who has been approved by the water undertaker for the area where a fitting is installed or used
  - a person who has been certified as an approved contractor by an organisation specified in writing by the regulator.
- **The Regulator:** in England the Regulator is the Secretary of State; in Wales it is the National Assembly for Wales.
- **Material change of use:** a change in how premises are being used, so that after they are changed they are used:
  - as a dwelling
  - as an institution
  - as a public building
  - for storage or use of substances that mix with water to make a category 4 or 5 fluid.
- **Supply pipe:** as much of any service pipe that is not vested in the water undertaker.

Activity

Give some examples of a material change of use.

The Water Supply (Water Fittings) Regulations 1999	
Do apply to:	**Do not apply to:**
Every water fitting installed or used where the water is supplied by the water undertaker	• Water fittings installed or used for any purpose not related to domestic or food production, so long as: o the water is metered o the supply does not exceed 1 month (3, with written consent) and o no water is returned to any pipe vested in a water undertaker. • Water fittings that are not connected to water supplied by a water undertaker. • Lawful installations used before 1 July 1999 (these do not have to be replaced).

Table 3.3: Summary of Regulation 2, which lists the water fittings that are covered by the Water Regulations

Water supplied from a private source

A private water supply may be defined as any water supply that is not provided by a statutory water undertaker and where the responsibility for its maintenance and repair lies with the owner or person who uses it. A private water supply can serve a single household and provide less than one cubic metre of water per day, or it can serve many properties or commercial or industrial premises and provide 1000 m^3 per day or more. The water source could be a borehole, well, spring, lake, stream or river.

The condition and purity of the water should be considered. Chemical and bacterial analysis is advisable before putting the source into use. Approval is needed from the local public health authority for drinking water supplies, and a licence to extract may be required from the water undertaker.

- If a private and a public supply are taken to a single property, the water undertaker must be informed and the Water Regulations must be complied with. Water from a private source must not be connected to a supply pipe served from the water undertaker's main.
- The water undertaker may require the private supply to be metered.

Notification requirements

Remember

With notification requirements, the onus is on you to actively notify the authorities. If you fail to do so, it will be your fault!

The Regulations also require you to notify various bodies when you are undertaking work on wholesome and recycled water systems. As with building permission, there can be serious ramifications if you fail to notify the appropriate authorities, and the work you have done could even have to be redone.

Notifying the water undertaker

Important aspects of Part II

WSR Part II, Reg 3

Key term

Workmanlike manner – working in line with appropriate British and European Standards, to a specification approved by the Regulator or the water undertaker.

Part II of the Regulations contains information about the quality or standard of water fittings and their installation. The aims of the Water Industry Act, Section 74(1), are given in detail in Regulation 3. Regulation 3 also states that any work on water fittings is to be carried out in a **workmanlike manner**.

WSR Part II, Reg 5

Regulation 5 requires a person who proposes to install certain water fittings to notify the water undertaker, and not to commence installation without the undertaker's consent. The undertaker may withhold consent or grant it on certain conditions.

This requirement does not apply to some fittings that are installed by a contractor who is approved by the undertaker or certified by an organisation specified by the Regulator.

The installation of the following water fittings and systems requires notice to the water undertaker, except those items in bold italics, if carried out by an approved contractor:

- the erection of a building or other structure not being a swimming pool or pond
- the extension or alteration of any water system in a building other than a house
- a material change of use of any premises
- the installation of:
  - a bath having a capacity of more than 230 litres
  - ***a bidet with ascending spray or flexible hose***
  - a single shower unit, not being a drench shower for health and safety reasons, approved by the Regulator
  - a pump or booster pump drawing more than 12 litres a minute
  - a unit that incorporates reverse osmosis
  - a water treatment unit that uses water for regeneration or cleaning
  - ***an RPZ valve assembly or other mechanical device for backflow protection from fluid category 4 or 5***
  - a garden watering system, unless designed to be operated by hand
  - any water system laid outside a building less than 750 mm or more than 1350 mm underground.

Where an Approved Contractor installs, alters, connects or disconnects a water fitting, they must provide a certificate to the person who commissions the work stating that it complies with the Regulations.

Remember

The aim is to prevent waste, misuse, undue consumption or contamination, or false measurement of the water supplied.

WSR Part II, Reg 6
WSR Part III, Regs 7 to 14

Important aspects of Part III

A brief description of the Regulations in Part III is given in Table 3.4.

Regulation	Content of Regulation
7 & 8	Provide for a fine not exceeding level 3 on the standard scale for contravening the Regulations. **It is a defence to show that the work on a water fitting was done by or under the direction of an Approved Contractor, and that the contractor certified that it complied with the Regulations.** This defence is extended to the offences of contaminating, wasting and misusing water under section 73 of the Water Industry Act 1991 (reg 8)
9	Enables water undertakers and local authorities to enter premises to carry out inspections, measurements and tests for the purposes of the Regulations
10	Requires the water undertaker to enforce the Regulations (this is done by the Regulator or the Director General of Water Services)
11	Enables the Regulator to relax the requirements of the Regulations on the application of the water undertaker
12	Requires the Regulator to consult water undertakers and organisations representing water users before giving an approval for the purpose of the Regulations, and to publicise approvals
13	Provides for disputes arising under the Regulations between a water undertaker and a person who has installed or proposes to install a water fitting to be referred to arbitration
14	Revokes the existing Water Byelaws made by water undertakers under section 17 of the Water Act 1945

Table 3.4: Summary of Regulations covered by Part III of the Water Industry Act 1991

Column 1	Column 2
Type of work	*Person carrying out work*
1. Installation of a heat-producing gas appliance.	A person, or an employee of a person, who is a member of a class of persons approved in accordance with regulation 3 of the Gas Safety (Installation and Use) Regulations 1998.
2. Installation of heating or hot water service system connected to a heat-producing gas appliance, or associated controls.	A person registered by the Gas Safe scheme in respect of that type of work.
3. Installation of: a. an oil-fired combustion appliance which has a rated heat output of 100 kilowatts or less and which is installed in a building with no more than 3 storeys (excluding any basement) or in a dwelling; b. oil storage tanks and the pipes connecting them to combustion appliances; or c. heating and hot water service systems connected to an oil-fired combustion appliance.	An individual registered by Oil Firing Technical Association Limited, NAPIT Certification Limited or Building Engineering Services Competence Accreditation Limited in respect of that type of work.
4. Installation of: a. a solid fuel burning combustion appliance which has a rated heat output of 50 kilowatts or less which is installed in a building with no more than 3 storeys (excluding any basement) or; b. heating and hot water service systems connected to a solid fuel burning combustion appliance.	A person registered by HETAS Limited, NAPIT Certification Limited or Building Engineering Services Competence Accreditation Limited in respect of that type of work.
5. Installation of a heating or hot water service system, or associated controls, in a dwelling.	A person registered by Building Engineering Services Competence Accreditation Limited in respect of that type of work.
6. Installation of a heating, hot water service, mechanical ventilation or air conditioning system, or associated controls, in a building other than a dwelling.	A person registered by Building Engineering Services Competence Accreditation Limited in respect of that type of work.

Figure 3.1: Extract from Schedule 2A Document L1

Building control or self-certification

In almost all cases of new works being undertaken on a new build it will be necessary to notify the Building Control Body (BCB) in advance of any work starting. There are two exceptions to this:

1 where work is carried out under a self-certification scheme as listed in Schedule 2A of Document L1 of the Building Regulations
2 where work is listed in Schedule 2B of Document L1 of the Building Regulations.

Competent person self-certification schemes

Schedule 2A

Under Regulation 12(5) of the Building Regulations it is not necessary to notify a BCB in advance of work that is covered by Approved Documents if the work is of a type as set out in Column 1 of Schedule 2A (see Figure 3.1) and is carried out by a person registered with a relevant self-certification (competent persons) scheme, as set out in column 2 of Schedule 2A. In order for that person to join such a scheme they must demonstrate their competency to carry out the type of work that the scheme covers.

Installer and user responsibilities

Under the Water Regulations 1999 there is no differentiation between the installer and the **user**: they both have the responsibility to provide evidence to demonstrate compliance to the reasonable satisfaction of the water supplier. This applies when the installer or user is claiming that a product complies with the Regulator's specifications.

If notification plans contain non-compliant fittings or installation details, consent – whether deemed or granted – does not remove the obligation on the installer, owner or occupier to ensure that the water system as installed complies with the Regulations.

Key term

User – the Water Regulations 1999 states that the user is the owner or occupier.

PROGRESS CHECK

1 What are the aims of the Water Regulations?
2 Who provides guidance on the Water Regulations?
3 Which regulation revokes the Water Byelaws?
4 What do the Water Regulations **not** apply to?
5 Who is the Water Regulator in England?
6 Regulation 9 of the Water Regulations enables the water undertaker and which other body to enter premises?
7 Which Act of Parliament enables the Secretary of State to enforce the Water Regulations?
8 The installation of which type of bidet would require notification to the water undertaker?
9 What is the definition of a private water supply?

Safety tip

Technology changes rapidly. Make sure that you keep up to date with the latest developments in plumbing components, so that you can keep your work as safe as possible.

Key terms

Stop valve – a valve, other than a servicing valve, used for shutting off the flow of water in a pipe.

Service valve – a valve for shutting off, for the purpose of maintenance or service, the flow of water in a pipe connected to a water fitting.

Supply pipe – the length of the service pipe between the boundary of the part of the street in which the water main is laid, and any terminal fitting directly connected to it and under mains pressure.

2. Know the types of cold water system layout used in multi-storey dwellings

Component layout features

It is essential that you understand each of the components that you are using in a cold water system layout. Knowing how the legislation works in this area will help you make the best choices possible, for the work and for the client.

Stop valves to premises

Paragraph 10(1) of the Regulations describes both **stop valves** and **service valves**. Every **supply pipe** or distributing pipe providing water to separate premises shall be fitted with a stop valve located to enable the supply to those premises to be shut off without shutting off the supply to any other premises. The Regulation deals generally with stop valves that are required to be fitted to control the whole supply of water to premises and those that are required to isolate individual sections of an installation.

Where a supply pipe or distributing pipe provides water in common to two or more premises, it shall be fitted with a stop valve to which each occupier of those premises has access. Every supply and distributing

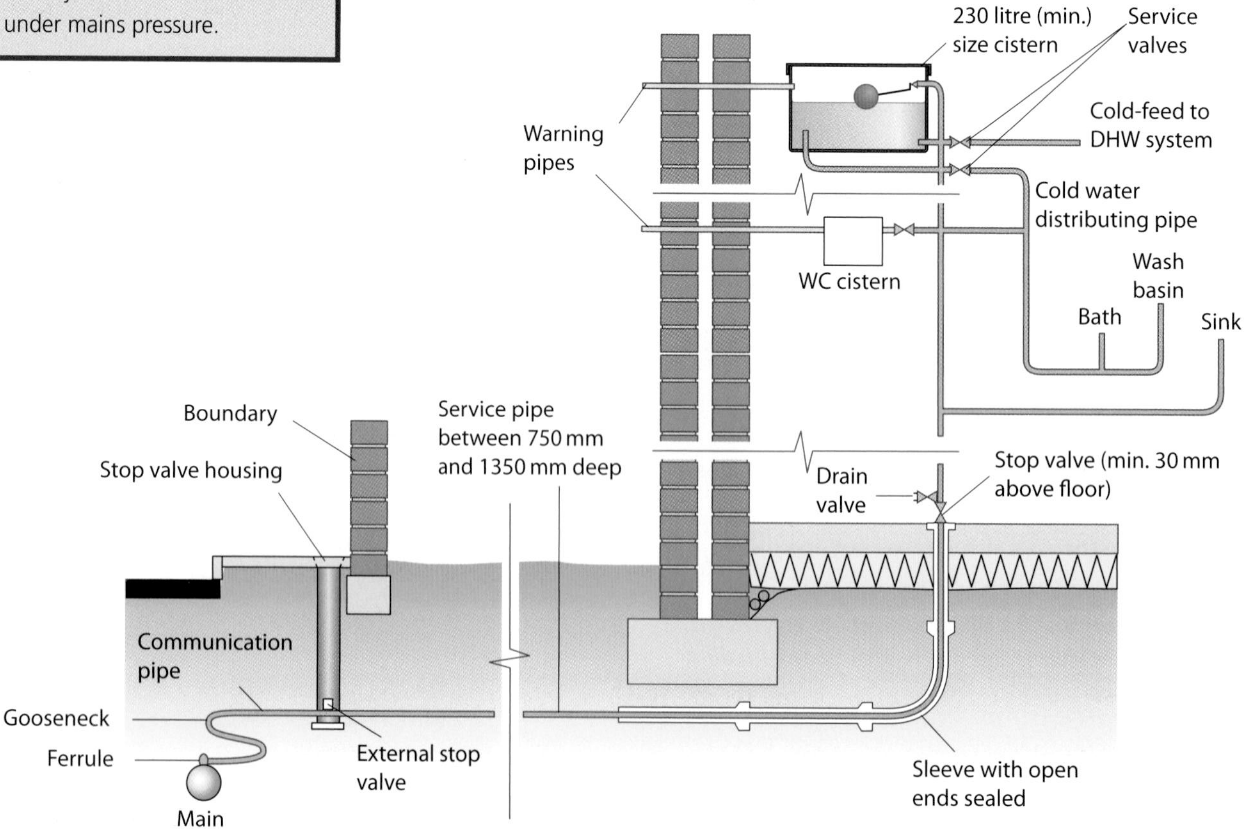

Figure 3.2: Location of stop valves

pipe providing water to premises must be of adequate size to suit the needs of the building and should be fitted with a stop valve located at the boundary of the premises, or elsewhere, to enable the supply to be shut off without shutting off the water supply to other premises. Stop valves to premises must be accessible so that the occupier, in the event of leakage or for other reasons, can isolate the supply.

WSR Schedule 2, para 10(1)

The stop valve should ideally be positioned on the supply pipe, inside the premises, above floor level and as close as possible to the point of entry.

WSR Schedule 2, para 10(2)

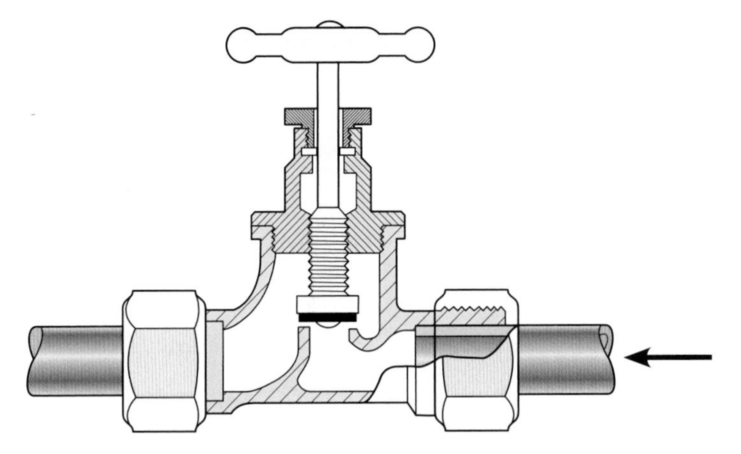

Figure 3.3: Internal workings of a stop valve

Activity

How could you keep yourself and those who work for you up to date with the latest plumbing technology? Make a list of useful ways of learning and sources of information.

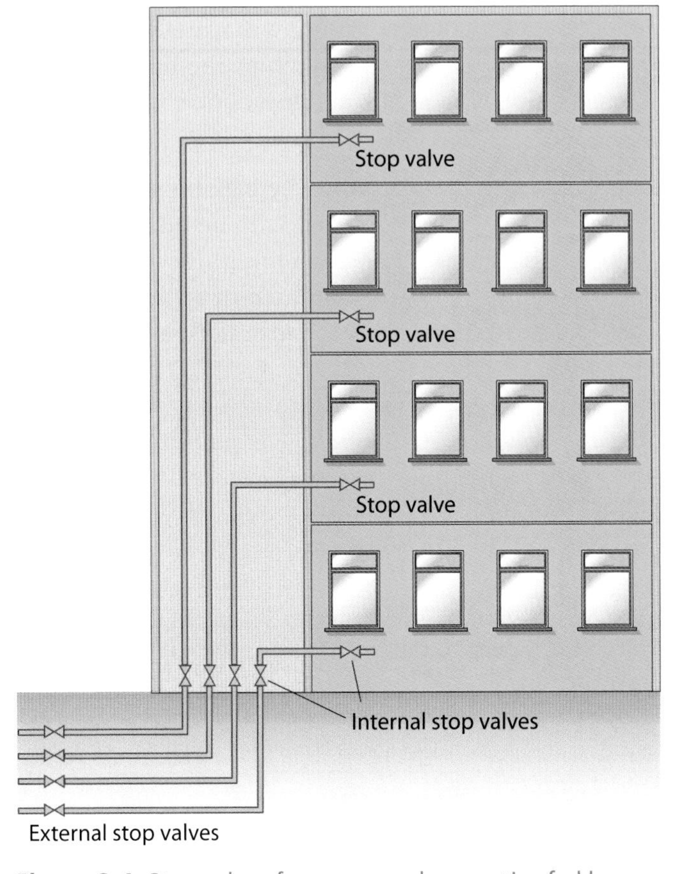

Figure 3.4: Stop valves for communal properties fed by separate supply pipes

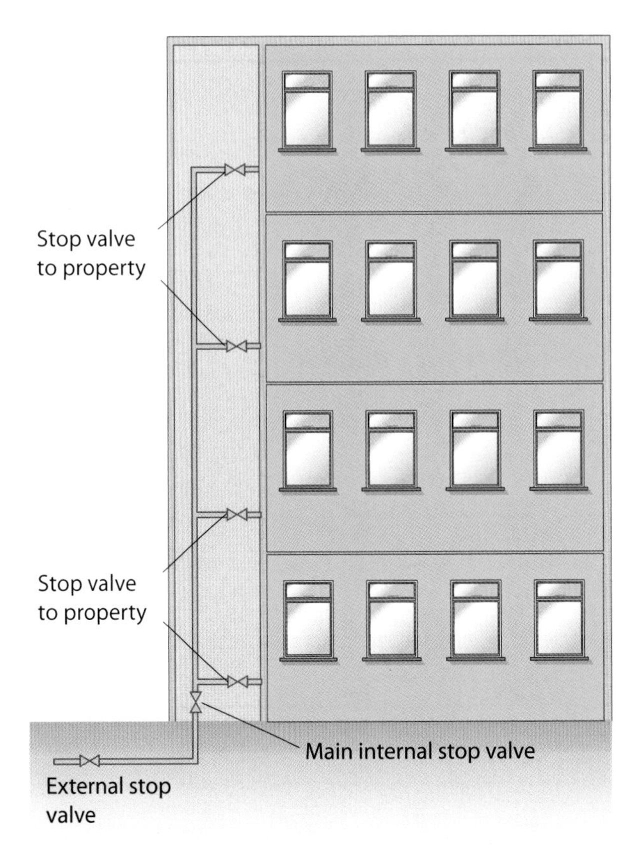

Figure 3.5: Recommended locations of stop valves for a block of flats fed from a common supply pipe

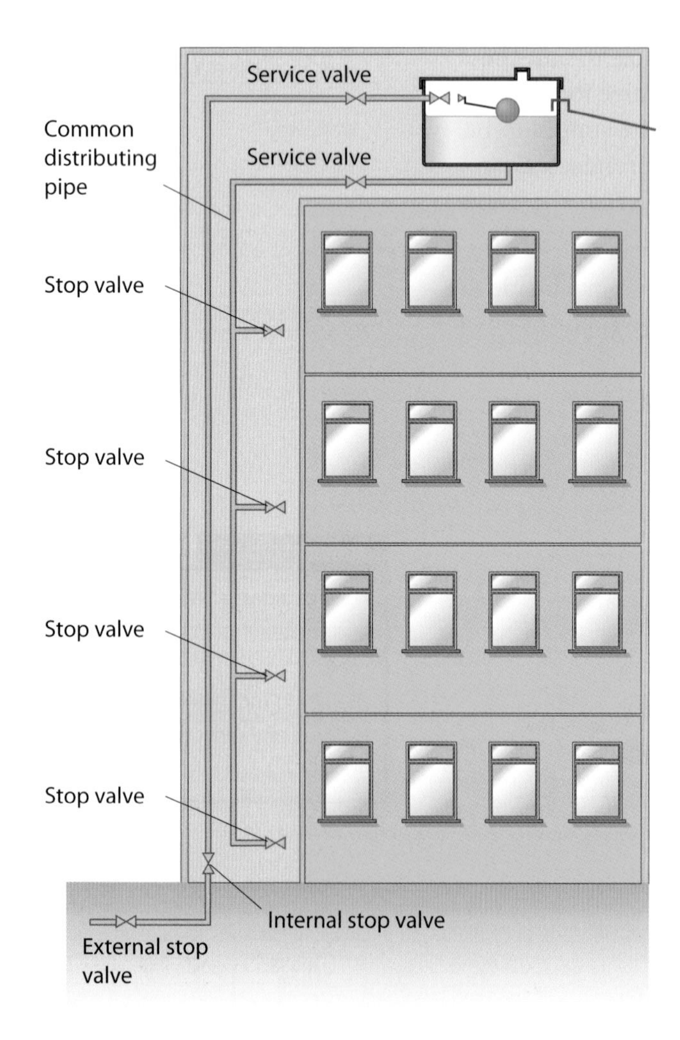

Figure 3.6: Recommended locations of stop valves for a block of flats fed from a common distributing pipe

Figure 3.4 on page 57 shows the recommended locations of stop valves in a block of flats where individual flats are fed by separate supply pipes. Separate internal stop valves are provided on entry to the property to ensure that the individual supply to the property can be isolated easily without having to gain access to the property.

Where a block of flats is fed from a common supply pipe, the recommended positions of the stop valves are as shown in Figure 3.5 on page 57.

Where a block of flats is fed from a common distributing pipe from a cistern, the recommended positions of stop valves are as shown in Figure 3.6 opposite.

Other examples are existing terraced houses fed by a common supply pipe, shown in Figure 3.7. This type of installation would today be allowed only under exceptional circumstances.

Where distributing pipes supply separately chargeable premises from a common storage cistern all separate premises are fitted with stop valves in similar positions to those supplied by a common supply pipe. These will usually be tall buildings that have fittings above the limit of the mains supply, working in conjunction with booster pumps.

Water supply systems must be capable of being drained down, and be fitted with an adequate number of servicing valves and drain taps so as to minimise the discharge of water when water fittings are maintained or replaced. A sufficient number of stop valves should also be installed for isolating

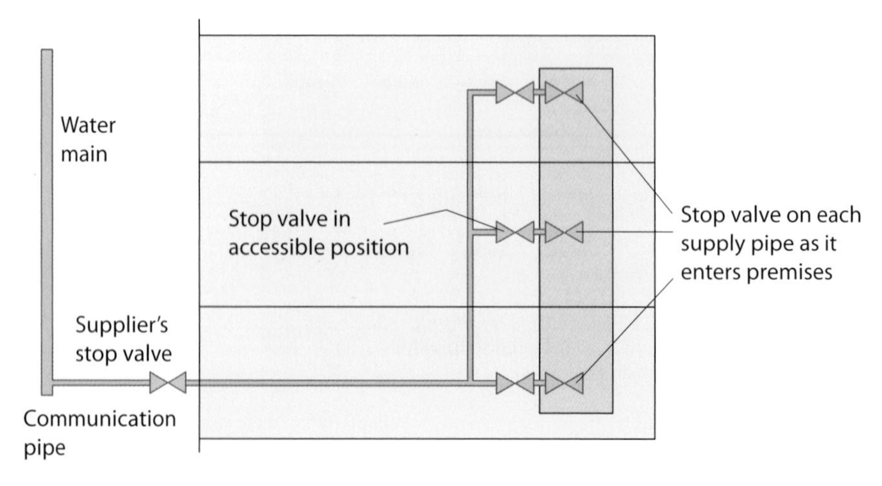

Figure 3.7: Stop valve locations for existing terraced houses (multiple premises) fed by a common supply pipe

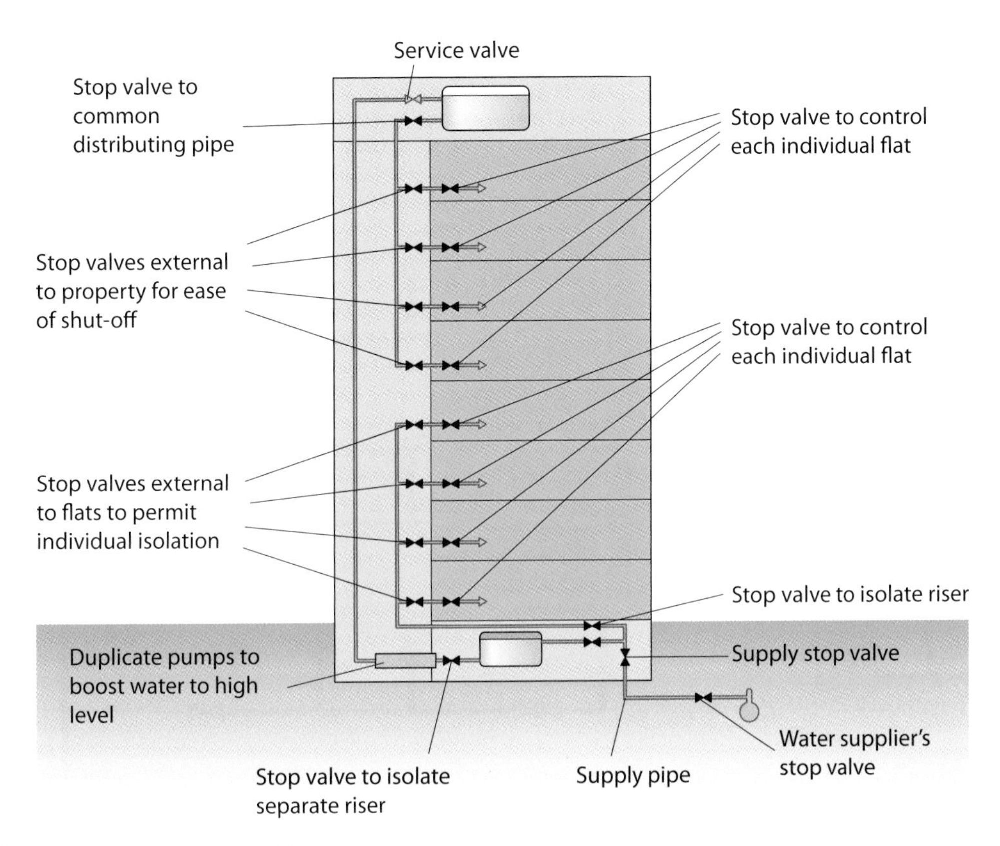

Figure 3.8: Requirements for tall buildings

parts of the pipework. Complying with this requirement will give full control over the installation, allowing sections of pipework or individual appliances to be isolated and drained down, without isolating the supply to other parts of the building.

WSR Schedule 2, para 11

The provision of service valves also applies to mechanical backflow-prevention devices. Servicing valves should be fitted as close as is reasonably practicable to float-operated valves or other inlet devices of an appliance, and they should be readily accessible.

Activity

What does 'reasonably practicable' mean? If you cannot remember, reread page 31.

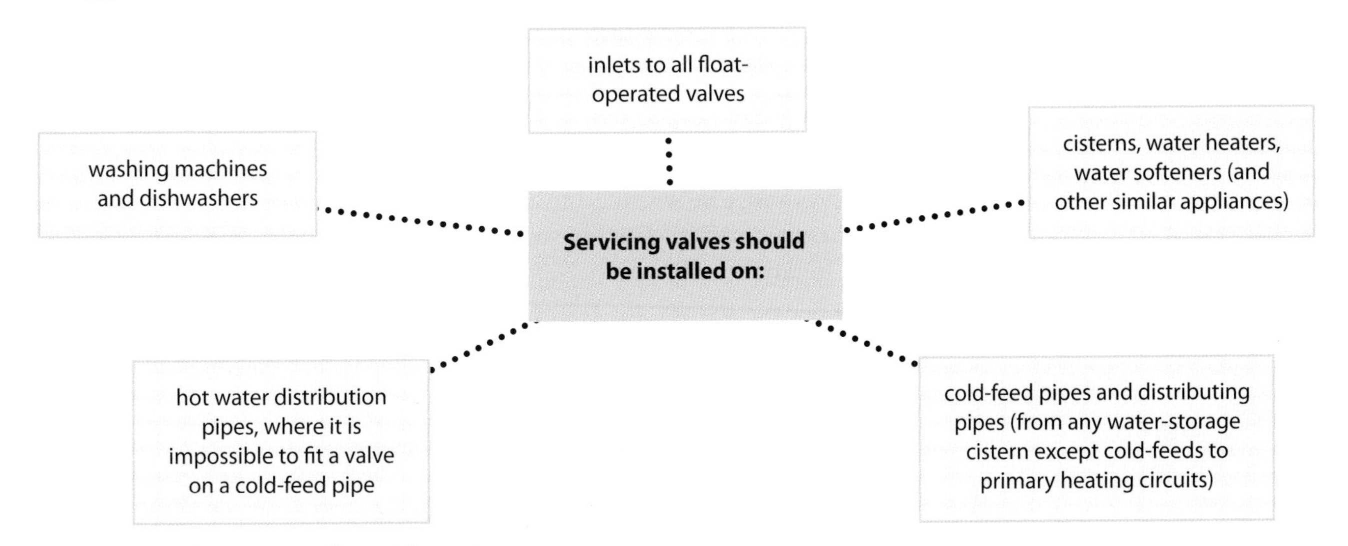

Figure 3.9: Where to install servicing valves

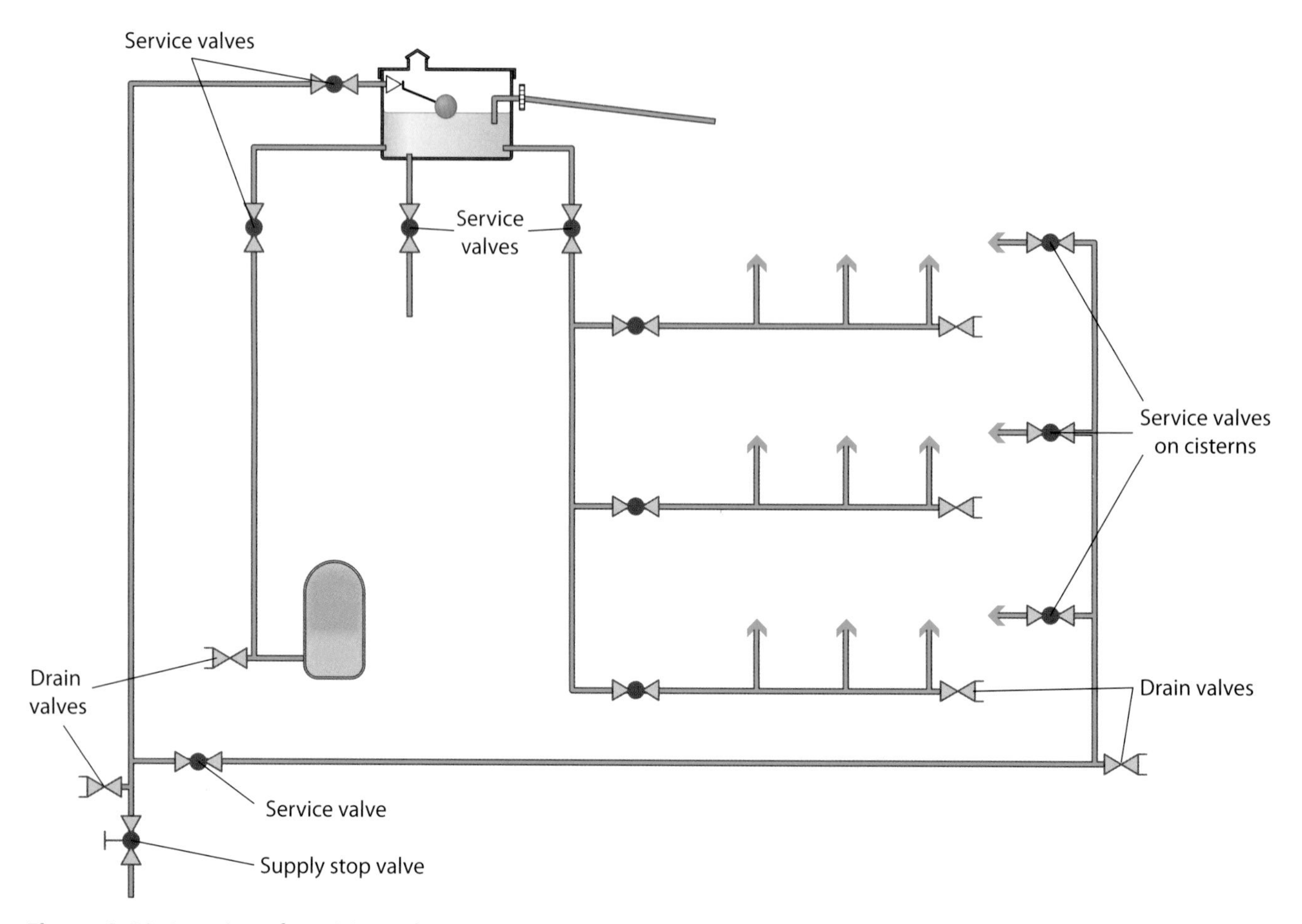

Figure 3.10: Location of servicing valves

Servicing valves should be installed on:

- inlets to all float-operated valves
- cisterns, water heaters and water softeners (and any other similar appliance)
- cold-feed pipes and distributing pipes from any water-storage cistern (except cold feeds to primary heating circuits)
- hot water distribution pipes, where it is impossible to fit a valve on a cold-feed pipe
- washing machines and dishwashers.

Stop valves must be installed to isolate parts of pipework for maintenance and for isolating sections of the supply should leakages occur. On larger installations servicing valves or stop valves should be fitted to:

- isolate pipework on different floors
- isolate various parts of an installation
- isolate branch pipes to a range of appliances.

BS EN 806:2

Drinking-water supplies and points

WSR Schedule 2, para 26

Paragraph 26 concerns the supply of water for domestic purposes through at least one tap, conveniently situated for supplying drinking water (wholesome water). In a house, a drinking-water tap should be situated over the kitchen sink, connected to the incoming supply pipe. In premises where a water softener is used, an un-softened 'drinking-water' tap must be provided.

The Water Industry Act 1991 refers to water for domestic purposes as water used for:

- drinking
- washing
- sanitary purposes
- cooking
- central heating.

The watering of gardens and washing of vehicles are included in domestic purposes if not done using a hosepipe.

WSR Schedule 2, para 27

Where it is not possible to provide the drinking-water tap with water from the supply pipe, the tap should be supplied from a cistern containing water of drinking quality. This appropriately takes us to the requirements of Paragraph 27, which states that a drinking-water supply shall be supplied with water from:

- a supply pipe
- a pump delivery pipe drawing water from a supply pipe, or
- a distributing pipe drawing water exclusively from a storage cistern supplying wholesome water.

In instances where it is not possible to supply water directly off the supply pipe due to there being insufficient water pressure available, it may be necessary to install pumps or a booster system.

If the amount of water required is less than 0.2 litres per second it is permissible to pump direct from the supply pipe. In cases where a greater flow capacity is required to serve the premises, written consent from the water supplier will be required for the direct or indirect pumping (via a cistern or closed vessel) from the supply pipe. If an indirect pumping system is installed, it must be of a type that minimises the possibility of the water quality deteriorating.

Figure 3.11 on page 62 shows an installation in which water is boosted from a break cistern. A pneumatic pressure vessel is included so that the pump is not continually shutting on and off; the pump is controlled by a low- and high-pressure switch sited in the pressure vessel to activate the pump on and off. The principles of this system are used in domestic properties fed from wells or boreholes; if you understand them, you should be able to transfer your knowledge to domestic installations.

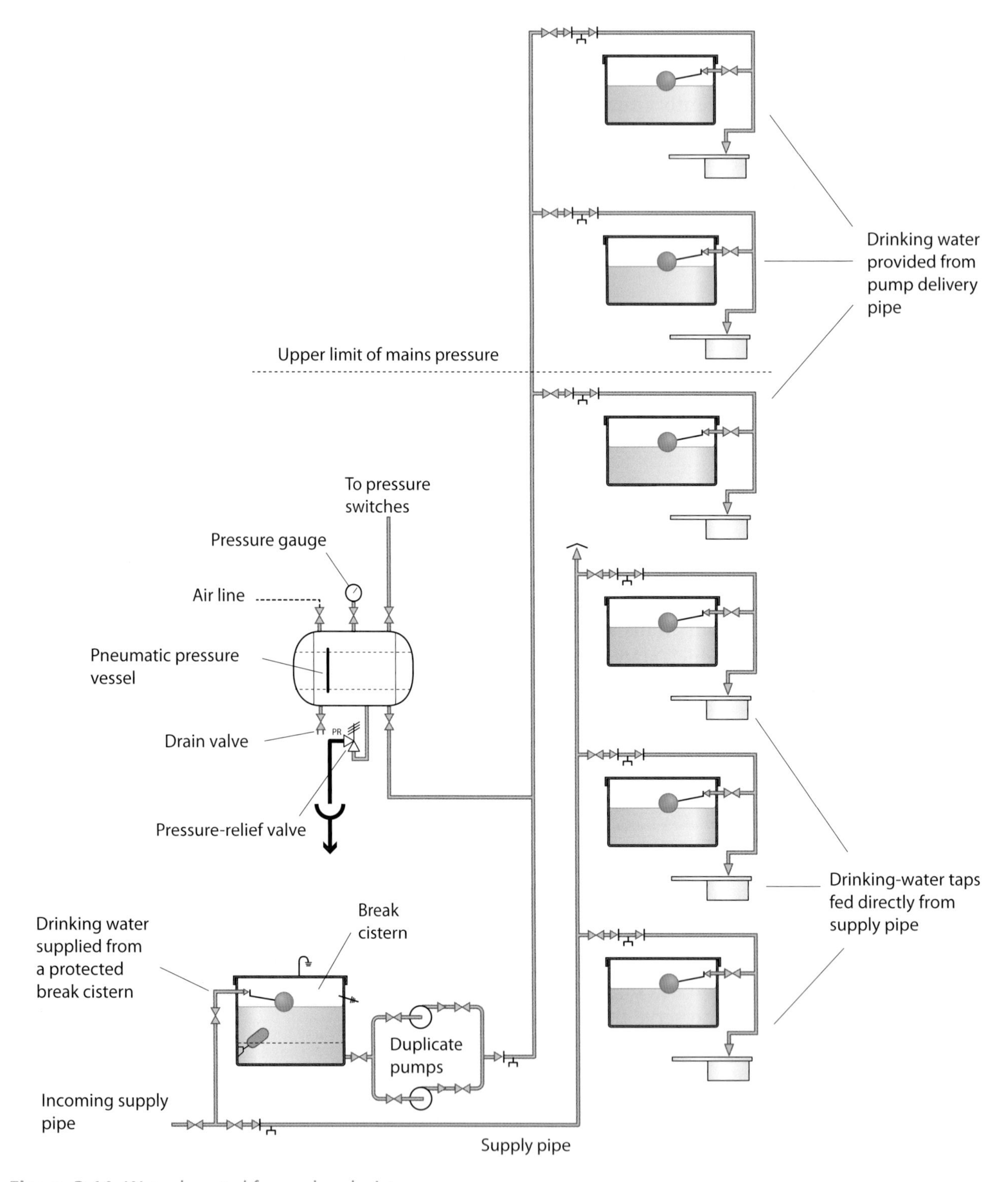

Figure 3.11: Water boosted from a break cistern

When drinking water is supplied direct from a storage cistern, it is recommended that:

- the interior of the cistern is kept clean
- the quantity of stored water is restricted to the minimum essential amount required, so that the throughput of water is maximised

- the stored water temperature is kept below 20°C, taking into consideration that cisterns sited in roof spaces or voids can be subjected to varying temperatures
- the cistern is insulated and ventilated, and fitted with a screened warning/overflow pipe
- the cistern is regularly inspected and cleaned internally.

Figure 3.12 shows an alternative method of providing drinking water using a large header pipe in place of the pneumatic pressure vessel.

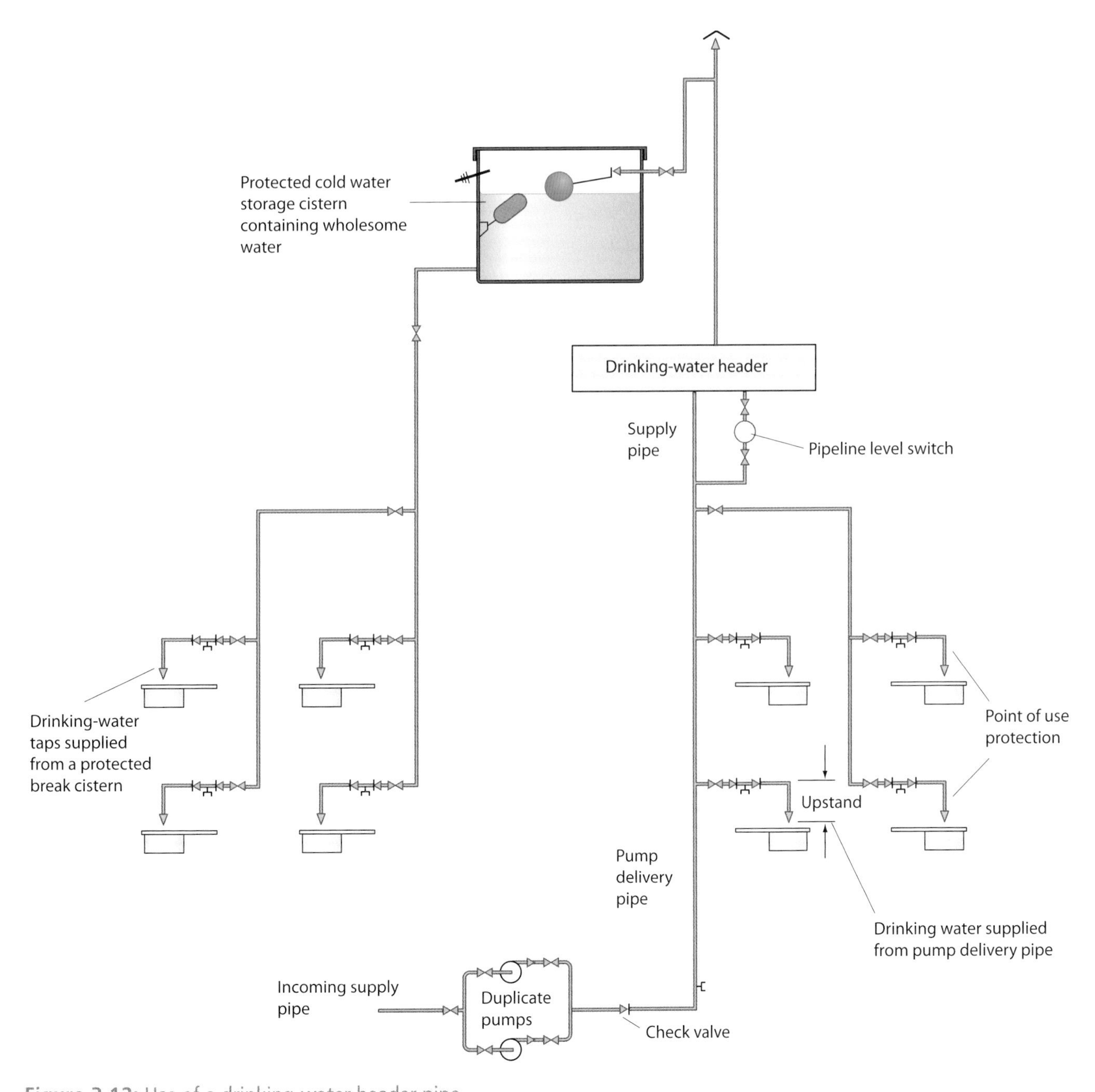

Figure 3.12: Use of a drinking-water header pipe

System layout features for large-scale storage cisterns

In this section you will look at cold water services and the requirements concerned with cold water storage cisterns. This includes the control of incoming water, overflow pipes and warning pipes, and preventing waste and contamination in cisterns.

WSR Schedule 2, para 16

Paragraph 16 of Schedule 2 ensures that water supplied by the water undertaker for domestic purposes remains wholesome. It also looks at the provision of servicing valves on inlet and outlet pipes to cisterns, and the requirements of thermal insulation to minimise freezing and undue warming.

Paragraph 16 states that:

(1) Every pipe supplying water connected to a storage cistern shall be fitted with an effective adjustable valve capable of shutting off the inflow of water at a suitable level below the overflowing level of the cistern.

(2) Every inlet to a storage cistern, combined feed and expansion cistern, WC flushing cistern or urinal flushing cistern shall be fitted with a servicing valve on the inlet pipe adjacent to the cistern.

(3) Every storage cistern, except one supplying water to the primary circuit of a heating system, shall be fitted with a servicing valve on the outlet pipe.

(4) Every storage cistern shall be fitted with:

  (a) an overflow pipe, with a suitable means of warning of an impending overflow, which excludes insects;

  (b) a cover positioned so as to exclude light and insects; and

  (c) thermal insulation to minimise freezing or undue warming.

(5) Every storage cistern shall be so installed as to minimise the risk of contamination of stored water. The cistern shall be of an appropriate size, and the pipe connections to the cistern shall be so positioned as to allow free circulation and to prevent areas of stagnant water from developing.

Storage cisterns should:

- be fitted with an effective inlet-control device to maintain the correct water level
- be fitted with servicing valves on inlet and outlet pipes
- be fitted with a screened warning/overflow pipe to warn against impending overflow
- be supported to avoid damage or distortion that might cause them to leak
- be installed so that any risk of contamination is minimised, and arranged so that water can circulate and stagnation will not take place

- be covered to exclude light or insects and insulated to prevent heat losses and undue warming
- be corrosion-resistant and watertight and must not deform unduly, shatter or fragment when in use
- have a minimum unobstructed space above them of not less than 350 mm.

In situations where two or more cisterns are used to provide the required storage capacity, the cisterns should be connected in parallel. To avoid stagnation, the float-operated valves should be adjusted so that they all operate to the same maximum water level. The cisterns must be connected so there is an equal flow of water through each cistern.

Remember

The entire base of all cold water storage cisterns must be adequately supported to avoid distortion or damage. They must be installed in a position where the inside may be readily inspected.

WSR Schedule 2, para 16(1)

Figure 3.13: Requirements for protected cisterns used to store drinking water conforming to fluid category 1

Did you know?

The clearance space allows any float-operated valve or other control to be installed, repaired, renewed or adjusted and also aids frost protection.

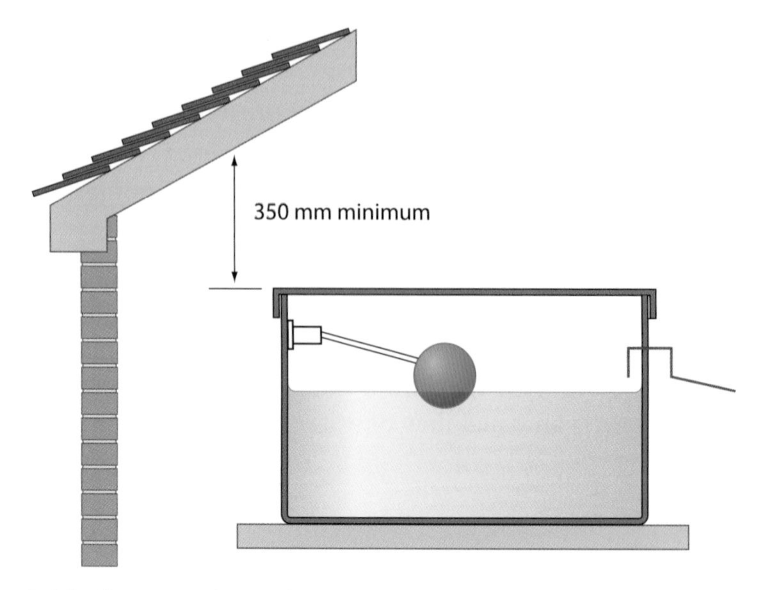

Figure 3.14: Clearance above cistern

Cistern inlet controls

Every pipe supplying water to a storage cistern shall be fitted with an effective, adjustable shut-off device that will close when the water reaches its normal full level below the overflowing level of the cistern. Generally, the device will be a float-operated valve, although larger cisterns may be fitted with a float switch, connected to an electrically operated valve or pump.

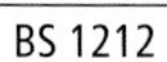
BS 1212

Where float-operated valves are used, they should comply with one of the following standards (which cover valves up to 50 mm in diameter):

- BS 1212 – Part I – Portsmouth type
- BS 1212 – Part II – diaphragm valve (brass)
- BS 1212 – Part III – diaphragm valve (plastic)
- BS 1212 – Part IV – compact-type float-operated valve.

Safety tip

Unless there is a suitable backflow-prevention device such as a double check valve directly upstream of the float valve, BS 1212 Part I valves are not acceptable in a WC cistern or in any location where any part of the valve may be submerged when the overflow pipe is in operation.

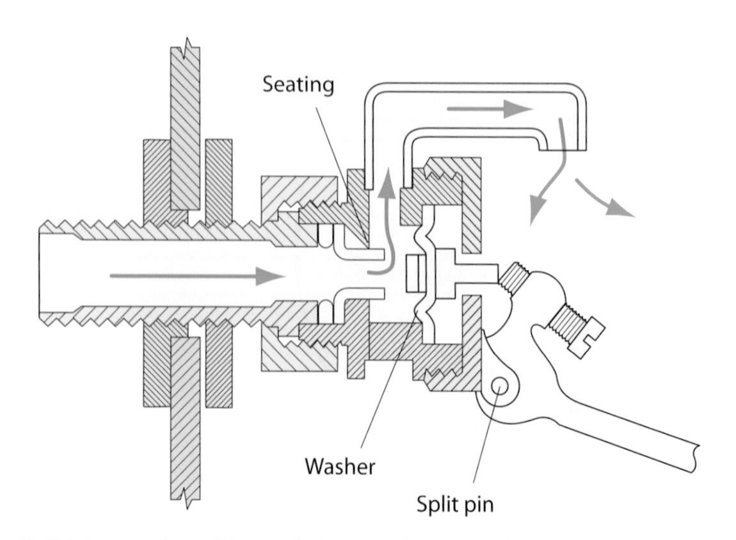

Figure 3.15: Internal workings of cistern inlet controls

Float-operated valves used in WC cisterns should comply with these or with Part IV, which are specially designed for use in WC cisterns.

There are many float valves available that do not comply with the requirements of BS 1212. Look in the WRAS Water Fittings and Materials Directory for other acceptable types.

If installing valves above 50 mm in diameter you will need to ensure that they meet Water Regulations standards. This can be done by checking in the Water Fittings and Materials Directory, asking the Water Regulations Advisory Scheme for advice or contacting the local water undertaker.

Cistern control valves

Every inlet to a storage cistern, combined feed and expansion cistern shall be provided with a servicing (isolation) valve on the inlet pipe adjacent to the cistern. This requirement also applies to WC and urinal cisterns.

WSR Schedule 2, para 16(2)

The servicing valve should be fitted as close as is reasonably practical to the float-operated valve or other device. This does not apply to a pipe connecting two or more cisterns with the same overflowing levels.

The requirements of Paragraph 16.3 state that every cistern (except one supplying water to a primary circuit of a heating system) shall be provided with a servicing (isolation) valve on the outlet(s). The valve should be fitted as close as is reasonably practical to the cistern.

Warning and overflow pipes

Every storage cistern shall be fitted with an overflow pipe, with a suitable means of warning of an impending overflow. This requirement excludes urinal-flushing cisterns.

WSR Schedule 2, reg 16(4)(a)

The requirement for overflow and warning pipes will vary depending on the water-storage cistern capacity. Cisterns up to 1000 litres or less actual capacity require only a single warning/overflow pipe. Where a cistern has a greater actual capacity than 1000 litres, it is recommended that a warning pipe and an overflow pipe should be provided.

Remember

The warning pipe must be installed so that it discharges immediately when the water in the cistern reaches the defined overflowing level.

All warning pipes must discharge in a conspicuous position, and the water overflow pipe should discharge in a suitable position elsewhere.

To ensure that you have the correct understanding of an overflow pipe and warning pipe, look at the distinctions between the two.

- An overflow pipe is a pipe from a cistern in which water flows only when the water level in the cistern reaches a predetermined level.
- The overflow pipe is used to discharge any overflowing water to a position where it will not cause damage to the building.
- A warning pipe is a pipe from a cistern that gives warning to the owners or occupiers of a building that a cistern is overflowing and requires attention.

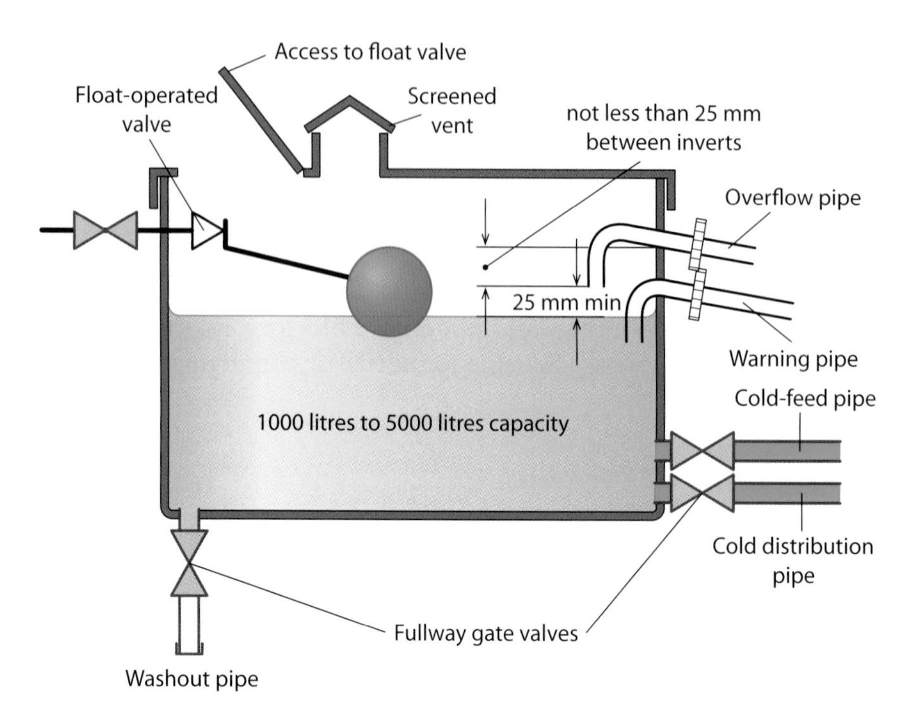

Figure 3.16: Overflow and warning pipes

Either pipe must not have an internal diameter of less than 19 mm, and the diameter of the pipe installed must be capable of taking the possible flow in the pipe arising from any failure of the inlet valve.

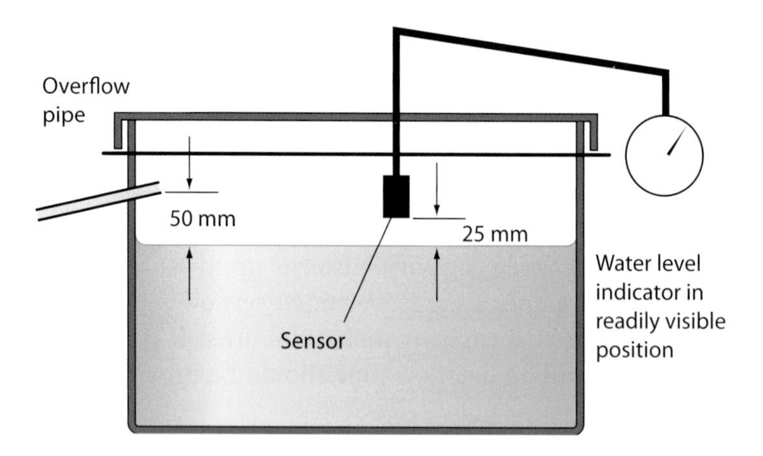

Figure 3.17: Cistern water level indicator

Larger cisterns

Cisterns that have an actual capacity greater than 5000 litres should be provided with an overflow that operates when the water level is 50 mm above the set shut-off level. It is acceptable to omit the warning pipe, but a level indicator should be provided, and the installation must include an audible or visible alarm that operates when the water reaches 25 mm below the opening of the overflow.

Alternative methods are a float-operated water level indicator and overflow pipe, as shown in Figure 3.18.

In situations where the cistern is larger than 10 000 litres capacity, the cistern must be fitted with either:

- a warning pipe and an overflow pipe (same criteria as for medium and large cisterns)
- an audible or visual alarm (electrically operated) that clearly indicates a rise in water level to within 50 mm of the cistern overflowing level
- a hydraulic audible or visual alarm that clearly indicates when the water level rises to within 50 mm of the overflowing level.

Installation factors

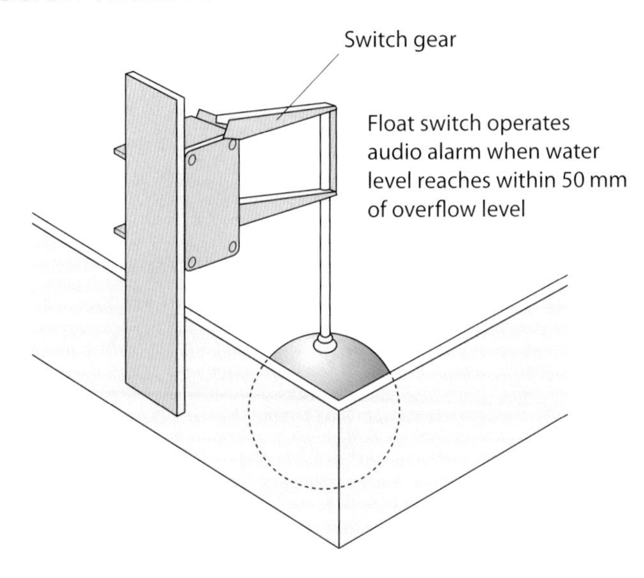
Switch gear
Float switch operates audio alarm when water level reaches within 50 mm of overflow level

Figure 3.18: Large-capacity cistern warning device

There are several important installation factors to consider when installing overflow/warning pipes.

- The overflow/warning pipe must be capable of removing the excess water from the cistern without the inlet device becoming submerged in the event of an overflow.
- Warning pipes are to discharge in a conspicuous position, preferably in an external location.
- Warning/overflow pipes must fall continuously from the cistern to the point of discharge.
- Feed and expansion cisterns must have separate warning pipes from those serving cold water cisterns.
- Warning pipes and overflow pipes must be fitted with some means of preventing the ingress of insects, etc. (usually in the form of screens or filters).
- When the installation consists of two or more cisterns, the warning or overflow pipe must be arranged so that the cistern cannot discharge one into the other.

WSR Schedule 2, para 16(4)

Paragraph 16.4(b) requires that cisterns are to be fitted with a cover and be positioned so as to exclude light and insects, and 16.4(c) requires that insulation shall be fitted to minimise freezing or undue warming – this includes insulation to the overflow and warning pipe.

Contamination of stored water

WSR Schedule 2, para 16(5)

Paragraph 16.5 requires that cisterns are to be installed to minimise the risk of contamination of stored water. This can be achieved to a certain extent by installing a 'protected cistern', especially in cases where water is supplied for domestic purposes.

Further requirements of the paragraph are that the cistern must be of an appropriate size and connections positioned to allow circulation and prevent areas of stagnation from developing. To reduce the potential risk of contamination in cisterns, the following factors should be considered.

- Cistern outlet connections should be connected as low as possible; this will allow sediment to pass through the taps rather than settle in the base of the cistern.
- Cisterns must be adequately sized and not oversized, thus reducing the risk of legionella and ensuring that there is a speedy replenishment of fresh water when stored water is being drawn off.
- Cistern outlet connections should be installed so as to allow movement of water throughout the entirety of the cistern. This can be achieved by connecting at least one outlet pipe to the appropriate end of the inlet connection.
- In instances where storage cisterns are linked together they should be installed in such a manner that they can be drained and cleaned easily. Cisterns that are connected in series should also be installed in such a manner as to allow a good throughput of water to reduce the risk of stagnation.

Remember

Make sure the connections are made into the bottom of the cistern and not the side wall.

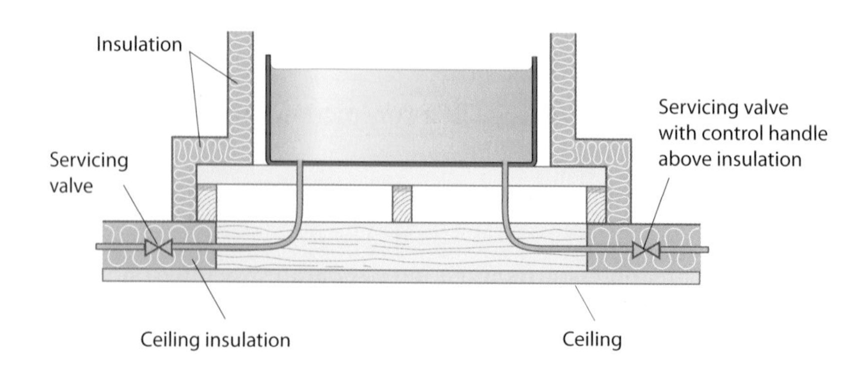

Figure 3.19: Cistern connections

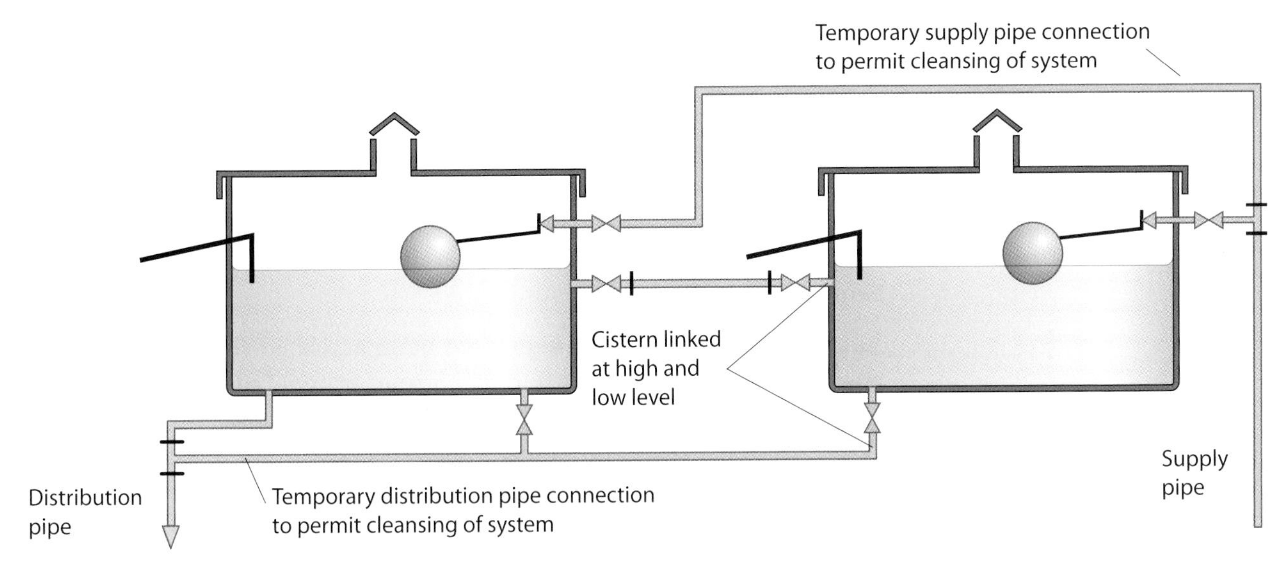

Figure 3.20: Connecting cisterns in series

Float switches and solenoid valves

Because of the advancement in electronics and electrical engineering, systems are available to control the flow of water into components such as cisterns. This then removes the float valve from the system and, along with it, the mechanical faults that occur with float valves.

One method of controlling the flow of water into a cistern is with the use of a **solenoid** valve.

Key term

Solenoid – generic term for a coil of wire used as an electromagnet; a device that converts electrical energy to mechanical energy using a solenoid.

Did you know?

Common applications of solenoids are to power a switch, such as the starter in an automobile, or a valve, such as in a sprinkler system.

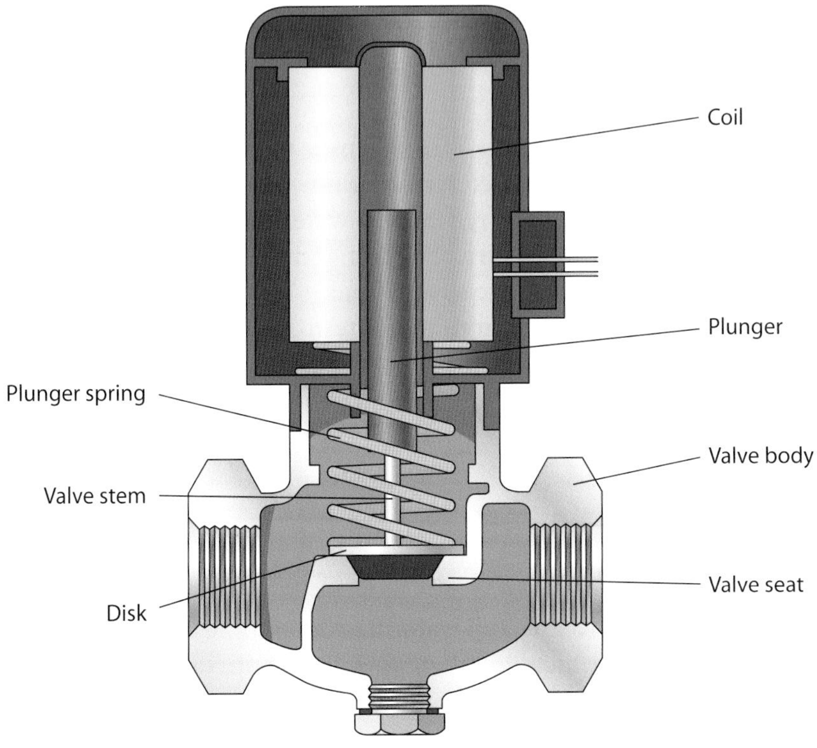

Coil
Plunger
Plunger spring
Valve body
Valve stem
Valve seat
Disk

Figure 3.21: Inside a solenoid valve

How a solenoid works

A solenoid is a coil of wire in a corkscrew shape wrapped around a piston, often made of iron. A magnetic field is created when an electric current passes through the wire, so it can be used as an electromagnet. Electromagnets have an advantage over permanent magnets in that they can be switched on and off by the application or removal of the electric current, which is what makes them useful as switches and valves and allows them to be entirely automated.

As in all magnets, the magnetic field of an activated solenoid has positive and negative poles that will attract or repel material sensitive to magnets. In a solenoid, the electromagnetic field causes the piston to either move backward or forward, which is how motion is created.

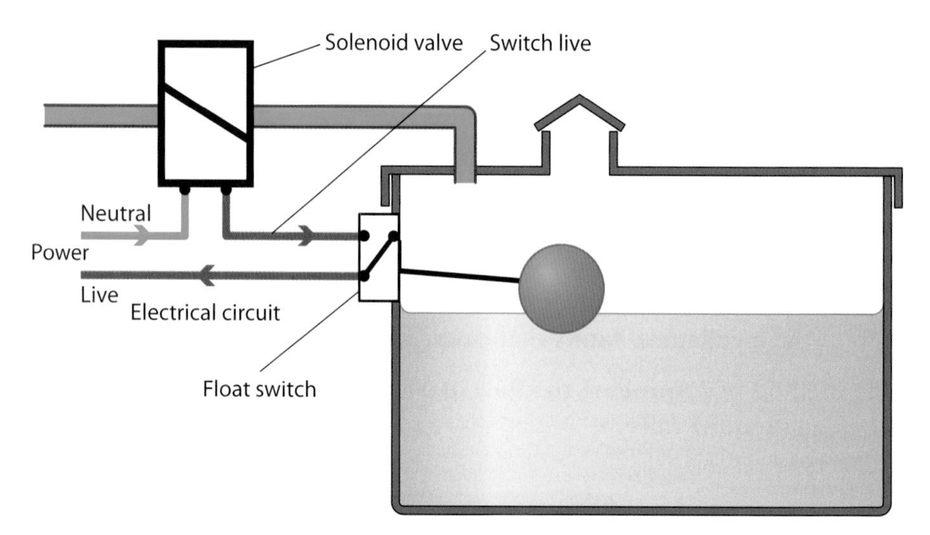

Figure 3.22: Float switch with solenoid valve

The solenoid valve needs some sort of sensor to enable it to work. The sensor acts as a switch that is either on or off. In Figure 3.22, you can see that, when the cistern has reached its water level, the float switch disconnects power to the electrical circuit and the solenoid closes; when the water level drops, the switch 'makes' (closes) and the solenoid opens, allowing water to flow into the cistern.

One of the drawbacks with this type of system is that, when the solenoid valve closes, it does not close slowly and this may cause shocks or water hammer in the pipework.

Specialist inlet valves

The Arclion®

The Arclion® is used to give full water flow at all times. It cuts out 'dribble conditions', which are often associated with conventional float valves. This valve is often used in automatically pumped and boosted water systems.

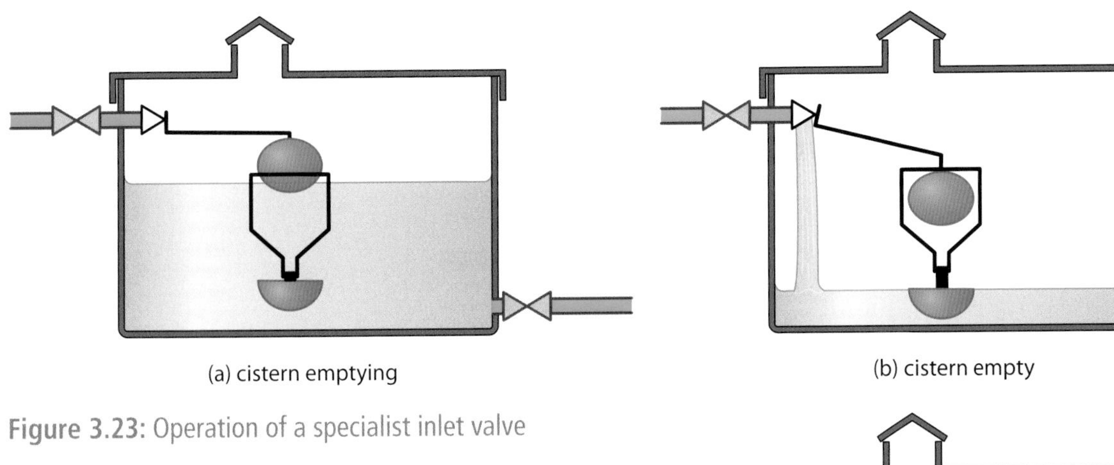

Figure 3.23: Operation of a specialist inlet valve

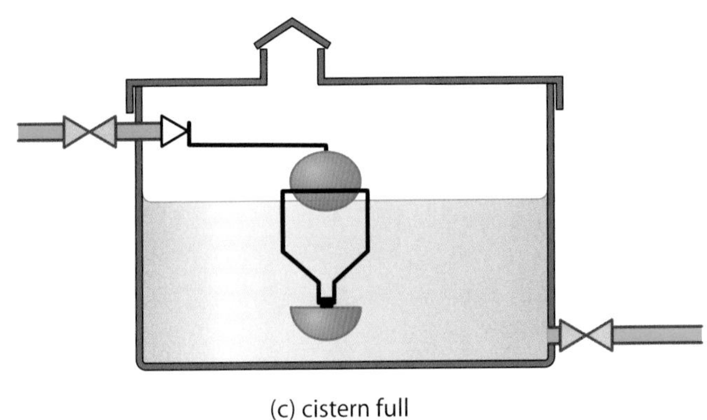

(a) **Cistern emptying**

As water is drawn from the storage cistern, the canister valve stays closed, which stops the main float-operated valve opening.

(b) **Cistern empty**

When the cistern water level reaches a preset depth, the canister valve opens. This makes the main float-operated valve open quickly and fully.

(c) **Cistern full**

When the cistern starts to fill, the canister valve closes, letting the water rise until it floods over the top of the canister. This causes the main float-operated valve to shut off quickly.

The Aylesbury delayed-action float valve

Aylesbury valves are designed to provide an accurate and efficient method of controlling the level of stored cold water in cisterns with and without raised float valve chambers. The valves are easy to install with an 'up and over' discharge arrangement which assists in facilitating Type AA, AB, AF or AG air gap requirements under the Water Regulations. The Aylesbury range is ideal for pumped systems as the open to closed 'on/off' valve operation avoids pump hunting and water hammer. This type of Aylesbury valve can be used for deep cisterns and cisterns with a minimum depth of one metre.

The benefits of fitting this type of valve to large cisterns are:

- no water hammer
- full flow during the fill
- able to maximise the capacity of water in the cistern
- suitable for all types of air-gap protections
- no valve bounce as with float-operated ball valve
- no water dribble as valve is closing
- has a delayed closing action
- minimises cistern wall stress.

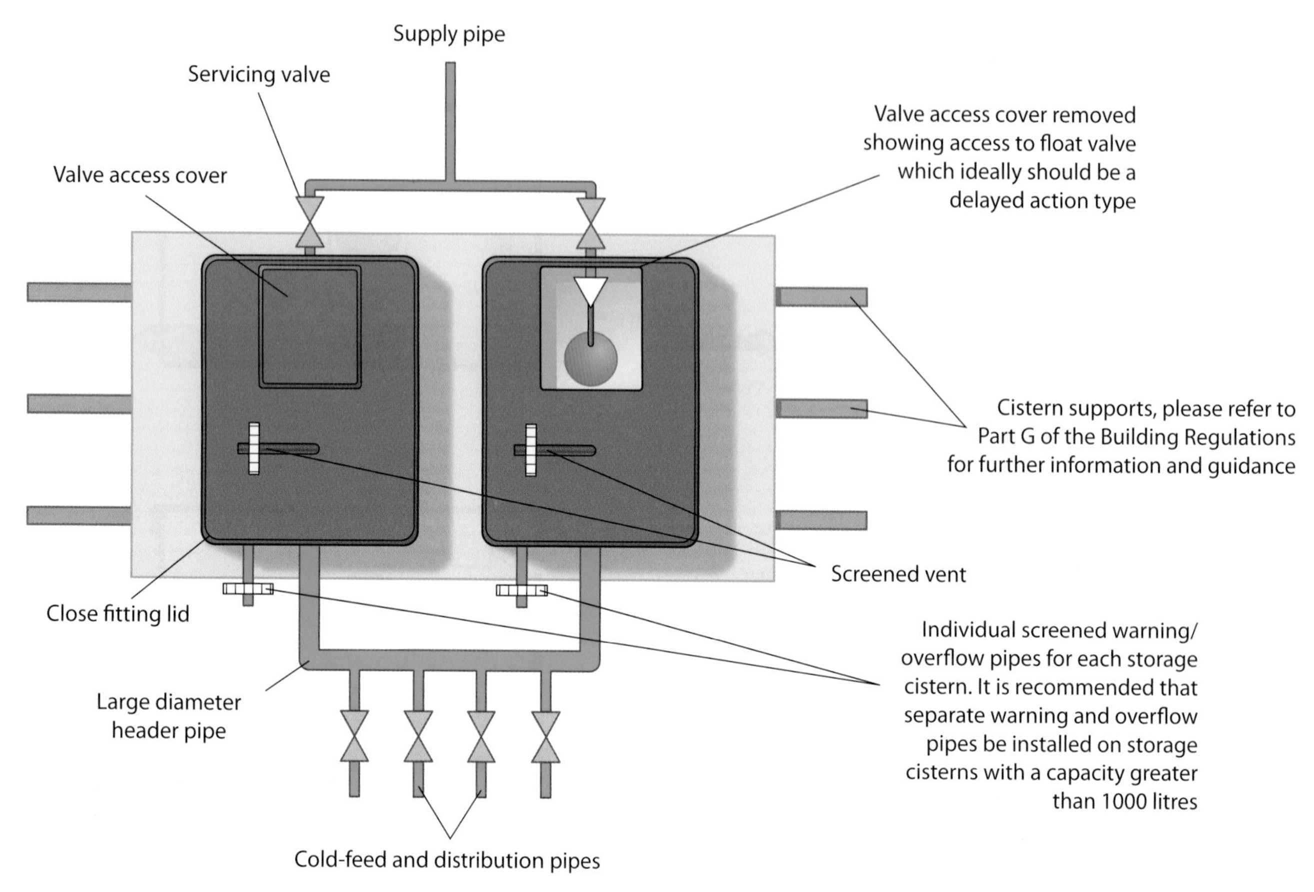
Supply pipe
Servicing valve
Valve access cover
Valve access cover removed showing access to float valve which ideally should be a delayed action type
Cistern supports, please refer to Part G of the Building Regulations for further information and guidance
Screened vent
Close fitting lid
Individual screened warning/ overflow pipes for each storage cistern. It is recommended that separate warning and overflow pipes be installed on storage cisterns with a capacity greater than 1000 litres
Large diameter header pipe
Cold-feed and distribution pipes

Figure 3.24: Plan view of linked cisterns to prevent stagnation

Interlinking multiple cisterns

Cisterns can be linked together in a way that complies with the Water Regulations, as is shown in Figure 3.24. This system will help avoid water stagnating, which can lead to health problems.

Use of sectional cisterns

Where larger cisterns are required for storage these would be of the sectional construction type. There are three types of sectional cistern:

1 externally flanged cistern
2 internally flanged cistern
3 totally internally flanged cistern.

They are manufactured to BS EN 13280: 2001 *Specification for glass fibre reinforced cisterns of one-piece and sectional construction, for the storage, above ground, of cold water.*

Space allowance

Externally and internally flanged cisterns must have a minimum of 500 mm access to the sides and base and 750 mm above, for access to the float valve and to gain access into the cistern for cleaning purposes.

Totally internally flanged cisterns can be used where space on the outside is at a premium, only requiring 25 mm clearance around the sides and 750 mm above.

The same requirements apply to large sectional cisterns as to small cisterns. Even when installed outside, the temperature of the water must not be allowed to rise above 20°C. This is to be controlled by the use of insulation installed during the manufacturing process of the GRP panels.

Break cisterns

Water supplies to buildings vary greatly in pressure and in the quantity available, as supplied by the water undertaker. In multi-storey developments, this can give rise to intermittent supplies at times of greatest use. In these situations there is the need for the use of a pumped boosted supply.

Boosted supplies can be divided into direct boosted supplies or indirect boosted supplies. The latter is the most preferred by the water undertakers as pumping direct from the main may reduce the main pressure to other users and can also increase the risk of backflow into the main water supplies.

In circumstances where water pressure is available in the supply pipe and the demand is less than 0.2 litres per second, or if the demand is greater and the water undertaker agrees, drinking water may be pumped directly off the supply pipe.

WSR Schedule 2, para 6(1)

When booster pumps are used they can cause excessive aeration. Although this does not cause deterioration of water quality, for some consumers the turbid (opaque) appearance of the aerated water can cause concern.

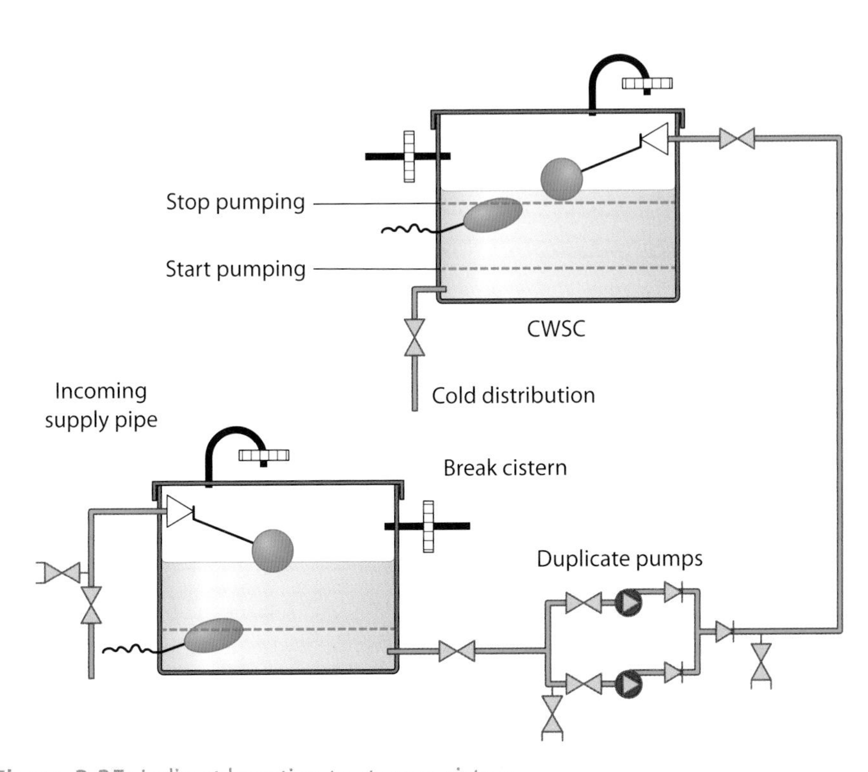

Figure 3.25: Indirect boosting to storage cistern

Within the booster pump system, sampling taps on the outlet side of the pumps are usually recommended for testing water quality.

Where boosted water systems are not directly connected to the mains they will be fed from a break cistern.

Indirect boosting to storage cistern

Where the water undertaker insists on a break cistern being incorporated in the installation, the pumps should be fitted to the outlet from the break cistern. Sizing of the break cistern should be decided after all aspects of the system requirements and location within the building have been decided. The break cistern should have a capacity equal to

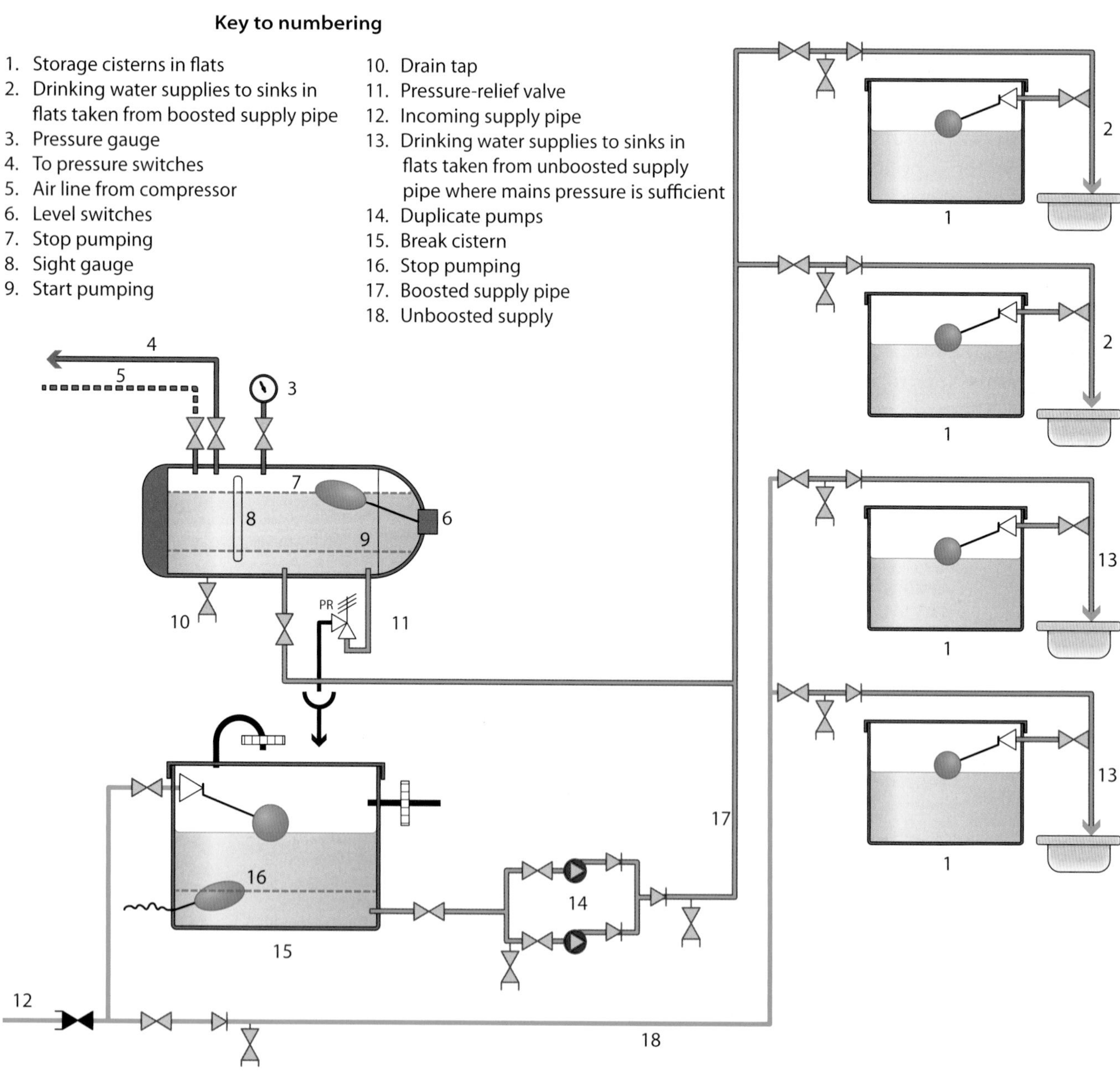

Figure 3.26: Indirect boosting with pressure vessel (note that this figure does not show any additional backflow-prevention devices that might be required in accordance with 5.6 of BS 6700: A1: 2009)

no less than 15 minutes of the pump output. The cistern should also not be oversized as this could lead to stagnation of the water.

The water level in the storage cistern(s) is controlled by means of water level switches that control the pumps. When the water level drops to a predetermined level, the pumps start and switch off when the water level reaches a point approximately 50 mm below the float-operated valve shut-off.

A water level switch should also be positioned in the break cistern to cut out the pumps when the level of water drops to approximately 225 mm above the suction connection in the break cistern. This will ensure that the pumps do not run dry (see Figure 3.26).

Clarke CBM25055 Boosted Cold Water Pump

Indirect boosting with pressure vessel

In buildings where a boosted supply serves a number of delivery points or storage cisterns at various levels (for example, in a block of flats), it might not be practicable to control the pumps by means of a number of level switches. An alternative would be to use a pressure vessel that contains both air and water under pressure. The pressure vessel, pumps and air compressor would usually be purchased as a packaged pressure set together, with all the necessary control equipment.

How does it work?

With the boosted water supply, if there were no pressure vessel, the pumps would be cycling continuously as taps or float valves are opened. Now when the water level falls, the float switch float also falls to a predetermined level in the pressure vessel, contacts are 'made' in the control box, and the pumps will activate.

Booster pumps

Booster pumps are available in two forms: either as a pre-assembled set with integral controls, or as sets to assemble yourself.

Sets with integral controls

For domestic properties the ideal boosted pump equipment comes as a pre-assembled set with all electrical controls, pumps and valves packaged together and ready to be integrated into a system. With these sets it is also advisable to install a cold water supply bypass into the system pipework, for those times that the unit is being serviced or has broken down. The electrical supply would normally be a single cable to either a plug or switched fuse spur. Information on the size of fuse required for the unit would be included in the manufacturer's instructions.

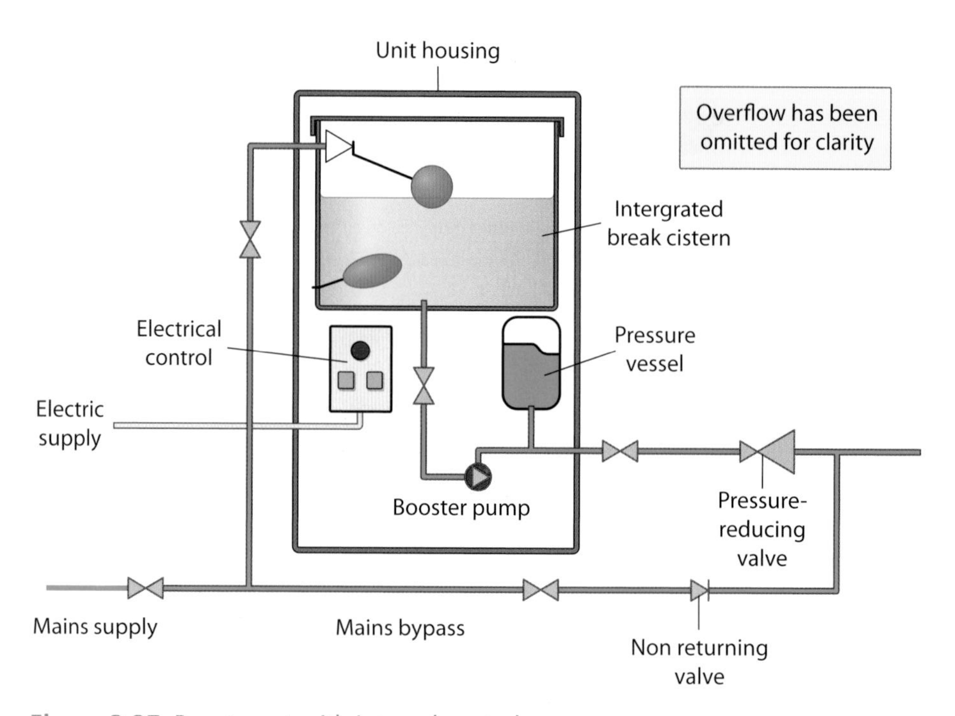

Figure 3.27: Booster set with integral controls

Self-assembled sets

Many of the larger systems may be bespoke systems designed by mechanical service designers or pump manufacturers. Because of their size it would not be feasible to pre-assemble them, so on-site assembly would be required. Instructions are usually provided and should be followed. One example would be a booster set such as the indirect boosting with pressure vessel, shown in Figure 3.26 on page 76.

The electrical controls would need to be wired in separately by an approved contractor. The manufacturers for this type of system would also commission the system on completion.

Pressure/expansion vessels

Pressure/expansion vessels are for use in buildings where the pump cannot be controlled by individual level switches due to the various levels having a number of storage cisterns. They can be charged with either air or nitrogen; if charged with nitrogen, then nitrogen must be used when recharging the vessel. When used for expansion of cold water systems, they would be used on the cold water supply to an unvented hot water cylinder. They take up pressure that has been built up within the unvented system.

An accumulator increases the flow rate of the water

Accumulator

An accumulator is a component to be used as a means of storing water at mains pressure. An accumulator usually consists of a sealed, steel shell that is spherical or near spherical in shape. The insides are split into hemispherical shells with a flexible diaphragm. The purpose of an accumulator on these systems is not to increase the pressure of the water, but to increase the flow rate of the water.

An accumulator would be used where there is sufficient pressure from the mains but not the flow rate.

How does it work?

The water pressure is balanced by air pressure between the accumulator wall and diaphragm. When any tap or shower is turned on, the stored water from the accumulator will supplement water from the incoming main, thus increasing the flow rate at that outlet. This will be maintained until empty, where the performance of the cold water supply is returned back to the main supply. The accumulator will refill once the outlets close.

Pressure switch (transducer)

Pressure switches are normally fitted within the system pump by the manufacturer. Without them, the pump would be operating at the slightest drop in pressure. Pressure switches detect a drop in the pressure of the water; as pre-determined by the manufacturer during the design of the booster pump.

Float switch

The purpose of a float switch is to open or close a circuit as the level of water rises or falls. Most float switches are 'normally closed', meaning that the two wires coming out of the switch complete a circuit when the switch is at its lowest point, and the component which it is controlling will activate.

PROGRESS CHECK

1. What is the term given by the Water Regulations to a cistern that is used to provide water for domestic purposes?
2. Where a sectional cistern is installed outside, what should the temperature of the water never exceed?
3. Why is an accumulator used on cold water systems?
4. What is the purpose of a float switch?
5. What is a solenoid valve?
6. Where should a service valve be fitted?
7. How would you know if a valve or fitting complies with the Water Regulations?
8. Where a boosted cold water system supplies delivery points or cisterns, which is the best option for controlling the pump?
9. What is the minimum space required above a cistern?
10. Within what distance should a service valve be fitted to a float-operated valve?

3. Know the types of cold water system layout used with single-occupancy dwellings fed by private water supplies

Providing private water supplies

The water you deal with in people's homes all starts as rainwater, but water supplies are collected from a variety of sources. It is important that water is collected and treated with care so that it reaches users in a clean, safe condition.

Pumped from wells and boreholes

Householders who live in remote areas away from mains-supplied water may have access to a private well or could be supplied from a stream, springs or ponds. Where the water supply is provided from these sources, other methods of distributing and storing the water at source will have to be taken into consideration. Before first use, the water must be tested within a laboratory, with a comprehensive report of the findings. The tests will check for all known contaminants that can be in the water supply.

Most deep wells should be usable with little or no treatment prior to use because of the type of strata as shown in Figure 3.28. Surface water cannot penetrate the impervious strata and therefore surface contaminants (such as chemicals which have been spread on the land) will not get into this water supply.

Where the well does not penetrate the impervious strata then there is a risk of surface contaminants seeping into the water. Any surface water that is taken and used will also have the same possibility of contamination from chemicals, animal droppings and also dead animals which can die on farmland.

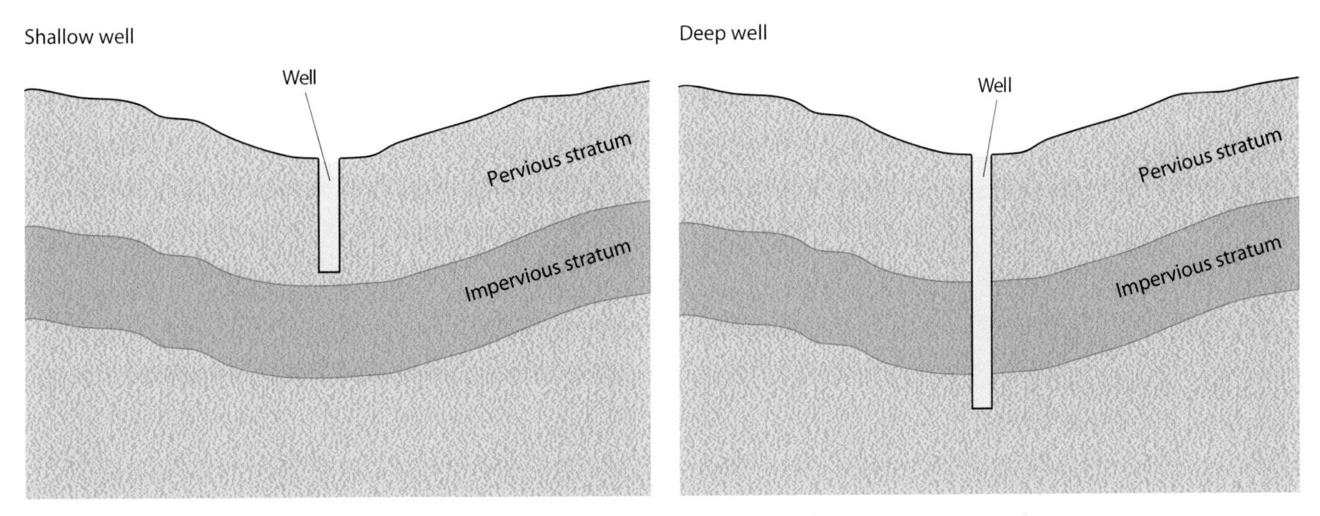

Figure 3.28: Shallow well does not penetrate impervious strata. Deep well penetrates impervious strata

The water when taken from any of these sources would first have to go through a break cistern. As with any cistern the installation would still have to comply with Schedule 2 Section 7 Paragraph 16 (1–5).

The sump pump is required for when the cistern is overflowing: the water needs to discharge to a point where it is visible and the user would have to take action to rectify the problem.

Insulation of pipework is essential and calculations for the thickness of insulation for pipework can be found in BS 5422. As the pipework is external and in a chamber, it would be advisable to install low-temperature trace heating to prevent freezing of the pipework.

Schedule 2, para 4

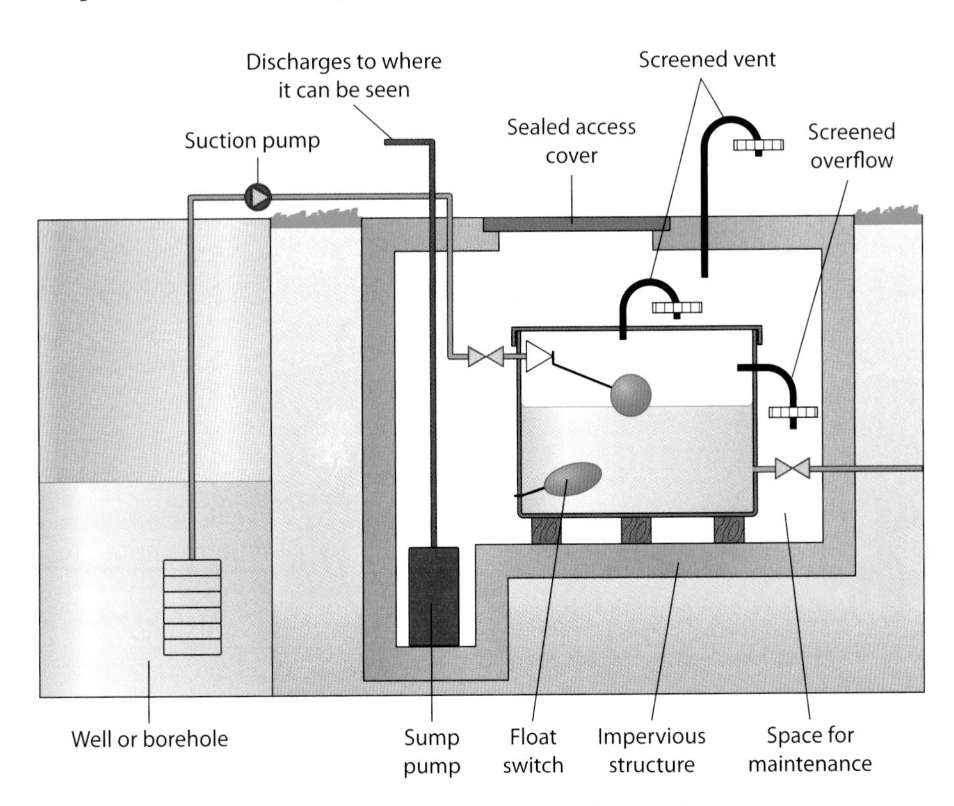

Figure 3.29: Supply to break cistern below ground from well or borehole

Collected from surface water sources – streams and springs

Did you know?

There are several methods of extracting water from wells or boreholes. The method used will depend on the depth of the well. A reciprocating positive displacement pump can be used in wells to a depth of 8 metres. Where the depth is greater than 8 metres the use of a submersible centrifugal pump is the preferred option.

Surface water is usually taken from streams and springs. Unlike groundwater, it is not protected from nature or human activities, so treatment is always necessary. Surface water level and quality will vary over the seasons: for example, after heavy rainfall or snowmelt, lots of solids and sand are washed downstream. These sharp and abrasive minerals, as well as biodegradable materials, need to be settled or screened off before pump intake to avoid negative effects on the final water-treatment process. Submersible pumps are ideal for applications with periodic uncontrollably high-water levels. Note that power cables and electric equipment must be elevated to permanently dry locations. Before extracting water from streams, the customer must first obtain a licence from the Environment Agency.

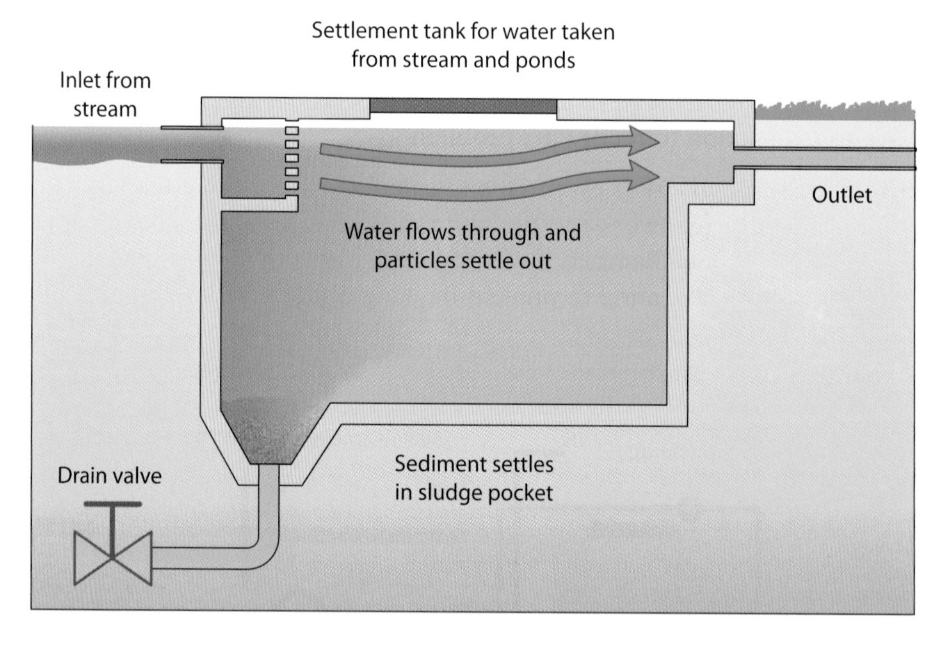

Figure 3.30: Settlement tank

Treating water

Once you have extracted the water from the spring or well, you need to filter the water to remove any objects and prevent contamination, to allow for a clean supply.

Localised water filtration units

In single-occupancy dwellings there are several means of filtering water for use. The methods vary from filter cartridges for single outlet demand to reverse osmosis units that can be either for a single outlet or whole-house demand.

Reverse osmosis is a process in which dissolved inorganic solids (such as salts) are removed from a solution (such as water). This is accomplished by household water pressure pushing the tap water through a semi permeable membrane. The membrane (which is about as thick as cellophane) allows only the water to pass through, not the impurities or contaminates. These impurities and contaminates are flushed down the drain.

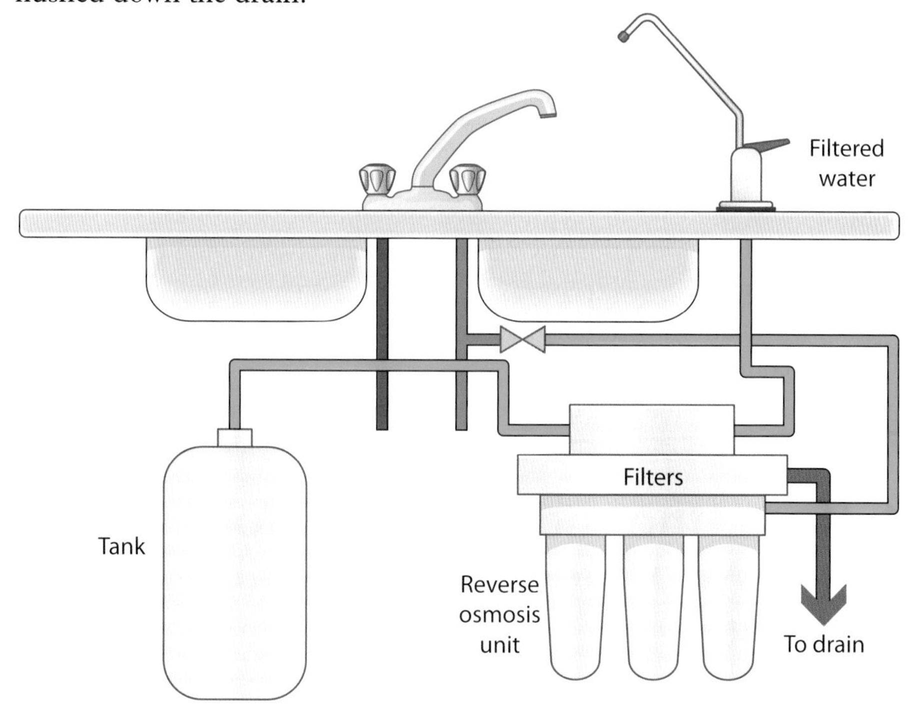

Figure 3.31: Reverse osmosis unit for single tap outlet

Localised water treatment units – ultraviolet

Ultraviolet disinfection of the water is the easiest to install. You can purchase one for single-point application or you can have them for the whole house. There is no need to store water; the system disinfects as the water passes the ultraviolet light. The bulbs are easy to change, just like changing a fluorescent tube. It is advisable to filter the water before passing it through the ultraviolet filter to remove suspended solids, as these solids can affect the effectiveness of the filter.

Did you know?

There are many methods of filtering water at source to remove particulates. One way is to use a settlement tank in which large particles settle out of the water. The other, more effective, method is the slow sand filter. The water passes through the Schmutzdecke layer. This layer provides the effective purification in potable water treatment with the underlying sand providing the support medium for this biological treatment layer.

Advantages of UV disinfection

- More effective against viruses than chlorine
- Environmentally and user-friendly, with no dangerous chemicals to handle or store, and no risks of overdosing
- Low initial capital cost as well as reduced operating expenses when compared with similar technologies such as ozone, chlorine, etc.
- Immediate treatment process, with no need for holding tanks, long retention times, etc.
- Extremely economical: thousands of litres may be treated for each penny of operating cost
- No chemicals added to the water supply, so no by-products (chlorine + organic compounds = trihalomethanes)
- No change in taste, odour, pH, conductivity, nor the general chemistry of the water
- Automatic operation without special attention or measurement, making it operator-friendly
- Simplicity and ease of maintenance, with periodic cleaning (if applicable), annual lamp replacement, and no moving parts to wear out
- Easy installation, with only two water connections and a power connection
- Compatible with all other water purification processes (reverse osmosis, filtration, water conditioning and softening, BioSand)

4. Know the requirements for backflow protection in plumbing systems

A backflow-prevention device stops contamination of drinking water by backflow. Devices can be mechanical, non-mechanical or an arrangement in the pipework system preventing backflow (for example, check valves, double check valves or air-gap arrangements).

Before installing a backflow-prevention device, you will need to know the correct type suitable for the particular installation, and your selection will also depend on the fluid risk category (see opposite). Guidance for these two important factors is given in the Regulator's specification on the prevention of backflow, where several tables are provided. Table A.1 in Appendix A is a list of non-mechanical backflow-prevention devices acceptable under the WSR. Table A.2 lists mechanical backflow-prevention devices.

The five fluid risk categories

WSR Schedule 1

Schedule 1 of the Water Regulations recognises and implements a five fluid risk category list, based on that developed by the Union of Water Supply Associations of Europe (EUREAU 12), and which is also currently used in North America and Australia. These fluid risk categories describe water based on how drinkable it is and how dangerous to health it may be, depending on impurities. Table 3.5 outlines the description and application of each category.

Remember

The type of backflow-prevention device needed will depend on the severity of the risk.

Fluid category	Description	Application
1	Wholesome water supplied by a water undertaker complying with the requirements of the Regulations made under Schedule 67 of the Water Industry Act 1991	
2	Water that would be classed as fluid category 1 except for odour, appearance or temperature. These changes in water quality are aesthetic changes only and the water is not considered a hazard to human health	(a) water heated in a hot water secondary system (b) mixtures of fluids from categories 1 and 2 discharged from combination taps or showers (c) water that has been softened by a domestic common salt regeneration process
3	These fluids represent a slight health hazard and are not suitable for drinking or other domestic purposes	(a) in houses or other single-occupancy dwellings, water in: (i) primary circuits and heating systems, whether additives to the system have been used or not (ii) wash basins, baths or shower trays (iii) washing machines and dishwashers (iv) home dialysing machines (v) handheld garden hoses with flow-control spray or shut-off control (vi) handheld garden fertiliser sprays (b) in premises other than a single-occupancy dwelling (c) where domestic fittings such as wash basins, baths or showers are installed in premises other than a single-occupancy dwelling (that is, commercial, industrial or other premises) these appliances may still be regarded as a fluid category 3, unless there is a potentially higher risk. Typical premises that justify a higher fluid risk category include hospitals and other medical establishments (d) house, garden or commercial irrigation systems without insecticide or fertiliser additives, with fixed sprinkler heads not less than 150 mm above ground level (e) fluids that represent a slight health hazard because of the concentrations of substances of low toxicity include any fluid that contains: – ethylene glycol, copper sulphate solution or similar chemical additives or – sodium hypochlorite (chloros and common disinfectants)

Fluid category	Description	Application
4	These fluids represent a significant health hazard and are not suitable for drinking or other domestic purposes. They contain concentrations of toxic substances	(a) water containing chemical carcinogenic substances or pesticides (b) water containing environmental organisms of potential health significance (microorganisms, bacteria, viruses and parasites of significance for human health which can occur and survive in the general environment) (c) water in primary circuits and heating systems other than in a house, irrespective of whether additives have been used or not (d) water treatment or softeners using other than salt (e) water used in washing machines and dishwashing machines for other than domestic use (f) water used in mini-irrigation systems in a house garden without fertiliser or insecticide applications such as pop-up sprinklers, permeable hoses or fixed or rotating sprinkler heads fixed less than 150 mm above ground level
5	Fluids representing a serious health risk because of the concentration of **pathogenic organisms**, radioactive or very toxic substances, including any fluid which contains: (a) faecal material or other human waste, or (b) butchery or other animal waste, or (c) pathogens from any other source	(a) sinks, urinals, WC pans and bidets in any location (b) permeable pipes or hoses in other than domestic gardens, laid below ground or at ground level with or without chemical additives (c) **greywater** recycling systems (d) washing machines and dishwashers in high-risk premises (e) appliances and supplies in medical establishments

Table 3.5: The fluid risk categories

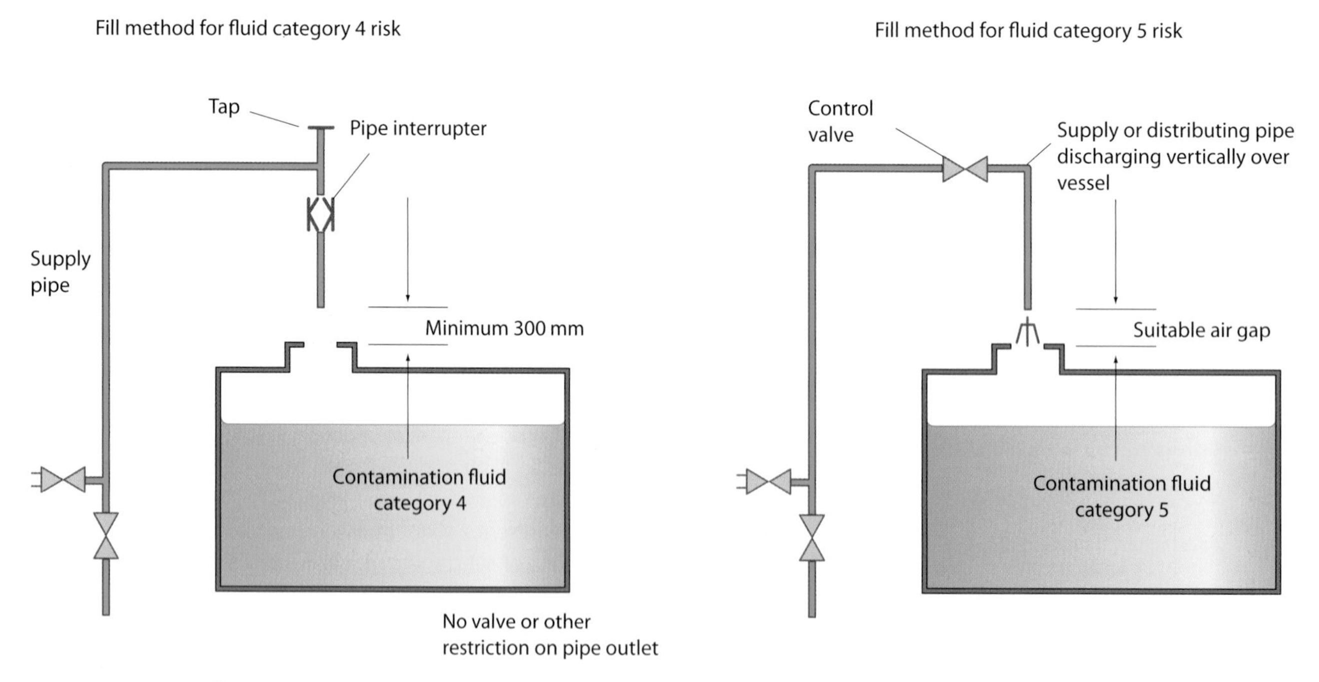

Figure 3.32: Backflow-prevention arrangements. These can be used for commercial or agricultural applications, such as farms or chemical plants

Professional Practice

Marlon is about to install a commercial washing machine (for clothes) in a laundry. This represents a fluid category 4 risk. What backflow-prevention device should he install?

Installation

After determining the fluid category and the appropriate backflow-prevention device, you will need to look at any installation requirements that may apply.

Backflow-prevention devices are normally **point-of-use devices**, which means that they protect against individual risk at or near the point of supply or at the point where the risk is likely to occur with an individual appliance.

Whole-site and zone protection are covered on pages 99–102.

Further advice on the installation of backflow-prevention devices is given in the Regulator's Specification on the Prevention of Backflow:

- G15.2 Backflow prevention can be achieved by good system design and the provision of suitable backflow-prevention arrangements and devices.
- G15.5 Where practicable, systems should be protected against backflow without relying on mechanical backflow protection; preferred protection is by the use of tap gap or air gap at the point of use.
- G15.6 Permanently vented distributing pipes will provide good 'secondary' protection (whole-site or zone protection) in many cistern-fed installations.
- G15.7 Mechanical backflow-prevention devices, which, depending on the type of device, may be suitable for protection against **back pressure** or back siphonage or both, should be installed.
  - (a) They should be readily accessible for inspection, operational maintenance and renewal, and
  - (b) except for types HA and HUK1 (backflow-prevention devices for protection against fluid categories 2 and 3) they should not be located outside premises;
  - (c) they should not be buried in the ground;
  - (d) vented or verifiable devices or devices with relief outlets are not to be installed in chambers below ground or where liable to flooding;
  - (e) if used for category 4 devices, they should have line strainers fitted **upstream** (before the backflow-prevention device) and a servicing valve upstream of the strainer; and

Key terms

Pathogenic organisms – microorganisms such as bacteria, viruses or parasites that are capable of causing illness, especially in humans: for example, salmonella, vibrio cholera and campylobacter.

Greywater – waste water generated by baths, showers, basins, washing machines and dishwashers, which can be recycled and used for flushing WCs and garden irrigation.

Point-of-use device – device to protect against individual risk at or near the point of supply or at the point where the risk is likely to occur with an individual appliance.

GD G15.2 and G15.5–7

Remember

The type of backflow-prevention device needed will depend on the severity of the risk.

Key terms

Back pressure – the reversal of flow in a pipe caused by a decrease in pressure in the system.

Upstream – the supply pipework before the backflow-prevention device.

Remember

Some devices are suitable for back pressure and for back siphonage, but they may not be suitable for the same fluid category.

Key term

Downstream – the supply pipework after the backflow-prevention device.

Reduced pressure zone valve

(f) the lowest point of the relief outlet from any reduced pressure zone valve assembly or similar device should terminate with a type AA air gap located not less than 300 mm above the ground or floor level.

The advantage of air gaps is that, because they are non-mechanical, they require virtually no maintenance – but they must be correctly installed.

Mechanical devices

There are a number of different types of mechanical backflow-prevention device.

Back pressure devices are devices or arrangements for prevention of back pressure, and are those where the outlet control valves or taps are positioned **downstream** of the backflow-prevention device; such devices are B, C and E families. Typical examples are reduced pressure zone (RPZ) valves, non-verifiable disconnectors and pressurised inlet valves.

Back siphonage devices are devices or arrangements for prevention against back siphonage, and are those where the control valve is located prior to the device; such devices are type A, D or H families. A typical example is a tap gap, a pipe interrupter or anti-vacuum valve.

Unlike with non-mechanical backflow-prevention devices, it is very important that mechanical devices are periodically inspected and tested to maintain correct operation. This is a particular requirement when RPZ valves are fitted. The RPZ valve is fairly new to the UK and is ideal for protection against fluid risk 4 applications. The installation of an RPZ valve requires a contractor's certificate, and the water undertaker must be notified. Approved contractors are not required to notify the water undertaker.

WSR Part II, Reg 5

Did you know?

The RPZ valve must only be installed by a competent person and the device must be tested every year. To install, commission and maintain this type of valve you will need to go on a special course. Furthermore, special test equipment needs to be used to commission this type of valve. When installing, altering or disconnecting an RPZ, approved contractors must send a signed certificate of competence to the customer, with a copy to the water undertaker.

The use of backflow-prevention devices

Backflow can be prevented by good system design or by using suitable backflow-prevention devices or arrangements.

WCs and urinals

WC pans and urinals are considered to be a fluid category 5 risk – a serious health hazard irrespective of whether they are installed in a domestic dwelling or industrial or commercial premises. There are two suitable backflow-prevention devices to protect against the risk:

1. an **interposed** cistern type AUK 1 – this means a siphonic or non-siphonic flushing cistern that may be used in premises of any type (see Figure 3.35 on page 90)
2. a pipe interrupter with a permanent atmospheric vent type DC installed to the outlet of a manually operated pressure-flushing valve. It may be connected to a supply pipe or distributing pipe (excluding domestic dwellings).

There should be no other obstruction between the outlet of the pipe interrupter and the flush-pipe connection to the appliance.

> **Key term**
>
> **Interposed** – A type of backflow-prevention device in which a water container is separated from the supply pipe feeding it via a storage cistern that includes an air gap arrangement. A correctly configured water closet fed by a flushing cistern is an example of an interposed cistern.

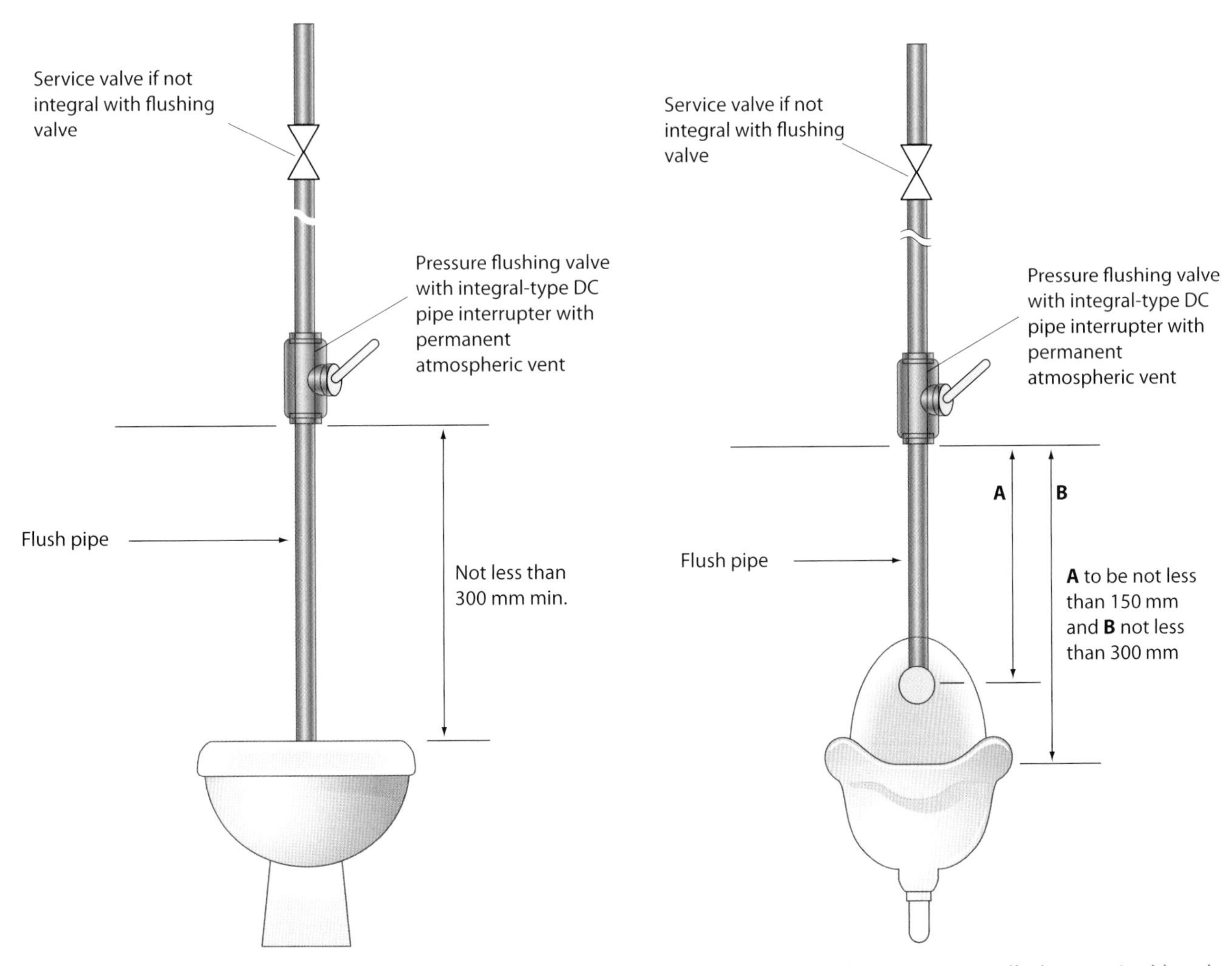

Figure 3.33: Pipe interrupter installed to WC

Figure 3.34: Pipe interrupter installed to a urinal bowl

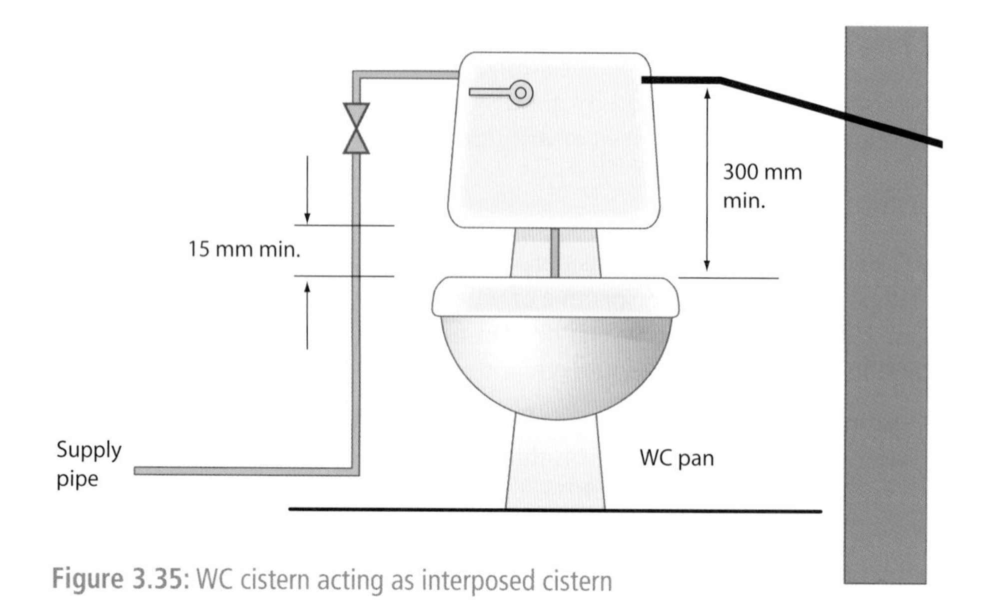

Figure 3.35: WC cistern acting as interposed cistern

Bidets

An over-rim type bidet

Bidets are another type of sanitary appliance with a high fluid risk category. 'Bidets' include WCs adapted as bidets with flexible hose and spray handset fittings or with submerged water inlets. Bidets can be divided into two groups, each of which requires different protection:

1 over-rim types – supplied from a tap at the back edge of the appliance
2 ascending-spray or submerged-inlet types – with a water spray jet situated below the rim of the spill-over level.

Included in the second group are bidets supplied from a tap that uses flexible spray attachments.

Over-rim bidets

Bidets installed in domestic dwellings that are of the over-rim type, having no ascending spray or flexible hose spray, can be supplied with cold and hot water through individual or combination tap assemblies,

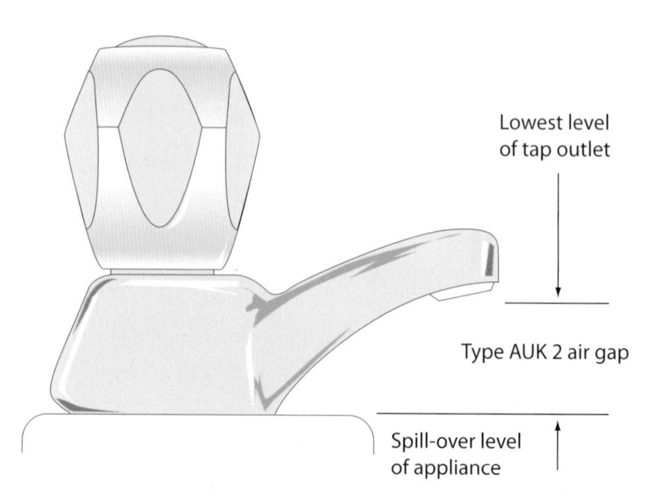

Figure 3.36: Air-gap requirements for over-rim bidets with no spray attachments

from either a supply or distribution pipe, providing that a type AUK 2 air gap is maintained between the outlet of the water fitting and the spill-over level of the bidet.

Ascending-spray bidets

When making connections to ascending-spray bidets, you must remember that they are not permitted to be connected directly from the supply pipe, but must be supplied from a storage cistern. Also, the hot and cold connections must be taken from independent, dedicated distributing pipes that do not supply other appliances.

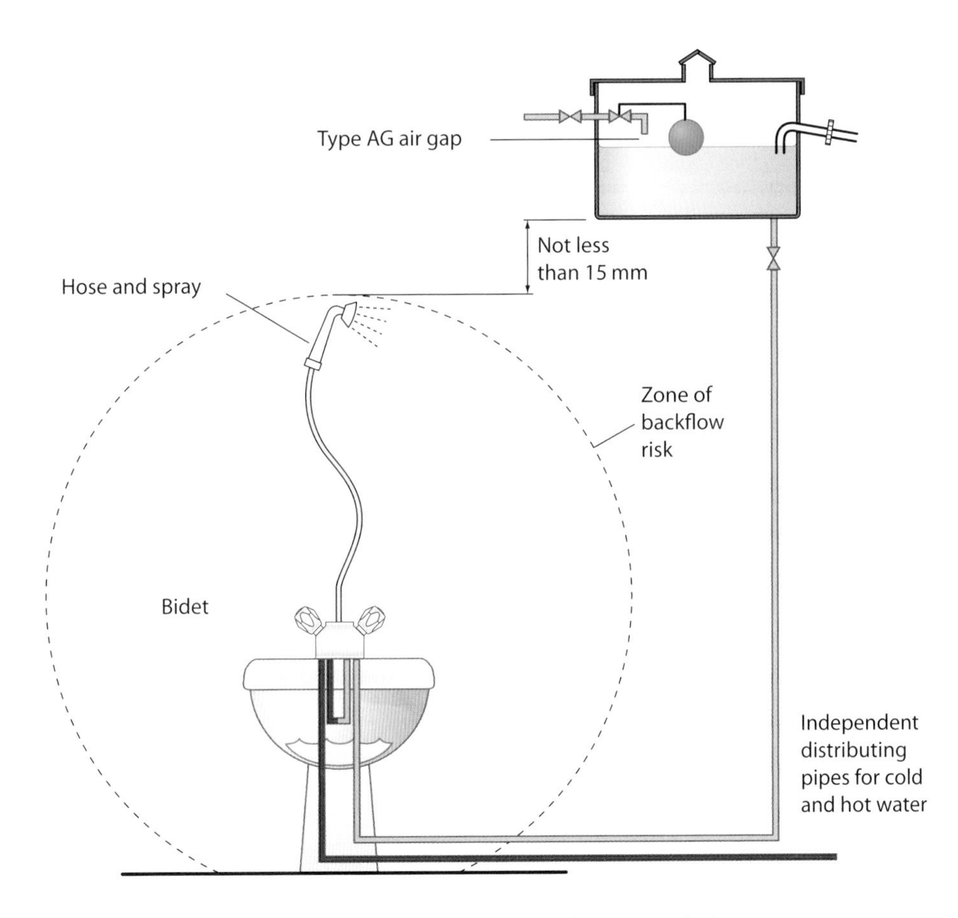

Figure 3.37: Installation requirements for ascending-spray bidet

Exceptions to this can be made for the following cases:

- the common distributing pipe serves only the bidet and a WC urinal flushing cistern
- the bidet is the lowest appliance served from the pipe, with no likelihood of any other fittings being connected at a later date
- the connection to the distribution pipe is not less than 300 mm above the spill-over level of the bowl
- for an over-rim bidet with a flexible spray connection, the connection should not be less than 300 mm above the spill-over level of any appliance that the spray outlet may reach.

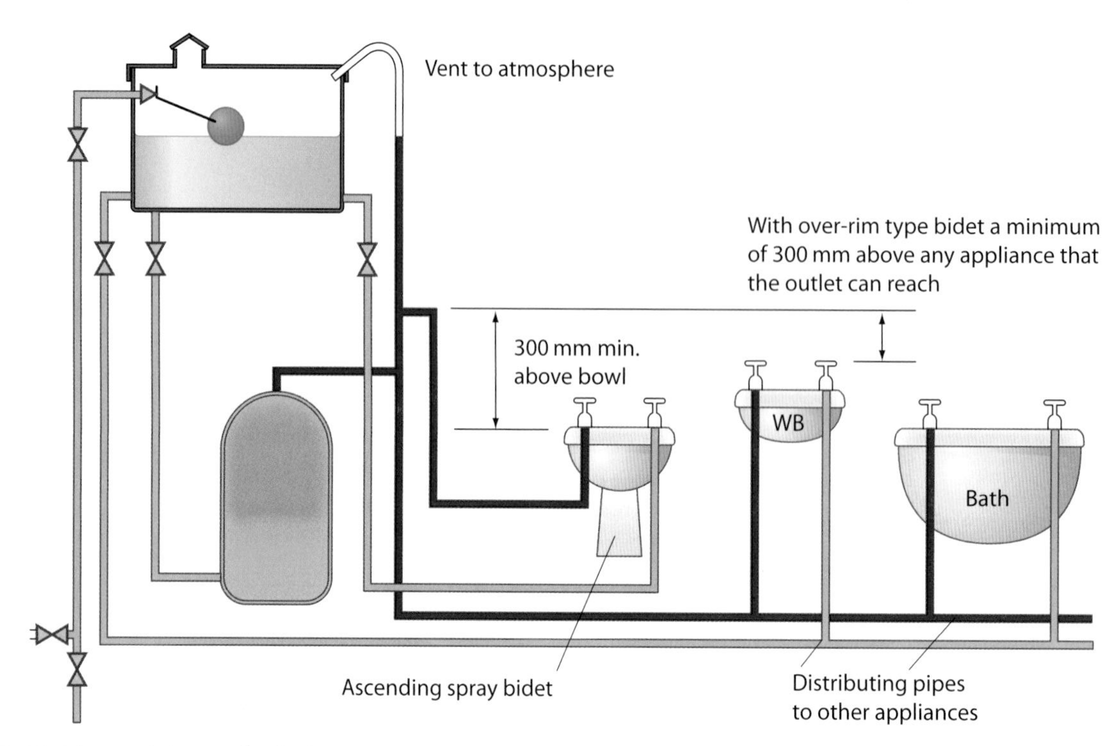

Figure 3.38: Pipework layout for bidet

Shower heads and tap inlets

Except where suitable additional backflow protection is provided, the Regulations require that all single tap outlets, combination tap outlets, fixed shower heads terminating over wash basins, baths or bidets in domestic situations should discharge above the spill-over level of the appliance with a tap gap type AUK 2.

Sinks in domestic and non-domestic situations are considered to be a fluid category 5 backflow risk and the minimum protection should be a type AUK 3 air gap/tap gap. Generally with sinks this is not a problem, as sinks require additional space for access to work and for the filling of buckets, and so on.

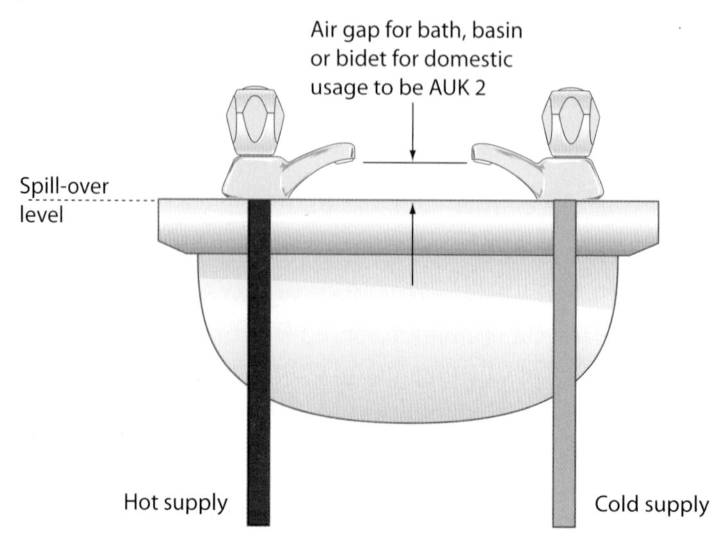

Figure 3.39: Air gap requirements for baths, basins or bidets for domestic usage

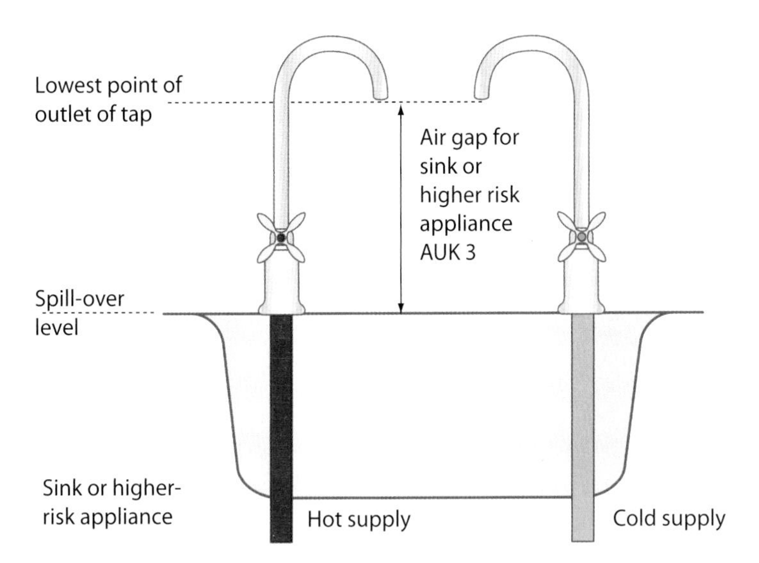

Figure 3.40: Air gap requirements for sinks

Baths and wash basins

Baths and wash basins fitted in domestic dwellings that have submerged tap outlets are considered to give a fluid category 3 risk, and should be supplied with water from a supply or distributing pipe through double check valves.

Submerged tap outlets to baths or wash basins in non-domestic situations are considered to be a fluid category 5 risk, and appropriate backflow protection is required for this higher risk level.

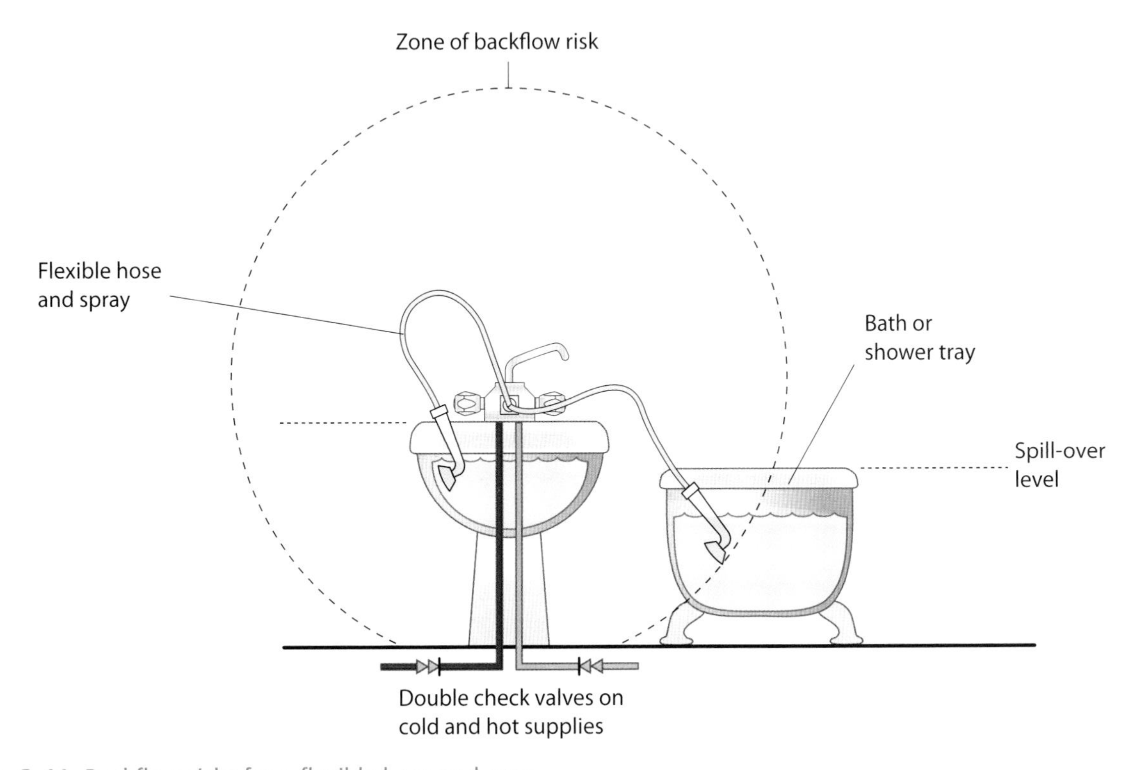

Figure 3.41: Backflow risks for a flexible hose and spray

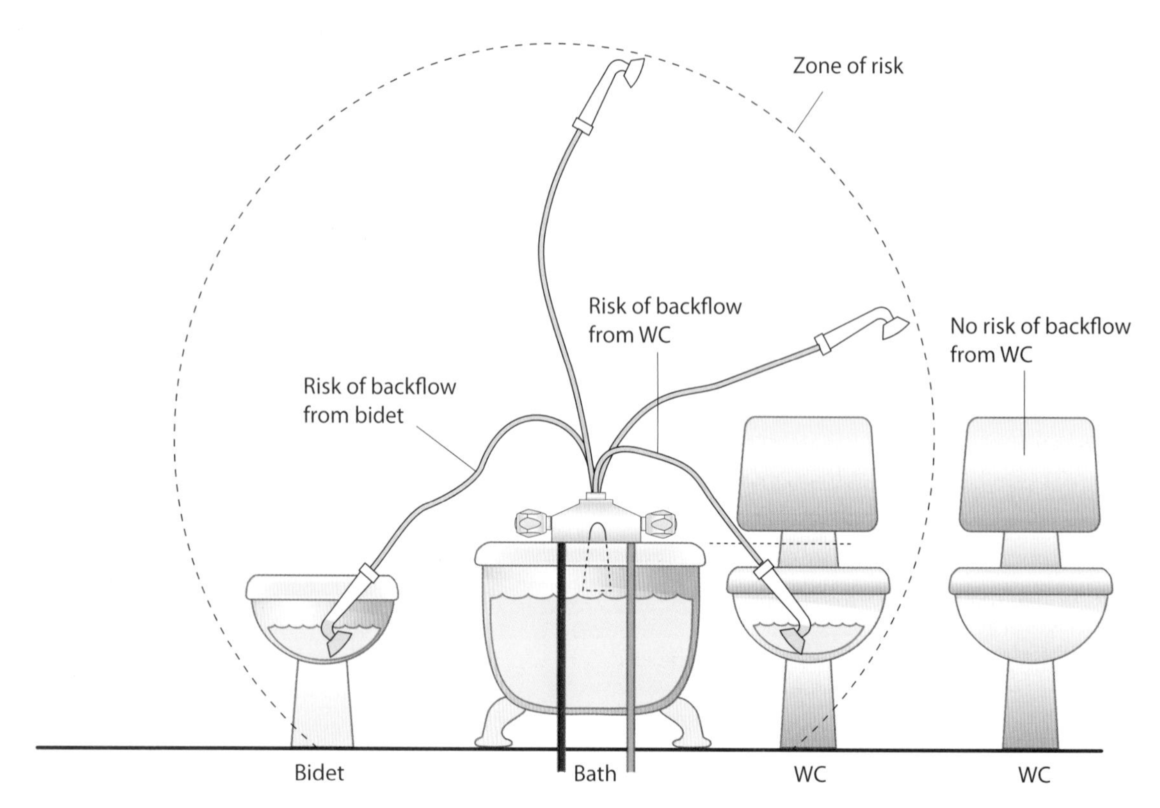

Figure 3.42: Backflow fluid category 5 risk for a bath mixer fitting

Where installations consist of a spray or jet served from a tap or combination tap assembly or mixer fitting located over a wash basin, bath or shower tray, you must ascertain the zone of backflow risk. If the spray or jet on the end of the hose is capable of entering a wash basin, bath or shower tray located within the zone of backflow risk, then a fluid category 3 prevention device such as a double check valve must be fitted on each inlet pipe to the appliance types EC or ED.

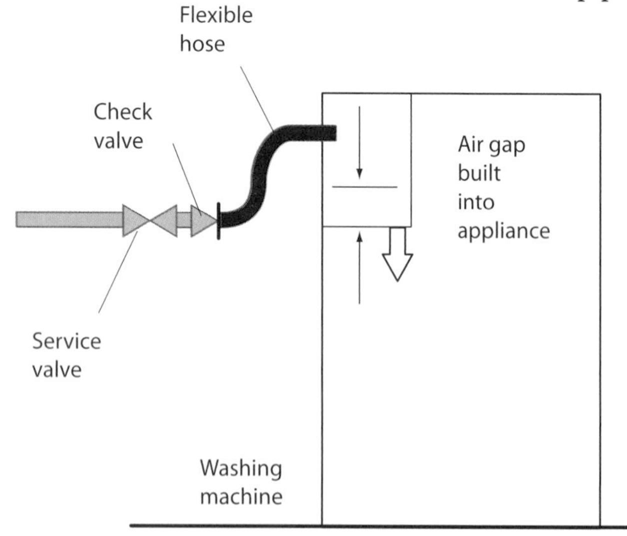

Figure 3.43: Connection to washing machine. A single check valve is shown because non-approved hoses are used, which may result in the possibility of tainting the water supply, leading to a category 2 risk

Washing machines

Domestic household washing machines, including washer-dryers and dishwashers, are manufactured to satisfy a fluid category 3 risk. Before installing the appliance, you should consult the Water Fittings and Materials Directory. If the hoses are approved, they will be listed under the WRAS. In instances where the hose is not approved, you must fit an appropriate fluid category 3 risk check valve.

Commercial machines such as those used in hotels, restaurants and launderettes are a fluid category 4 risk. Machines that are used in healthcare premises and hospitals are classed as a fluid category 5 risk and higher protection will be required. Most machines now incorporate a type AD device to guard against any type of risk.

Drinking fountains

Drinking fountains have their own requirements. These should be designed so that there is a minimum 25 mm air gap between the water delivery jet nozzle and the spill-over level of the bowl. The nozzle should be provided with a screen or hood to protect it from contamination.

Drinking fountains have their own requirements

Domestic garden installations

Handheld hosepipes for garden or other use must be fitted with a self-closing mechanism at the hose outlet. This will reduce the risk of backflow into the supply pipe if the end is dropped on the ground, and additionally promotes water conservation.

Any garden tap that enables a hose connection to be made to it must be:

- fitted with a double check valve
- positioned where it will not be subject to frost damage.

The installation of a double check valve is also adequate protection for handheld hosepipes used for spraying fertilisers or domestic detergents in domestic garden situations.

Remember

The correct protection device for installation of external taps and hosepipes will depend on the fluid risk category and the type of hose equipment.

Activity

What fluid category risk are spray fertilisers and domestic detergents?

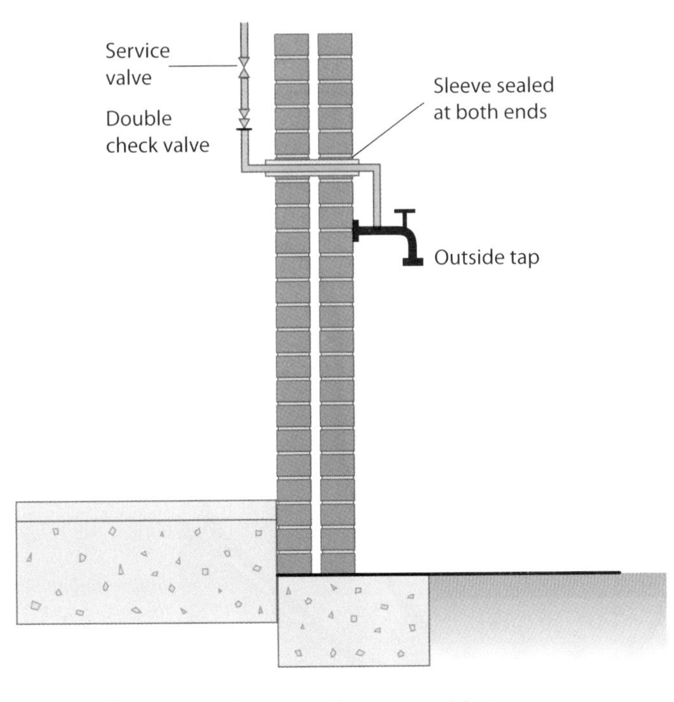

Figure 3.44: Installation requirements for an outside tap

Irrigation and porous-hose systems

Irrigation and porous-hose systems, laid above or below ground, are a serious potential backflow risk, and the pipe supplying such systems must be protected against a fluid category 4 or 5 risk. Mini-irrigation systems and smaller applications are category 4, whereas commercial systems are category 5.

Irrigation in action

Irrigation systems that consist of fixed sprinkler heads located no less than 150 mm above ground level, which are not intended to be used in conjunction with insecticides, fertilisers or additives, are considered as a fluid category 3 risk.

Figures 3.45 and 3.46 show how pop-up sprinklers or porous hoses should be installed in a domestic situation (category 4 risk).

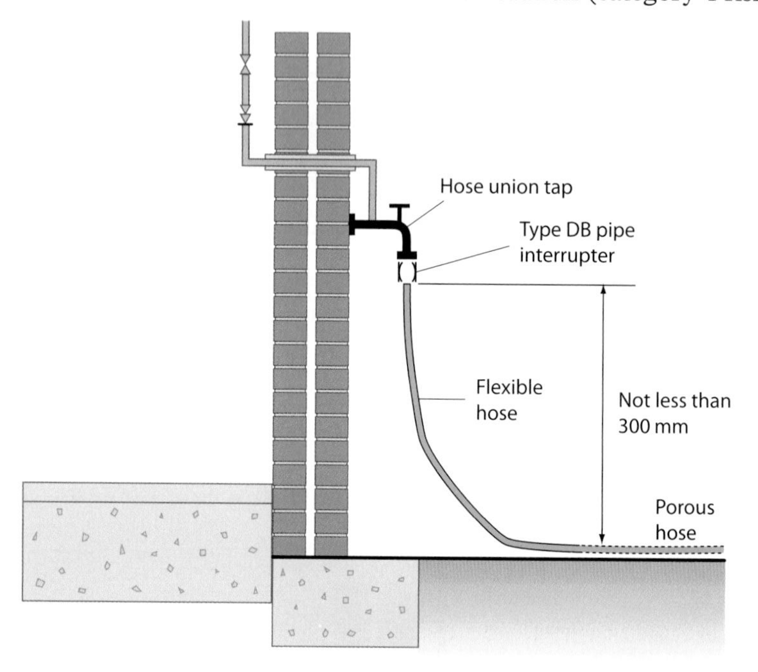

Figure 3.45: Installation details for mini irrigation or porous hose where ground surface is level or falling away from house

GD G15.23

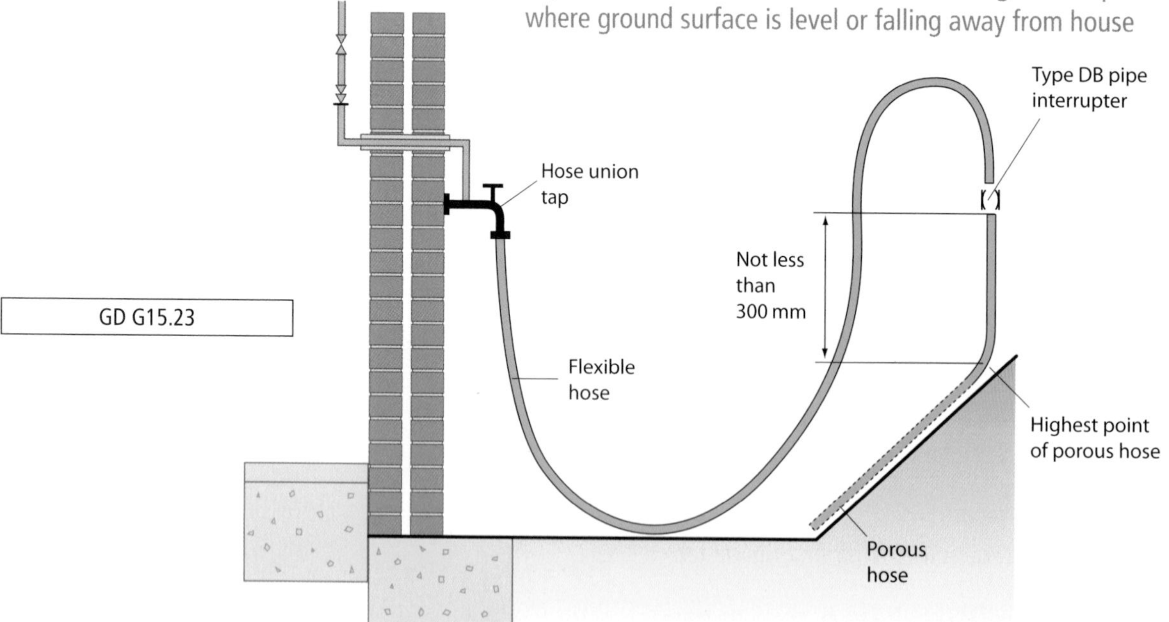

Figure 3.46: Installation details for mini irrigation or porous hose where ground is rising away from house

Where installations that have chemical additives are added, then the fluid category must be a category 4 and a verifiable backflow preventer with a reduced pressure zone (type BA device) or no less effective device must be fitted.

Existing garden-tap installations

While the Water Regulations are not retrospective, appropriate measures must be taken against any known situation where there is a potential backflow risk from hoses. Theoretically, if a tap was installed legally under previous byelaws (before 1 July 1999) and has a hose connected to it, it remains legal. However, as soon as the hose is disconnected and reconnected, the installation becomes illegal unless appropriate steps are taken.

GD G15.21

Where an external tap is being replaced the following factors apply:

- if practicable, a double check valve should be provided on the supply to the tap type EC or ED. This should be installed inside the building or, where it is not practicable to locate a double check valve within a building, the tap could be replaced with:
  - a hose union tap that incorporates a double check valve type HUK1, or
  - a tap that has a hose union backflow preventer type HA or a double check valve type EC or ED, fitted and permanently secured to the outlet of the tap.

External taps and systems in commercial situations

Taps and fittings used for non-domestic applications such as commercial, horticultural, agricultural or industrial purposes (this may include small catering establishments and hotels) must be supplied with backflow-protection devices appropriate to the downstream fluid category and, where appropriate, an additional zone protection system.

Animal drinking troughs or bowls

The water inlet to an animal or poultry drinking trough should be provided with a float-operated valve or other similar effective device. The inlet device should be of type AA or AB air gap installed to prevent backflow from a fluid category 5 risk and contamination of the supply pipe. The inlet device and backflow arrangements must be protected from damage. The installation arrangements of the trough will be accepted as being satisfied providing they comply with the requirements of BS 3445: *Fixed agricultural water troughs and water fittings*.

A service valve should be provided on the inlet pipe to every drinking appliance for animals or poultry.

Where a number of animal drinking troughs are supplied with water from a single trough, the spill-over levels of other troughs must be at a higher level than the initial drinking trough where the water inlet device is located.

Animal drinking troughs must comply with BS 3445

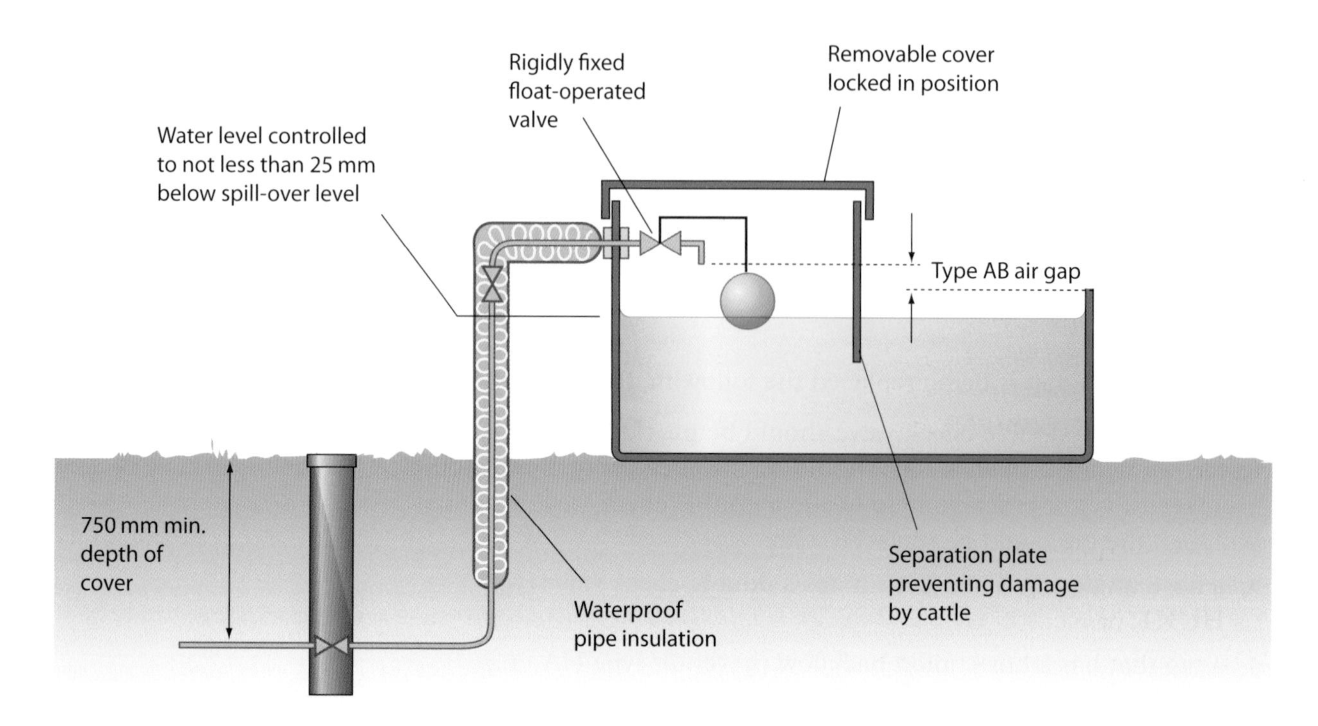

Figure 3.47: A cattle trough installation with a Type AB air gap

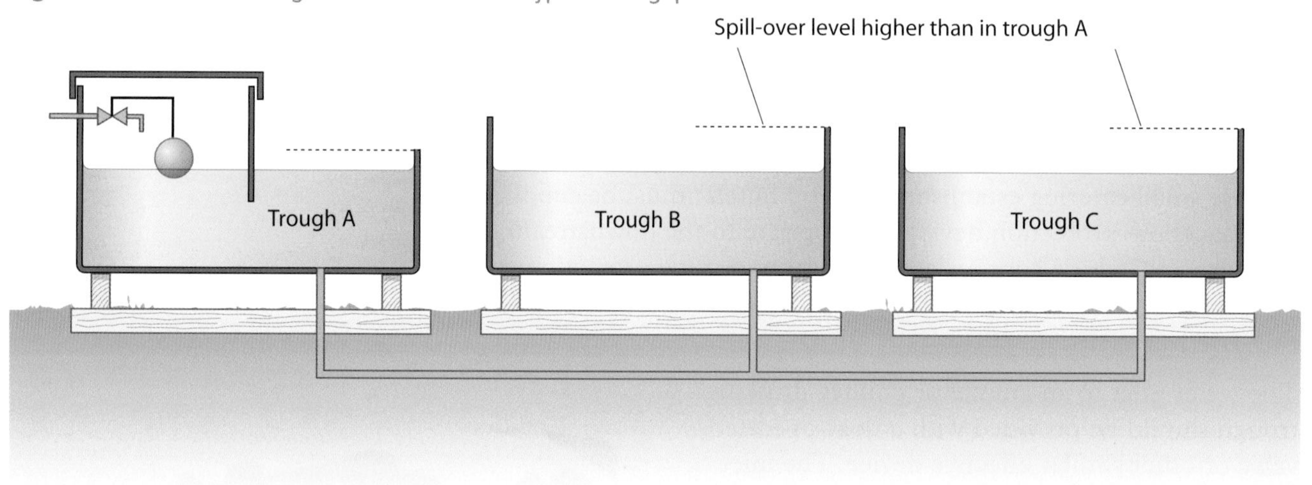

Figure 3.48: A combination of interconnected cattle troughs

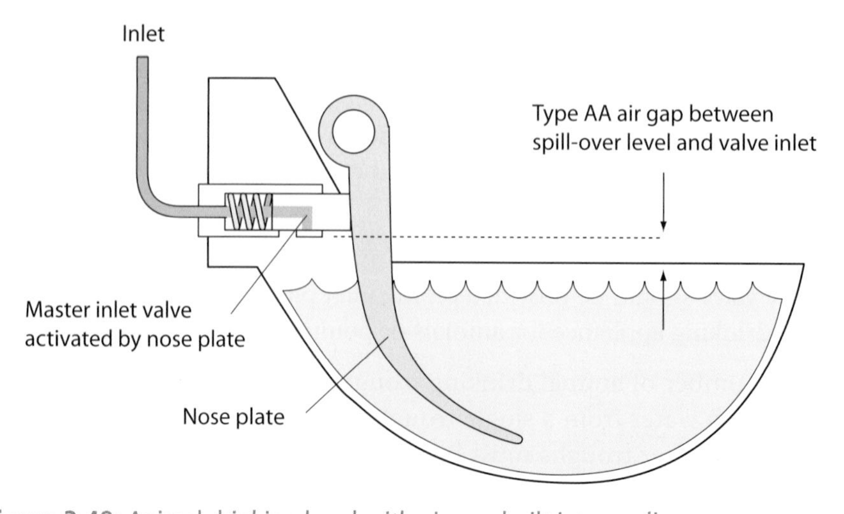

Figure 3.49: Animal drinking bowl with air gap built into appliance

Where animal drinking bowls are to be installed, the source of the water supply will depend on the type of bowls being installed.

The type of bowl shown in Figure 3.49 has a spring-return or float-valve device which operates when depressed by the animal's mouth. These may be connected directly from a supply pipe or distributing pipe, providing the type AA air gap is maintained and that the animals' mouths cannot come into direct contact with the outer nozzle.

Where the outlet nozzle is below the spill-over level or is likely to be contaminated by animals' mouths, the bowl must be supplied from a dedicated distributing pipe that only supplies similar appliances.

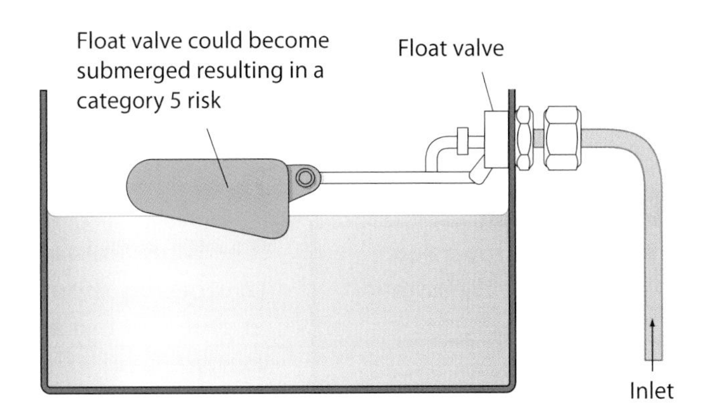

Figure 3.50: Animal drinking bowl with submerged water outlet or inadequate air gap at the appliance

Whole-site and zone protection

The WSR Guidance Document states that whole-site or zone backflow-prevention devices should be provided on the supply pipe, such as a single check valve, or double check valve, or other no less effective

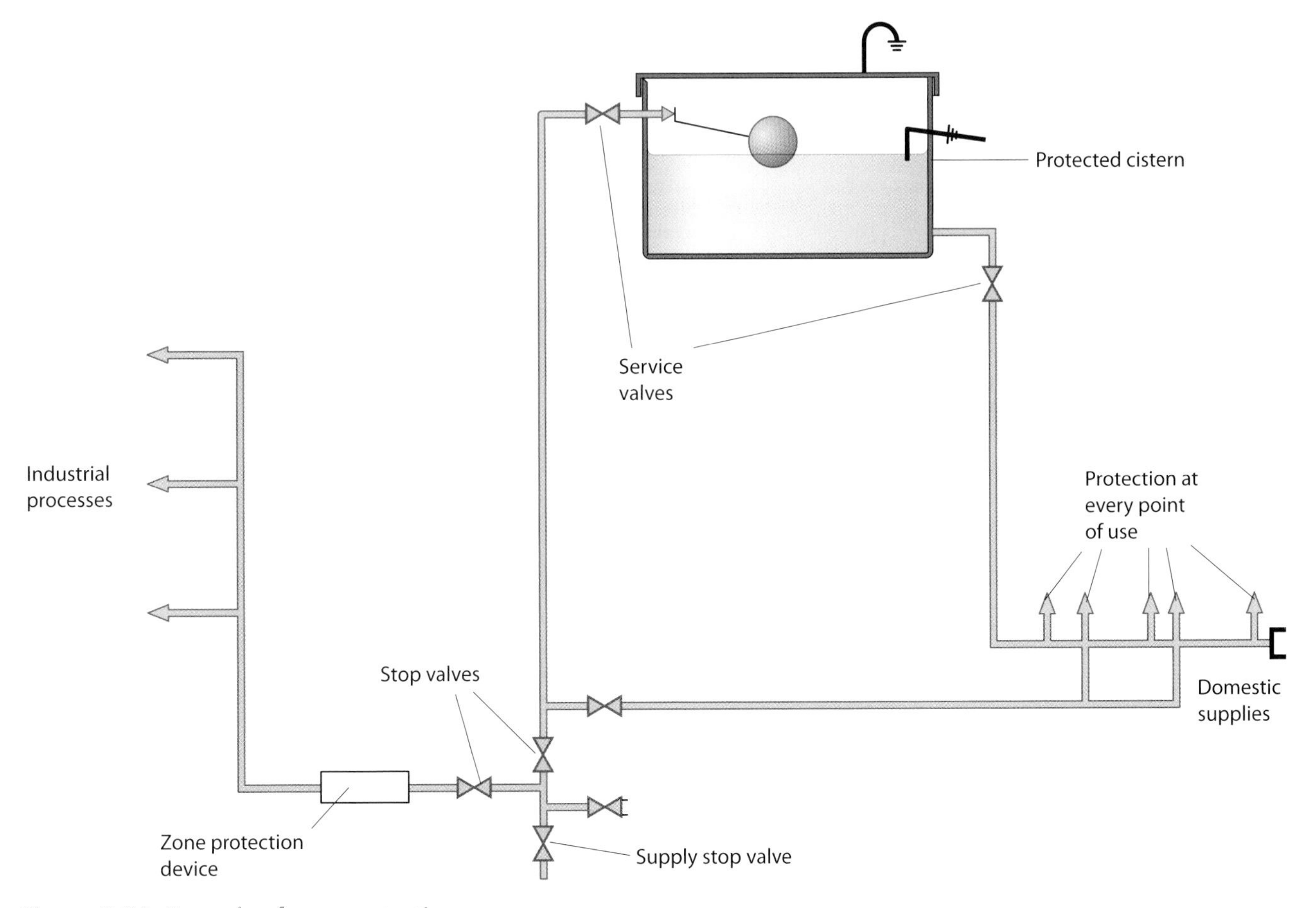

Figure 3.51: Example of zone protection

Did you know?

Whole-site protection used to be termed 'secondary protection'.

GD G15.24

Key terms

Whole-site protection – used to protect one building from another.

Zone protection – used to protect one part (zone) of a building from another part.

GD G15.25

GD G15.27–29

A sprinkler head

backflow-prevention device, according to the level of risk as deemed by the water undertaker, where:

- a supply or distributing pipe conveys water to two or more separately occupied premises
- a supply pipe conveys water to premises that are required to provide sufficient water storage for 24 hours of ordinary use.

The WSR Guidance Document requires that **whole-site** or **zone protection** should be provided in addition to individual requirements at points of use and within the system. The use of zone protection is extremely important in premises where industrial, medical or chemical processes are undertaken alongside the supply of water for domestic purposes such as drinking.

Backflow protection to fire systems

Fire-protection systems require backflow protection to suit the level of risk. Wet sprinkler systems (that contain no additives), fire-hose reels and hydrant landing valves are considered to be a fluid category 2 risk and will require the minimum protection of a single check valve.

Wet sprinkler systems in exposed situations often have additives in the water to prevent freezing at low ambient temperatures, and these are considered to be a fluid category 4 risk. Also included in this risk category are systems that contain hydro-pneumatic pressure vessels; the system requires either a verifiable backflow preventer (RPZ or type BA) or it must be fitted with a suitable air gap (type AA, AB, AD or AUK 1).

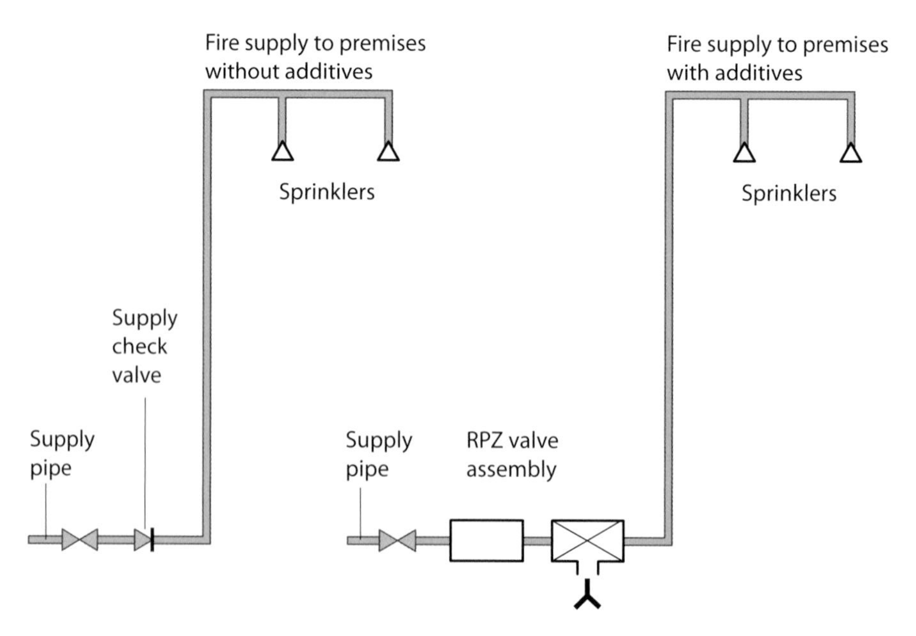

Figure 3.52: A sprinkler system (no additives used) with a single check valve installed, and a system (additives used) with a type BA (RPZ) valve

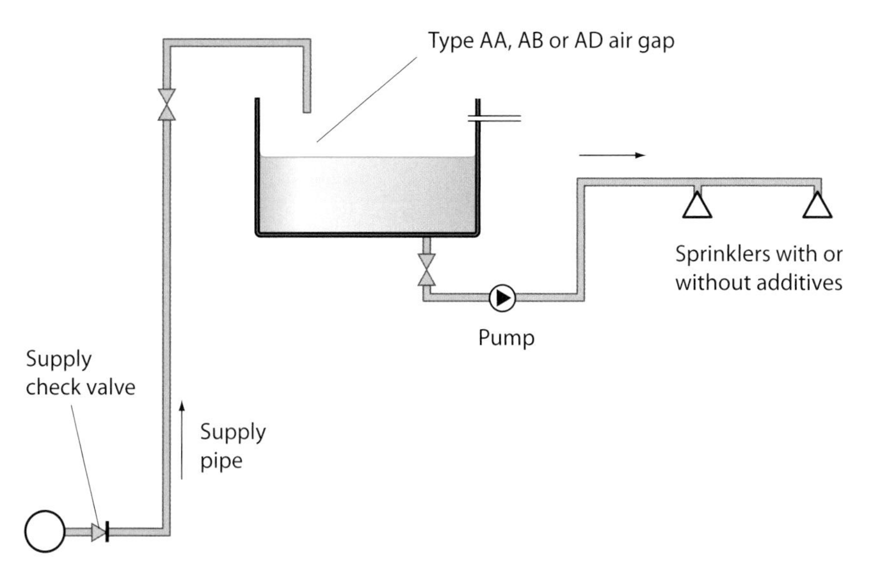

Figure 3.53: Sprinklers supplied with water pumped from storage

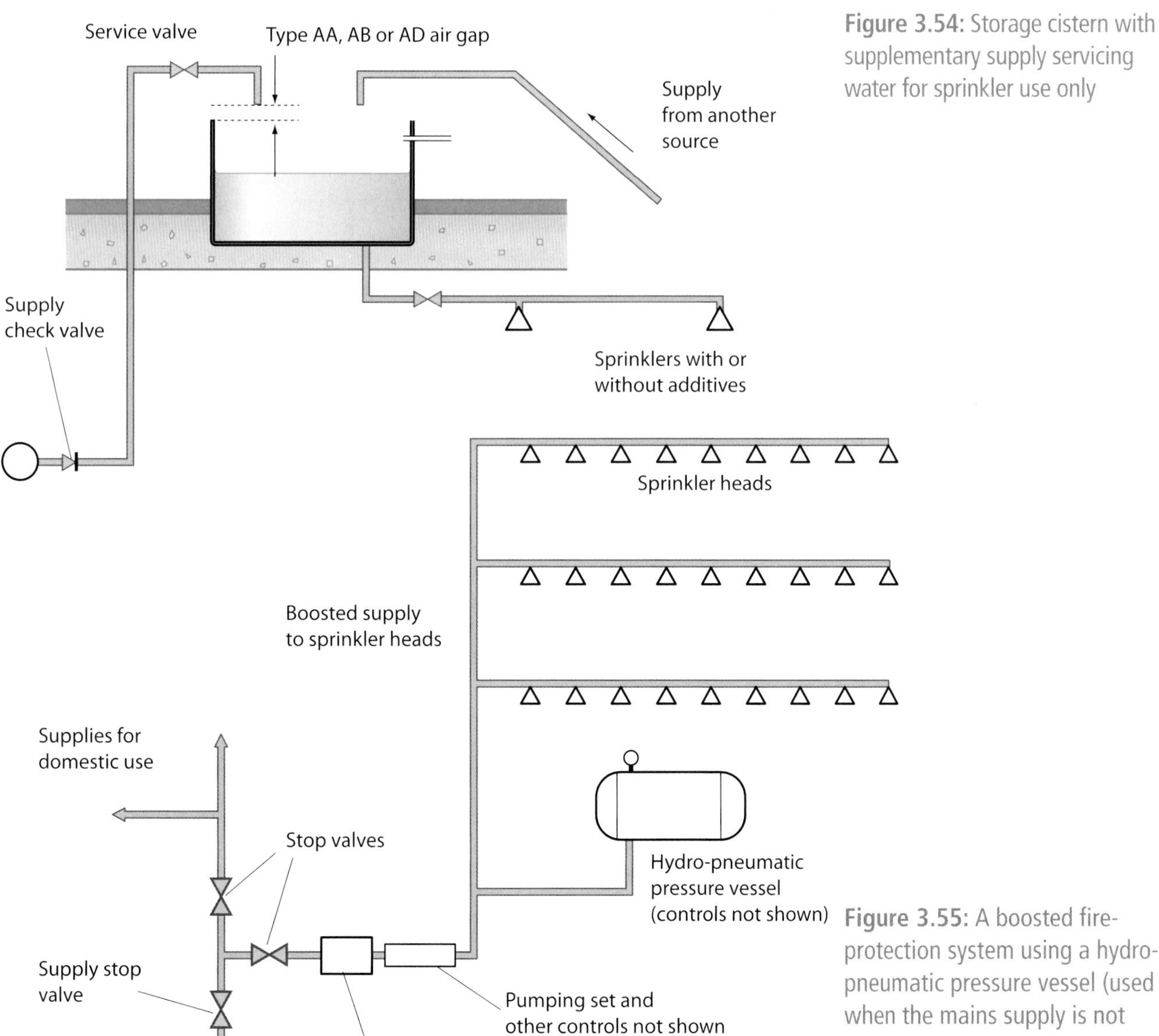

Figure 3.54: Storage cistern with supplementary supply servicing water for sprinkler use only

Figure 3.55: A boosted fire-protection system using a hydro-pneumatic pressure vessel (used when the mains supply is not sufficient and water storage cannot be provided)

Safety tip

Some water suppliers may insist on independent service pipes for domestic supplies and fire protection.

In situations where fire-protection systems and drinking-water systems are served from a common domestic supply, the connection to the fire system should be taken from the supply pipe directly on entry to the building, and an appropriate backflow-protection device must be installed.

Backflow-protection devices: mechanical and non-mechanical

These are outlined in detail in Appendix A.

Preventing cross-connection to unwholesome water

The Water Industry Act places duties and responsibilities on suppliers of water that it should be clean, free from impurities and fit for drinking. As the installer, it is your duty not to contaminate the water supplied. This also applies to water users under the Water Supply (Water Fittings) Regulations 1999.

WSR Schedule 2, para 14(1)

The Water Regulations Advisory Committee has made recommendations for requirements, suggesting that the use of recycled water could make significant contributions to water conservation, and considers that systems making use of recycled water in future years could become more common. This pre-empts the fact that, as the use of these waters increases, so will the risk of cross-connection and backflow.

Identification

Pipes can be identified by colour-coded pigmentation incorporated in plastic pipes, permanent marks or labels, or colour painting of the pipes themselves.

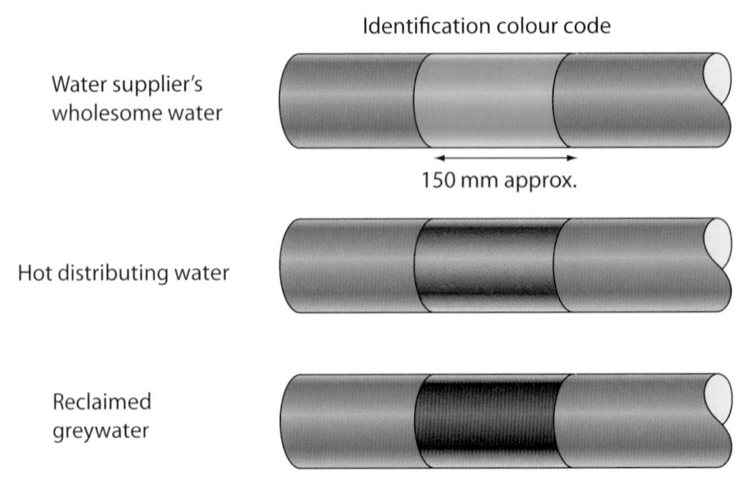

Figure 3.56: Colour-coding of pipes for non-potable water should follow this scheme

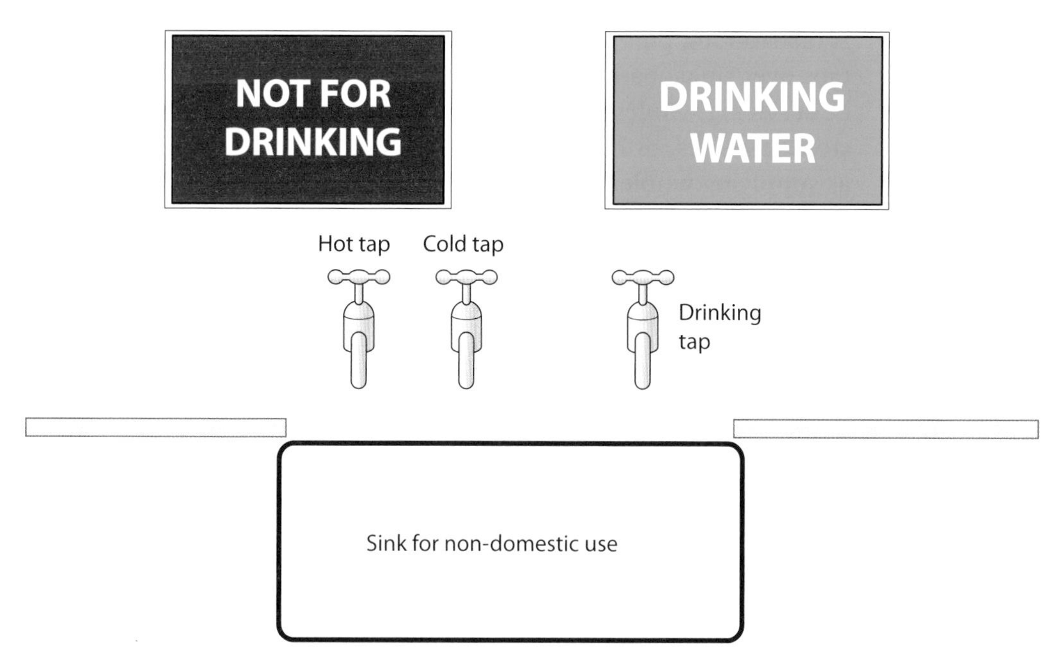

Figure 3.57: Labelling requirements for wholesome and unwholesome water

Pipes located above ground within buildings should be colour-coded to BS 1710 to distinguish them from others. Generally, this applies only to industrial and commercial buildings but, in cases where a house or small building uses water other than wholesome water supplied by the undertaker, colour-coding of the pipes will be required.

Colour identification should be fitted on pipes at junctions, inlets and outlets of valves, and service appliances where a pipe passes through a wall (both sides).

Ideally, on larger installations, accurate pipe-layout drawings should be handed over to the customer, identifying the locations of pipes, within and below ground level, feeding the building.

Any pipe carrying fluid that is not wholesome water must not be connected to a pipe that is carrying wholesome water, unless a suitable backflow-prevention device is fitted.

The requirement would be satisfied if wholesome water (fluid category 2) feeding into a cistern containing greywater was fed into the cistern via a backflow-prevention device or an arrangement suitable for protection against a fluid 5 category risk (examples are Type AA, AB and AD air gaps).

It is very important that you remember that water derived from a supply pipe is considered to be wholesome water, but water derived from a distribution pipe may not be. This depends on the quality of the water contained in the cistern supplying the distribution pipe and on what the distribution pipe is serving.

Activity

Find out the colour-coding used for a cold distributing pipe and a pipe carrying water for firefighting purposes.

Remember

The identification requirement also applies to all water fittings, including cisterns and valves. It is particularly important that taps should be labelled, identifying those that are suitable for drinking purposes and those that are not.

WSR Schedule 2, para 14(2)

A distributing pipe from a cistern containing wholesome water, servicing taps over sinks, baths, wash basins and showers, could be considered to be servicing wholesome water. A distribution pipe servicing hot water storage vessels, or hot water distribution pipes, should not be considered as supplying wholesome water and should not have any connections made into it for drawing wholesome water.

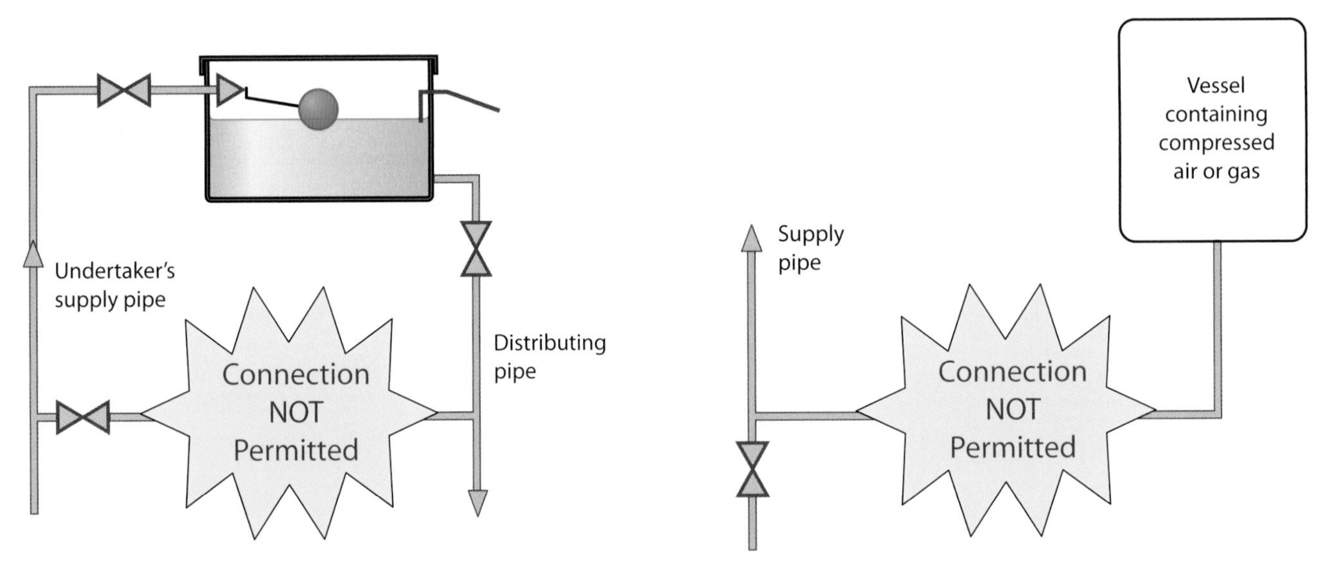

Figure 3.58: Unlawful connection between a supply pipe and distributing pipe (left) and an unlawful connection between a supply pipe and a vessel containing compressed air or gas (right)

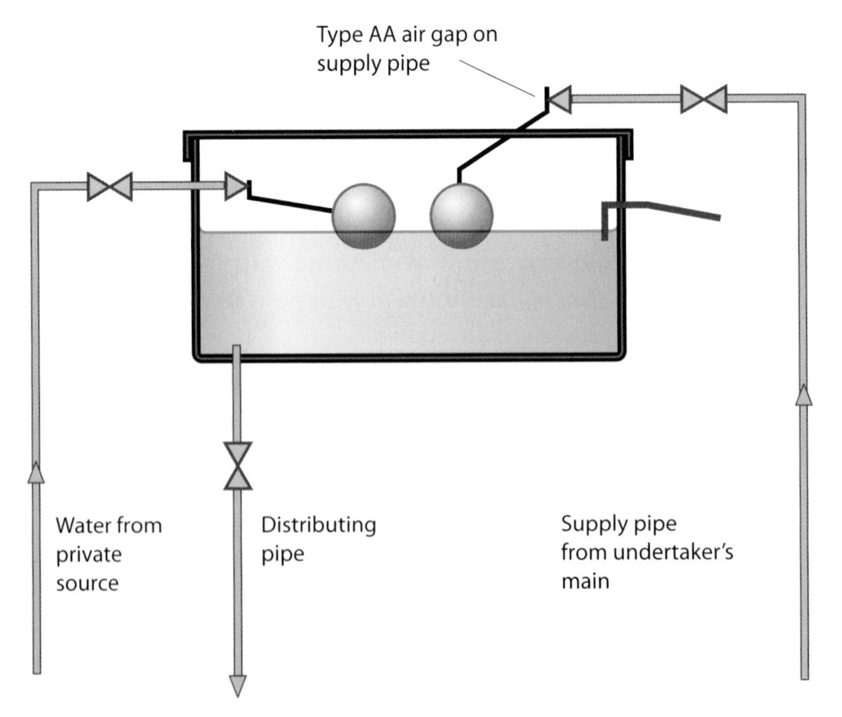

Figure 3.59: Correct connection for a cistern containing reserve water for firefighting or other industrial use

PROGRESS CHECK

1. What is the purpose of a backflow-prevention device?
2. Under which fluid category would hot water be considered a risk and why?
3. Which type of backflow device should be used for a kitchen sink?
4. Which backflow device should be used for whole-site protection?
5. On which type of system should a type DB pipe interrupter be installed?
6. If a wash basin has a mixer tap that mixes the water within the body, which type of backflow-prevention device must be installed?
7. What are the installation requirements for a bidet with an ascending spray?
8. On which sanitary appliance would you find an AUK 1 air gap?
9. If a pipe had identification labels of green, blue, green which type of water would it be conveying?
10. Which type of backflow device should be fitted to a central heating filling loop?

5. Know the uses of specialist components in cold water systems

How cold water system components work

As a Level 2 plumber, you had to deal with the basic components that make up the most common cold water systems. Now as you move up to Level 3, you will encounter an ever-expanding range of specialist components, of which you will need to develop a solid, working knowledge.

Here you will look at just some of these specialist components.

Infrared operated taps

Infrared taps are being used more and more, due mainly to the spread of bacteria. Their main places of use are:

- schools
- hospitals
- food preparation areas
- food production areas
- users with disabilities.

Nowadays they are being specified in domestic dwellings more for aesthetic looks and as a water-saving device.

How do they work?

- An object (usually your hands) approaches the sensor eye.
- The infrared proximity is triggered or disrupted once an object enters its infrared sensing zone.

- The proximity sensor zone is live once powered (typical sensor range 20–26 cm wide).
- The sensor eye part beams out an infrared signal.
- The sensor signal wire transfers or sends an electronic signal to the solenoid valve to OPEN or CLOSE.
- The solenoid valve acts as a latching mechanism that restricts or allows water to flow through it.
- It opens up and releases water through the flexible hose as soon as an electronic signal is received from the sensor.
- The solenoid valve is always in a CLOSED position, and opens up once an electronic signal is received; it goes back to the CLOSED position when the object leaves the infrared sensing zone.
- Water exits – water let through the solenoid valve comes out.

Non-concussive taps (self-closing taps)

BS EN 8160

These taps should conform to BS EN 8160, and should be capable of closing against 2.6 times the working pressure. When new, these taps are very effective in saving water in places such as washrooms, offices, schools and public toilets. New taps are effective but do require regular maintenance: if maintenance is ignored, they have a tendency to remain in the open position after being pressed. For this reason, such taps should only be fitted in buildings where maintenance is carried out on a regular basis.

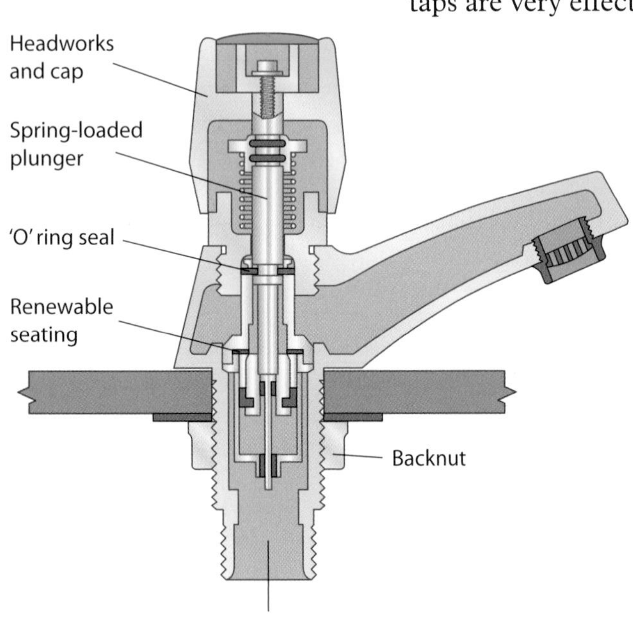

Figure 3.60: Non-concussive tap

Non-concussive taps limit the amount of water delivered to a tap, and can be timed to anywhere between 1 to 15 seconds. As well as being used for water-saving in situations such as public conveniences, they are installed as anti-vandal devices on wash basins and showers; many of them have devices integrated into the taps that prevent them from constantly allowing water to flow in the event of the top being removed during an act of vandalism.

Combination bath tap and shower head

The main use of this component is as a bath filler with a shower. Many variations are available: some are a simple diverter valve arrangement, while others have a thermostatic cartridge that has its own flow and temperature control.

Combination bath taps will require equal pressures; if supplies are being sourced from a cistern, they decrease the risk of flow and pressure variations. As the shower is fed from a cistern, this eliminates the risk from backflow.

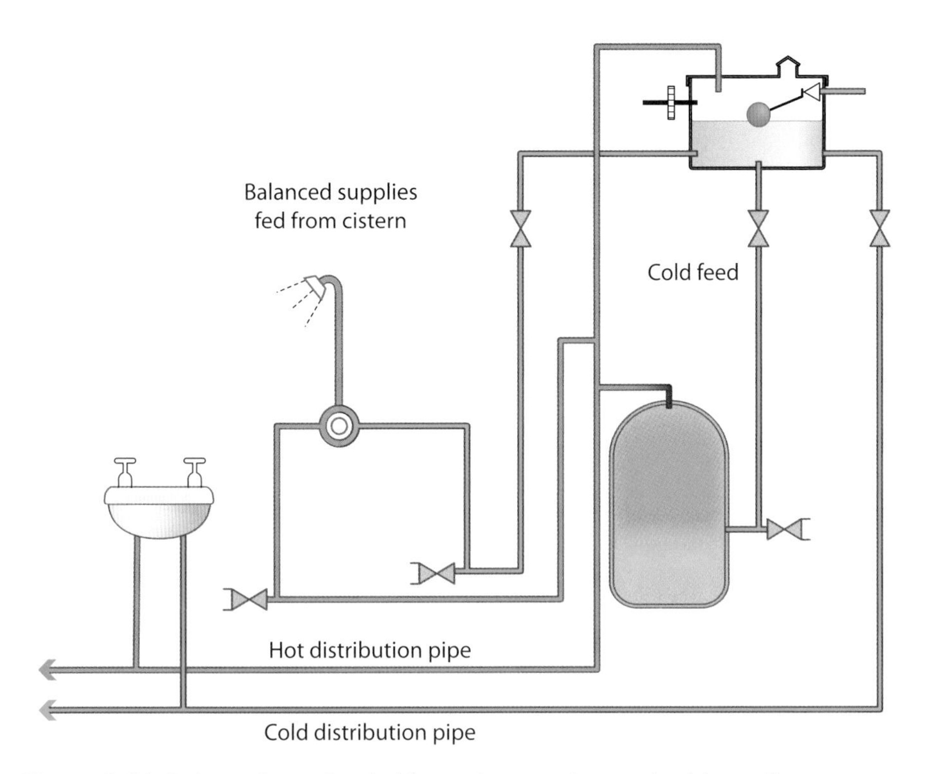

Figure 3.61: Balanced supplies fed from cistern to hot and cold supplies

When combination bath taps are fed from a multipoint, combination boiler or unvented system, the shower outlet has to have a double-check valve (either type EC (verifiable) or ED (non verifiable)) installed in the flex outlet of the shower, because the bath is a category 3 backflow risk (see Figure 3.62).

G15.14

To save water from the bath/shower mixer, you can fit flow limiters into the shower hose. These vary from 7 lt/m to 12 lt/m, depending on the manufacturer and the requirements of the customer.

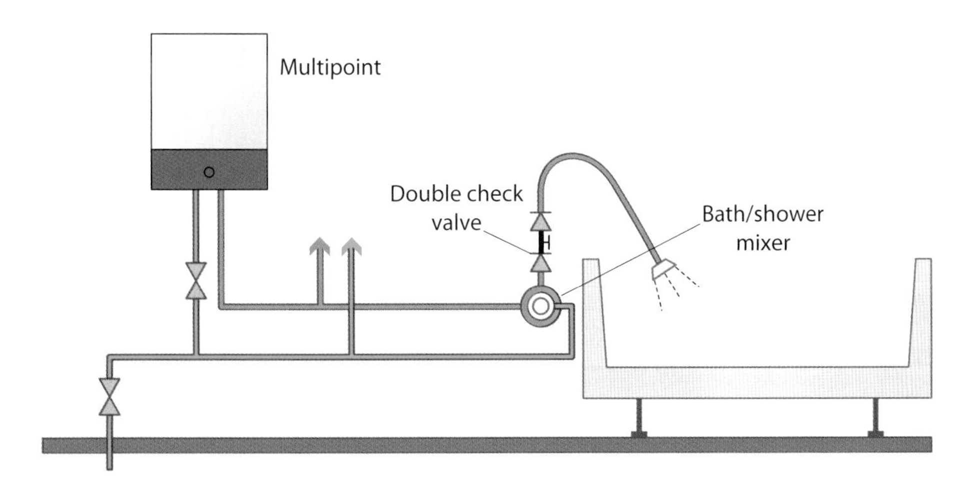

Figure 3.62: Pipework from multipoint/combination boiler showing double check valve

Flow-limiting valves

Instead of limiting flow to individual appliances, you can limit flow to all appliances, if required, by using an inline flow-limiting valve. These valves look similar to service valves, but a removable cover on the side of the valve gives you access to the flow-limiting valve. These cartridges can be changed depending on the rate of flow you require, and for maintenance purposes.

Limiting the flow can save water and assist with balancing, preventing some appliances (such as a shower) consuming all the available water while other appliances at a higher level or further downstream are starved of water.

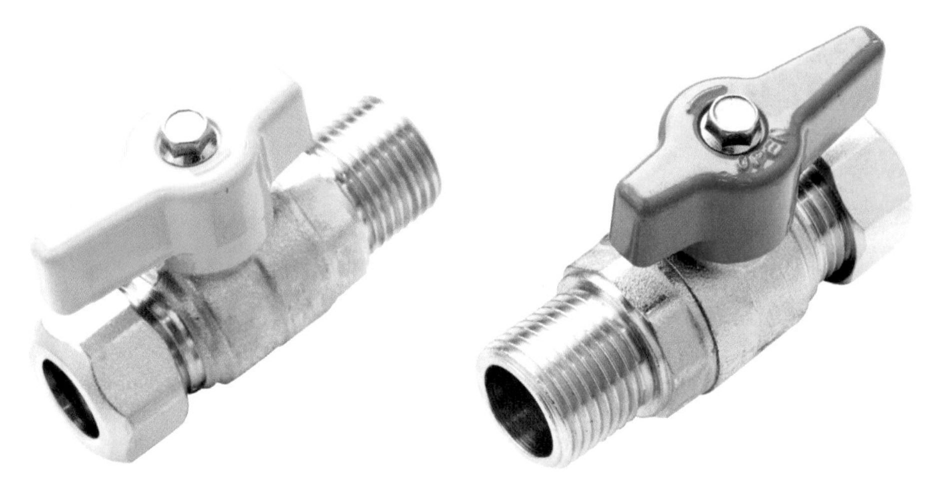

Double check valves with integral flow limiter

Basin spray taps can save up to 80% in water usage

Spray taps

Spray taps are also used to minimise water usage, giving a saving of up to 80 per cent. You can convert the outlet of taps with a conversion kit. Good locations for spray taps are in washrooms or toilets, or incorporated into a non-concussive tap. They do, however, have several disadvantages.

- They should not be used where basins are subject to heavy soiling from dirt and grease.
- They require regular maintenance (the spray block).
- They are only suitable for hand rinsing.
- Self-cleansing velocities for the waste water may not be achieved and residues of soap and grease can accumulate in the waste pipes.

Urinal – water conservation controls

Flushing urinals

Flushing devices for urinals should be designed so that they do not supply more water than is necessary to effectively clear the urinal and replace the trap seal. The acceptable flushing methods to meet the Regulations are:

- by flushing cistern, operated manually or automatically
- by flushing valve, operated manually or automatically.

Paragraph 25(4) sets out the maximum volumes of water permitted for urinal flushing.

WSR Schedule 2, para 25(4)

Flushing cisterns

Automatically operated

These should supply no more water than:

- 10 litres per hour for a single urinal bowl or stall
- 7.5 litres per hour, per urinal position, for a cistern servicing two or more urinal bowls, stalls or per 700 mm slab position.

Manually operated (chain-pull or push-button) to a single urinal bowl

These are required to flush no more than 1.5 litres each time the cistern is operated.

Pressure-flushing valves, operated manually or automatically

These should not flush more than 1.5 litres each time the valve is operated.

Pressure-flushing valves may be fed with water from either a supply pipe or a distributing pipe. The outlet of the pressure-flushing valve should either be provided with a pipe interrupter with a permanent atmospheric vent, or incorporate the flushing valve being installed, so that the level of the lowest vent aperture is not less than 150 mm above the **sparge outlet** and not less than 300 mm above the spill-over level of the urinal.

Unless a servicing valve is integral with the pressure-flushing valve, it is recommended that a separate servicing valve be provided on the branch pipe to each pressure-flushing valve.

Key terms

Sparge outlet – the outlet that spreads the water across the face of a urinal for cleaning and flushing purposes. They can be in the form of a bar with a series of holes drilled, when used for stall urinals, or individual outlets when used on bowl urinals.

Sparge pipe – the pipework that connects the urinal cistern to the urinal outlets.

Water-saving valves

WSR Schedule 2, para 25(1)

Paragraph 25(1)(j) of the Regulations focuses on water-saving controls for urinals, and states that any urinal supplied either manually or electronically from a flushing cistern must have a time-operated switch (and a lockable isolating valve) fitted to its incoming supply, or some other equally effective automatic means of regulating the periods during which the cistern may fill.

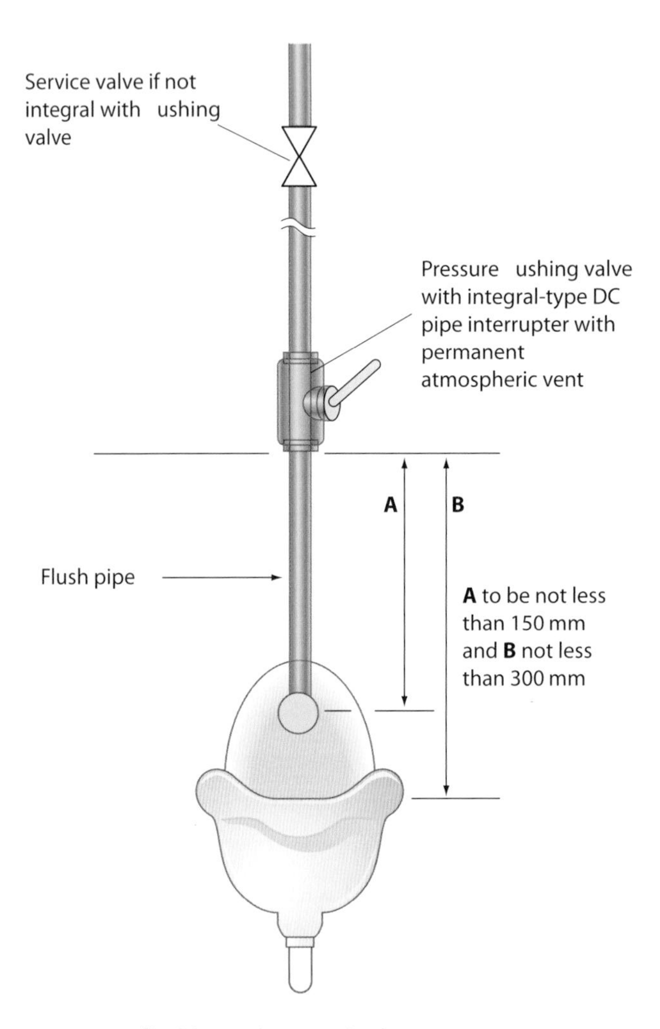

Figure 3.63: Pressure-flushing valve to urinal

The prevention of water flow to urinal cisterns during periods when the building is not occupied can be achieved in several ways:

- by incorporating a time-operated switch controlling a solenoid valve, which cuts off the water supply when other appliances are not used
- by an 'impulse'-initiated automatic system that allows water to pass to a urinal cistern only when other appliances are used
- by proximity or sensor devices (infrared).

Figure 3.64 shows an example of a system containing a timing device and an automatic isolation valve. Figure 3.65 on page 112 shows the urinal operation controlled by a hydraulic valve. Table 3.6 gives volumes and flushing intervals for urinals.

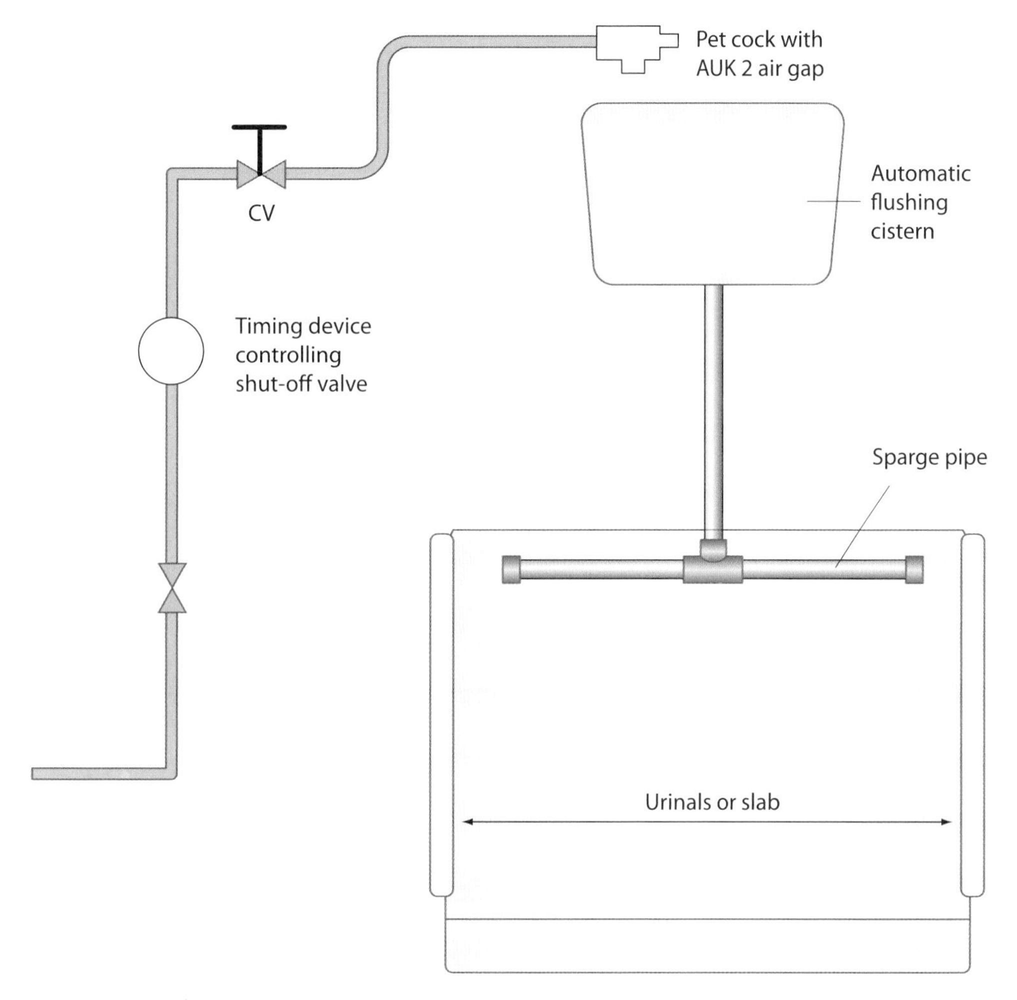

Figure 3.64: System timing device

Number of bowls, or stalls, per 700 mm of slab	Volume of automatic flushing cistern				Maximum fill rate in litres per hour
	4.5 litres	9 litres	13.5 litres	18 litres	
	Shortest period between flushes in seconds				
1	27	54	81	108	10
2	18	37	54	72	15
3	12	24	36	48	22.5
4	9	18	27	36	30
5	7.2	14.4	21.6	28.8	37.5
6	6	12	18	24	45

Table 3.6: Volumes and flushing intervals for urinals

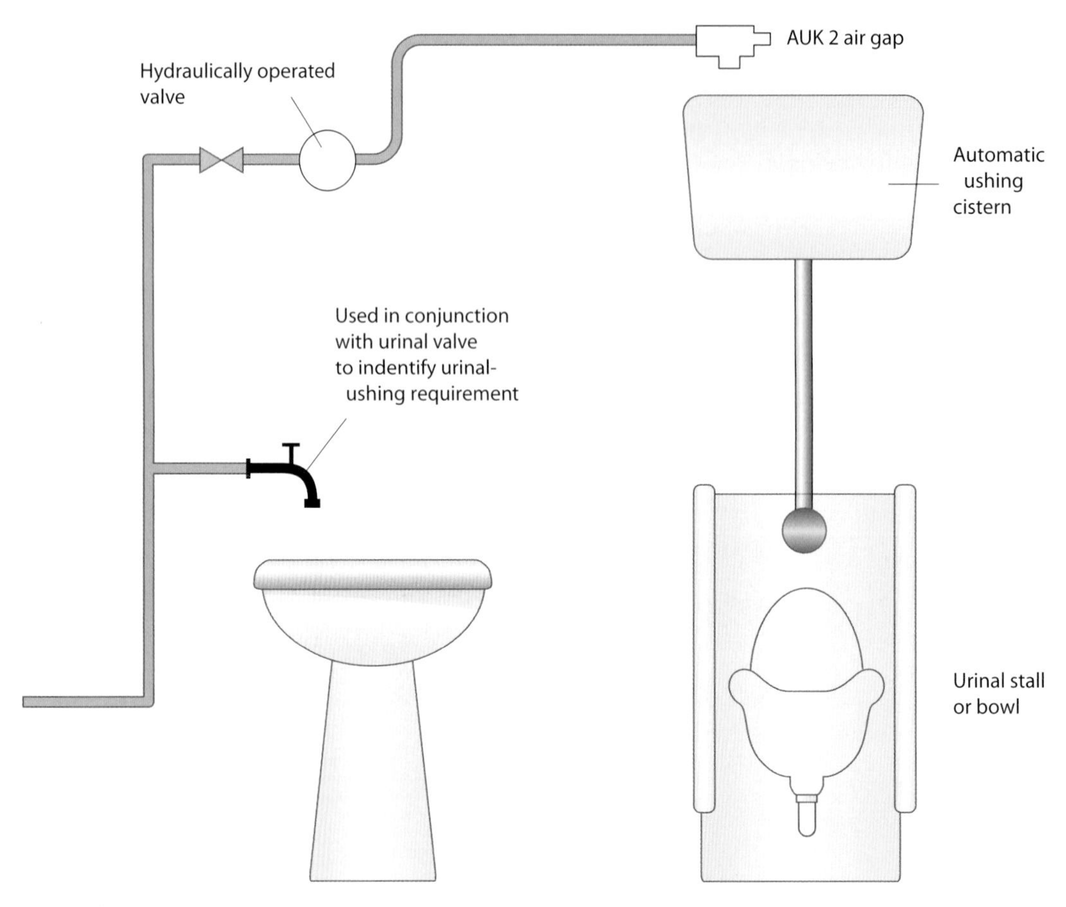

Figure 3.65: Hydraulic valve

Shower pumps – single and twin impellor

Where the head of water is less than one metre you would need to install a shower pump. Installation of the pump is straightforward, providing you follow these instructions.

- Position the pump at the same level as the base of the cylinder.
- Ensure that anti-vibration feet are fitted.
- Connections to the pump must be flexible (to prevent noise in the pipework).
- Fit strainer onto feed(s) into the shower pump (normally supplied by pump manufacturer).
- Supplies must be balanced supplies – that is, not taken from a mains supply.
- For maintenance purposes, fit full-bore service valves.
- To activate the pressure switch you will need to have a minimum of 150 mm head of water above the shower head.
- If the head of the shower is higher than the level of water in the cistern, there are several ways to overcome this problem, by installing:
  - a negative head kit (available from pump manufacturers)
  - a negative head shower switch, fixed to ceiling in shower area
  - a negative head shower pump.

  You should check the manufacturer's instructions for details.

Surrey flange

- A dedicated supply for the hot supply is preferable. This can be achieved by:
  - a boss being pre-installed during the manufacture of the cylinder
  - using a Surrey flange in the top of the cylinder
  - installing an 'Essex' boss into the side of cylinder near to its top.
- For any other methods of connection of the pump into the system, you should check the manufacturer's instructions.

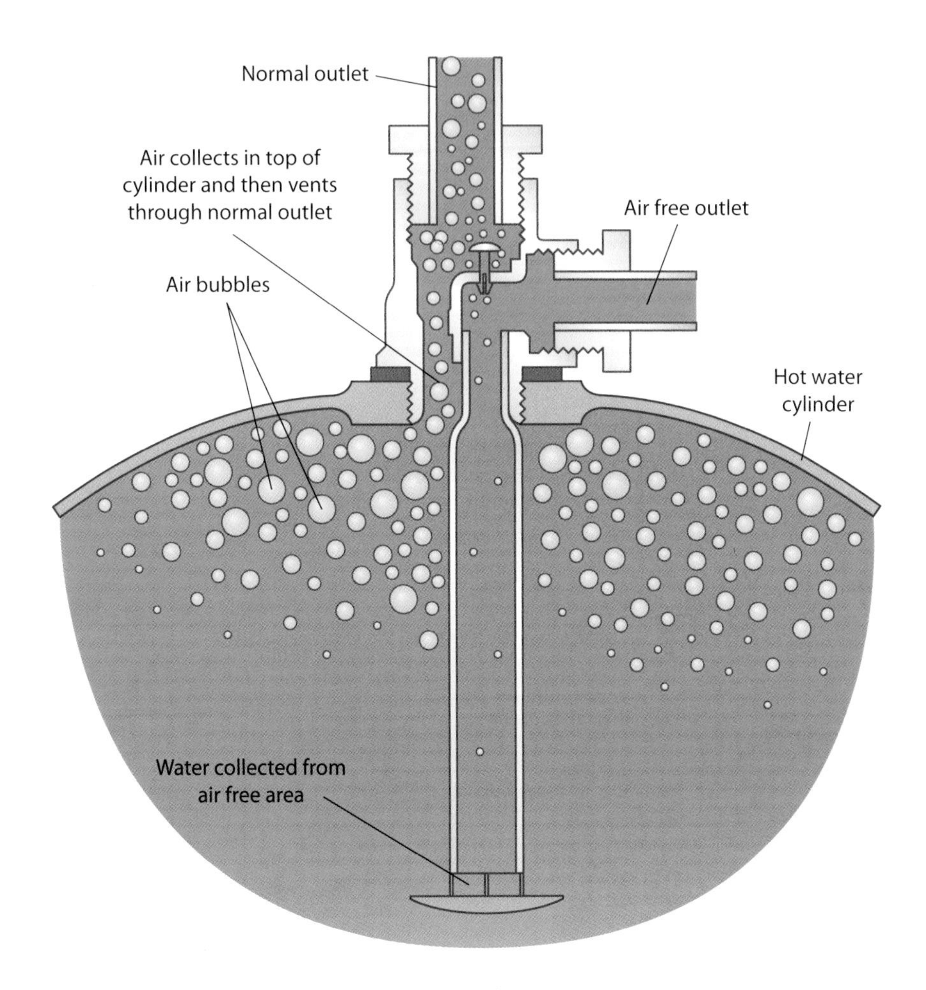

Figure 3.66: Working principle of a Surrey flange

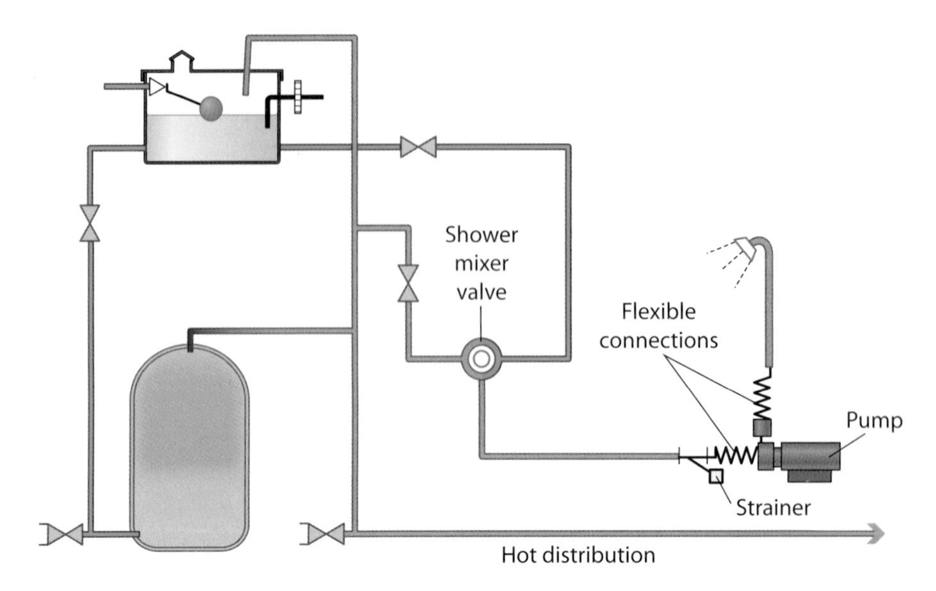

Figure 3.67: Single-ended pump installation

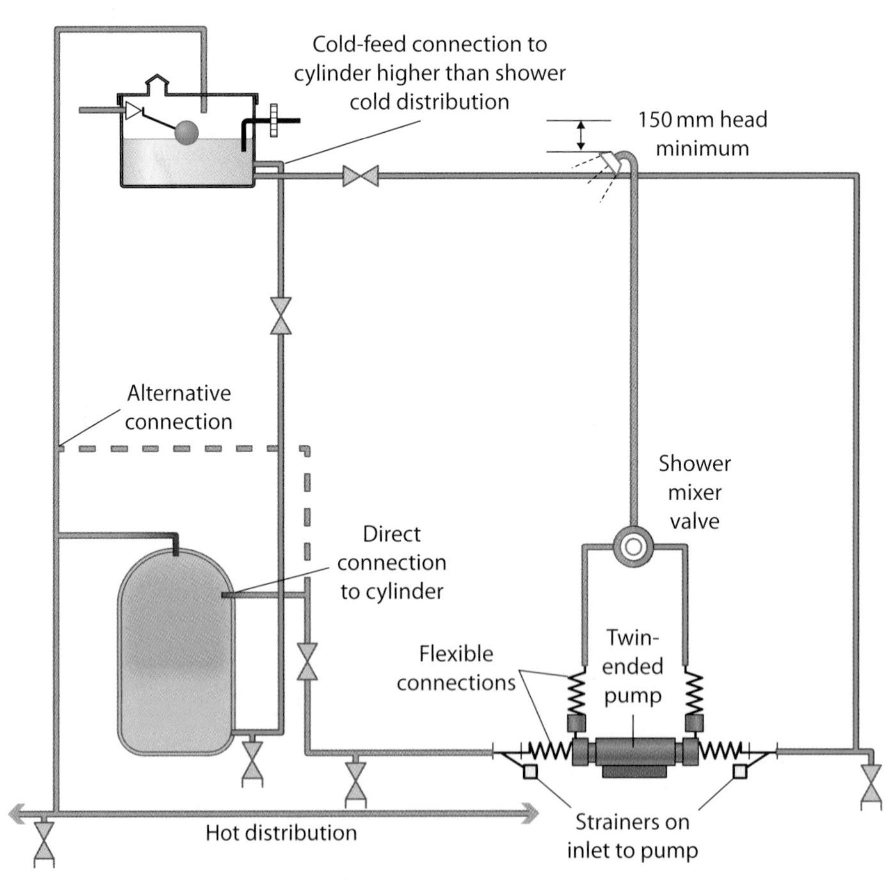

Figure 3.68: Twin-ended pump installation

Key term

No-flow condition – when there is pressure within the system but there is no water flowing from an outlet.

Pressure-reducing valves

Under **no-flow conditions**, the downstream (outlet) pressure acts on the diaphragm and overcomes the spring pressure. The diaphragm moves up and the linkage that joins the diaphragm to the seat holds the seat closed, so that downstream pressure cannot increase (see also photo of a pressure-reducing valve on page 178.

Under flow conditions, the downstream (outlet) pressure decreases until the spring can overcome the pressure. The diaphragm moves the linkage down and so opens the seat and water flow through the valve. When the outlet is closed, pressure builds up until the spring pressure is overcome and the seat is closed again.

Remember

Valves used on unvented systems are pre-set by the manufacturer, so you do not need to set or adjust them.

Balanced pressure valves

These operate in basically the same way except that there is an additional 'piston' of the same area as the main seat. This gives better control under low- and high-flow conditions.

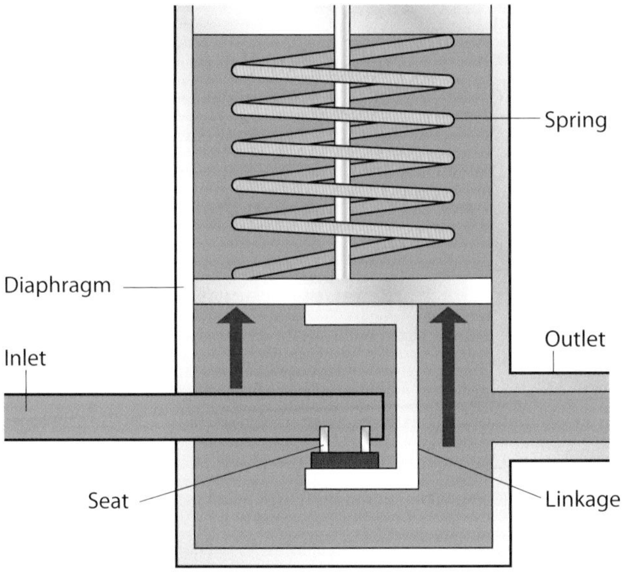

Figure 3.69: Pressure-reducing valve closed

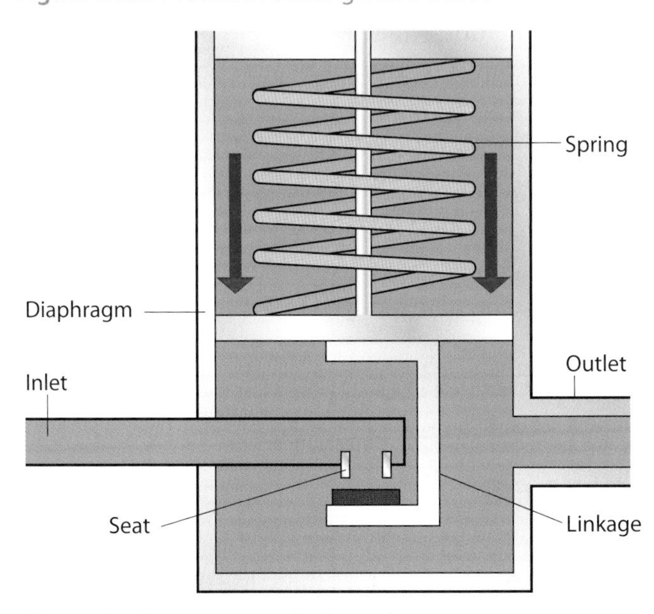

Figure 3.70: Pressure-reducing valve open

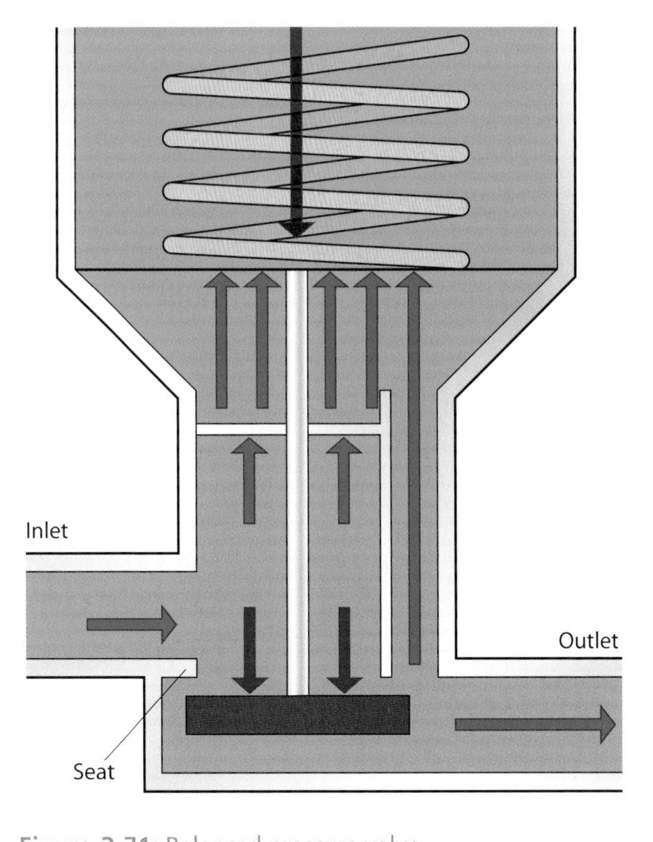

Figure 3.71: Balanced pressure valve

Shock arrestor

Pipe (or water) hammer and reverberation are both caused by shock waves running through the water system, evidenced by the production of noise or, in extreme cases, violent pipe movement. Hammer can be a one-off shock wave, while reverberation is a series of shock waves in quick succession. The pressure wave in such circumstances can be up to three times greater than the standing pressure. Reverberation can feed on itself (positive feedback) and, under extreme conditions, can go on increasing until the pipe bursts.

Hammer occurs when the flow rate is changed suddenly: for example, when a valve or terminal fitting is closed suddenly. Reverberation occurs when system components have moving parts, which respond to the initial shock wave by trying to open or close.

Curing pipe hammer is not an exact science. While it can be described scientifically, and it is possible to use a mathematical model to design a system that should not hammer, an existing system cannot be treated in the same way. The following methods of fault finding are given as a guide; as each system has its own peculiarities, these methods cannot guarantee success.

The first step is to find out when the hammer/reverberation occurs and under what circumstances. This will give the main clue as to how to solve the problem.

One of the most common problems is when water hammer occurs when a new tap or valve closes. Now that quarter-turn valves and taps are more commonly used, the speed of the tap being closed, combined with the pressure of the water flow and the suddenness of the quarter turn, can cause a shockwave through the system, creating water hammer. Where the tap or valve has a washer, the problems can be due to a loose or faulty washer. Where an installation whether it be old or new has insufficient clipping of the pipework this can also cause water hammer.

Mini expansion vessels

Where unvented hot water units of less than 15 litres are fitted and there is no means of expansion provision, whether it is internal or within the cold water pipework, you would need to use a mini expansion vessel. This can be seen in more detail in hot water systems, pages 179–81 of Unit 4.

Overcoming temperature and pressure effects of backflow-prevention devices

Where a backflow device would be under the effects of temperature and pressure, an expansion valve would be fitted to allow for the excessive pressure and temperature, which expand the water. This would be most common with unvented hot water systems.

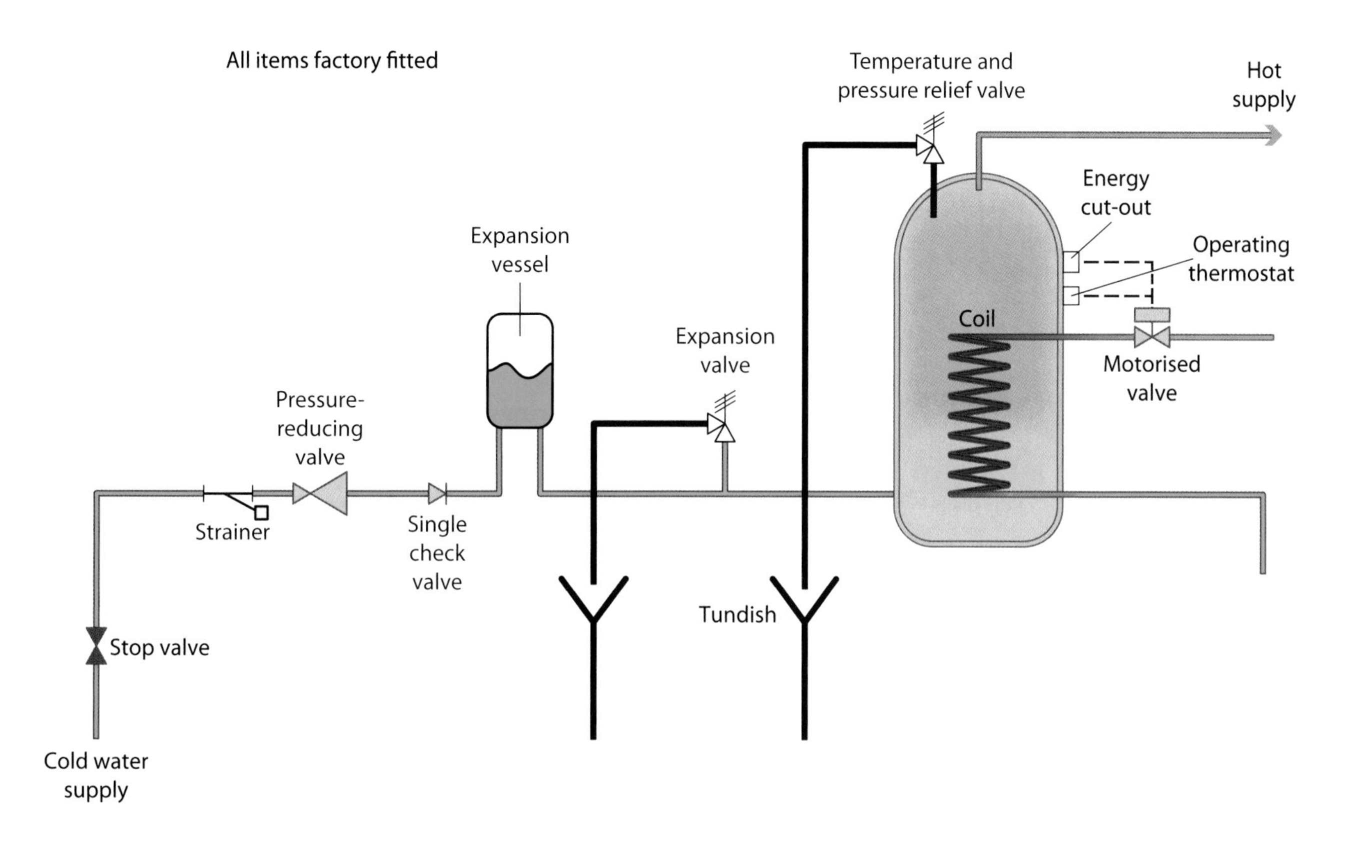

Figure 3.72: The package

PROGRESS CHECK

1. When would a Surrey flange be used?
2. How does an infrared tap work?
3. What is the maximum flush for a single bowl urinal?
4. What could be installed to overcome water hammer in a system?
5. What is the purpose of a concussive tap?
6. When a bath shower mixer is fed with water from a combination boiler what should be fitted and where?
7. When would a mini expansion vessel be fitted?
8. State the disadvantages of using spray taps.
9. What is the minimum head required for a shower pump to operate?

6. Know and be able to apply the design techniques for cold water systems

Plumbing information sources

Plumbing information covers a range of sources, including plans, drawings and specifications, manufacturers' literature, Codes of Practice and legislative documentation. You will use this information in designing or specifying systems installations, so you need to know how to use it. You will also use plans and drawings in construction, so you should be familiar with how to read them.

The information sources in this section are ones that you will need to draw on again and again, and not just when dealing with cold water systems. You will find reminders to look back at this section in each of the later units, when you deal with hot water, electrics, central heating and sanitation.

Block plan

While a block plan does not provide details of a specific system installation, it is classified as part of the general documentation that is required for a new housing development. A block plan records a number of uses of space. Developers will use it to show the proposed site's relationship to amenities such as schools, shops and leisure facilities, together with the transport infrastructure – meaning road and rail networks.

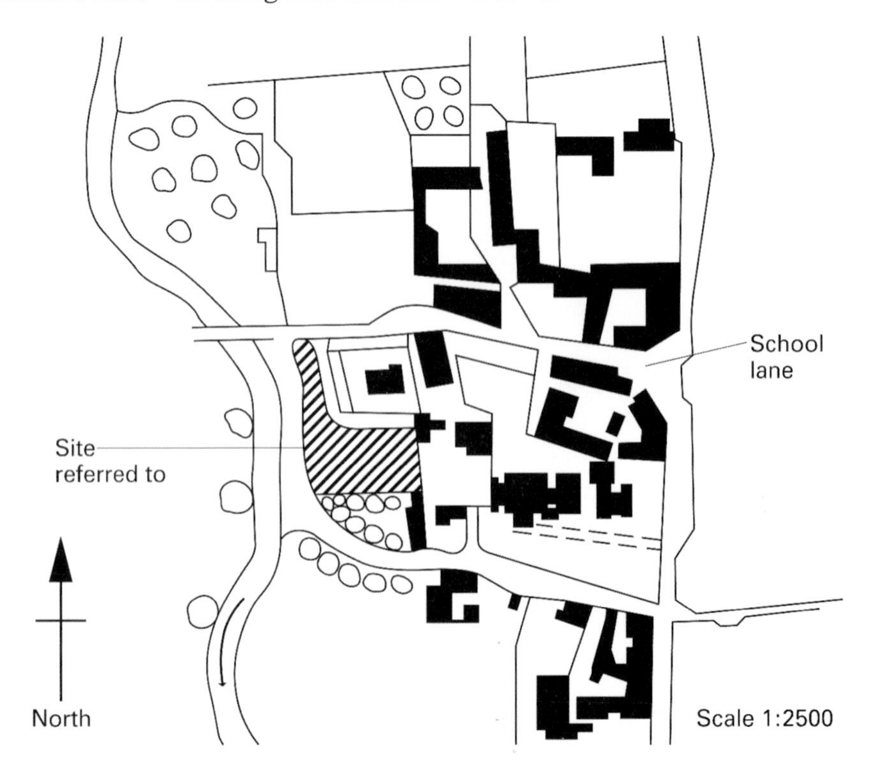

Figure 3.73: Block plan

Site plan

The site plan shows the proposed development in much greater detail. The road access to the site is shown, together with the position of each property. Drainage and sewerage requirements are also included.

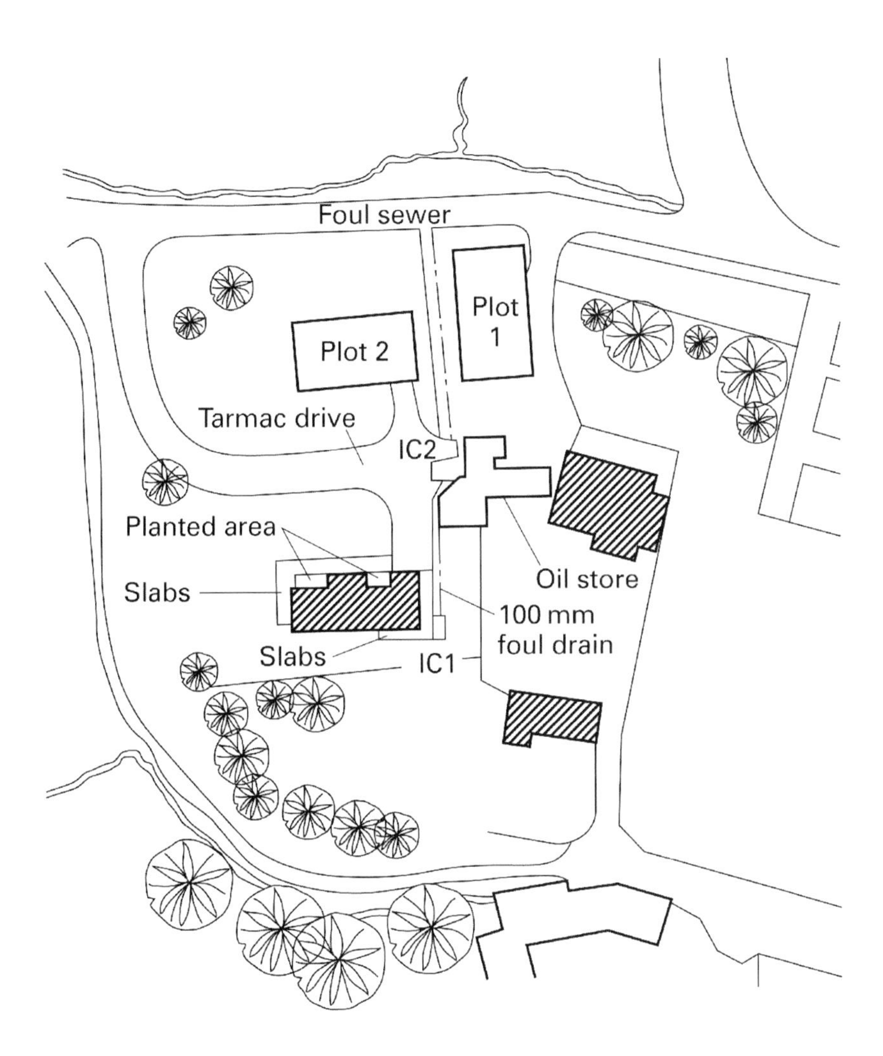

Figure 3.74: Site plan

Location drawing

A location drawing is used to show overall sizes, levels and references to assembly drawings. A location drawing could be:

- block plans
- site plans
- floor plans
- foundation plans
- roof plans
- section through the building
- elevations.

Assembly drawings

Assembly drawings are used to show how a building is erected on site, and will show a detailed section through each aspect of the construction.

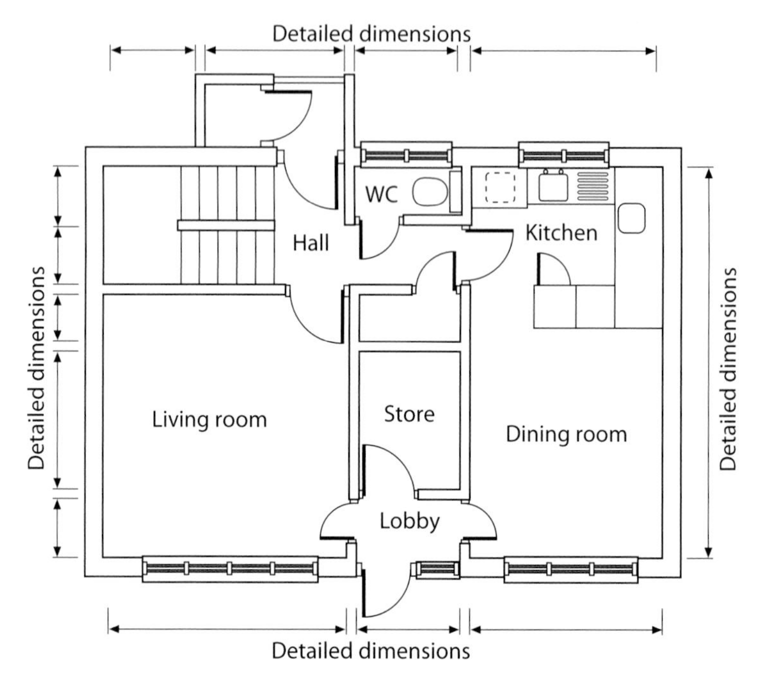

Figure 3.75: Location drawing showing a ground-floor plan

Installation drawings

Generally, on one-off installations, the pipe runs will be determined by the plumber, and the location of appliances and components agreed

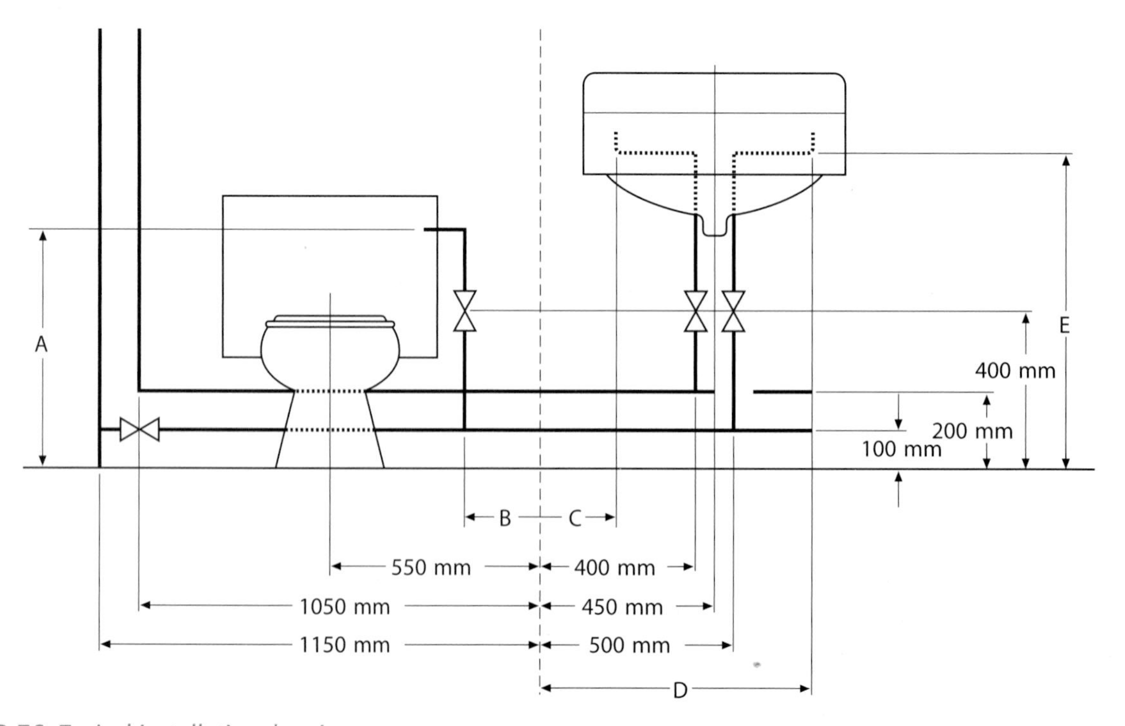

Figure 3.76: Typical installation drawing

in consultation between the customer or client and the plumber. On some jobs, however, installation drawings may be provided, showing the position of appliances and components, pipe sizes and pipe runs.

Design principles for cold water systems

At Level 3, you are not expected to have the capability for a full system design. The aim of this section is to give you the basic skills and information you will need to deal with problems as they arise, and in particular to deal with small component problems or faults.

Pipe sizing

For systems in larger buildings (including larger detached domestic properties) that require several outlets over a greater area, or on several floors, the pipework will need to be sized correctly, ensuring that there is adequate pressure and flow at all the draw-off points, without any excessive system noise problems. Correct pipe sizing will ensure adequate flow rates at the appliances, preventing any associated problems caused by over-sizing or under-sizing. In this section we will look at the design principles required to help you achieve this.

Pipe-sizing procedure

You will need access to BS 6700, which provides a series of tables and charts for pipe sizing. Pressure is not always measured in the same units on the charts, so you will need to be able to convert between values.

BS 6700

Remember

Use this rule of thumb for converting different values:

1 metre head = approx 10 kPa (kN/m2) = approx 0.1 bar

Determining the flow rate

In most buildings, it is unlikely that all appliances installed will be in use at the same time. As the number of outlets increases, the probability of them all being used at the same time decreases. It is therefore more economical to design the system for peak flow rates, which are based on the probability theory, using 'loading units' instead of using possible maximum flow rates.

- A loading unit is a factor or number given to an appliance; it relates to the flow rate at the terminal fitting, to the period of time in use and frequency of use.
- Table D1 from BS 6700 gives the loading units for various appliances.

By multiplying the number of each type of appliance by its loading unit, then adding the results together, you will find the total units for the installation. You can then covert the loading units into litres per second by using the conversion chart in Table 3.7.

In this example, a building with five floors has on each floor one bath, two WC cisterns, two wash basins with taps and one shower. At this stage, you are only dealing with the cold water; the hot water will be dealt with separately, as it comes from a distribution system (storage) –

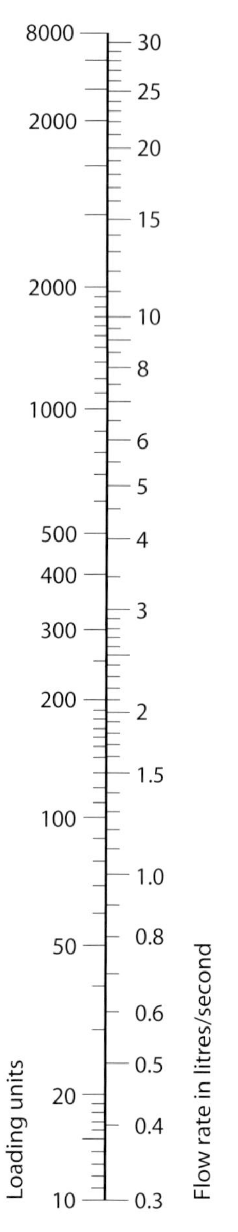

Figure 3.77: Loading units to flow rate calculator

unless it is an unvented system (or a combi), in which case you will need to look at sizing aspects of the hot and cold systems together.

Outlet fitting	Design flow rate (l/s)	Minimum flow rate (l/s)	Loading units
WC cistern – dual or single flush	0.13	0.01	2
WC trough cistern	0.15 per WC	0.01	2
Wash basin – ½" size	0.15 per tap	0.01	1.5 to 3
Spray tap or spray mixer	0.05 per tap	0.03	–
Bidet	0.20 per tap	0.01	1
Bath tap – ¾" size	0.30	0.20	10
Bath tap – 1" size	0.60	0.40	22
Shower head	0.20 hot or cold	0.10	3
Sink tap – ½" size	0.20	0.10	3
Sink tap – ¾" size	0.30	0.20	5
Washing machine – ½" size	0.20 hot or cold	0.15	3
Dishwasher – ½" size	0.15	0.10	3
Urinal flushing cistern	0.004 per position	0.002	–

Table 3.7: Loading units for different appliances

Using the conversion chart, you get a figure of 0.45 l/s for the section of pipework. Therefore, the five floors being supplied have a total loading unit rating of 20 x 5 = 100, converted to litres per second, gives 1.25 l/s.

- **Continuous flows** Appliances such as automatic flushing cisterns must be considered as having a continuous flow rate, and instead of applying the probability theory for using loading units, you must use the full design flow rate for the outlet fitting.
- **Design flow rate** The design flow rate for a pipe is the sum of the flow rate determined from 'loading units' and the 'continuous flows'.

Effective pipe length

Because valves and fittings create a resistance to the passage of water, you must convert the resistance created by them to an equivalent length of straight pipe run:

Effective pipe length = actual pipe length + equivalent pipe length (valves and fittings)

This calculation also benefits from a ready-made chart. Table 3.8 shows the equivalent pipe lengths for various types of fittings and pipe sizes for use on copper, plastic and stainless steel pipework.

Bore of pipe (mm)	Equivalent pipe length			
	Elbow (m)	Tee (m)	Stop valve (m)	Check valve (m)
12	0.5	0.6	4.0	2.5
20	0.8	1.0	7.0	4.3
25	1.0	1.5	10.0	5.6
32	1.4	2.0	12.0	6.0
40	1.7	2.5	16.0	7.9
50	2.3	3.5	22.0	11.5
65	3.0	4.5	–	–
73	3.4	5.8	34.0	–

Table 3.8: Equivalent pipe lengths for various types of fittings and pipe sizes

Note that:

- for tees only the change of direction should be considered
- the pressure loss through gate valves can be ignored.

From Table 3.8 you can see that using a 20 mm stop valve is equivalent to adding another 7 m of pipe run.

Pressure loss

Pressure loss through outlets

To size your pipework you will also need to know the pressure loss across any outlet fittings. BS 6700 provides you with standard data for some common fittings; for more specialist fittings, such as shower valves, the manufacturer will be able to provide this information.

For the system to work, the pressure at the inlet to the tap should be more than the pressure loss across it from inlet to outlet. Pressure loss through taps can be calculated using Table 3.9.

Nominal size of tap	Flow rate (l/s)	Loss of pressure (kPa)	Equivalent pipe length (m)
1/2"	0.15	5	3.7
1/2"	0.20	8	3.7
3/4"	0.30	8	11.8
1"	0.60	15	22.0

Table 3.9: Calculating pressure loss through taps

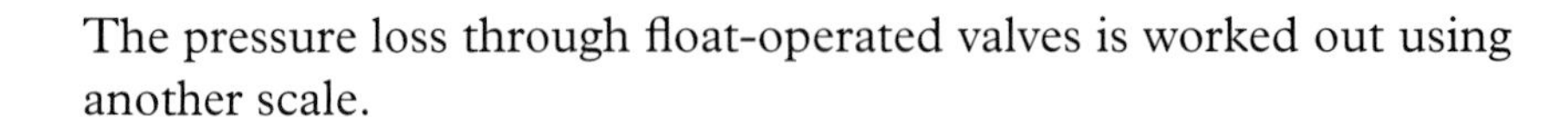

The pressure loss through float-operated valves is worked out using another scale.

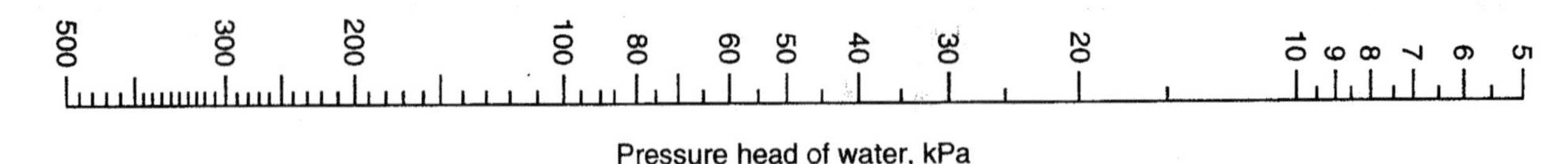

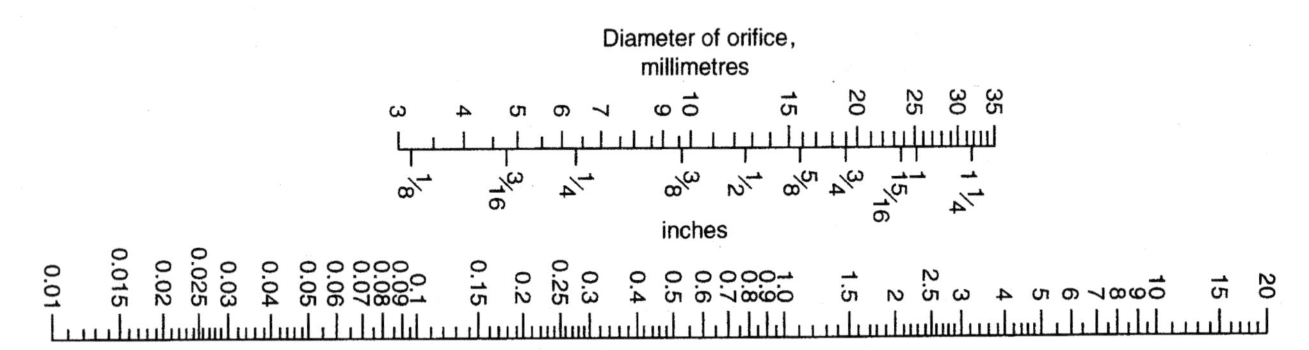

Figure 3.78: Scale for working out pressure loss through float-operated valves

To establish the pressure-head loss through the float-operated valve, you need to know the flow rate through it (you should already know this from looking at design flow rates) and the size of the orifice through which it discharges.

A standard ½″ float-operated valve has an orifice size of 3 mm, so if the flow rate required is 0.05 l/s, then by projecting a line across from flow rate 0.05 l/s through diameter of orifice 3 mm, you arrive at a pressure-head loss of approximately 45 kPa.

Remember

The absolute maximum velocity (speed) of water flow as recommended by BS 6700 is 3.0 m/s, so the example you have just looked at is only just acceptable.

Pressure loss per metre run of pipe

The final factor that you will need to be able to determine is the pressure-head loss per metre run of pipe. You determine this using another scale.

To determine the pressure loss per metre, use the suggested pipe size and project it across to the flow rate you require through the section of pipework. For example, if the pipe size (OD) is 38 mm and the desired flow rate is 3.5 l/s, then the pressure loss per metre run of pipe will be approximately 2.25 kPa.

Finally, you need to check the velocity and record it at the same time. There is a limit to the velocity at which we want water to travel through pipework. This is because, at high velocity, water is noisy and can cause wear and damage to fittings and components. In the domestic sector you often see small pipes working at quite high velocity. However, if you are to do the job properly, you should size the pipe appropriately.

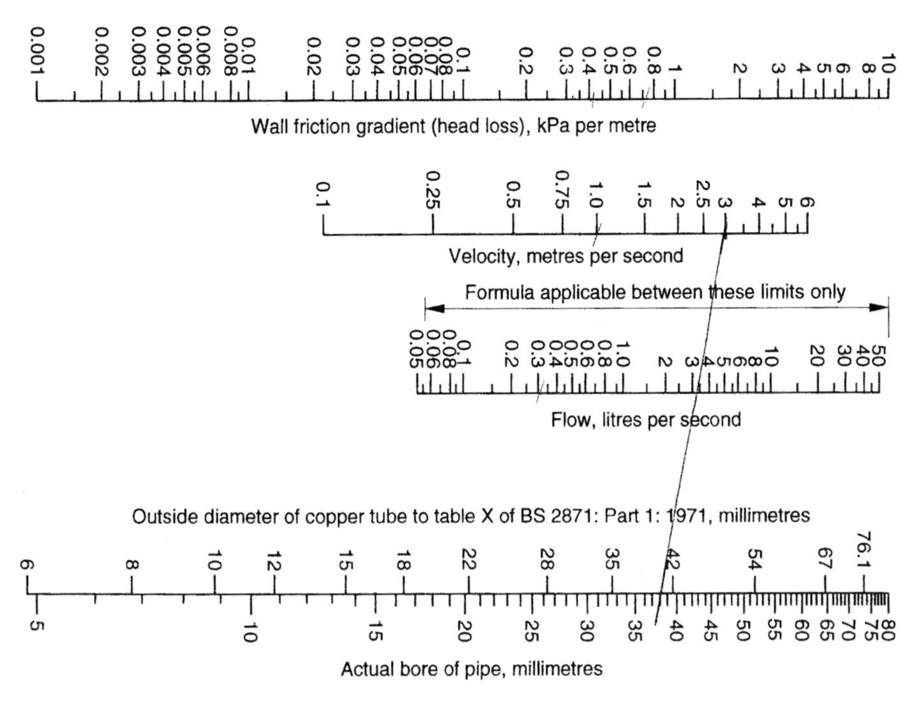

Figure 3.79: Scale for pressure loss per metre

Worked example: Cold water pipe size

First make a drawing of the installation and break it down into pipework sections. Then identify the flow rate to each appliance, which for this example will be:

- bath 0.30 l/s (10 loading units)
- basin – 0.15 l/s (1.5 loading units)
- WC – 0.05 l/s (2 loading units) – 3 mm orifice
- sink – 0.20 l/s (3 loading units)

A stop valve is included in section 1–2.

A stop valve and check valve are included in section 2–3.

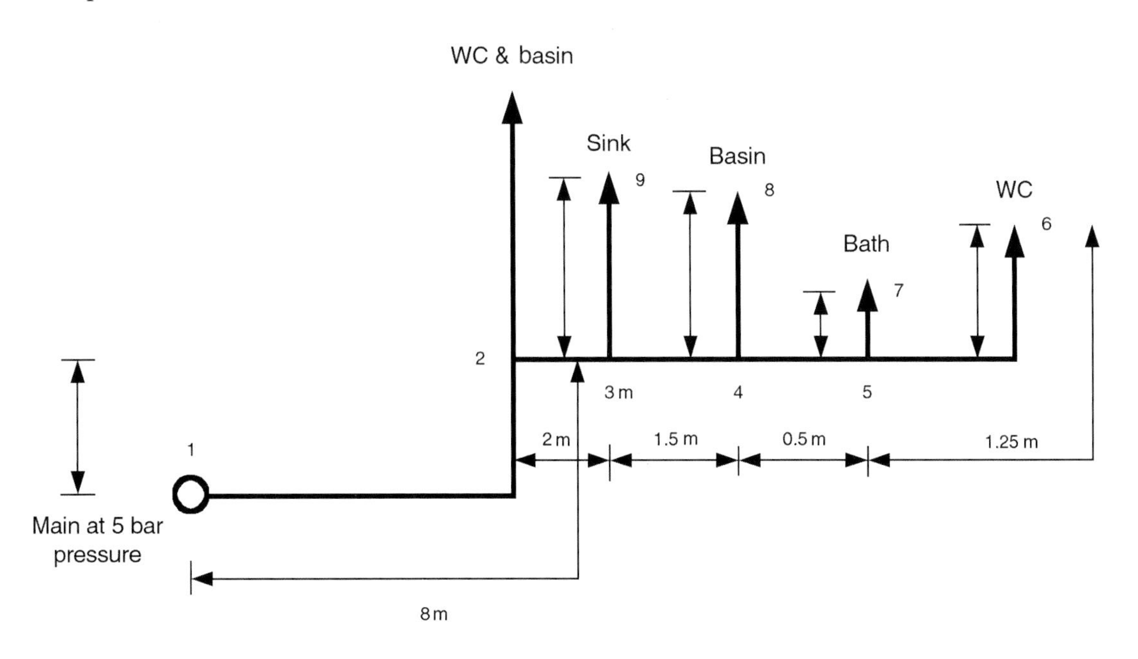

From the drawing you can establish the pressure at the main is 5 bar. This should be based on the minimum pressure usually available, not the maximum.

For supply pipework sizing, if you do not know what the pressure is at the main, you will need to put the system on test. You can do this by putting a pressure gauge onto the system and taking a reading. You do, however, need to know the height in metres of your test point in relation to other parts of the system, so that you can work out the pressure-head differences.

So the 5 bar pressure converts to 500 kPa or 50 m head. Use a chart to make the work easier and follow a step-by-step plan of how to size the pipes.

1	2	3	4	5	6	7	8	9	10	11	12	13	14	15	16	17
	Flow rate							Pipe length		Head loss				Residual head		
Pipe reference	Total (LU)	Design (l/S)	Pipe size (mm)	Velocity (m/s)	Head loss (kPa/m)	Drop– Rise+ (kPa)	Available head (7+14) (kPa)	Actual (m)	Effective (m)	Pipe (10x6) (kPa)	Valves (kPa)	Total (11+12) (kPa)	Available (8-13) (kPa)	Fitting type	Required (kPa)	Surplus (kPa)
1-2	20	0.45	15	3.4	10	-20	480	8.0								
1-2	20	0.45	22	1.5	1.5	-20	480	8	8.8	13.2	70	83.2	396.8			

Table 3.10

1	2	3	4	5	6	7	8	9	10	11	12	13	14	15	16	17
	Flow rate							Pipe length		Head loss				Residual head		
Pipe reference	Total (LU)	Design (l/S)	Pipe size (mm)	Velocity (m/s)	Head loss (kPa/m)	Drop– Rise+ (kPa)	Available head (7+14) (kPa)	Actual (m)	Effective (m)	Pipe (10x6) (kPa)	Valves (kPa)	Total (11+12) (kPa)	Available (8-13) (kPa)	Fitting type	Required (kPa)	Surplus (kPa)
2-3	16.5	0.4	15	3.0	7.5	0	396.8	2.0	2.6	19.5	65	84.5	312.3			
3-4	13.5	0.36	15	2.6	7.0	0	312.3	1.5	1.5	10.5	0	10.5	301.8			
4-5	12	0.33	15	2.5	6.0	0	301.8	0.5	0.5	3.0	0	3.0	298.8			

Table 3.11

1	2	3	4	5	6	7	8	9	10	11	12	13	14	15	16	17
	Flow rate							Pipe length		Head loss				Residual head		
Pipe reference	Total (LU)	Design (l/S)	Pipe size (mm)	Velocity (m/s)	Head loss (kPa/m)	Drop– Rise+ (kPa)	Available head (7+14) (kPa)	Actual (m)	Effective (m)	Pipe (10x6) (kPa)	Valves (kPa)	Total (11+12) (kPa)	Available (8-13) (kPa)	Fitting type	Required (kPa)	Surplus (kPa)
5-6	2	0.05	15	0.4	0.2	-6	292.8	1.25	1.75	0.4	0	0.4	292.4	Float valve	43	249.4
3-9	3	0.20	15	1.4	2.4	-10	302.3	1.0	1.0	2.4	0	2.4	299.9	0.2l/s tap	8	291.9
4-8	1.5	0.15	15	1.2	1.5	-7.5	294.3	0.75	0.75	1.2	0	1.2	293.1	0.15l/s tap	5	288.1
5-7	10	0.30	15	2.3	5	-5	298.8	0.5	0.5	2.5	0	2.5	296.3	0.2l/s tap	8	288.3

Table 3.12

Activity

You are required to size the indirect cold water pipework from the cold water storage cistern shown. The details of the system are similar to those you have seen before:

- bath 0.30 l/s (10 loading units)
- basin – 0.15 l/s (1.5 loading units)
- WC – 0.05 l/s (2 loading units) – 3 mm orifice
- sink – 0.20 l/s (3 loading units)

A gate valve is included in the pipework, but you can ignore this for pressure loss.

The head from the base of the cistern to the lowest point on the pipework is 4 metres.

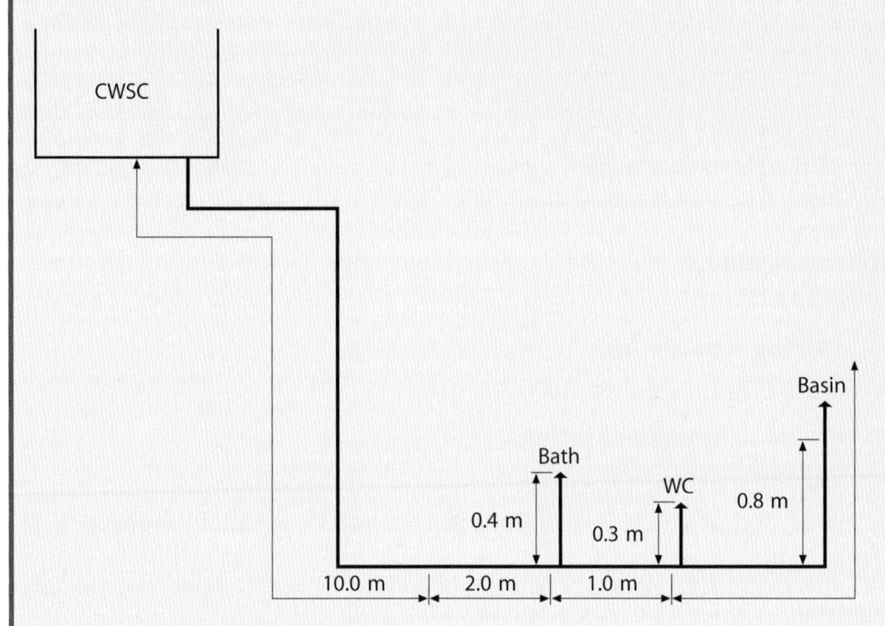

Figure 3.80: Activity scenario

Hint: Section and label the pipework first, and remember that this time you start off gaining pressure from 4.0 metres head at the outlet of the cistern.

Use Table 3.13 on page 128 to carry out the calculation.

Cold water storage cisterns

Traditionally, cold water in domestic dwellings was stored to provide a reserve in the event of cold mains failure. However, over recent years there has been a steady decline in the use of indirect systems, mainly because of the increase in the use of combination boilers and unvented hot water storage vessels that do not require a cold water storage vessel.

BS 6700 gives a recommended minimum storage capacity for domestic dwellings of approximately 230 litres for indirect systems. Cisterns that are supplying cold water only should have a minimum capacity of approximately 100 litres.

1	2	3	4	5	6	7	8	9	10	11	12	13	14	15	16	17
Pipe reference	Flow rate		Pipe size (mm)	Velocity (m/s)	Head loss (kPa/m)	Drop– Rise+ (kPa)	Available head (7+14) kPa)	Pipe length		Head loss			Residual head			
	Total (LU)	Design (l/S)						Actual (m)	Effective (m)	Pipe (10x6) (kPa)	Valves (kPa)	Total (11+12) (kPa)	Available 8–13) (kPa)	Fitting type	Required (kPa)	Surplus (kPa)

Table 3.13: Pipe size chart

In larger premises the cold water storage capacity will depend on the following factors:

- building type and use
- number of occupants
- number and types of fittings
- rate and pattern of use
- likelihood of an interruption or breakdown of the mains supply.

These factors have been taken into consideration, and Table 3.14 gives recommended guidance for the storage of water in a variety of different property types.

Type of building occupation	Minimum storage (litres)
Hostel	90 per bed space
Hotel	200 per bed space
Office premises – with canteen facilities without canteen facilities	 45 per employee 40 per employee
Restaurant	7 per meal
Dayschool nursery primary	15 per pupil
Dayschool secondary technical	20 per pupil
Boarding school	90 per pupil
Children's home or residential nursery	135 per bed space
Nurses' home	120 per bed space
Nursing or convalescent home	135 per bed space

Table 3.14: Minimum storage recommended for different property types

The minimum cold water storage shown in Table 3.14 also includes water used to supply hot water outlets.

Imagine that a boarding school with 200 residential pupils has opened a new day nursery annexe. The nursery contains provision for a further 40 pupils, and you need to determine the amount of cold water storage required.

Storage capacity = number of pupils x storage per pupil

Day school nursery = 40 pupils x 15 litres per pupil = 600 litres

Boarding school = 200 pupils x 90 litres per pupil = 18 000 litres

Add the two figures together and the total storage capacity is:

600 +18 000 = 18 600 litres.

Remember

The cold-feed cistern must have a capacity at least equal to that of the hot water storage vessel it is supplying.

Sizing booster pumps

On some contracts you may be required to size the booster pump that is required for the installation. The type of pump that is required would depend upon the application for which it is being used (for example, pumping well water).

To size a booster pump you will need two factors:

- fluid flow rate
- pressure to be developed.

The pressure that the pump should develop should equal the pressure drop (frictional resistance) in the system and be capable of overcoming the static head of water. The pressure drop can usually be found by using pipe-sizing tables (nanograms) which are available in BS 6700.

You can also determine the fluid flow rate from nanograms, which you would have already used when pipe sizing.

You should add a 20 per cent margin to the pump pressure to allow for any future extensions to the system and also to allow for a drop in efficiency of the pump over time (wear and tear).

Worked example

If the static head required is 30 m then you can calculate the pressure required from the pump to delver the water to 30 m.

Pressure (Pa) = density of water x acceleration due to gravity x head (m)

Or **P = p x g x H**

Where P = Pump pressure (Pa)
p = Density of water approx 1000 kg/m^3
g = Acceleration due to gravity 9.81 m/s^2
H = Head (m)

P = 1000 x 9.81 x 30
P = 294300 (Pa)

Convert Pa to Kilo pascals 294300/1000
= 294.3 kPa

You can find the head that a pump can deliver by using the following calculation.

$$H = (P / p) / g$$
$$H = (P / 1000) / 9.81$$
$$H = P / 9810$$
$$H = (294300 / 1000) / 9.81$$
$$H = 2943 / 9.81$$
$$H = 30 \text{ m}$$

So a pump with a delivery pressure of 294300 Pa will pump the water to a head of 30 metres.

Now you can consult a pump catalogue to choose a suitable pump. The operating point of the pump can be superimposed using a graph. The graph shows pressure (head) against the flow rate in kg/s or l/s.

When selecting a pump for your system it is best to choose a pump with the operating speeds at or near the lower end of the performance curve, otherwise known as the Q/H curve, so that it is not operating at its maximum capacity; this allows a little room for error or margin. A typical pump sizing curve is shown in Figure 3.81.

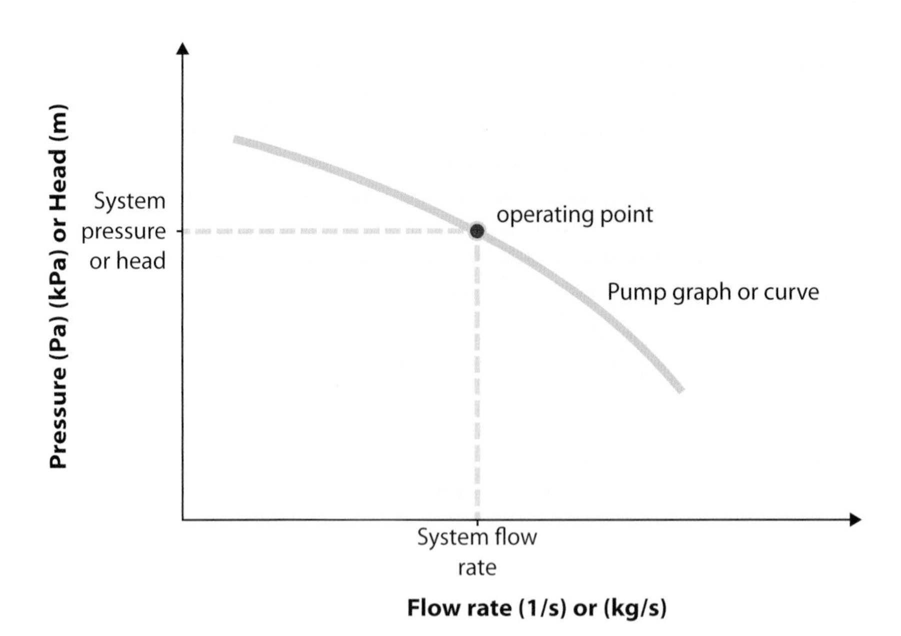

Figure 3.81: Q/H curve for pump

Where you have a multi-speed or 3-speed pump, it would be best to choose a pump that operates at the lower speed if possible, to prolong the life of the pump.

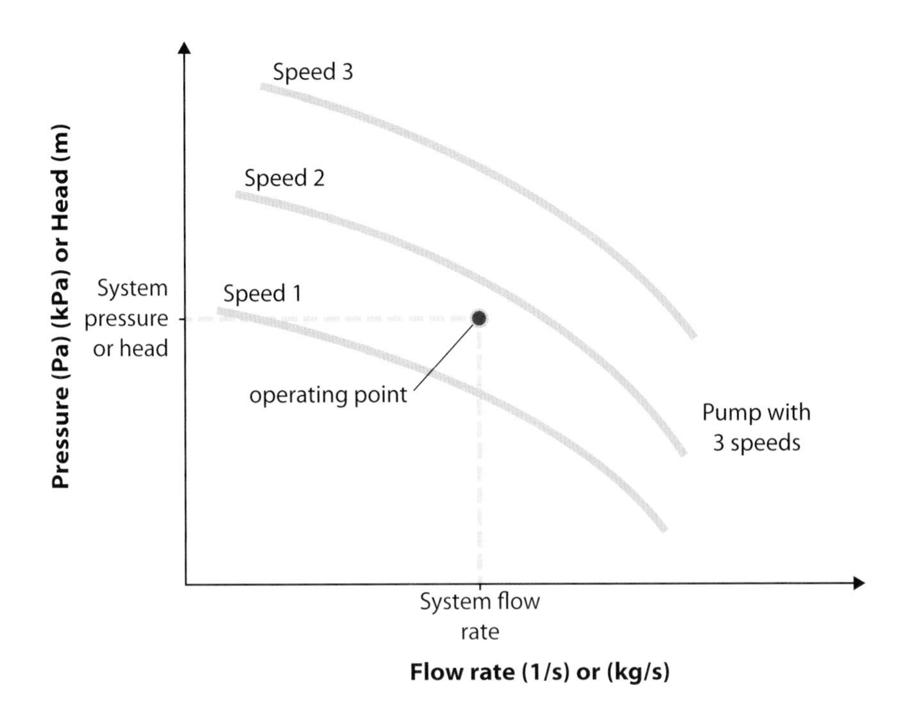

Figure 3.82: Q/H curve for 3-speed pump

Pump power and efficiency

Pump power

Pump power calculations are based on the physics of work done relative to time.

$$\text{Power of a pump} = \frac{\text{Work done}}{\text{time (seconds)}}$$

$$= \frac{\text{Newton x metres}}{\text{seconds}}$$

$$= \frac{\text{force x total head}}{\text{seconds}}$$

$$= \frac{\text{kg x 9.81 x head}}{\text{seconds}}$$

$$= \frac{\text{9.81 x litres x head}}{\text{seconds}}$$

Pump power = 9.81 x l/sec x m

$$\text{Watts} = \frac{\text{Nm}}{\text{s}}$$

Note:

1 kg/s = 1 l/sec
1 Newton = 1 kg x m/sec^2 = force
1 Newton = kg x 9.81

When sizing up the pump, you must include the measured pump head, plus all pipes and fitting resistance plus the velocity head.

Total head = L + Lf + 1/2 mV^2

Key term

Pump power – the total amount of energy required to drive a pump or a pumping set.

Efficiency

The efficiency of a pump is the ratio of the input power to the brake power. The brake power is the power absorbed by the pump output.

$$\frac{\text{output}}{\text{input}}$$

Efficiency % = x 100

Input or power to be provided:

$$\text{Input in kW} = \frac{\text{output in kW}}{\text{efficiency \%}}$$

Owing to the combined effectiveness of the pump and motor, overall efficiencies of about 50% are quite normal.

Work done is applied force through distance moved. The unit of measurement is the Joule (the work done when a 1 Newton force acts through 1 metre distance, i.e. 1 Joule = 1N x 1 m).

Time is expressed in seconds. By combining work done over a period of time:

Power = work done - time
= (force x distance) / seconds
= (Newtons x metres) / seconds [J/s] where 1 J/s = 1 Watt

Force in Newtons = kg mass x acceleration due to gravity [9.81 m/s^2]
Power expressed in Watts = (mass x 9.81 x distance) / time

Worked example

Power = (mass x 9.81 x distance) / time
= (5 x 9.81 x 30) / 1
= 1471.5 Watts

Allowing for the pump efficiency: 1471.5 x (100 ÷ 75) = 1962 Watts
Pump rating: 2 kW at 5 l/s (1962 Watts rounded up to nearest kW)

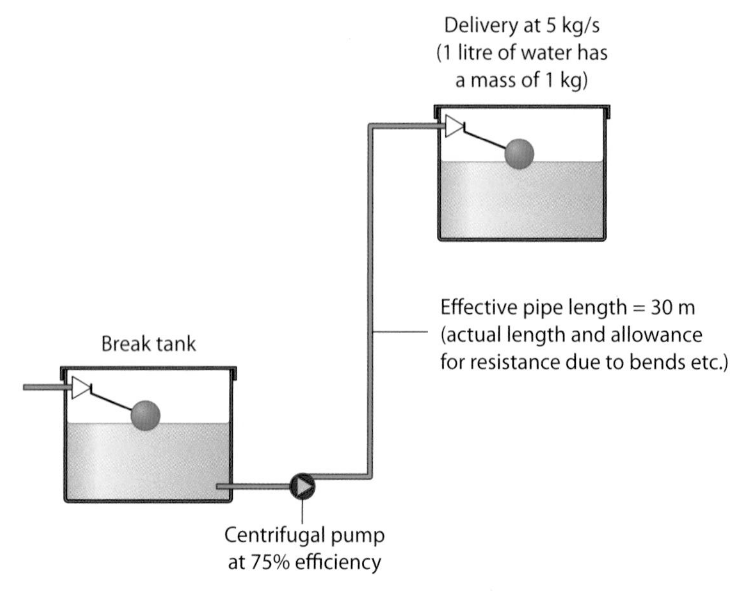

Figure 3.83: Pump sizing two cisterns

In normal applications, with the exception of maintenance and repair, the components of a water pump will remain unchanged during use. If a pump proves unsuitable for purpose, the complete unit is usually replaced with a pump of better specification.

Pumping definitions

- **Capacity:** the flow rate discharged by a pump, usually expressed in m^3/h or l/s.
- **Static head:** the vertical height of the liquid being pumped.
- **Velocity pressure:** the pressure needed to set the liquid in motion.
- **Friction head:** the head necessary to overcome resistance to flow of the liquid. Friction head must not exceed the total delivery head.
- **Negative suction lift:** this exists when the pump is above the liquid to be pumped and is the vertical distance from the centre line of the pump down to the free surface of the liquid.
- **Positive suction head:** this exists when the pump is below the liquid to be pumped and is the vertical distance from the centre line of the pump up to the free surface of the discharge liquid.
- **Static delivery head:** the vertical distance between the centre line of the pump and the free surface of the discharge liquid.
- **Total negative suction lift:** the static suction lift plus the friction head and velocity head in the suction pipe system.
- **Total positive suction head:** the static suction head minus the friction head and velocity head in the inlet pipe system.
- **Total delivery head:** the static delivery head plus the friction head and velocity head in the delivery pipe system.
- **Self-priming:** when a pump is operating under suction lift (negative) conditions, self-priming is the characteristic which enables the pump to evacuate air from the suction line, thus creating a vacuum which allows the atmospheric pressure to push the liquid through the suction pipe to the pump.
- **Velocity head:** this is the pressure required to set a liquid in motion and is generally of practical importance only in the case of pumps of large capacity or where the suction lift is near the limit.

  Velocity head $= 0.5 \times M \times V^2$
  Where $M = kg$
  $V =$ velocity in m/second
- **Cavitation:** occurs when the static pressure somewhere locally within the pump falls below the pressure of the liquid (the word originates from the Latin for 'hollow').

Cavitation is a rather complex and undesirable condition that may occur and can be recognised by a metallic knocking which may vary from very mild to very severe.

Cavitation is normally associated with centrifugal pumps, but can arise with any pump installation unless care is taken with pump selection and good planning of the installation.

When planning, the following conditions should be avoided:

- suction lift higher or suction head lower than recommended by manufacturer
- liquid temperature higher than that for which the system was originally designed

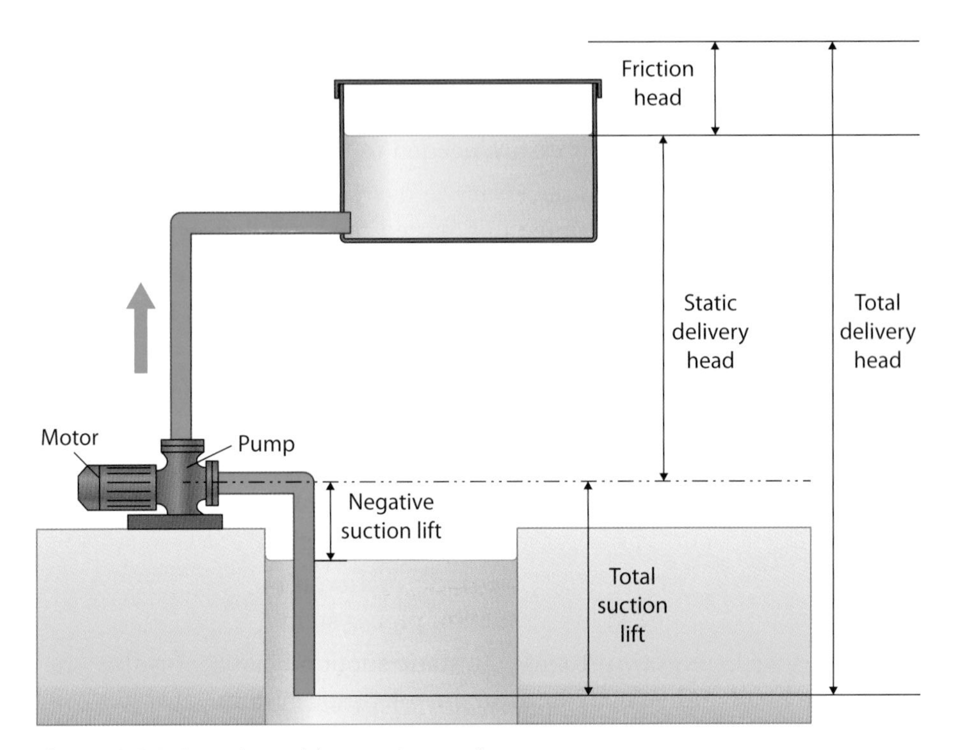

Figure 3.84: Pumping with negative suction

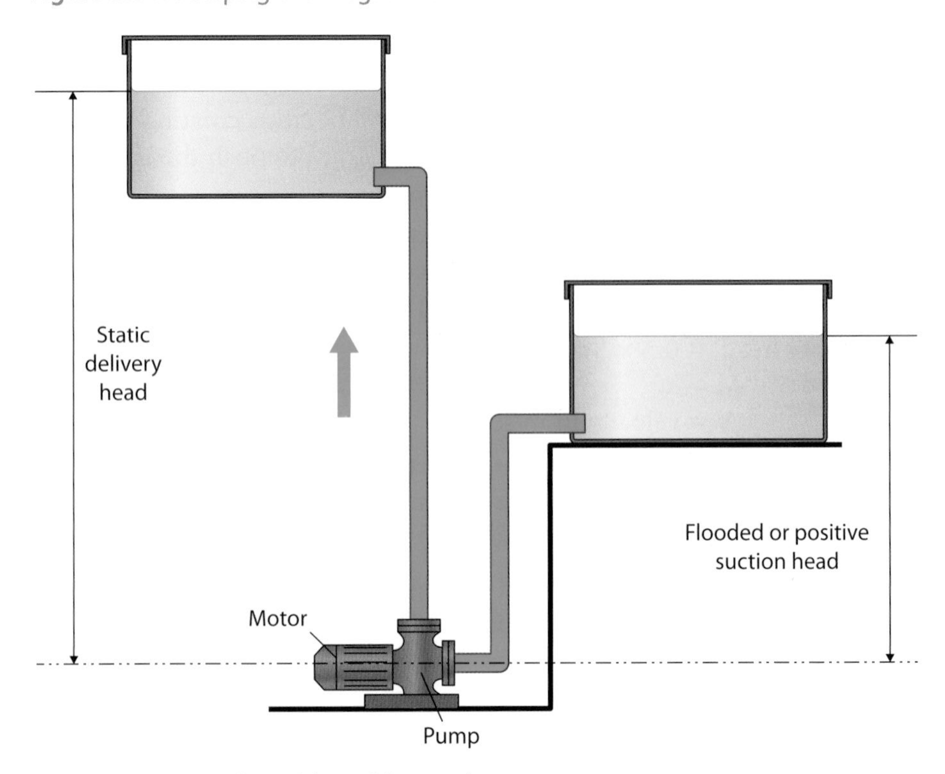

Figure 3.85: Pumping with positive suction

- speeds higher than manufacturer's recommendation. For centrifugal pumps care should be taken that they are not operated with heads much lower than the head for peak efficiency or with capacities much higher than the capacity for peak efficiency, particularly if there are other adverse conditions present which may tend to promote cavitation.

Types of pump

There are two main classifications of pump:

- centrifugal
- positive displacement.

Centrifugal

This is the simplest form of a booster pump as it consists of an impellor and a volute. This type of pump has to be filled completely with the liquid it is pumping to operate. The way that the pump works is that the impellor 'throws' the pump liquid to the outside of the volute. This action within the pump imparts kinetic energy, which is why the pump can provide a certain head for a given speed of the pump.

The relationship between capacity and head is expressed by a 'characteristic curve'. This is often referred to as the Q/H curve, where Q is the quantity (flow rate) and H is the head.

The main characteristics of centrifugal pumps are:

- capacity varies with head
- capacity is proportional to pump speed
- head is proportional to the square of the pump speed
- non self-priming
- suitable for low-viscosity liquids.

Positive displacement pump

Positive pumps normally operate by the liquid being positively transferred from suction to discharge port. Because these pumps are manufactured to such fine clearances, most positive pumps are self-priming. These pumps deliver an almost constant capacity irrespective of variations in head. When drawn for this type of pump, the usual Q/H curve is almost a vertical straight line. It is not usual to provide Q/H curves for positive pumps.

The main characteristics are:

- capacity independent of head
- capacity proportional to speed
- self-priming
- suitable for viscous liquids.

Applications

Pumped systems or booster pump sets will be required where any one or a number of the following conditions prevail:

- where there is no piped public supply
- where pressure from the public supply pipe is insufficient to reach the top-most point in the system
- where pressure is fine for normal domestic applications, but not sufficient for other more specialised equipment
- where the water undertaker cannot guarantee a constant pressure to the supply in the foreseeable future.

Be able to apply design techniques for cold water systems is a practical outcome, so you will be learning about this in your workplace. Activities and further information can also be found on the *Level 3 NVQ Diploma Plumbing Tutor Resource Disk.*

7. Know and be able to apply the fault diagnosis and rectification procedures for cold water systems and components

General principles

Information below applies whatever system you are working on, whether for hot or cold water, sanitation or central heating. You should follow this guidance whatever work you are carrying out.

Discussing faults with clients

The best ways of communicating with clients were covered in Unit 2. Look back at pages 14–17 to remind yourself before reading on.

When fault finding, it is important that you get as much information from the end-user as you can. This is essential, as the end-user may be able to give you key information about when and under which situations the fault occurs: for example, if the fault only happens during the evening, this may be because the water pressure increases during this time, leading to excessive noise in the system or even water hammer.

Getting the right information when repairing faults in components

When installing, commissioning, servicing and fault finding systems it is important that you have the correct information available.

The information can be available in many forms, including:

- manufacturer's brochure
- computer-based information

- manufacturer's installation instructions
- British Standards
- Building Regulations.

When trying to fault find on a component or appliance, it is important that you have the manufacturer's instructions, which should have been left when the job was completed. If these are not available, many manufacturers now have them on their websites as downloads.

These instructions would normally carry either a fault finding flow chart or table, and would give details of the possible rectification methods that you would have to undertake. With some faults, the outcome may be that you would have to renew the whole product.

Activity

Using the Internet or other means, research the checks to make when servicing and maintaining the following components:

- shower pump
- booster pumps
- float switches
- accumulators.

Make notes and ask your tutor to check them.

Routine checks and diagnostics

Checking operating pressures and flow rates

You should check the pressure and flow rate by taking measurements from various outlets on the system. In multi-storey dwellings, such as flats, you would need to check pressure and flow rates at varying levels.

You would check pressure using pressure gauges, which you can attach to tap outlets; the flow rates would be taken using a weir cup. Make sure that you note all measurements taken and, when you have finished testing, compare these against the performance specification in the installation manual.

Remember

When servicing a component, make sure that you have the technical data available: for example, pressure and flow rates when servicing pumps.

Cleaning components

Periodic servicing and cleaning of component working parts is important to increase the component's working life. As with all servicing and fault finding, you would need manufacturer's instructions for this. Manufacturers make service kits available for their components. These can comprise of replacement 'O' rings, washers and sometimes grease. With parts like these, it is important that you purchase manufacturer-specific items, and do not try to 'make do' with components you have picked off the shelf. There are many variations and sizes to these, and the metal or fibre that they are made from may not be to the correct specification for your job.

Isolating supplies

With water supplies, isolation is simple. If the component has local service valves, you should isolate at that point, locking off the valve. If not, you may have to isolate at the nearest point to the component. Check with the client or customer before isolating supplies, as other arrangements may need to be made for staff in a commercial environment for toilets and so on. Make sure you leave a notice to indicate that the system is isolated and should not be re-energised.

Dismantling the component

Drain down the component where necessary. If possible use a hose, but otherwise make sure that you catch any water and do not damage the client's property.

In the servicing section there is normally a blow-out drawing of the component with all parts identified. This will also have a dismantling and a reassembly sequence, which need to be followed to the letter. When dismantling, make sure that you keep all screws, bolts and other small items safe.

If you have to disconnect wiring, make a note of the connections. This is important: if you reconnect wrongly, you could cause damage to circuit boards or, in the case of pumps, make them work in the wrong direction.

Lay out the components in a logical order on your dust sheet, labelling where necessary. Check the condition of washers, O-rings and wiring. Make sure that the insulation on the wiring is still OK, with no exposed cabling. Check that none of the connections is shorting out – indications for this could be a smell of burning, or the connecting block having soot around it.

Reassembling the component

Checking with the service instruction part of the manual and the service kit instructions, reassemble the component in reverse order, replacing those parts in the service kit that need replacing.

After reassembly, connect the water supply and test for water tightness, which may involve undertaking a pressure test. When all is OK, reconnect the electrical supply, making sure that you follow any notes that you have made. After connection, check that continuity is OK for the earthing and also check for polarity of phase and neutral. Replace all covers that expose electrical cabling and re-energise the component.

Checking component operation

After servicing the component, you should now check that all components are working correctly as per the manufacturer's specifications.

Pumps

To check for correct operation of the pump, ask the following questions:

- Is it pumping in the correct direction?
- Can you hear the pump working?
- Is the pressure on the outlet side correct (use a pressure gauge at the test point on the delivery side of the pump)?
- Does the pump stop pumping at the correct pressure?

Pressure switches (transducers)

Without the pressure switches, which are usually fitted within the system by the manufacturer to the pump, the pump would be operating at the slightest drop in pressure.

You will have learned about pressure switches in Outcome 2. Look back at page 79.

When checking for correct operation, you will need to ensure that the system is up to pressure then, using the information from the manufacturer, drop the pressure by opening an outlet and let the water flow until the pressure drops to the lower pressure at which the pump should activate.

Float switches

You will have learned about float switches in Outcome 2. Look back at page 79.

To check that the float switch is working, empty the water out of the vessel in which the switch is fitted, then turn on the component that it operates and check that this activates. If the component does not operate, there is a fault. To check that the switch is not the faulty component, use a multimeter to check that there is a voltage across both of the cables that come out of the switch. If there is no voltage, the float switch may be faulty. Before deciding this, check the rest of the circuit(s).

Safety tip

Modern float switches use a reed switch, but older versions used mercury to control them. If you come across one of these, be extremely careful not to get mercury on your skin as it is extremely poisonous.

Expansion and pressure vessels

Checking expansion vessels is normally straightforward. You will need a tyre pressure gauge to check the pressure. This is normally done at the Schrader valve situated on the top of the vessel. Check the pressure against the data plate or manufacturer's instructions.

Pressure vessel pre-charge pressure is normally supplied at 1.5 bar, but under normal operating conditions this must be adjusted to a value of 90 per cent of the cut in pressure of the pump.

Here is an example:

Required cut-in pressure	2 bar
Required cut-out pressure	3.5 bar
Therefore tank pressure	= (0.9 x 2)
	= 1.8 bar

To recharge pressure in the vessel, make sure that the vessel is isolated, remove pressure from the vessel and then, using a foot pump connected to the Schrader valve, pump up to the required pressure. If the expansion vessel/pressure vessel will not hold the pressure, this could mean either a new diaphragm or a new vessel is needed.

Gauges and controls

Gauges

Pressure gauges can be calibrated by specialist companies. For basic checks of gauges, you need to make sure that the system is up to pressure. Check that:

- the glass is in place
- the needle has not become dislodged

Activity

Find information about gauges, in particular the Bourdon gauge, and how they work. Check the information with your tutor.

- there are no cracks or damage to the gauge body
- you are using a properly calibrated pressure gauge.

If you find any faults with the gauges, you need to change them for new ones. Most installations will have an isolation valve below the gauge, so it is a straightforward job to change gauges. Make sure that you have the correct size and that the pressure readings are the same.

Controls

For controls, check the expansion valves. Open the valve by twisting the cap or lifting the lever, release it and check that it re-seats and does not pass water. Also check that the valves operate correctly, and have not seized.

Checking treatment devices

Water filters

You can check if water filters are working correctly by using testing kits and equipment. These include testing strips, which you use to check if chemicals are in the system by comparing the colour on the strip to a colour in a chart. If the colours match, this indicates that there is an excess in the system and the filtration methods are not working, so would need to be changed for new filters. There are also other devices for checking **total dissolved solids (tds)**; these can be portable handheld devices with digital readout or electronic devices fitted in-line.

Key term

Total dissolved solids – these refer to any minerals, salts, metals, cations (positively charged ions) and anions (negatively charged ions) dissolved in water.

Water softeners

There are many products on the market for checking the softness of water. These range from kits that create foam to show that the water is soft, to kits that use reagents to change the colour of a sample of water.

As you should know from Level 2, water with a pH less than 7 is acidic, and acidic water is soft water; water with a pH greater than 7 is alkaline, and alkaline water is hard water. You can test for this using either testing strips or an electronic pH meter.

With the other type of test kit, you take a measured sample of the water from an outlet that is using the softened water. Pour this into a larger container and add the reagent solution, then stir the two together, making sure that the reagent has dissolved. Check the colour of the water: if it is red, the water is hard. This means that your softener will need recharging.

Locating faults

Plumbing is a complex area, involving a wide range of interdependent components and systems. This means that locating the source of any fault and identifying the exact problem can be difficult. Even when you are familiar with the most common faults, you will need to be logical and methodical in your approach to working out what is wrong on each occasion.

Booster (pump) set to a system

As with any component, you should check with the manufacturer's fault finding charts. Table 3.15 lists how to spot some of the most common faults, and how to correct them.

If you cannot find the fault in this table or in the manufacturer's chart, you will then need to contact the manufacturer's technical helpline.

Fault	Indication/cause	Corrective action
No lights on in controller box	• No electrical supply to boosted pump	1 Check mains isolator is turned on 2 Check fuse in plug or switch 3 Confirm that there is a 240V a.c. between live and neutral
Pump runs for a short time then stops	• Pump could be air-locked • Break cistern is empty	1 Check that pump is vented 2 Check water level in break cistern 3 Check incoming mains supply 4 Check float valve is working correctly
Pump running light lit in controller panel and pump not running	• Possible relay contacts defective • Pump shaft seized	Check electrical supply in pump motor terminal box for voltage between live and neutral
Pump stops then starts again after a few seconds	• Pre-charge pressure in vessel is incorrect	Check pressure in vessel and adjust to 90% of cut in pressure
Pump delivers correct pressure, but does not stop with no demand	• Pressure setting too high • Flow demand or possible leak	1 Check switch settings in pump controller 2 Close isolating valve; pump should stop. If pump does not stop, there is a system flow that is keeping the pump running on the flow switch

Table 3.15: How to spot and correct some common faults

Safety tip

When locating faults, safe isolation is essential – for your own safety and for the safety of others. Look at pages 409–415 now to remind yourself what is involved.

Remember

When locating faults:

- make sure that the water supply is on
- make sure that the electrical supply is switched on.

Isolating and dismantling a shower pump

Step 1: Isolate supplies (electrical and water), making sure the electrical isolation is safe.

Step 2. Leave warning notice where pipes are isolated.

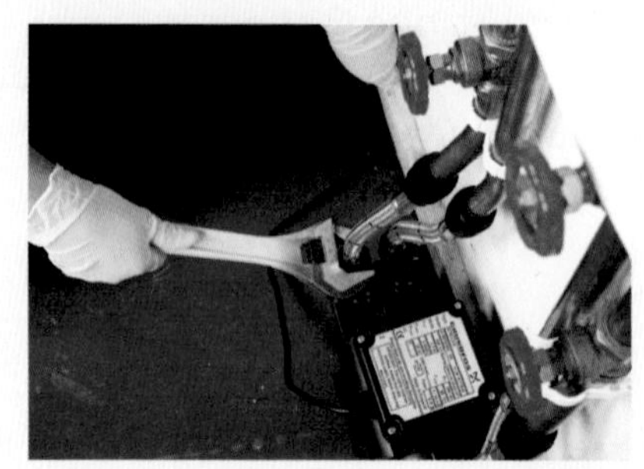

Step 3: Remove flexible connections ensuring that washers (if fitted) do not get lost or damaged.

Step 4: Check strainer on inlet to pump, clean if necessary under clean running water.

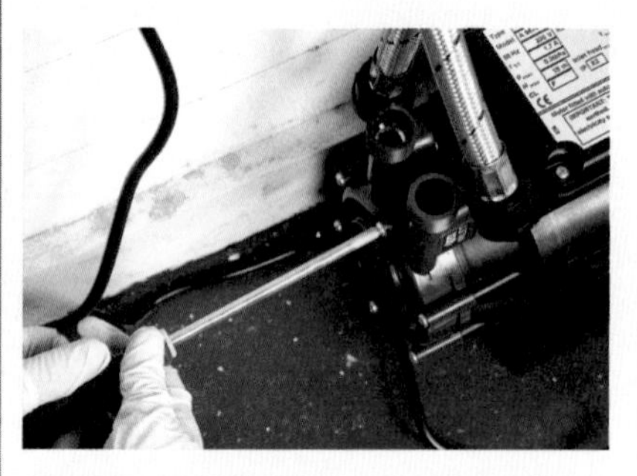

Step 5: Remove screws to impellor casing (this will vary from pump to pump, so read the manufacturer's instructions).

Step 6: Remove casing, taking care not to damage any parts, O rings, etc.

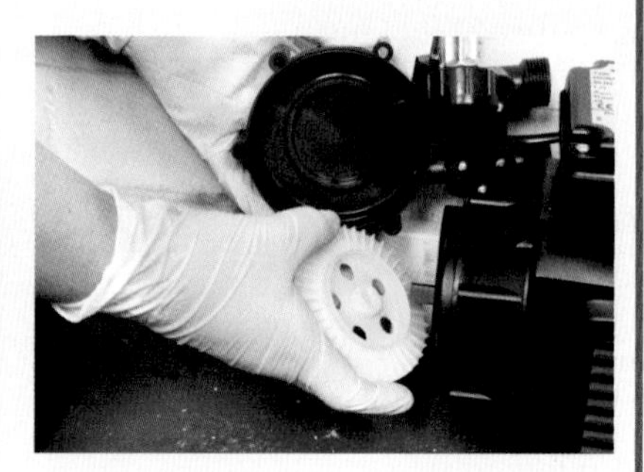

Step 7: Locate the impellor.

Step 8: Remove impellor (this one is sliding off easily, but with a circlip may need circlip pliers).

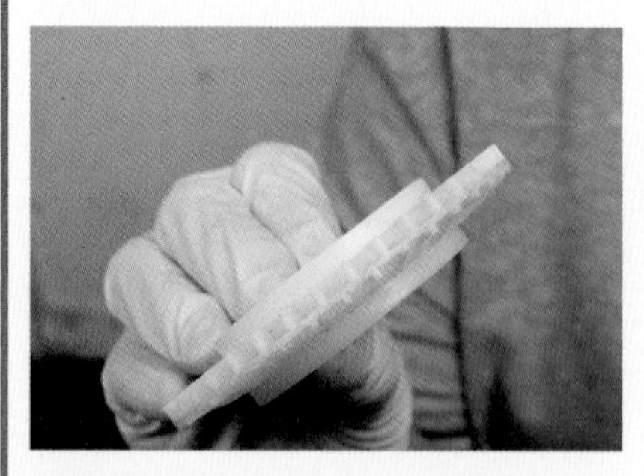

Step 9: Check for any damage, which could be causing noise.

Step 10: If damaged replace with new impellor.

Step 11: Reassemble in reverse order (it is advisable to replace washers to flexible connections).

Step 12: Turn on water supplies to check for water tightness, and remove air from system at shower.

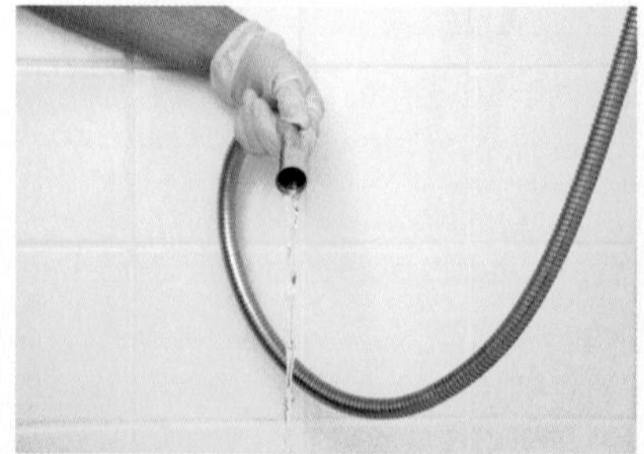

Step 13: Re-energise electrical connections and test shower.

Step 14: Confirm with customer on completion of job.

Backflow-prevention devices

With many backflow-prevention devices you would need to isolate by the servicing valve closest to the device that is being worked on. See pages 84–105 for more on these devices.

Diagnosing and preventing pipework corrosion

Electrolytic corrosion

Water fittings need to be immune to, or protected from, galvanic action.

The further apart metals are in the electrochemical series, the more likely it is that corrosion will take place. If two dissimilar metals are placed in contact with each other, the metal at the lower base end of the scale will be the one to corrode.

A typical example of this corrosion can be seen in galvanised steel cisterns (coated with zinc) that are connected to a copper pipework system. From Table 3.16 you will see that copper and zinc are some distance apart, and that zinc is the metal that will corrode.

Metal	Chemical symbol	Electrode potential (volts)
Silver	Ag	+ 0.80 cathode
Copper	Cu	+ 0.35 noble end
Lead	Pb	– 0.12 anodic
Tin	Sn	– 0.14 base end
Nickel	Ni	– 0.23
Iron	Fe	– 0.44
Chromium	Cr	– 0.56
Zinc	Zn	– 0.76
Aluminium	Al	– 1.00
Magnesium	Mg	– 2.00
Sodium	Na	– 2.71

Table 3.16: The electrochemical series

Another example of galvanic corrosion occurs when connecting copper pipe directly into lead pipe. The lead, being at the lower base end of the scale, will corrode, resulting in it being taken into solution, contaminating the water. The lead will also be weakened by the corrosion, eventually resulting in leakage.

Sometimes cathodic protection can provide protection against galvanic action. A sacrificial anode can be put inside hot water vessels, cisterns and tanks, and on pipelines. The anode will corrode instead of the fitting that it protects.

Did you know?

To check that a verifiable double check valve is operating correctly, check pressure in the system, decrease pressure in the upstream pipework and verify the check valve does not pass any water back into the pipework by removing the verifiable screw. If the backflow device is not working then water from the downstream section will be passing back into the upstream section and the device will need to be changed.

WSR Schedule 2, para 3(a)

WSR Schedule 2, para 3(b)

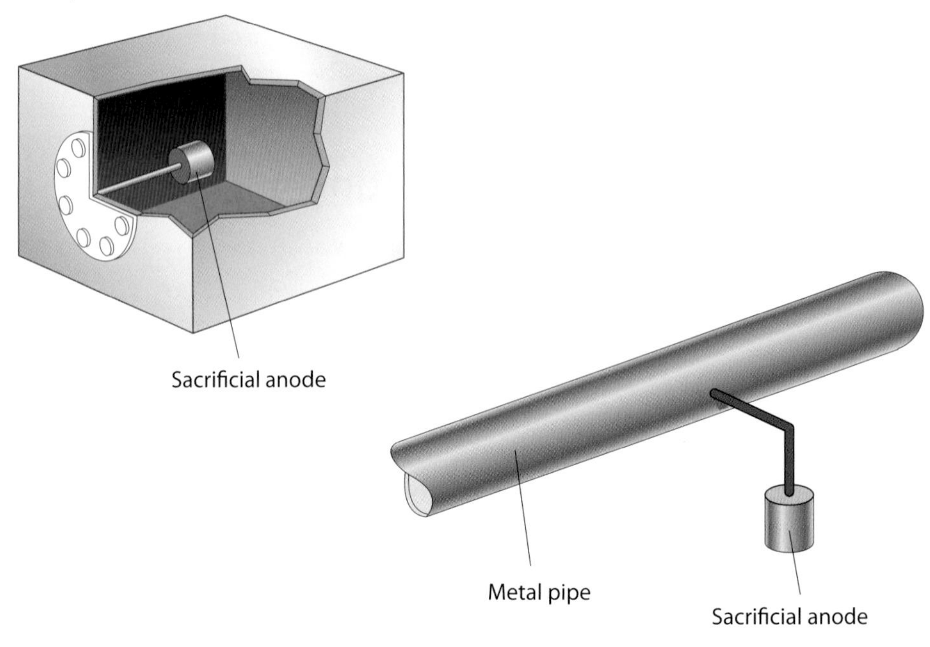

Figure 3.86: Tank protection

Blue water corrosion

Stagnation is suspected as one of the causes of blue water corrosion in newly installed copper pipework. This happens when the natural protective layer, which normally forms quickly on new copper pipework, fails to do so and the pipe starts to corrode. This releases copper into the water and can lead to a characteristic blue, cloudy appearance of water drawn from the outlet.

To minimise water quality problems caused by stagnation, the following actions are recommended.

(a) Where newly completed copper pipework is unlikely to be used within a few days after flushing or pressure testing, it should be drained down to prevent it being left with water standing in it.

(b) If this is impracticable, the water system should be flushed once or twice a week to prevent the water stagnating. It is also sensible to do this where existing copper pipework is unlikely to be used on a regular basis.

(c) In newly-occupied premises, each day for a fortnight occupants should ensure that the taps used for drinking purposes are run briefly until the water becomes noticeably cooler, in order to clear standing water from the pipes serving them.

8. Know and be able to apply the commissioning requirements of cold water systems and components

Interpreting information from different sources

Plumbing information covers a range of sources, including plans, drawings and specifications, manufacturers' literature, Codes of Practice and legislative documentation.

Plans and drawings

Plans and drawings take many forms. Here are some examples.

Specifications, materials schedules and component and appliance details

On larger installation contracts, a bill of quantities specifies how the work will be carried out, as well as the quality and quantity of the materials. A materials schedule contains similar information but is used as a working document on site. Component and appliance details are usually supplied in the delivery packaging.

These important documents were covered in Unit 2. Look back now at pages 37–38 to remind yourself.

Industry standards

There are two reasons for making sure that an installation meets industry standards:

- to ensure that the materials are of a satisfactory standard
- to ensure that the work is of a satisfactory standard and conforms to Regulations.

The main relevant legislation was covered in Unit 2. However, this is an important area now that you are starting some of the practical parts of the Level 3 qualification, so here is a reminder.

- Materials used in plumbing installations should be to the relevant EN or BS number. British Standards also make recommendations on design and installation practice: for example, BS 6700.

Remember

Legislation changes. Make sure that you are aware of the latest versions of industry standards, and that those you supervise understand changes that are relevant to their work.

WSR Schedule 2, paras 12 and 13

- In addition to British Standards, the following legislation places statutory responsibilities on plumbers:
  - Water Supply (Water Fittings) Regulations 1999
  - Gas Safety (Installation and Use) Regulations 1998
  - Electricity Supply Regulations 1998 and Electricity At Work Regulations 1989
  - Building Regulations 2000
  - Health and Safety at Work etc. Act 1974.

Commissioning

Key term

Commissioning – completing an installation, checking for faults, putting the system in use, and ensuring that it operates safely and efficiently and is to the customer's satisfaction.

Commissioning is an important part of your work that needs to be given time and care. The processes described below apply equally to hot water systems, which will be covered in Unit 4.

Commissioning a water installation includes:

- making a visual inspection of the installation
- soundness testing (testing for leaks)
- flushing and disinfection
- performance testing
- final checks/hand over.

WSR Schedule 2, para 12
BS 6700:2006+A1:2009

All the above procedures are considered to be good practice, but unfortunately they are not always carried out correctly and are sometimes completely missed.

Visual inspection

A visual inspection includes making sure all pipework and fittings are thoroughly examined to ensure that:

- they are fully supported, including cisterns and hot water cylinders
- they are free from jointing compound and flux
- all connections are tight
- terminal valves (sink taps, etc.) are closed
- in-line valves are closed to allow stage filling
- the storage cistern is clean and free from **swarf**.

Key term

Swarf – debris or waste from wear and tear on metal.

It is useful at this stage to advise the customer or other site workers that soundness testing is about to commence.

Checking cold water pipework

Checking for leakage

When testing for leaks you should follow this checklist.

Checklist

Testing for leaks

1. Slowly turn on the stop tap to the rising main.
2. Slowly fill, in stages, to the various service valves, and inspect for leaks on each section of pipework, including fittings.
3. Open service valves to appliances, fill the appliance and again visually test for leaks.
4. Make sure the cistern water levels are correct.
5. Make sure the system is vented to remove any air pockets before pressure testing.

Pressure testing

The **test pressure** applies to all tests and all installations. The requirement does not distinguish between installation sizes, new or replacement work or location (for example, above or below ground).

Key term

Test pressure – an internal water pressure of not less than 1.5 times the maximum pressure to which the installation or relevant part of it is designed to be subjected in operation (WSR Schedule 2, 12(1)).

Additional guidance on testing, flushing and disinfection of water installations can be seen in BS 6700:2006+A1:2009. The British Standard specifications give guidance only and are not law. However, if you carry out testing, flushing and disinfection in a workmanlike manner (Regulation A(5)) following the recommendations of BS 6700:2006+A1:2009, you are doing what is reasonably expected, to satisfy the law.

WSR Schedule 2, para 12(2)

Paragraph 12(2) sets out the 'test criteria', giving separate criteria for:

WSR Schedule 2, para 12(2)(9)

- systems that contain no plastics, such as copper, steel and cast iron
- systems that do contain plastic pipes and fittings.

Testing non-plastic systems

For non-plastic systems there are three test requirements:

- the installations shall be pumped up to test pressure or the maximum operating pressure, plus an allowance for any expected surge pressure (whichever is the greatest)
- the test shall be for one hour
- during this time there should be no visible leaks and no loss of pressure.

Safety tip

Whenever you use chemical agents for flushing and disinfection of installations and equipment, it is very important that you take into consideration the manufacturer's recommendations and advice.

Water-soundness testing should be carried out on all completed installations, including supply pipes, distributing pipes, fittings and components.

Remember

The final test of an installation is a crucial part of the commissioning procedure, and any buried or concealed pipework must be successfully tested before backfilling or encasing takes place.

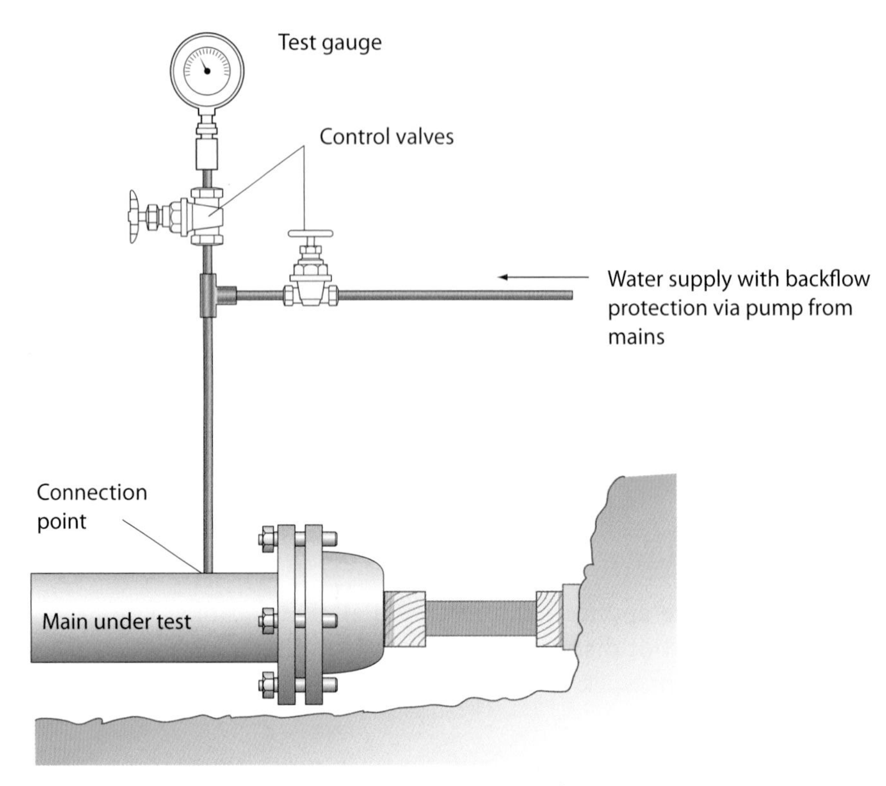

Figure 3.87: Testing equipment and requirements for a water main

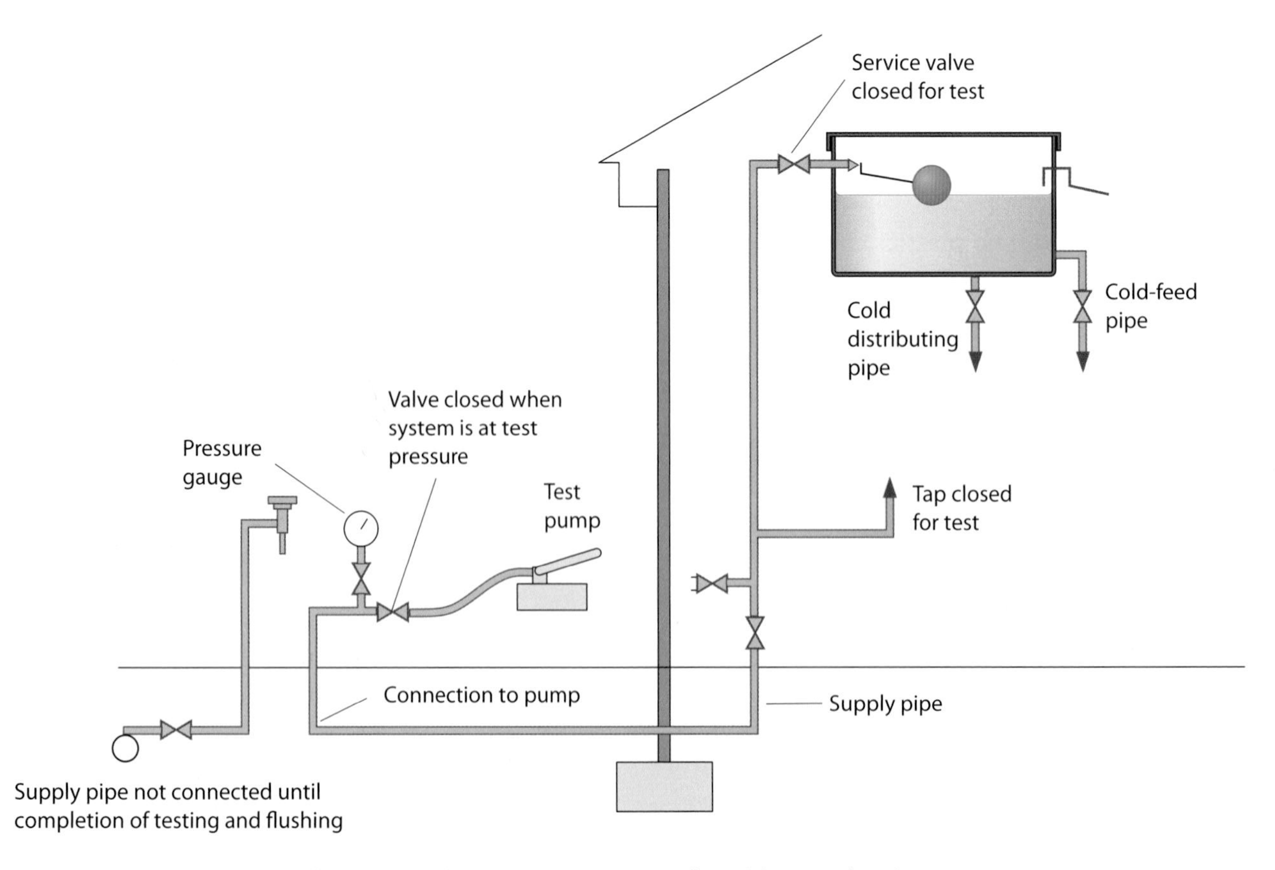

Figure 3.88: Testing a supply pipe and the soundness testing of a cold water distributing system

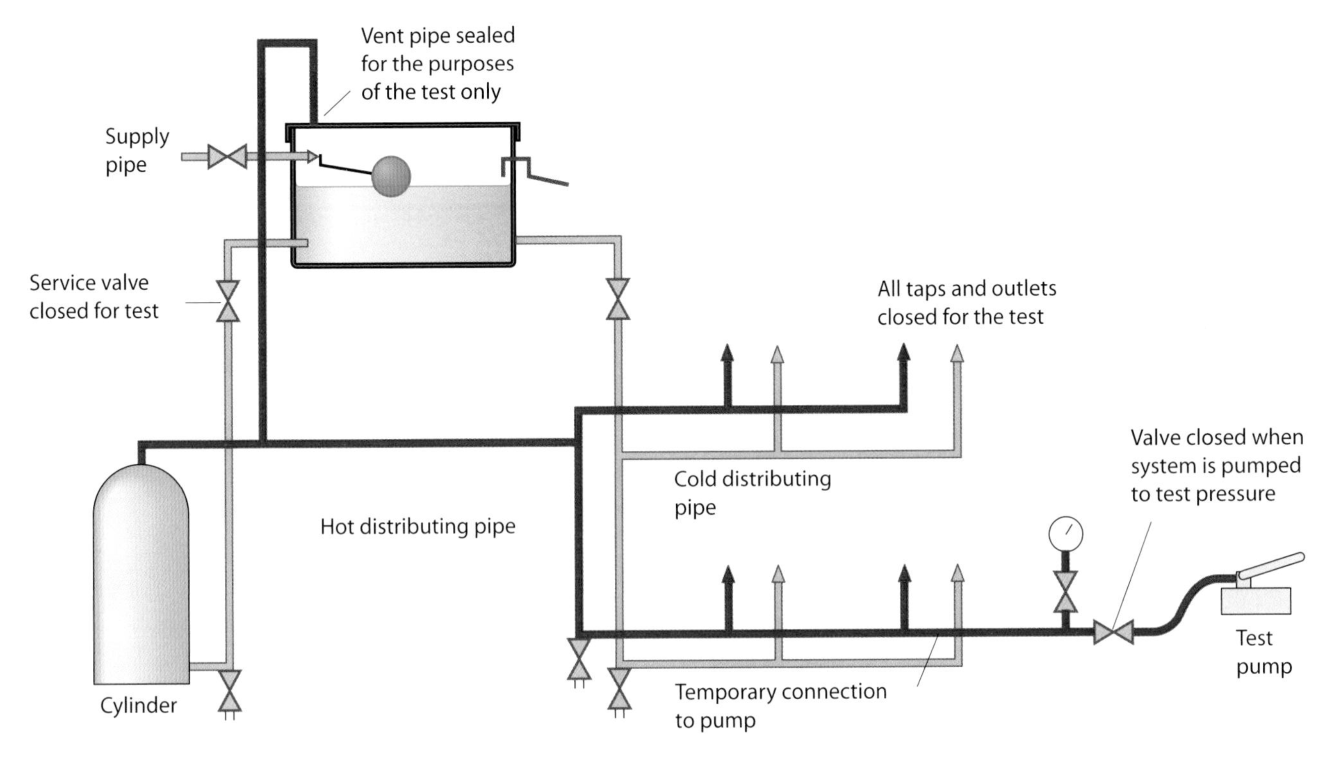

Figure 3.89: Soundness testing of a hot water distributing system

Testing systems containing plastics

WSR Schedule 2, para 12(2)(b)

There are two tests for installations containing plastics. The Guidance Document states that systems containing plastic pipes should be tested in accordance with the recommendations of BS 6700:2006+A1:2009.

Did you know?

The testing procedure varies slightly between BS 6700:2006+A1:2009 and the Water Regulations. The variation takes into consideration the fact that some plastic materials, when subjected to a test pressure, suffer stress that can be retained in the pipe material once the test is over. This can result in pipe failure at a later date, so for this reason pipes of an **elastomeric** material are allowed to be subjected to a less severe test.

Key term

Elastomeric – having elastic properties: for example, a polybutylene pipe.

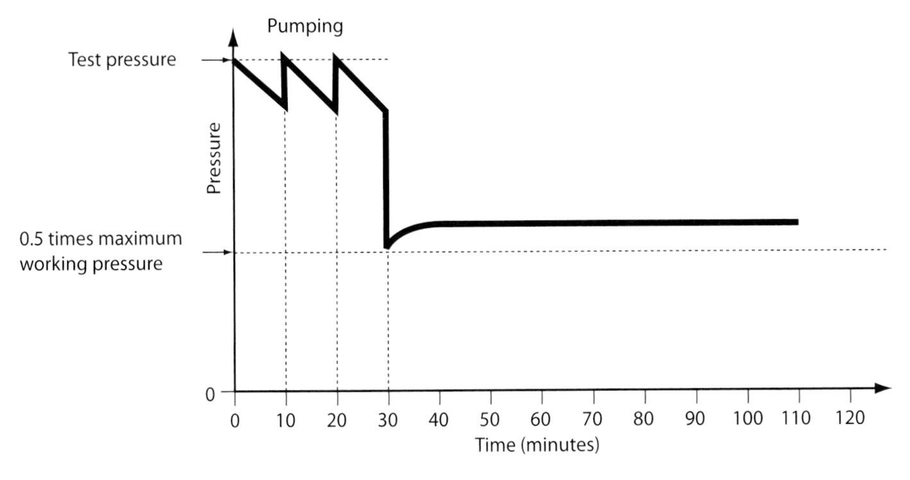

Figure 3.90: Test A procedure for plastic pipes

Test A

The whole system is subjected to the test pressure by pumping for 30 minutes, after which the pressure is noted and the test continues without further pumping.

a) The pressure is reduced to one-third of the test pressure.
b) The pressure does not drop over the following 90 minutes.
c) There is no visible leakage throughout the test.

Test B

The whole system is subjected to the test pressure by pumping for 30 minutes, after which the pressure is noted and the test continues without further pumping.

a) The drop in pressure is less than 0.6 bar (60 kPa) after a further 30 minutes, and less than 0.2 bar after the next 120 minutes.
b) There is no visible leakage throughout the test.

Remember

Acceptable test results are shown in Test A on the previous page and Test B above.

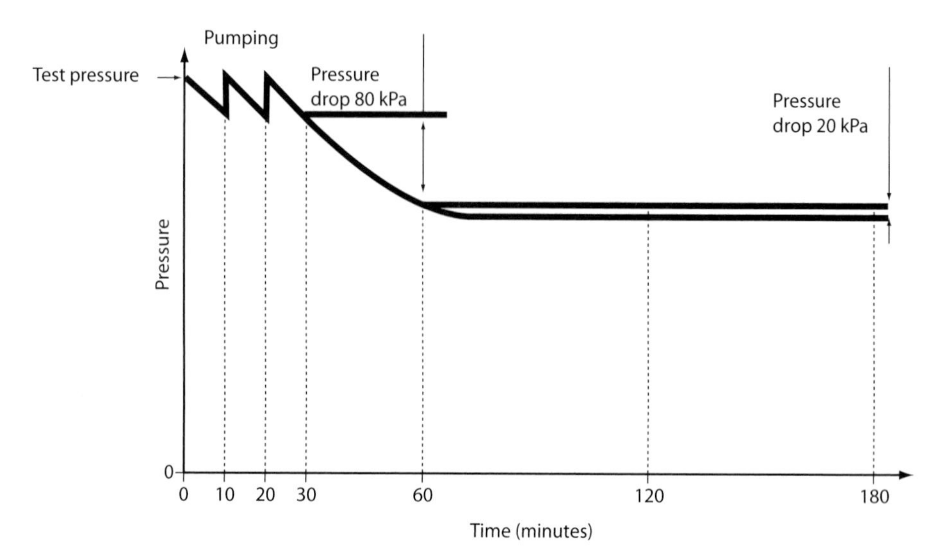

Figure 3.91: Test B procedure for plastic pipes

System flushing

WSR Schedule 2, para 13

Every water system shall be flushed out before it is first used. This applies to new installations, alterations, extensions or any maintenance to existing installations. This makes sure that any debris, including excessive flux that may have collected in the pipework during installation, is removed from the system.

System disinfecting

Every water system shall where necessary be disinfected before it is first used. The Guidance Document recommends that, after system flushing, the system should be disinfected. This applies to:

- new installations, except private dwellings occupied by a single family
- major extensions or alterations except private dwellings occupied by a single family
- underground pipework, except localised repairs or insertion of junctions
- where contamination may be suspected, such as fouling by sewage, drainage, animals, insects and vermin
- after physical entry by personnel for interior inspection, painting or repairs where a system has not been in regular use and not regularly flushed. Regular use means periods of up to 30 days without use, depending on the characteristics of the water.

Activity

Why does the pipework need to be free of debris? Research and make some notes.

Disinfection recommendations

BS 6700:2006+A1:2009

The Guidance Document refers to the recommendations of BS 6700:2006+A1:2009 for flushing and disinfection of systems, and following these procedures will satisfy the requirements of the Regulation.

BS 6700:2006+A1:2009 sets guidance on the sterilisation of pipework fittings.

- Where pipework is under mains pressure, or has a backflow device fitted downstream, the water undertaker should be notified.
- If water used for disinfection is to be discharged to a sewer, drain or water course, the authority responsible for the sewer, drain or water course should be notified.
- Chemicals used for disinfection of drinking-water installations must be chosen from a list of substances compiled by the Drinking Water Inspectorate, which is listed in the *Water Fitting and Materials Directory* published by the WRAS.

Unless a specific chemical disinfectant is specified, sodium hypochlorite (diluted chlorine) can be used. Other disinfectants are available in tablet form, and these are easily obtainable and safe to use.

Did you know?

Private dwellings include a normal house containing a single family. In instances where a larger house has been divided into separate units, such as bedsits or student flats, disinfection will be required.

Safety factors

- Before carrying out disinfection, the system must be taken out of use, marking all outlets 'DO NOT USE: DISINFECTION IN PROGRESS'.
- All operatives carrying out the disinfection procedure must receive appropriate health and safety training under the COSHH Regulations.
- No other chemicals (for example, sanitary cleaners) should be added to the water during disinfection as this could generate toxic fumes.
- All occupants within the premises must be notified that disinfection is taking place.
- Extreme care must be taken when using disinfectants: some can be hazardous, and operatives must wear safety goggles and protective clothing and refrain from smoking.

Safety tip

The correct sequence for system disinfection should follow the flow of water into the premises: first the water mains, then the supply pipe and cisterns, and finally the distribution systems.

The procedure for disinfection

This can be used for both hot water and cold water installations and is the general procedure for the disinfection of a system, whether chlorine or any other approved disinfectant is used.

- Thoroughly flush the system before disinfection.
- Introduce a disinfection agent at specified concentrations into the system, filling systematically to ensure a total saturation. If using chlorine, use initial concentrations of 50 mg per litre (50 parts per million (ppm)).
- Leave the system for a contact period of one hour. If using chlorine, check the free residual chlorine level at the end of the contact period. If this is less than 30 mg per litre, the procedure needs repeating.
- Immediately following successful disinfection, the system should be drained and thoroughly flushed with clean water until the residual chlorine level is the same as that of the drinking water supplied.

Did you know?

When disinfecting supply pipes or any hot or cold storage vessels that are connected directly to mains pressure, the chlorine solution should be injected into the lower end of the pipe near its point of connection to the communication pipe using a suitable injector point.

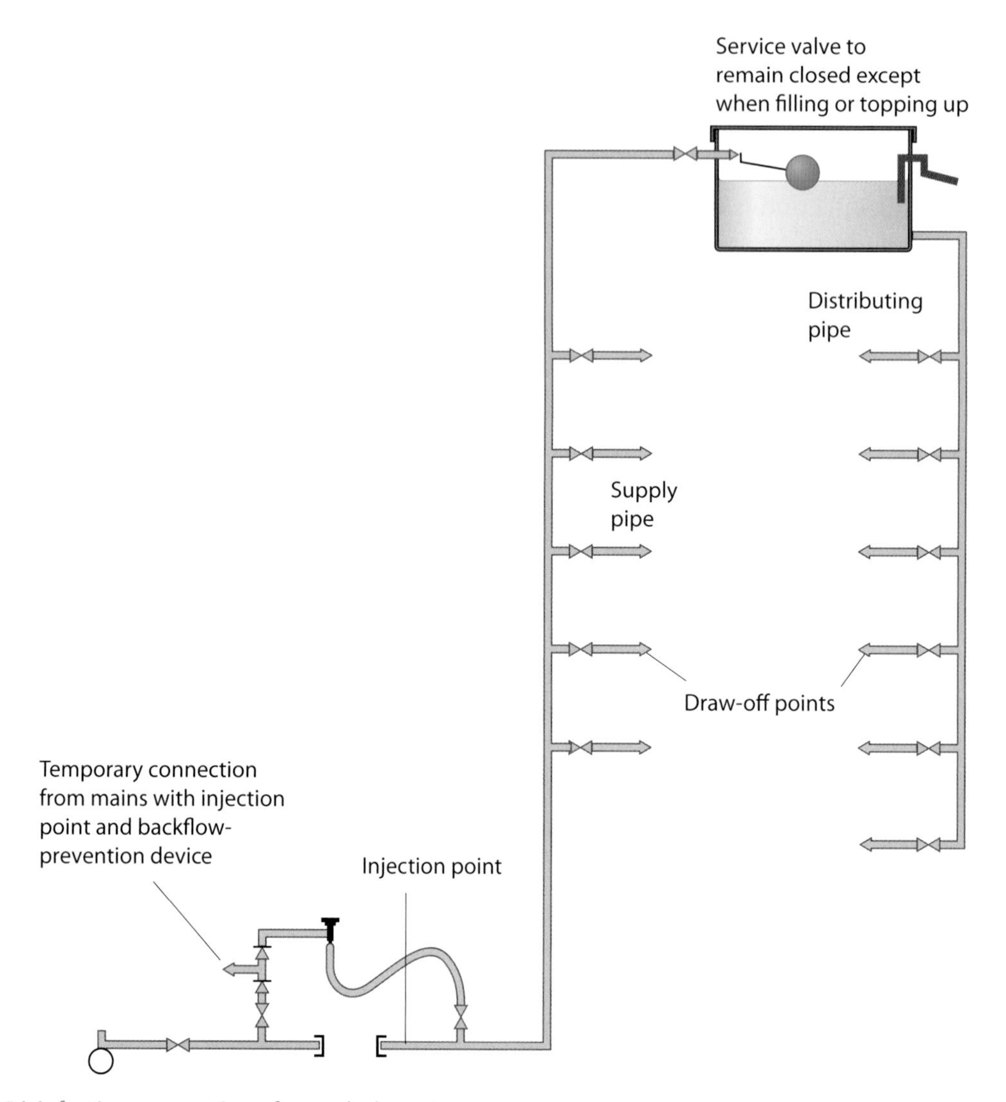

Figure 3.92: Disinfection connections for a whole system

Remember

On no account should high concentrations of disinfectant be discharged into the natural environment (such as into water courses via surface water drains). Seek advice from the Environment Agency before disposing of disinfection fluids.

After flushing, samples should be drawn off and taken for bacteriological analysis. If the test result proves unsatisfactory, the disinfection and sampling test procedures should be repeated.

Water should be introduced into the system by systematically opening individual taps, working away from the point of connection, until the whole of the system is filled with water of the specified concentration (50 ppm).

Disposal of disinfection fluid

Generally, water and other fluids that have been used for disinfecting water systems can be safely discharged into a public sewer. However, particularly in rural areas, the discharge may have an adverse effect on sewage treatment (for example, by affecting cesspools). Contact the sewage undertaker if you are in any doubt.

Notifying the relevant authority

For installations to be supplied with water from the public water supply, the following information must be sent to your local water supplier:

- the name and address of the person giving notice and, if different, the name and address of the person to whom the consent should be sent
- a description of the proposed work or any significant change of use of premises
- the location of the premises and their use or intended use
- a plan of those parts of the premises that relate to the proposed work and a diagram showing the pipework and fittings to be installed
- the plumbing contractor's name and address, if an Approved Plumber is to do the work.

Granting consent

There is no charge by your water supplier for dealing with notifications or granting consent. Work on installations must not start until consent has been given. Consent will not be withheld unreasonably, and may be granted subject to conditions, which must be followed. If, within ten working days of receipt of a valid notification by the water supplier, consent is neither granted nor refused, it is deemed to have been granted. This does not alter the obligation on the installer and owner or occupier to see that the Regulations are fully met.

Approved Plumbers

Approved Plumbers will certify that their installation work satisfies the Regulations. In the event of breaches of the Regulations in connection with the certified work, the owner or occupier can use the certificate as a legal defence against any resulting prosecution.

An Approved Plumber does not have to receive consent before starting work. However, on completion, as well as supplying a certificate to the person who asked for the work to be done, the plumber must send a copy of the compliance certificate to the water supplier.

Most UK water suppliers either operate their own Approved Plumbers Scheme or support the national Water Industry Approved Plumbers Scheme (WIAPS), whose members have demonstrated their experience of plumbing work and knowledge of the Regulations and have liability insurance cover. Contact your water supplier for a list of Approved Plumbers who are available for work or look on the WRAS website for details. The Institute of Plumbing and Heating Engineering, the Scottish and Northern Ireland Plumbing Employers Federation (SNIPEF) and the Association of Plumbing and Heating Contractors (APHC) also list Approved Plumbers.

Commissioning records

After completing commissioning, you must make a record and hand it over to the client customer. Figure 3.93 shows you the sort of information you would need to include.

Commissioning Record		
Address: 10 Well Lane Leek Staffordshire	Date of commissioning: 12th September 2012	Operative name: Andrew Pipe
Type of system	Cold water	
Working pressure (Bar)	2.5	
Test pressure (Bar)	3.75	
Date of test	12/09/2012	
System flow rate (L/s)	15 L/s	
System disinfected (Y/N)	N	
Additional components		
Booster pump	N/A	
Working pressure	N/A	
Test pressure	N/A	
Flow rate	N/A	
Notes		

Figure 3.93: Example of a commissioning record

Handing over to the end-user

On completion of the installation, before leaving the job you need to go over details of the system with the end-user. The details will depend on the type of system you have installed and which components you have used.

Here is a checklist for handing over to the end-user.

Checklist

Handing over

- Do a walk through the system.
- Explain how to set any programmers and timers.
- Identify where to isolate the water supplies.
- Identify where to isolate electrical supplies for components and controls.
- If a water softener is fitted, explain:
  - how to check the softness of the water
  - how to fill it with salt.
- Explain the need for maintenance of components.
- Leave manufacturer's instructions with the client or customer.

Dealing with defects revealed during commissioning

Systems that do not meet correct installation requirements

During the commissioning of some systems or components you may have identified that the installation is not correct, or you may be unable to achieve the correct design output for a given appliance. This could be due to poor design and/or calculations or pre-system checks not being carried out by the installer, particularly with unvented hot water systems.

Before the system can be handed over to the client or customer, these defects must be rectified. If the defect is due to design by a third party, that party must be notified and this could possibly incur costs and alterations. If the defect is by the installer who has not installed as stated in the specification, it would be the installer who would have to pay the costs.

Micro-biological contamination within a cold water system

Where testing of the water supply has indicated contamination, the system will need to be taken out of service, drained, flushed and disinfected. On completion of this process further testing of the water supply will need to be undertaken.

Remedial work associated with defective components

When a defective component is found, the first thing to check is whether it has been damaged during the installation stage. If so, it will be the responsibility of your company to replace the defective part.

If you can prove that it is the manufacturer's responsibility, you must contact them. As the units are new, they will either:

- send out one of their own engineers, or
- pay your company to replace the part, with the manufacturer supplying the replacement.

PROGRESS CHECK

1. State the procedure for testing metallic pipework systems.
2. What is blue water corrosion and what can be the cause of it?
3. When should the sewage undertaker be consulted?
4. What should be used to take flow readings for cold water systems?
5. What steps does commissioning of cold water include?
6. Where would the disinfection procedure for cold water systems be found?
7. What is the maximum permissible drop of pressure for a metallic system?
8. What type of chemical can be used for disinfection of a system?
9. What should be done to a cold water system on completion of testing?
10. When handing over a system to a client/customer, state what points you should cover.

Check your knowledge

1. Which of the following statements is correct in relation to water supply?
   - a Water byelaws have been replaced by the Water Supply (Water Fittings) Regulations 1999.
   - b The Water Supply (Water Fittings) Regulations 1999 replaced British Standard 6700.
   - c The Water Supply (Water Fittings) Regulations 1999 have been replaced by Water byelaws in England and Wales only.
   - d Water byelaws are in addition to the Water Supply (Water Fittings) Regulations 1999.
2. BS 6700 has test pressure procedures for plastic pipework and states a test pressure which should be maintained for an initial period. What is this period?
   - a Forever
   - b 24 hours
   - c 30 minutes
   - d 45 minutes
3. Which of the following is a requirement for industrial/commercial cold water systems?
   - a Disinfection of the system
   - b Visual inspection of the system
   - c Flushing of the system
   - d Interim testing of the system
4. According to the Water Regulations, what fluid category is given to fluids that are a slight health hazard because of the concentration of substances of low toxicity?
   - a Category 1
   - b Category 2
   - c Category 3
   - d Category 4
5. Which of the following apply to plumbers who are approved contractors?
   - a They can work for the HSE to check that installations meet safety regulations.
   - b They can give out certificates to confirm that work complies with the Construction Regulations.
   - c They can give out certificates to confirm that work complies with the Water Regulations.
   - d They can carry out plumbing inspections for the local authority.
6. On a new dwelling a hose union bib-tap must be fitted with a ____?
   - a Single check valve
   - b Double check valve
   - c RPZ valve
   - d PRV valve
7. The Water Regulations require that the type of air gap used when adjusting water levels in a WC cistern is type ___?
   - a AB
   - b AD
   - d AG
   - d AUK 1
8. Where a bath or wash basin is fitted in a domestic dwelling with submerged tap outlets, this is considered to be a fluid category 3 risk. What should be fitted to the supply and distributing pipes that supply the appliance?
   - a Type EC device
   - b Type EA device
   - c Type CA device
   - d Type AUK 3 device
9. What is the purpose of a transducer on booster pumps?
   - a To detect heat in the water
   - b To detect pressure in the water
   - c To detect flow of water
   - d To detect electrolysis in water
10. If carried out by an approved contractor, which one of the following would NOT have to be notified to the water undertaker?
    - a A bath having a capacity of more than 230 litres
    - b Any water system laid outside a building less than 750 mm or more than 1350 mm underground
    - c A bidet with ascending spray or flexible hose
    - d A unit that incorporates reverse osmosis

Preparation for assessment

Now that you have completed all of your learning outcomes it is time to prepare for assessment.

Use the following revision strategy for the underpinning knowledge test.

- Do not try to take in all of the information in one go; take your time and do a little each day.
- Reread Unit 3, *Understand and apply domestic cold water system installation, commissioning, service and maintenance.*
- Look in detail at backflow devices both mechanical, non-mechanical and the situations where they will be fitted.
- Go over the notes you have taken over the weeks of teaching.
- Research more information about anything which you are not sure about.
- Check any queries with your tutor.
- Do the practice questions in your book.
- Check your answers with your tutor.

You will also have to take practical assessments. These assessments are not about installation – you would have done that at Level 2 – the Level 3 practical cold water is primarily about the commissioning and fault finding of boosted cold water systems.

You will have to identify if the backflow device is correct for given situations which will be set up in the practical workshops.

Follow the practical sheets which are on the Training Resource Disc.

As part of the qualification you will also be undertaking a design assignment for cold water in a single occupancy dwelling. This will include being able to:

- use information sources when designing hot water systems
- calculate the size of hot water system components:
  - cistern
  - pipework
  - pump
  - pressure vessel
- Present the design calculations in an acceptable format:
  - a single line drawing, not to scale
  - details for insertion into a quotation or tender for work.

This assessment is in assignment format and must be completed as part of your unit award. On completion of this your course tutor will mark your work.

Good luck and we hope you do well.

4 Understand & apply domestic hot water system installation, commissioning, service & maintenance techniques

This unit focuses on hot water systems, both vented and unvented. The installation of unvented hot water systems must be carried out by properly qualified plumbers, as identified in Approved Document G of the Building Regulations. The systems provide potentially high water flow rates to plumbing systems and components such as power showers.

By the end of the unit, you should be able to pipe size hot water systems and calculate storage capacities for hot water. You will need to have read Unit 7 on electrical systems to fully grasp the related principles in this unit.

This unit covers the following learning outcomes:

- know the types of hot water system and their layout requirements
- know the uses of specialist components in hot water systems
- know and be able to apply the design techniques for hot water systems
- know and be able to apply the fault diagnosis and rectification procedures for hot water systems and components
- know and be able to apply the commissioning requirements of hot water systems and components.

Regulation requirements

The Regulation requirements for hot water services are generally concerned with the prevention of wasted water and the overall safety of the building and its occupants where services are installed.

WSR Schedule 2, paras 17–24

Paragraphs 17 to 24 of Schedule 2 mostly regulate the installation of the hot water service within the building, and look at the following issues:

- expansion in hot water systems
- the measures required to accommodate expansion in vented and unvented systems
- control of water temperature and safety devices
- the discharge from temperature-relief valves and expansion valves
- backflow prevention to closed circuits (filling loops, and so on).

You will find in this chapter that there are some cross-references back to Unit 3 (Domestic cold water) and forward to Unit 7 (Electrical principles). This is to avoid repetition of key concepts. Make sure you read these cross-referenced pages carefully.

1. Know the types of hot water system and their layout requirements

Types of hot water supply system

There are two main types of hot water system: vented and unvented.

- The vented system has a cold water storage cistern (CWSC) connected by a cold-feed pipe to a hot water storage cylinder (HWSC). The system also includes a vent pipe, which is a key safety component of the system.
- In the unvented system, water is stored in the system itself and is available at mains supply pipe pressure. This gives a much higher-performance hot water system with greater flow rates.

The vented system

This type of system has been installed in the UK for years; the main reason for its widespread use is that the system is very safe, if installed correctly. The vented system is open to the atmosphere and is not designed to work at pressures above atmospheric pressure, so the system has safety built into its design.

As water is heated in the system, it expands from the cylinder through the cold-feed pipe and into the cold water storage cistern. The volume of water in the system increases as the water heats, leading to the water level rising in the cold water storage cistern.

The vent pipe has a dual purpose. First, it acts as a vent to remove air from the system but, more importantly, it also provides a safety back-up

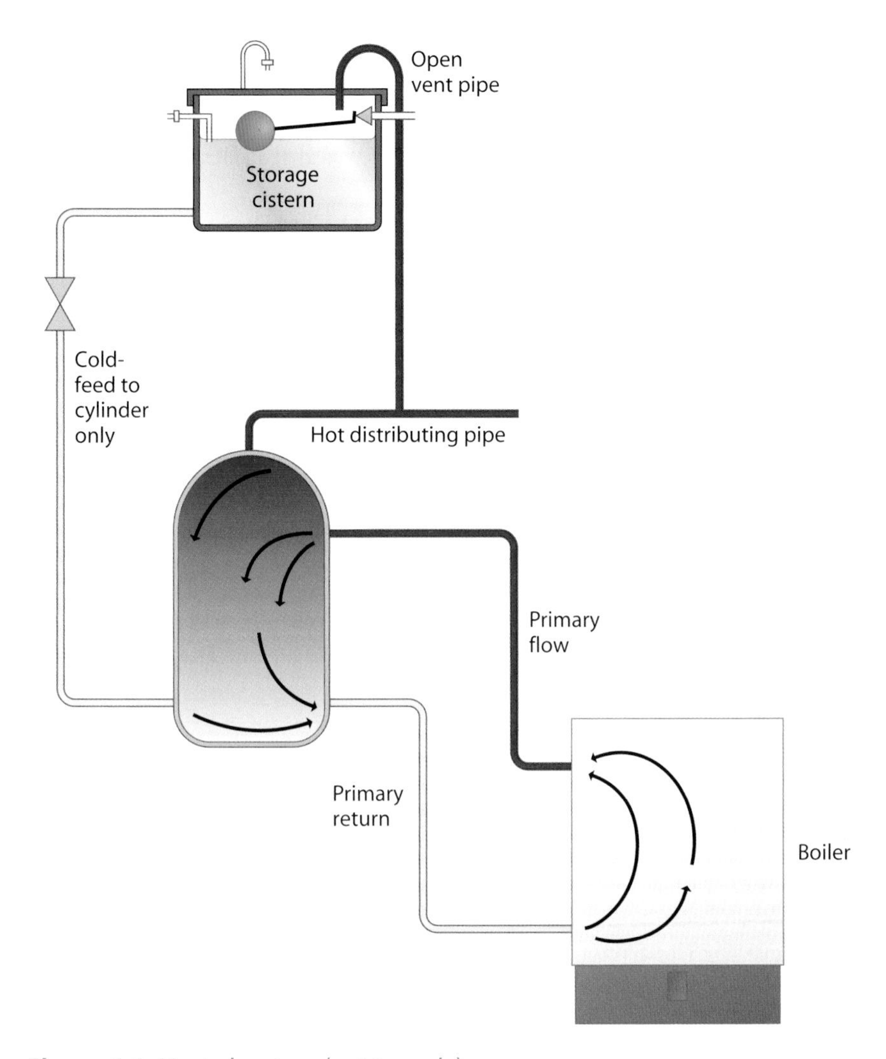

Figure 4.1: Vented system (not to scale)

if the cold-feed becomes blocked or does not work, offering a route to the atmosphere to safely relieve pressure in the system.

With the addition of thermostatic controls to the heater of the vented system (either by an immersion heater or indirectly heated by a boiler), the system is effectively protected against a build-up of temperature and pressure.

However, the flow rate from the hot water taps is restricted by the head of water generated over the tap by the cold water storage cistern, and can often be quite poor, especially as many of today's users want better performance from systems (such as more powerful showers).

There are three types of vented hot water system still installed in many dwellings around the UK. These systems are:

- direct
- single-feed indirect
- double-feed indirect.

These systems were covered at Level 2. Look back at pages 308–310 of *Level 2 NVQ Diploma Plumbing* to remind yourself.

Safety with hot water systems

When water is heated, it expands and its volume increases. Figure 4.2 represents the expansion of water when it is heated.

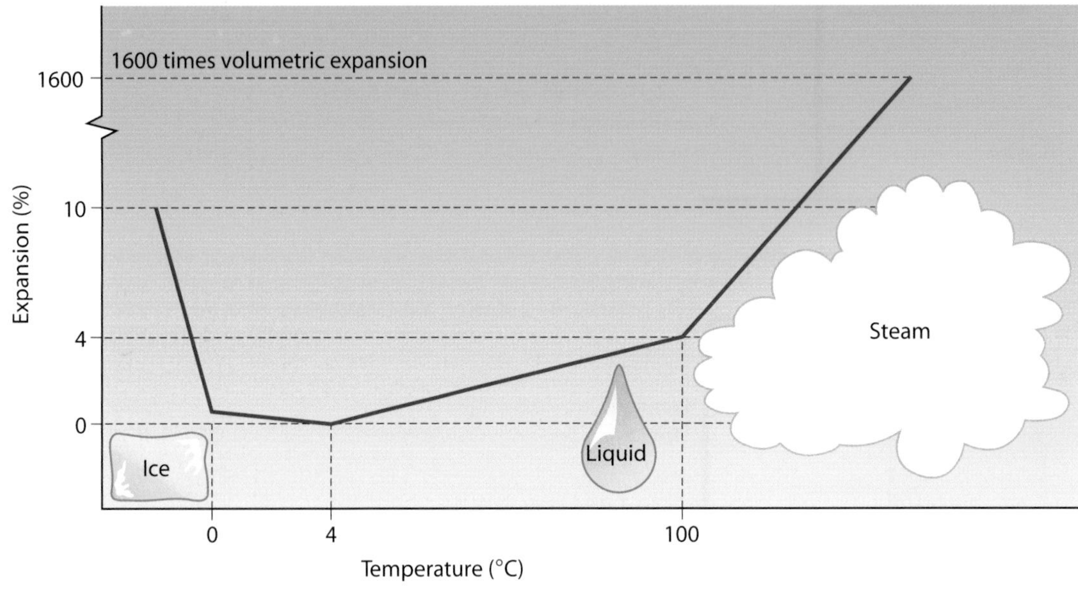

Figure 4.2: Water expansion

Below 0°C, water expands as it cools; this means that water in an enclosed space (such as a pipe) increases in volume when it freezes. With nowhere for the frozen water to go, the pipe ruptures. Similarly, when water is heated from 4°C to 100°C, it expands. This expansion has to be allowed for in any hot water system, to stop pressure build-up and prevent damage to the components.

A typical increase in volume for water heated in a 120-litre hot water cylinder is:

120 x 0.04 (4%) = 4.8 litre increase in volume

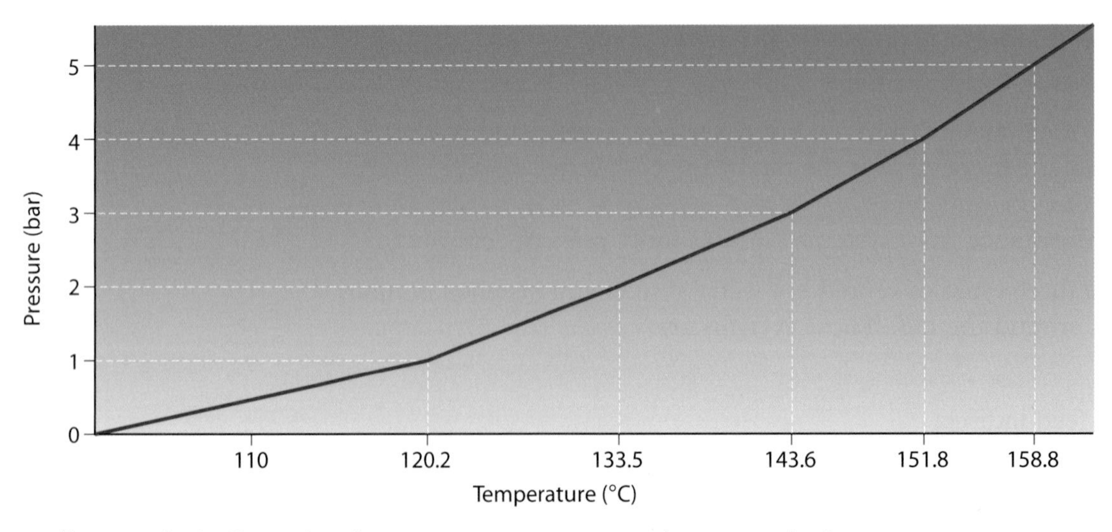

Figure 4.3: Effects on the boiling point of water at pressures greater than atmospheric pressure

Water expands by up to 4 per cent under normal system heating conditions and this must be catered for somewhere in the system.

The boiling point of water increases as the pressure in a system increases. For example, water at 2 bar pressure has a boiling point of 133.5°C. If the temperature in a closed vessel is increased, then the pressure in that vessel is increased.

Remember

It is extremely important that both the temperature and pressure are controlled in hot water systems.

The unvented system

By design, the system is not open to the atmosphere: it works at a much higher pressure, and must have a different range of safety control features from those of vented systems. Figure 4.4 shows an example of an unvented system, where you can see that cold water is supplied to the cylinder from the mains supply directly through the cylinder to the hot water tap. Hence the term 'unvented'; it may be described as a closed circuit.

When positioning a hot water system, it can be identified as either a centralised or a localised system.

- A centralised system is where the storage vessel is positioned in one central point of the hot water system and all outlets are fed from that vessel, whether it is unvented or a vented system.
- A localised system is where the hot water vessel is local to the appliance outlet that it is supplying hot water to.

As a Level 3 plumber you will need to identify the types of systems and be able to choose a system for a given situation.

An unvented cylinder

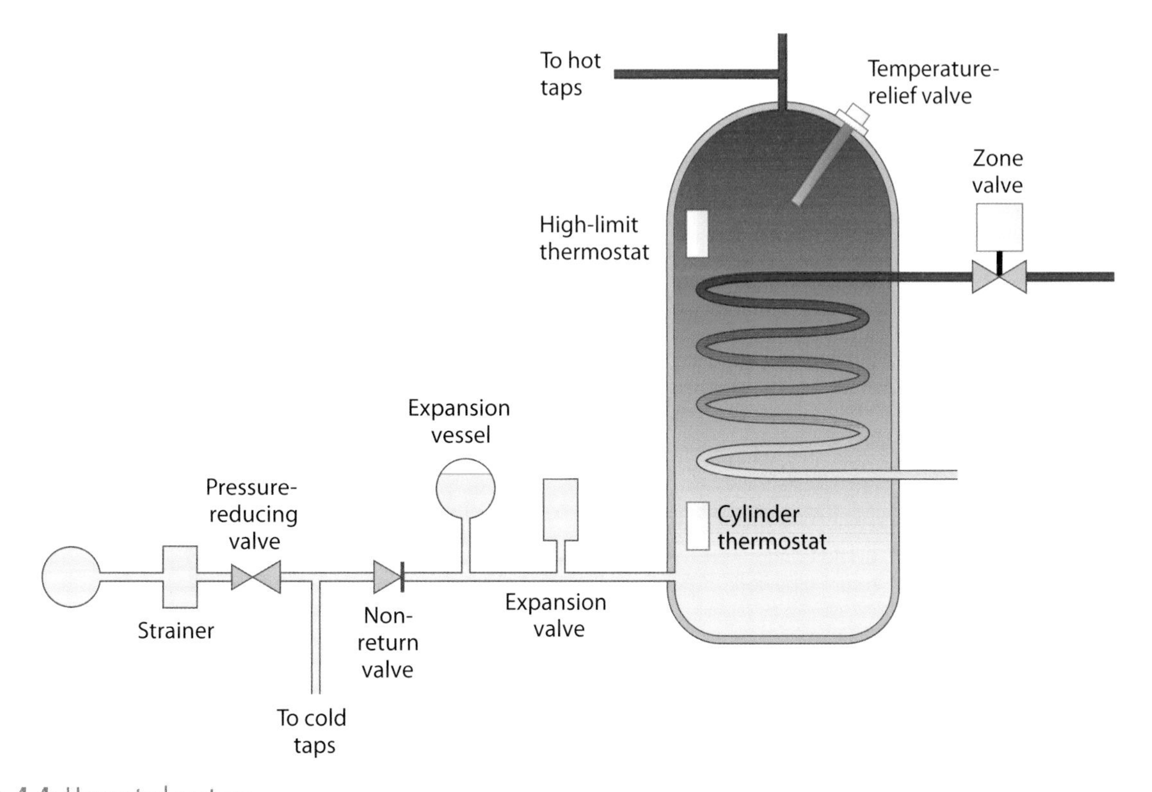

Figure 4.4: Unvented system

Centralised systems

Unvented hot water systems

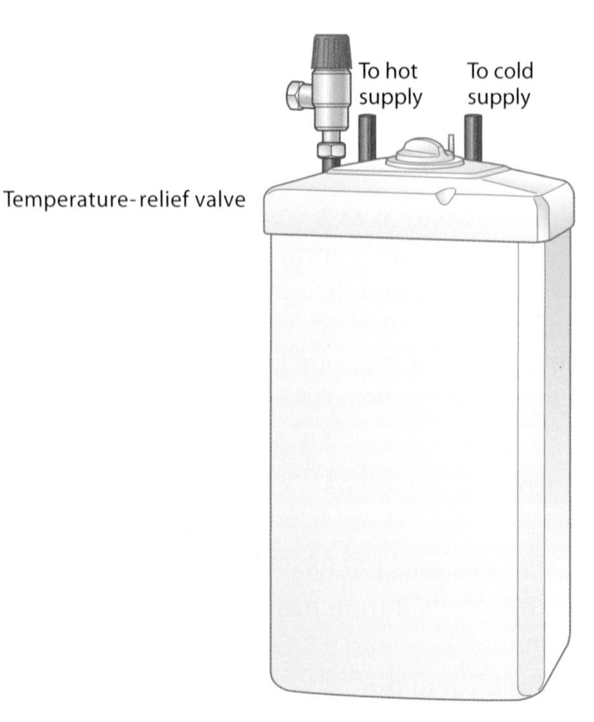

System: Unvented instantaneous heater
Capacity: Usually less than 15 litres
Notes: Gas or electric multipoint heater
Centralised

Feed and expansion cistern (can be built into unit)
Adjustable thermostatic mixing valve
Expansion chamber
Boiler
Hot taps
Cold taps
CH flow and return pipework
Supply pipe

System: Unvented water-jacketed tube heaters
Capacity: Usually up to 15 litres
Notes: Indirectly heated via an internal coil; known as thermal storage and combination boilers
Centralised

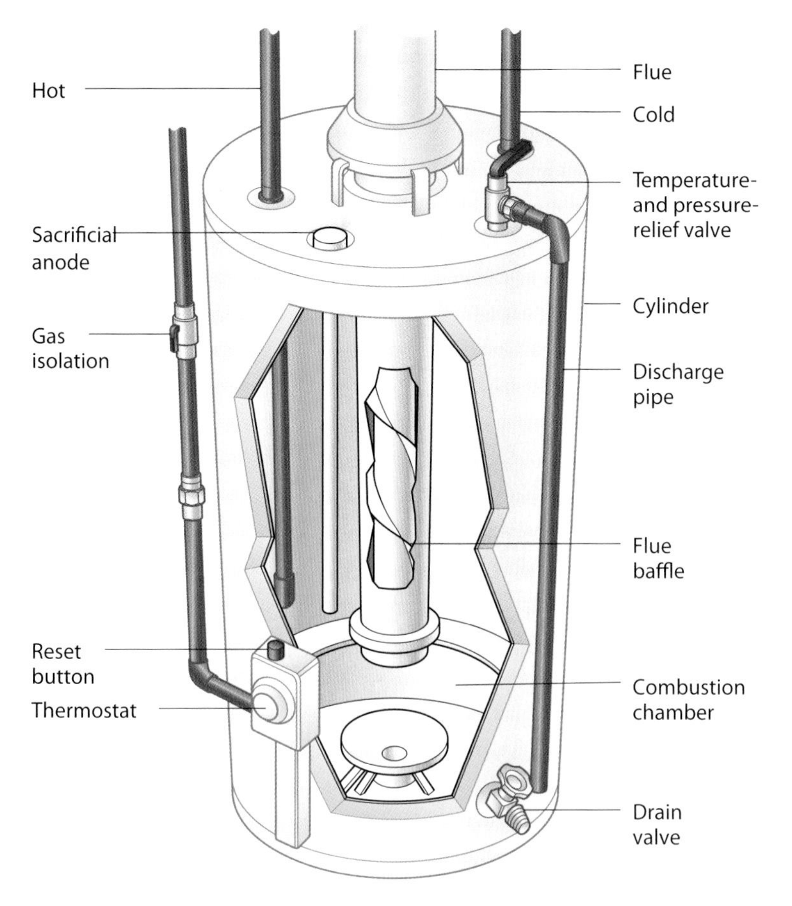

System: Unvented directly-heated storage heaters
Capacity: Over 15 litres capacity
Notes: Gas-fired
Centralised

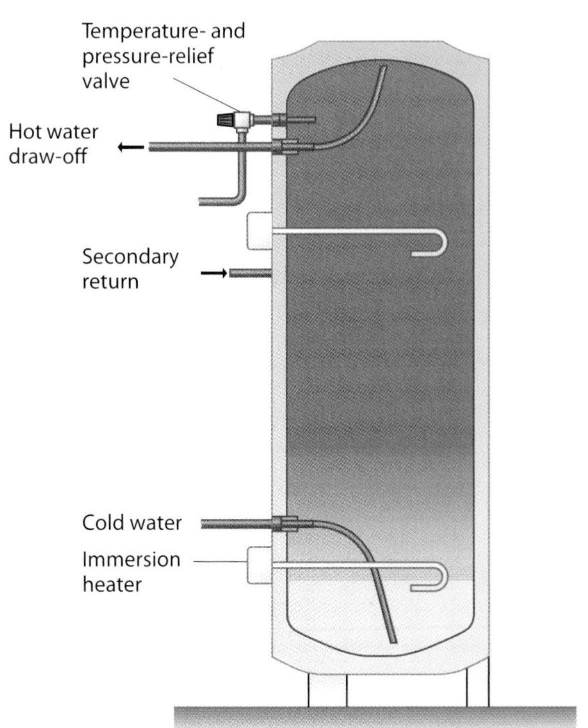

System: Unvented direct-heated storage heaters
Capacity: Over 15 litres capacity
Notes: Can be electrically heated or gas-fired
Centralised

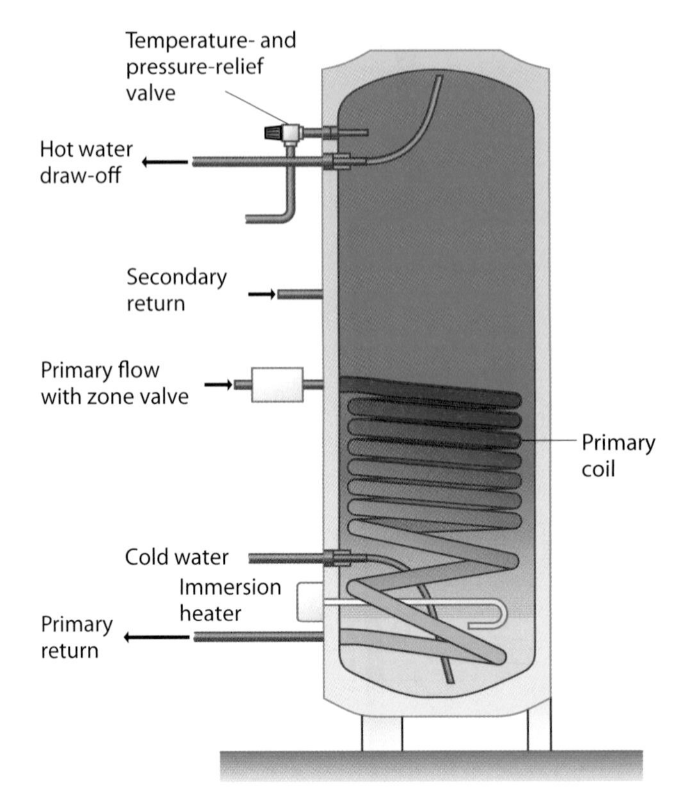

System: Unvented indirectly-heated storage heaters
Capacity: Over 15 litres capacity
Notes: External heat source provided via an internal coil
Centralised

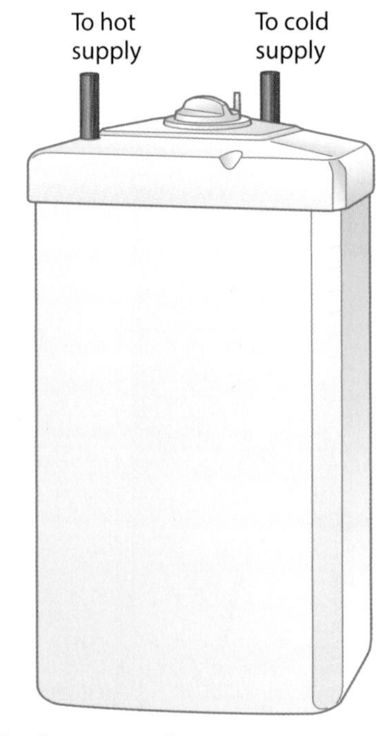

System: Unvented displacement heaters
Capacity: Up to 15 litres
Notes: Heated by an immersion heater; can be fitted above or below the sink; more popular in commercial properties
Localised

Open vented hot water systems

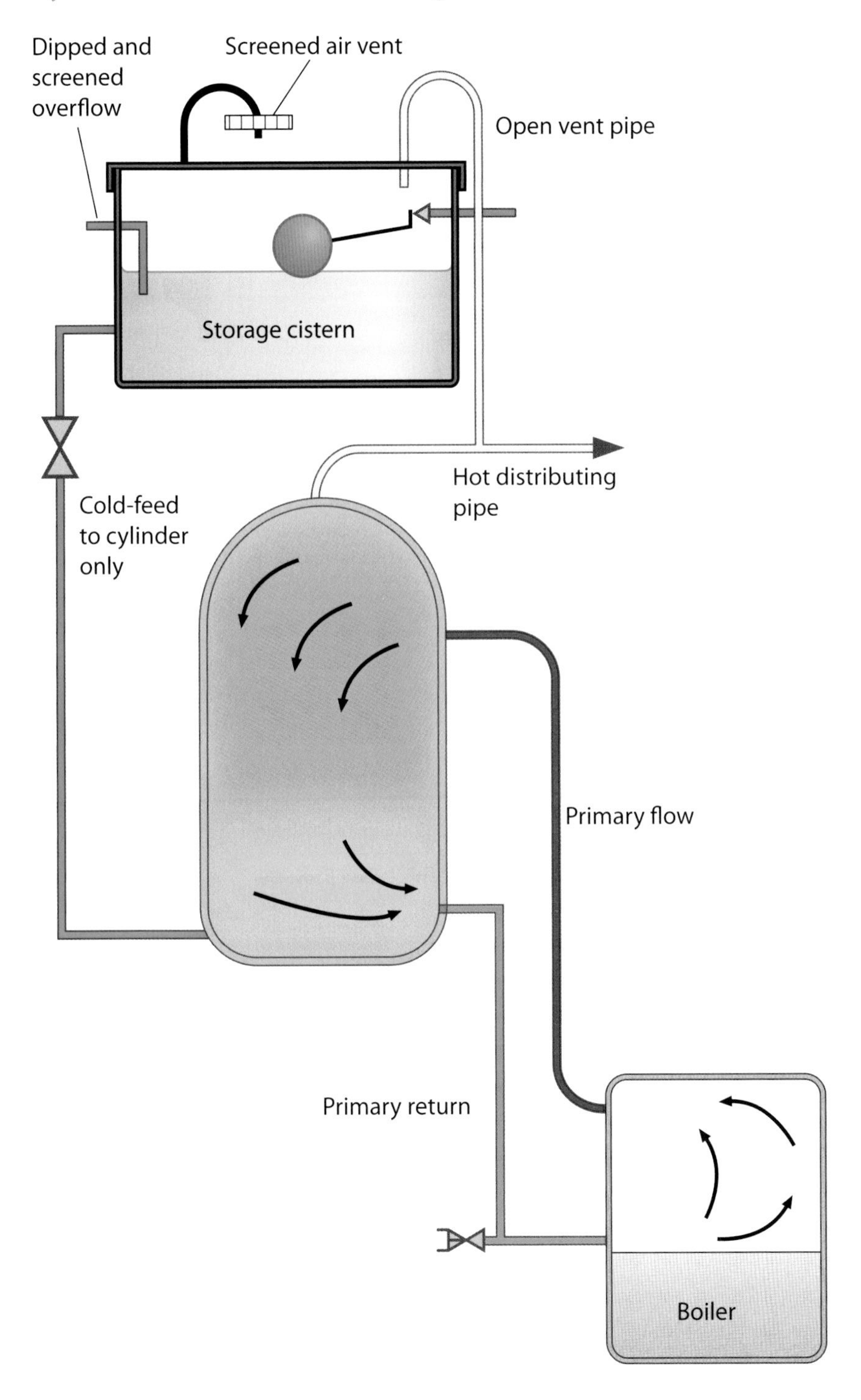

System: Direct system
Capacity: In dwellings the storage capacity should normally be based on 45 litres per occupant, unless pumped primary circuits or special appliances justify the use of smaller storage capacities **(BS6700:2006+A1:20095.3.9.3.2)**
Notes: Water heated directly from boiler or immersion heater; gravity circulation from boiler to cylinder.

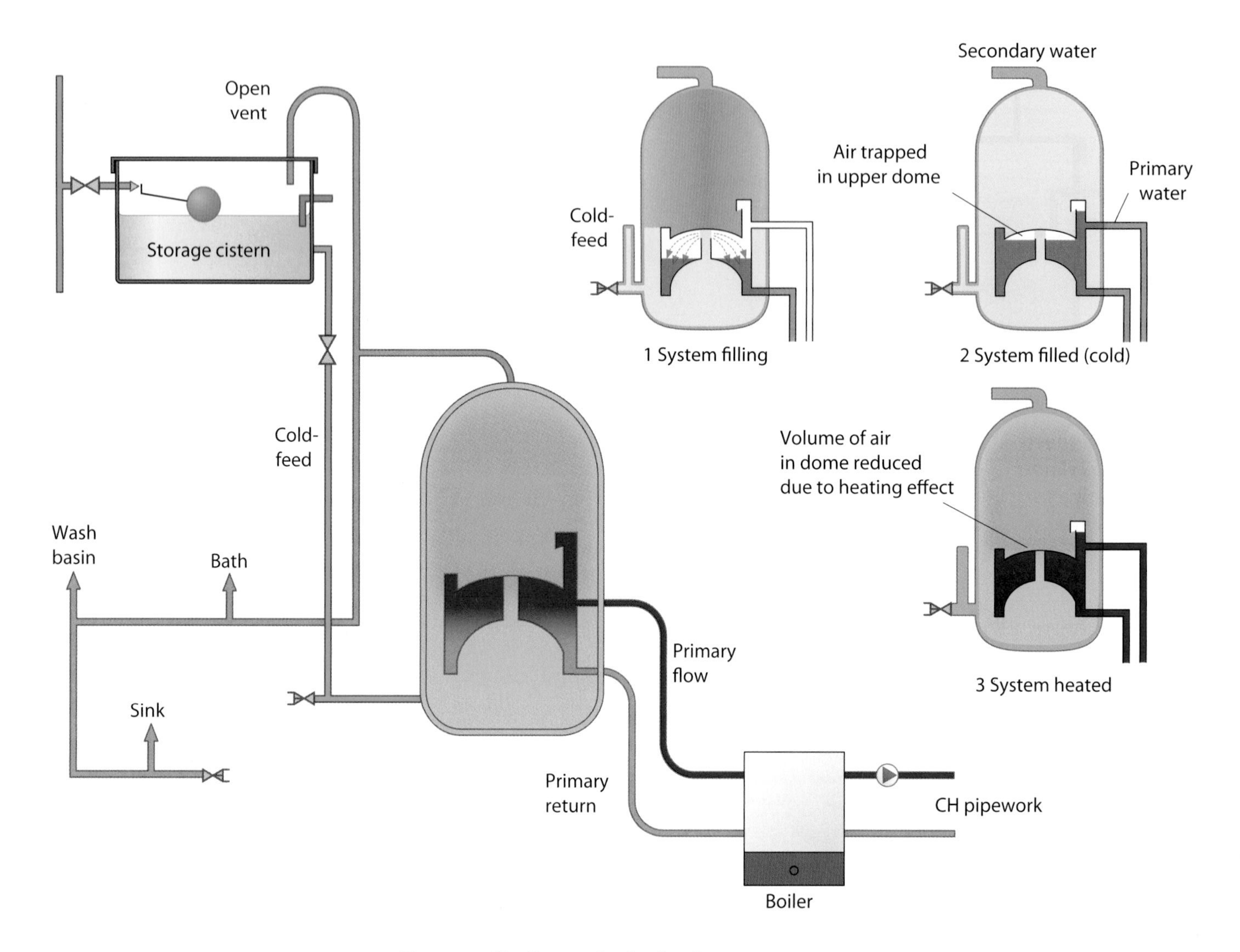

System: Indirect single-feed
Capacity: See Direct system
Notes: Water heated indirectly from boiler

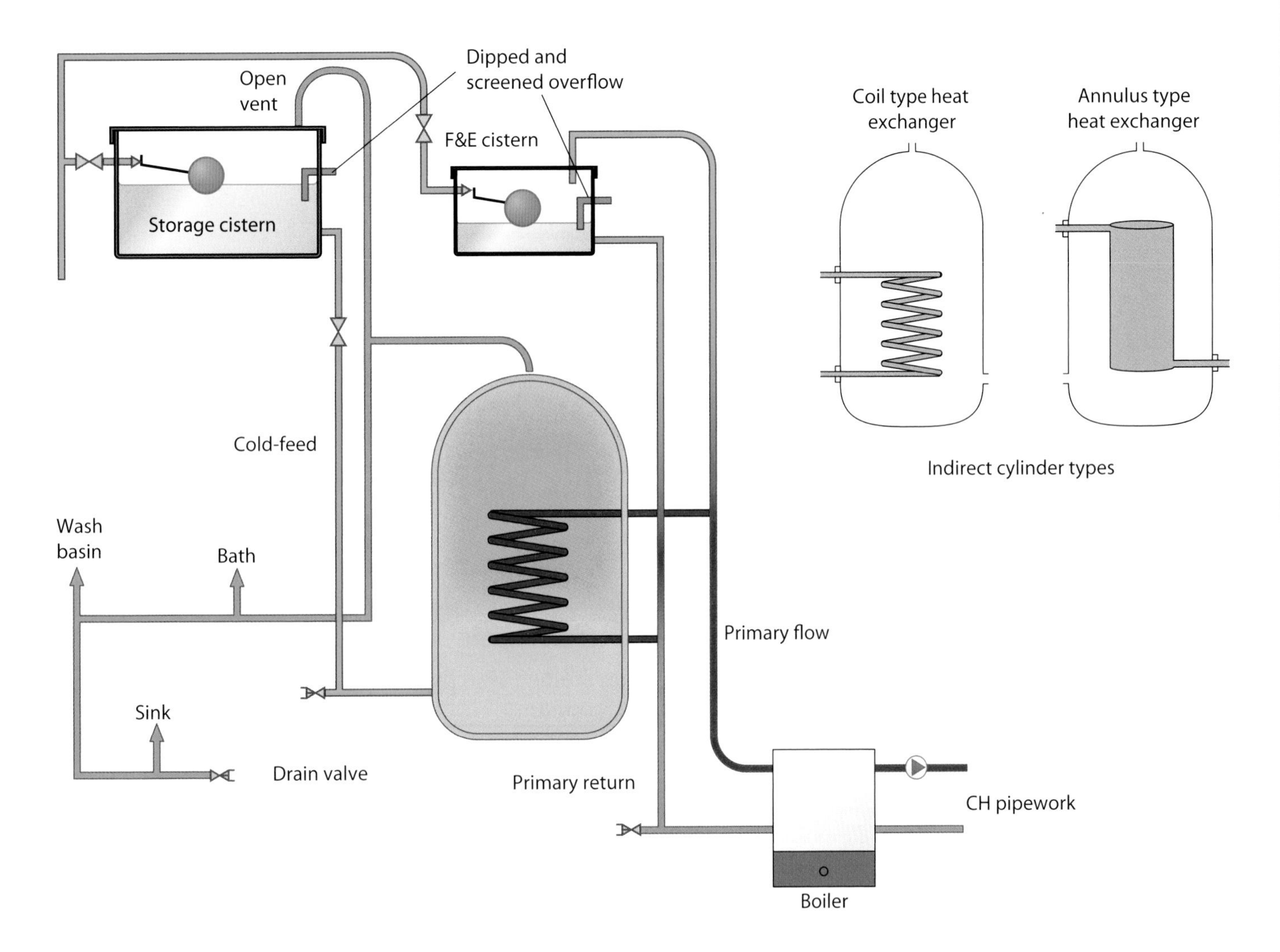

System: Indirect double-feed
Capacity: See Direct system
Notes: Open vent pipe may be connected to the primary flow pipe and the cold-feed pipe may be connected to the primary return pipe or fed separately into the boiler; where vent pipe is not connected to highest point in a primary circuit, an air release valve should be fitted; separate feed and expansion cistern needs to be provided to feed primary circuit (ensures that where double-feed cylinder is used, primary water stays separate from secondary hot water)

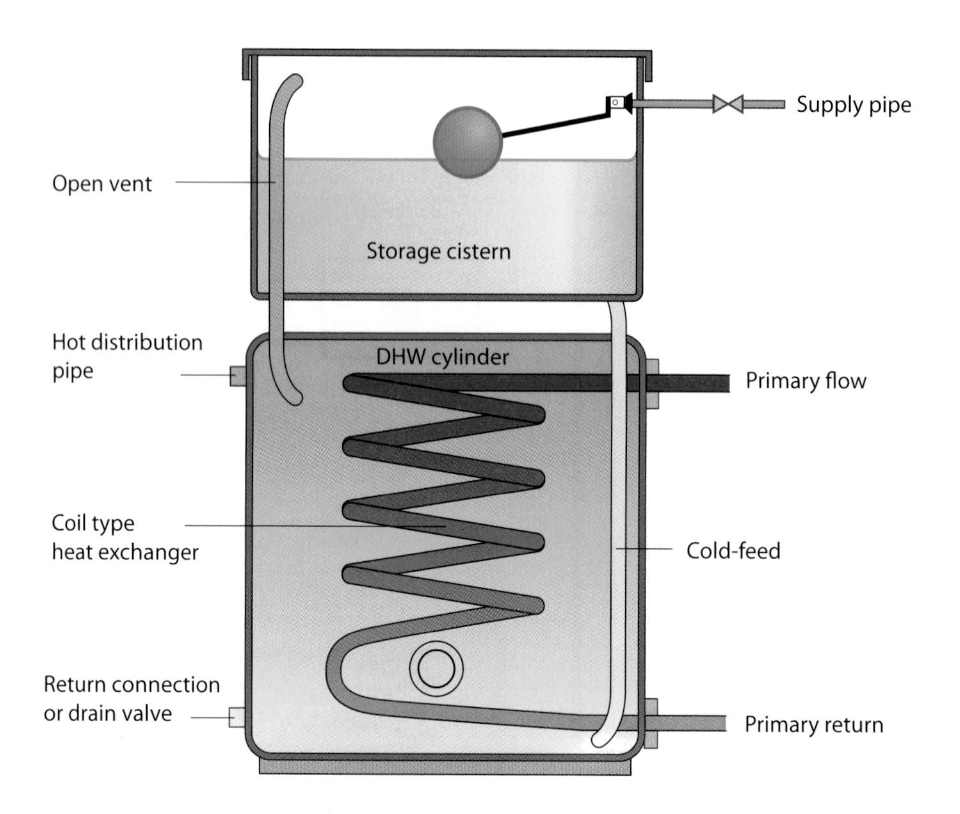

System: Combination storage system
Capacity: See Direct system
Notes: Vented system; can be indirectly heated by other means (boiler); directly heated by immersion heater; cistern combined with cylinder.

Localised systems

Instantaneous heaters

Small instantaneous water heater

An instantaneous heater, more commonly called a single-point instantaneous water heater, uses electricity or gas. Sited directly above the appliance, it is usually inlet-controlled, with the hot water delivered via a swivel spout. The electric single-point is a small tank of water with an electric heating element inside. Because of its low volume, the water quickly heats up as it is drawn through the heater. The temperature at the outlet will be related to the water flow rate and the kW rating of the heater.

You will find single-point heaters used in situations where a small number of hot water draw-offs are fed by individual heaters in a non-domestic building, and where the use of a centralised hot water system would be uneconomical: for example, in the WC of a small cafe.

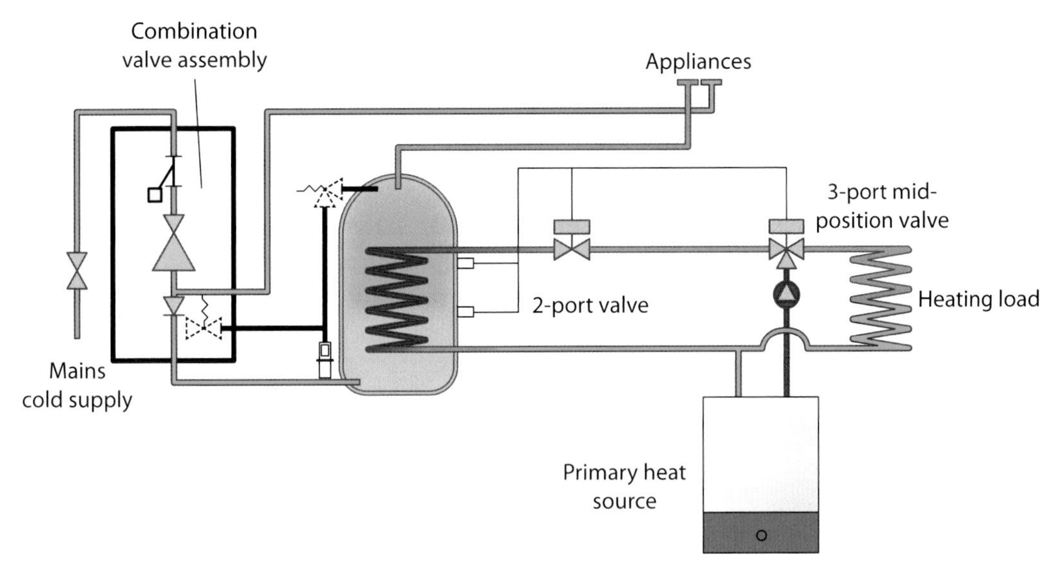

Figure 4.5: Centralised system

Hot water system pipework

Centralised unvented hot water systems

A centralised system is where the cylinder is located within a dwelling and the hot water supplies to the outlets are fed from a single source.

Larger systems requiring a secondary circulation system

To avoid **'dead legs'** and waste of water, larger systems of hot water will need to be installed with a secondary circulation circuit. With an open vented system cylinder, if you know **before** installation, you could

Key term

'dead leg' – a distribution pipe without secondary circulation.

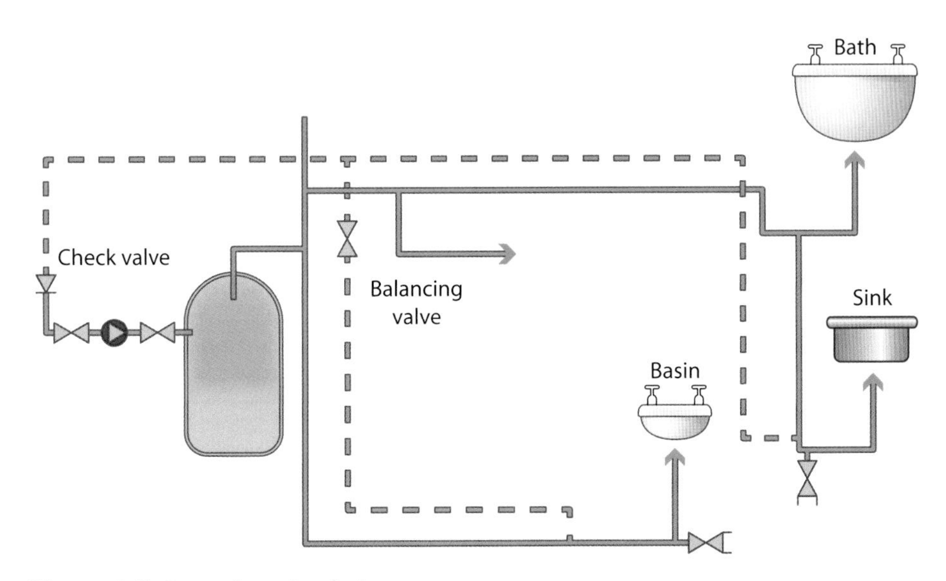

Figure 4.6: Secondary circulation system

Key term

Essex Boss – a purpose-made fitting to make a connection into a cylinder for a secondary return pipe or a shower hot water connection.

have a cylinder made to your specification, with the connections for a secondary circuit. If you are connecting to an existing cylinder that has no provision, you can make provision by installing an **Essex Boss** into the side of the cylinder, no more than one-third of the way down from the top of the cylinder.

The domestic heating compliance guide states that when secondary circulation is used, all pipes kept hot by that circulation should be insulated for the total length of the installation. This is to maintain a minimum of 50°C in the flow and return pipework temperature in the secondary system.

Recommended design temperatures

Some domestic hot water storage systems have the capability of exceeding 80°C under normal operating conditions. These types of vessel are those used as heat stores and those connected to solar heat collectors or solid fuel boilers. The outlet from these vessels should be fitted with a device such as an in-line hot water tempering valve in accordance with BS EN 15092:2008. This will ensure that the temperature supplied to the domestic hot water distribution system does not exceed 60°C.

BS EN 15092:2008

Hot water is responsible for the highest number of fatal and severe scald injuries in the home. Every year, around 20 people die as a result of scalds caused by hot bath water, and a further 570 suffer serious scald injuries. Those most at risk are young and old people, because their skin is thinner.

Safety tip

The temperature of hot water delivered to a bath outlet must not exceed 48°C. This temperature is a maximum and it is strongly recommended that temperatures are set lower.

Building Regulation Document G3 3.65 states: 'The hot water supply temperature to a bath should be limited to a maximum of 48°C by the use of an in-line blending valve or other appropriate temperature control device, with a maximum temperature stop and a suitable arrangement of pipework.'

In housing for older people, it is considered essential that all showers and hot water outlets are controlled by thermostatic mixing valves (TMVs). Valves are now available that have been certified to the new BuildCert TMV2 scheme, which recommends maximum hot water outlet temperatures for use in all premises. These valves maintain the preset temperature even if the water pressure varies when other appliances are in use. The TMV3 valve is made to a different standard for NHS applications.

Safety tip

Mixing valves to limit the temperature of hot water outlets should not be easy for the dwelling's users to alter.

Table 4.1 gives details of appliance and pipework hot water temperatures.

Appliance/outlet	Temperature	Notes
Hot water storage vessel	60°C	To prevent and kill the growth of legionella and prevent scale forming in hard water areas
Hot water distribution pipework	>50°C	Within 1 minute, legionella will be inactive between the temperatures of 50°–60°C
Hot water secondary pipework	>=50°C	As above
At point of use recommended maximum mixed outlet temperatures.		
Assisted bathing	48°C	A requirement of new dwellings. Mixed at point of use with thermostatic mixing valves
Bath	44°C	
Bidet	38°C	
Wash basin outlet	41°C	
Shower outlet	41°C	
Note: Temperatures should never exceed 48°C		

Table 4.1: Pipework and outlet hot water temperatures

Types of unvented hot water system

The main types of unvented hot water system involve indirect and direct storage systems; this refers to the way that the water is heated within the system.

Indirect storage systems

In an indirect storage system, the cylinder is heated by means of a remote heat source, such as a system boiler. The primary flow and return are connected to the cylinder and heat the secondary water through the coil, which is within the cylinder (see Figure 4.7).

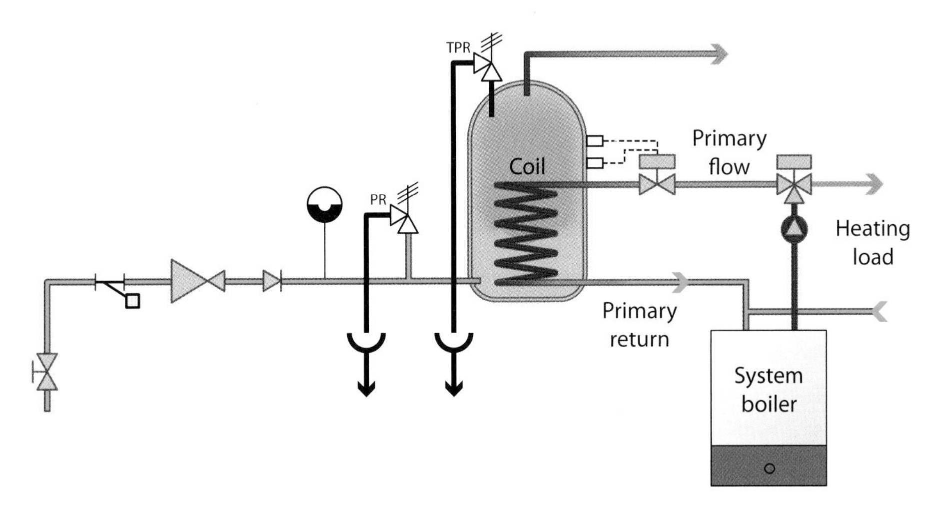

Figure 4.7: Indirect storage system

Indirectly heated systems

The safety devices listed for direct heating are also required for indirectly heated units and packages, but a number of further points must also be considered.

- The non-self-resetting thermal cut-out should be wired up to a motorised valve or some other suitable device to shut off the flow to the primary heater.
- If the unit incorporates a boiler, the thermal cut-out may be on the boiler. There are only a few manufacturers who provide unvented cylinder and boiler packages that must be fitted together. These take advantage of minimising the number of thermostats by only having one energy cut-out on the boiler.
- The temperature-relief valve should be sized and located, and the discharge pipe provided, as described for direct heating (see page 183).
- Where an indirect unit or package has any alternative direct method of water heating fitted, such as an immersion heater, a non-self-resetting thermal cut-out device will also be needed on the direct source(s).

Direct storage systems

Electrically heated

Under Regulation G3, a directly heated unit or package should have a minimum of two temperature-activated safety devices operating in sequence:

BS EN 60335-2
BS 4201
BS 6283 Part 3

- a non-self-resetting thermal cut-out either to BS EN 60335-2 for electrical controls or to BS 4201 for thermostats for gas-burning appliances, and
- one or more temperature-relief valves manufactured to BS 6283 Part 2 for temperature-relief valves or to BS 6283 Part 3 for combined temperature- and pressure-relief valves.

Both these devices are in addition to the standard control thermostat fitted to maintain the temperature of the stored hot water – hence the three-tier level of protection. The Building Regulations do state that it is permissible to use other forms of safety device, but they have to be approved by a testing body such as the British Board of Agrément (BBA) and provide an equivalent degree of safety, so is not a matter for the installer.

In both units and packages:

- the temperature-relief valve(s) should be located directly on the storage vessel, so that the temperature of the stored water does not exceed 100°C
- the valve(s) must be properly sized (this is for the manufacturer to do) in accordance with the requirements of BS 6283
- the valves should not be disconnected other than for maintenance or repair (so temporarily capping off a faulty valve on a working system while a new one is obtained is absolutely not allowed)

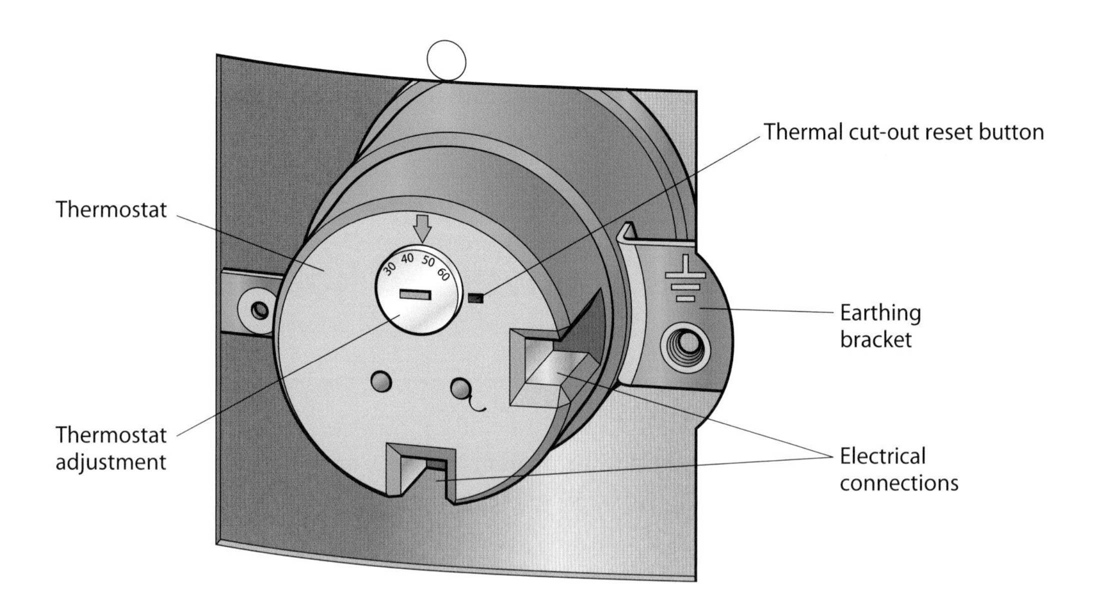

Figure 4.8: Immersion thermostat

- each valve should discharge via a short length of metal pipe (called D1, as covered on page 183) of a size not less than the outlet size of the temperature-relief valve, through an air break over a tundish located vertically as near as possible to the valve(s).

There are special requirements for immersion heaters used with unvented hot water storage systems. Figure 4.8 shows that the immersion heater has a special design and includes an energy cut-out device in addition to the control thermostat. Note that, when maintaining these, replacement parts may not be available off the shelf, so you may have to order them in specially.

Gas- or oil-fired

This type of heater can be gas- or oil-fired. If the heater is oil-fed then the installer must be OFTEC registered. This heater works by the burner at the base of the cylinder heating the plate within the combustion chamber. Heat is transferred by conduction and the secondary water is heated up from the bottom. Heat is also taken off from the flue which passes through the centre of the heater (see figure on page 165).

Safety tip

A gas-fired heater would have to be installed by a Gas Safe registered installer.

Using cold water accumulators in unvented hot water systems

An accumulator is a component used as a means of storing water under pressure. An accumulator normally consists of a sealed steel shell that is spherical or near spherical in shape. The insides are split into hemispherical shells with a flexible diaphragm. The purpose of an accumulator on these systems is not to increase the pressure of the water, but to increase the flow rate of the water. It would be used in instances where there is sufficient pressure from the mains, but insufficient flow rate.

A cold water accumulator works by balancing the water pressure by using air pressure between the accumulator wall and diaphragm. When any tap or shower is turned on, the stored water from the accumulator will supplement water from the incoming main, increasing the flow rate at that outlet.

Components in unvented hot water systems

System controls and devices

Safety devices are included to protect the user and the property. Functional devices are used to protect the supply of water.

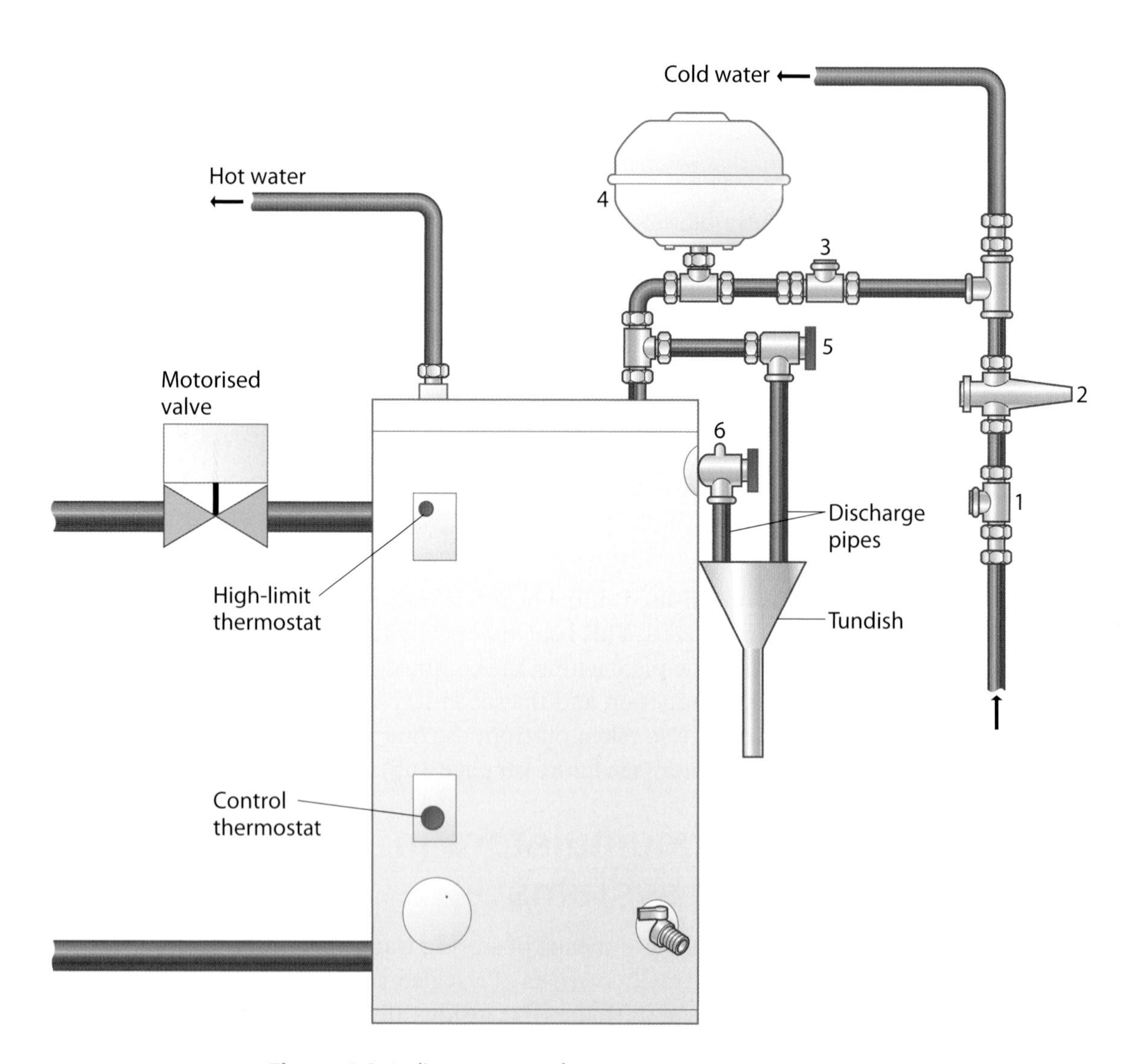

Figure 4.9: Indirect unvented storage system

Safety devices

There are three levels of safety protection to guard against overheating: the control thermostat, the high-limit thermostat (if the control thermostat fails) and the temperature-relief valve (if the high-limit thermostat fails).

Safety item 1	Control thermostat	Controls the water temperature in the cylinder to between 60°C and 65°C
Safety item 2	High-limit thermostat (energy cut-out device)	A non-self-resetting device that isolates the heat source at a temperature of around 80°C–85°C
Safety item 3	Temperature-relief valve (component 6)	Discharges water from the cylinder at a temperature of 90°C–95°C. Water is dumped from the system and replaced by cooler water to prevent boiling

Table 4.2: Safety devices in an unvented storage system

Device	Function
Line strainer (1)	Prevents grit and debris entering the system from the water supply, which can make the controls malfunction
Pressure-reducing valve (on older systems may be a pressure-limiting valve) (2)	Gives a fixed maximum water pressure set by the manufacturer
Single check valve (3)	Stops stored hot water entering the cold water supply pipe, which would be a contamination risk
Expansion vessel or cylinder air gap (4)	Takes up the increased volume of water in the system from the heating process
Expansion valve (5)	Operates if the pressure in the system rises above design limits of the expansion device, as the cylinder air gap or expansion vessel fail
Isolating (stop) valve (not shown in Figure 4.9)	Isolates the water supply from the system (for maintenance)

Table 4.3: Functional devices in an unvented storage system, often provided as composite valves

Key term

Composite valve – valve that can provide three functions in one unit: for example, line strainer, check valve and expansion valve.

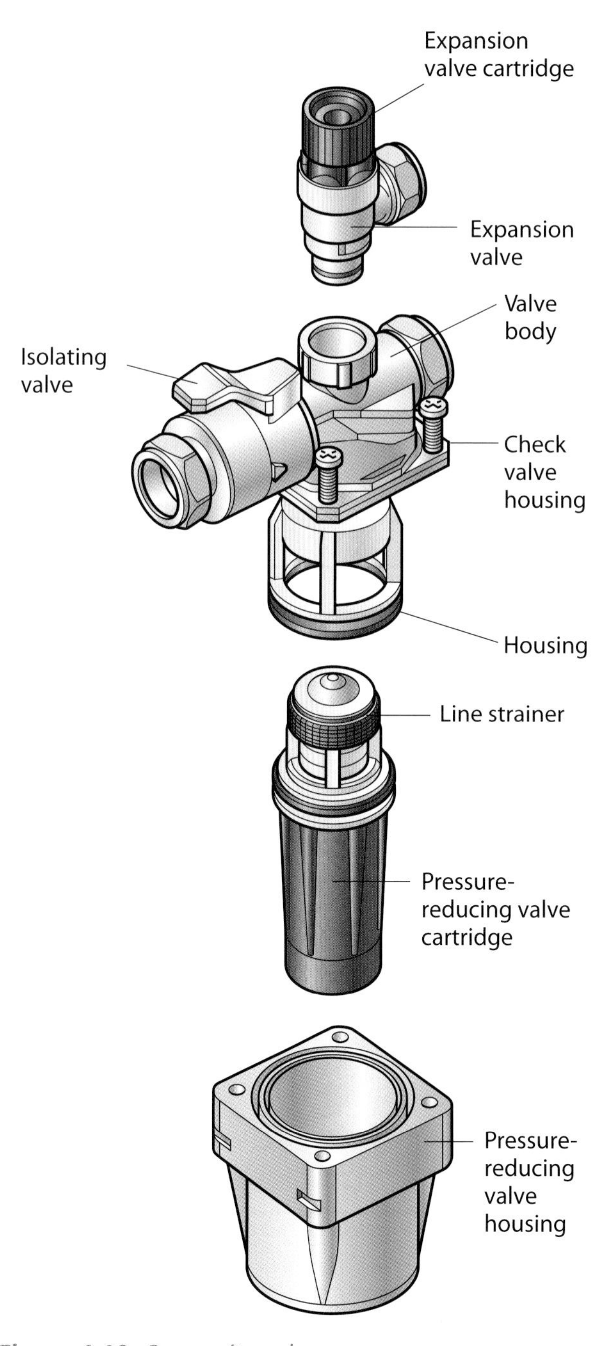

Figure 4.10: Composite valve

Full-bore spherical ball valve

BS 1010 stop valve

Remember

Valves used on unvented systems are preset by the manufacturer, so you do not need to set or adjust them.

Control components in detail

Isolating (stop) valve

To be able to maintain the system, the isolating (stop) valve needs to be isolated, by one of two means:

- full-bore spherical ball valve, or
- stop valve to BS 1010.

Line strainer

A filter must be provided to prevent particles and grit from the water supply passing into the system and affecting the correct operation of other expensive controls. In modern systems, the line strainer is usually an integral component of another valve. It has been shown as an individual item on Figure 4.10 (page 177) so that you understand where it is positioned in the system.

Pressure-control valve

A pressure-reducing valve reduces the incoming water pressure to that recommended by the manufacturer for the cylinder and also helps to prevent fluctuation of pressure within the system.

The pressure-control valve may be in the form of either a pressure-reducing valve (PRV) or a pressure-limiting valve. Pressure-limiting valves are not widely used on modern systems as they are not as effective as pressure-reducing valves. With pressure-limiting valves in high-pressure supply areas it was found that, if the incoming pressure in the system dropped slightly, the valve could provide a higher than required working pressure to the system and cause the expansion valve to discharge water.

A pressure-reducing valve

Single check valve

A non-return or single check valve is designed to prevent backflow of water from the system into the supply pipe or mains as the system water is heated up. The valve is located on the cold water inlet upstream of the connection to the expansion vessel, to prevent expanded water being discharged down the cold supply pipe.

Single check valve

Expansion device (vessel or integral to cylinder)

There are two methods of accommodating or taking up the expanded water in an unvented hot water storage system – a cylinder air gap, or an expansion vessel.

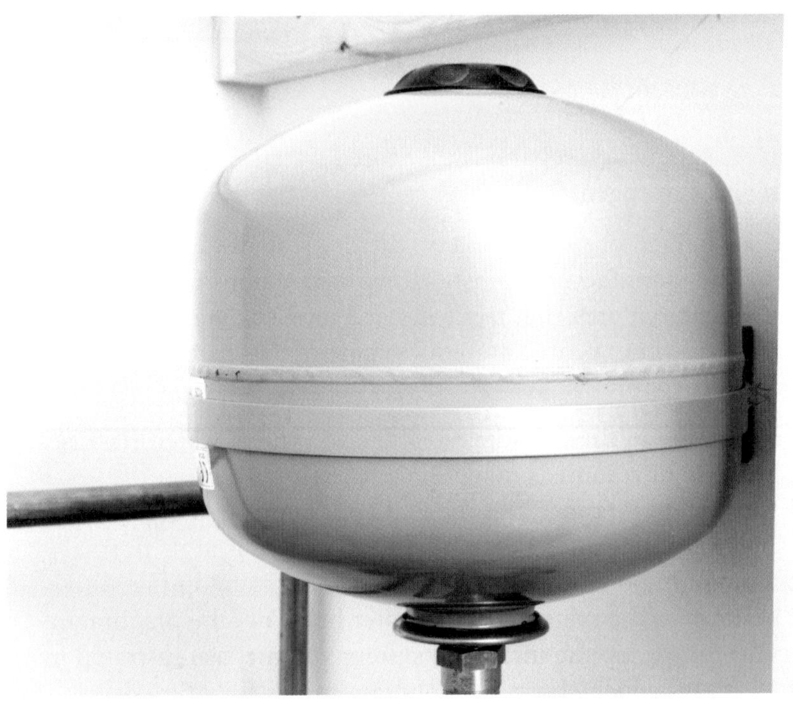

An expansion vessel takes up the volume of expanded water in the system

Figure 4.11: Cylinder air gap (bubble top)

Cylinder air gap

Cylinder air gaps accommodate the expansion of the heated water using an internal air bubble, which is generated and trapped at the top of the unit during commissioning. The size of the air bubble is determined by the cylinder manufacturer. In exceptional circumstances and with systems with extremely long pipe runs, you may have to consult the manufacturer to find out whether the expansion volume is sufficient for the system contents, or whether an additional expansion vessel needs to be provided. Figure 4.11 shows the air gap or expansion chamber built into the design of the cylinder. In modern cylinders of this type, the air gap varies in size, based on the pressure inside the cylinder as a result of a moving or floating **baffle**.

An expansion vessel can be used to take up the volume of expanded water in the system due to the heating process. The vessel contains a

Key term

Baffle – plate to slow down combustion processes, to extract more heat before the products are discharged.

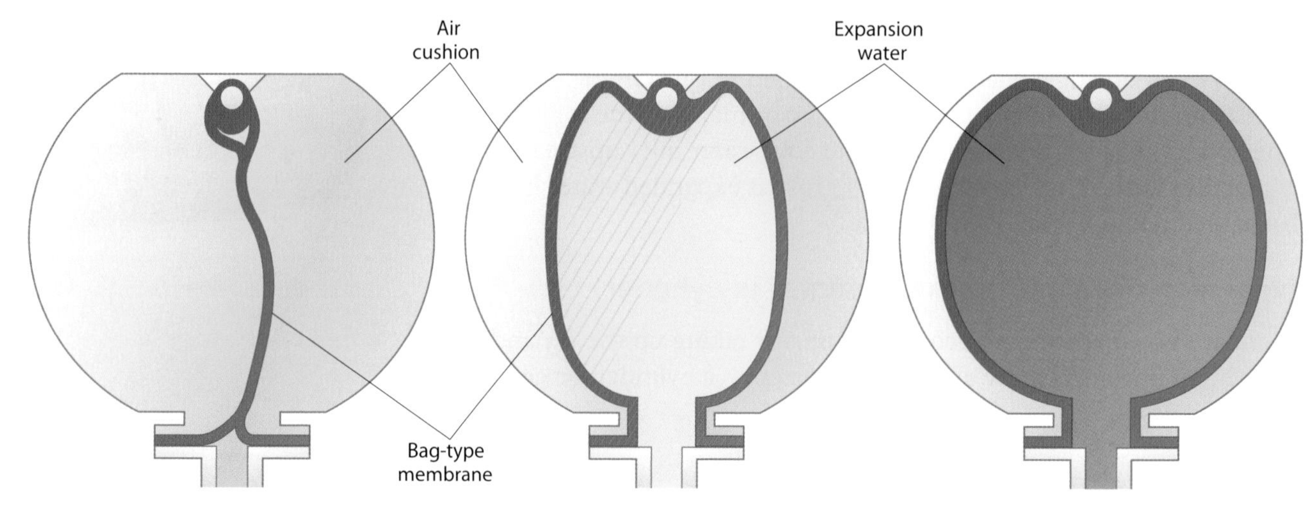

Figure 4.12: Operation of expansion vessel

flexible diaphragm which separates the stored water in the vessel from a cushion of air or nitrogen.

The vessel is charged or pressurised on the dry side of the flexible diaphragm. The manufacturer's installation instructions usually determine the charge pressure. You will need to check and charge expansion vessels on unvented systems as part of the installation and maintenance procedure.

As the temperature in the cylinder rises, the volume of water increases, and that increase in volume is taken up by the air cushion in the vessel. So when the system is empty, the vessel diaphragm is in a collapsed state, as shown in Figure 4.12, left. When the system is at its normal cold water operating pressure, the membrane has taken up the initial pressure built up in the system. As the water heats up, the diaphragm flexes further to take up the increased system volume, as illustrated in Figure 4.12, right, which shows a diaphragm that is fully flexed.

Did you know?

Contact with heated water reduces the life expectancy of the flexible membrane in an expansion vessel.

The size of the expansion vessel is usually determined by the cylinder manufacturer, but, as with the bubble top version, if there are excessive pipe runs resulting in an excessive increase in system volume, a larger than normal expansion vessel may be required. The best advice is to consult the manufacturer. You should fit the vessel on the cold side or inlet side of a storage cylinder, to prevent a build-up of scale in hard-water areas and to prevent deterioration of the flexible membrane.

The expansion vessel acts as a 'dead leg' in the circuit and the water contained in it is not often changed. In certain conditions, the water can begin to stagnate and bacteria can grow on the flexible membrane. The WRAS Guide to the Water Regulations identifies two ways to overcome this problem:

- use a throughflow expansion vessel – the cold water is constantly flowing through the vessel as it has an inlet and outlet connection to prevent possible stagnation

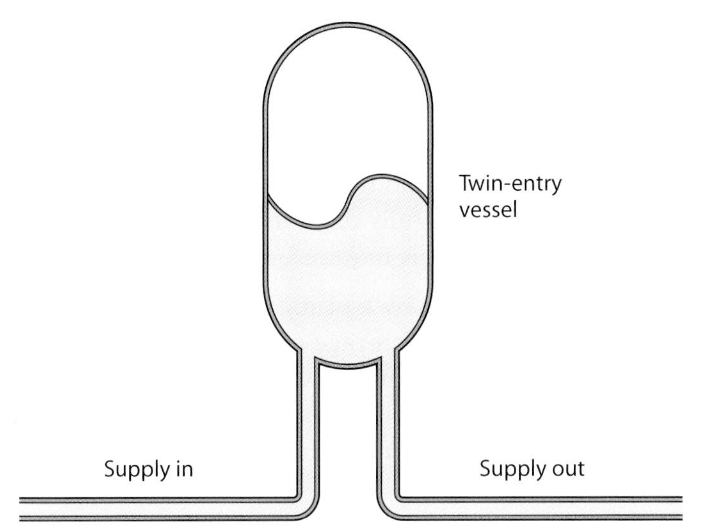

Figure 4.13: Throughflow expansion vessels

Did you know?

The Venturi effect is named after Giovanni Battista Venturi, who was a professor of geometry, philosophy and physics in eighteenth-century Italy.

- provide an anti-legionella valve – this is connected to the inlet of an expansion vessel to maintain circulation within it under flow conditions using a **Venturi effect**. The valve also incorporates an isolation and drain-down valve facility for easy removal of the vessel for maintenance or replacement.

Key term

Venturi effect – the reduction in fluid pressure that results when a fluid flows through a constricted section of pipe.

Currently there are hardly any manufacturers installing anti-stagnation devices in factory-built unvented hot water units. Instead the installer is left to provide and install what the manufacturer specifies.

Expansion valve

The expansion valve is fitted in the system to relieve excess pressure build-up in the system, which is usually due to a component failure. The cylinder manufacturer predetermines the pressure at which the valve begins to operate, and you should not adjust this. On modern systems, expansion valves typically tend to operate at pressures ranging from 6 to 8 bar. With earlier systems, the operating pressure for the expansion valve was usually 0.5 bar higher than the system operating pressure.

The two main features of the valve are to protect the system in the event of:

- failure of the pressure-reducing valve to maintain the inlet cold water pressure at the design operating pressure of the system
- failure of the expansion vessel or, in the case of a bubble top cylinder, a loss of expansion volume (when a loss of air pressure causes an imbalance of pressure between water and air in the device).

No other valve should be fitted between the expansion valve and the storage cylinder.

The valve can be either 'lever-top' pattern (operates by lifting a lever) or 'twist-top' pattern, as shown in the photo opposite.

Expansion valve and tundish

Temperature- and pressure-relief valve

Activity

What do you think is the purpose of an anti-vacuum valve? And why would it need to be installed in an unvented hot water system?

Key term

Anti-vacuum valve – valve designed to open a tap to atmosphere if the internal pressure drops.

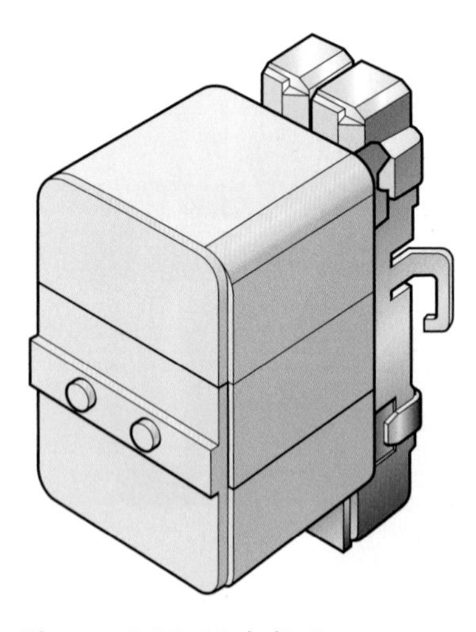

Figure 4.14: High-limit thermostat

The safety controls

Temperature-relief valve

This valve is usually supplied and fitted to the storage cylinder. It is preset by the manufacturer to fully discharge at a temperature of 90°C–95°C, so no adjustment is required.

The temperature activation is by a probe that is immersed in the hottest and highest part of the storage cylinder. The probe is filled with a temperature-sensitive liquid or wax that reacts to temperature change. The valve's key purpose is to protect the system in the event of failure of both the control thermostat and the energy cut-out device.

The valve is often supplied by the manufacturer with an in-built pressure-relief function – called a temperature- and pressure-relief valve. Temperature- and pressure-relief valves are commonly installed on continental systems that have different installation requirements. This pressure-relief function in-built in the valve is not a requirement for UK systems as we use an expansion valve, but a temperature- and pressure-relief valve may well be supplied by the manufacturer to eliminate the need for them to produce different valves for the UK market.

The pressure setting in a combined temperature- and pressure-relief valve is set higher than the expansion valve to ensure that the expansion valve opens first – hence not really providing a useful function.

A means of preventing the formation of a vacuum (**anti-vacuum valve**) was built into some early temperature-relief valves fitted mostly on copper cylinders in the UK market. Anti-vacuum valves tend not to be fitted on modern systems.

High-limit thermostat (non-self-resetting thermal cut-out)

An unvented hot water system must include high-limit thermostats or non-self-resetting thermal cut-out devices to BS EN 60335-2. This type of protection must be provided on all heat sources (hot-water zone feeding an indirect cylinder) or be built into the design of any immersion heaters on direct cylinders (a special immersion heater is required). With direct-fired heaters there must be an overheat thermostat alongside the normal control thermostat. This device must not be capable of automatically resetting itself. It will typically operate at around 80°C–85°C.

Control thermostat

This is the control thermostat/cylinder thermostat provided on the cylinder that can be adjusted by the end-user. The temperature setting on the thermostat will usually be around 60–65°C.

Remember

The control and high-limit thermostats are supplied to control the operation of a **motorised valve** on indirect systems as the means of isolating the heat source. A motorised valve will usually be supplied with an indirect unvented hot water cylinder.

Key term

Motorised valve – valve which is operated by an electrical device, such as a room thermostat or programmer.

Tundish arrangements

Discharge points

Unvented systems give great flexibility in terms of where they can be sited, but the main factor to consider is the provision of a discharge pipe from both the temperature-relief valve and the expansion valve to a safe position, which can sometimes be problematic.

Notice that the discharge pipe is divided into two sections, (D1) and (D2), and is separated by means of a tundish.

The Building Regulations relating to D1 say that the tundish should be vertical, should be located in the same space as the unvented hot water storage system, and should be fitted as close as possible to and within 600 mm of the safety device (temperature-relief valve).

A tundish separates the two sections of the discharge pipe

Temperature-relief valve

Metal discharge pipe (D1) from temperature-relief valve to tundish

600 mm max.

Tundish

300 mm min.

Metal discharge pipe with minimum fall from tundish (D2)

Discharge below fixed grating

Fixed grating

Trapped gully

Figure 4.15: Discharge pipe arrangements

Remember

Building Control has the right to monitor, inspect and control the installation of unvented systems under G3.

With early systems, the D1 section of pipe used to be manufacturer-supplied; it is now common for you to have to produce this section of pipe yourself. It commonly joins together both the outlets of the temperature-relief and expansion valves, as shown in Figure 4.16.

The requirement here is to make sure that you install to the D1 section, with the pipework falling to the tundish and no more than 600 mm between valve outlet and tundish. Also, when joining both valve outlets together, you will need to consider siting the cold water supply pipe and the expansion valve, to ensure that the outlet from that valve can fall continuously to the tundish with an AUK 3 air gap – it is not allowed to rise upwards!

The requirements for D2 are a bit more complicated. The discharge pipe (D2) from the tundish should:

- terminate in a safe place where there is no risk to people in the vicinity
- be of metal (although there are plastic materials available on the market which will cope with temperatures in excess of 100°C, and these can be suitable)
- be at least one pipe size larger than the outlet size of the safety device, unless its total equivalent hydraulic resistance exceeds 9 m: discharge pipes between 9 m and 18 m equivalent resistance would be at least two sizes larger than the outlet size of the safety

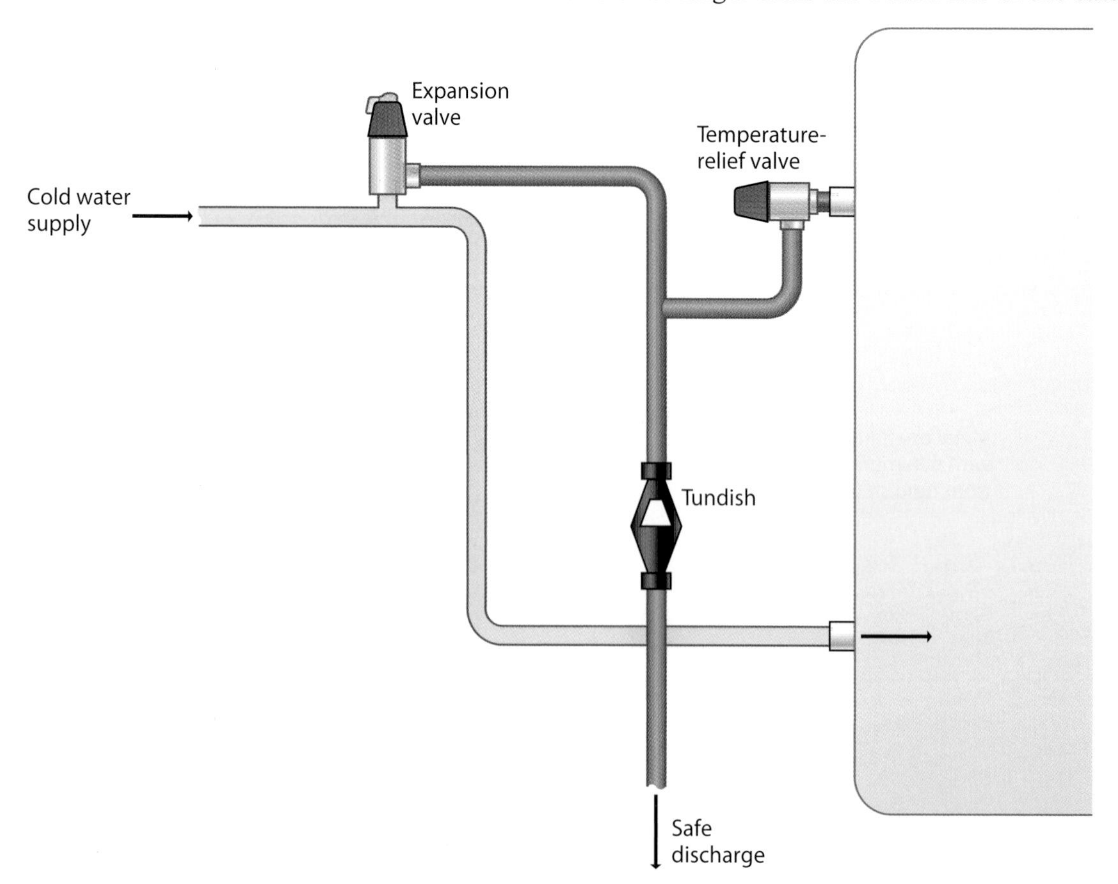

Figure 4.16: Requirements for D1

device, between 18 m and 27 m at least three sizes larger, and so on (note that bends must be taken into account when working out the resistance)

- have a vertical section of pipe at least 300 mm long below the tundish, before any elbows or bends in the pipework
- be installed with a continuous fall of 1:200
- have discharges visible at both the tundish and the final point of discharge; where this is not possible or is practically difficult, there should be clear visibility at one or other of these locations.

Here are some examples of acceptable discharge arrangements.

Downward discharge at low level
Low level below a fixed grating

Discharges at low level (up to 100 mm above external surfaces, such as car parks, hard standings and grass areas) are acceptable provided that, wherever children may play or otherwise come into contact with discharges, a wire cage or similar guard is positioned to prevent contact while maintaining visibility. Low level below a fixed grating is the ideal location as it is highly safe and visible, however it should not discharge below the water level in the gully in case of freezing.

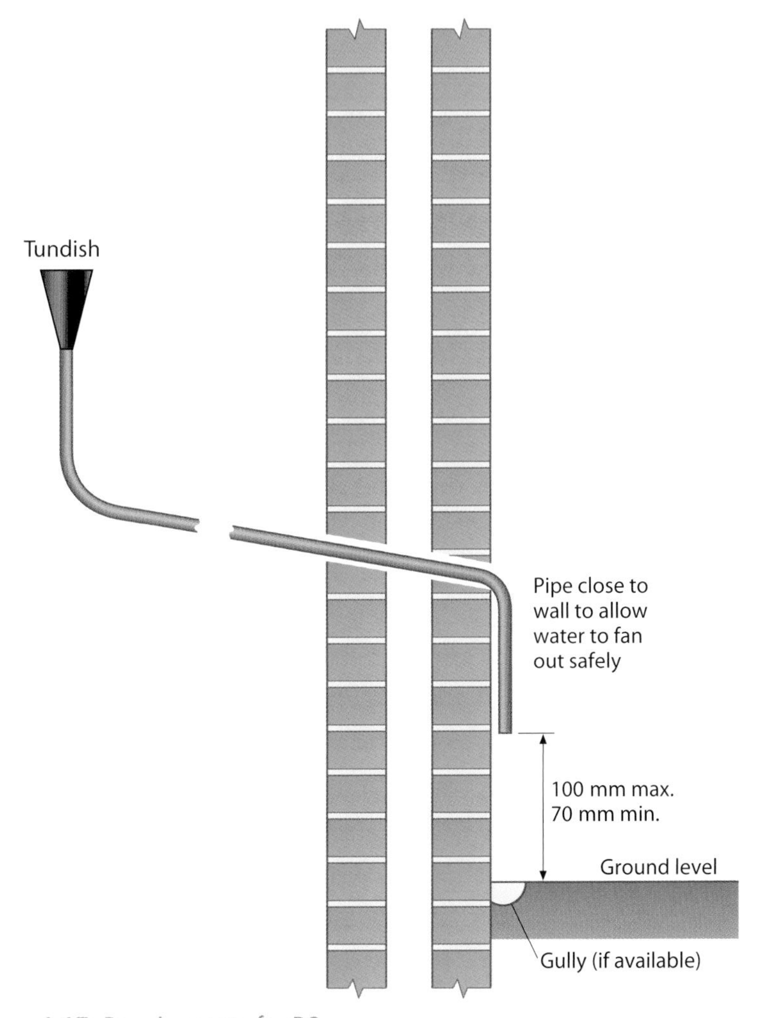

Figure 4.17: Requirements for D2

Discharge at high level into hopper

Discharge may be made to a metal hopper and metal downpipe with the end of the discharge pipe clearly visible (tundish visible or not), or onto a roof capable of withstanding high-temperature discharges of water and 3 m from any plastic guttering that would collect such discharges (tundish visible).

Where a single D2 pipe serves a number of discharges, such as in blocks of flats, the number of systems served should be limited to no more than six so that any installation discharging can be traced reasonably easily. The single common discharge pipe should be at least one pipe size larger than the largest individual discharge pipe (D2) to be connected.

If systems are to be installed where discharge is not apparent (for example, in dwellings occupied by the blind) you should consider installing an electronically operated device to warn when discharge takes place.

Safety tip

On no account must the pipe be able to discharge onto people. You should seek feedback from Building Control before attempting discharge at high level, to make sure they will approve it.

Remember

As a minimum, the pipe size must be one size larger than the valve outlet, up to 9 m in length, and between 9 m and 18 m it must be two sizes larger, and so on.

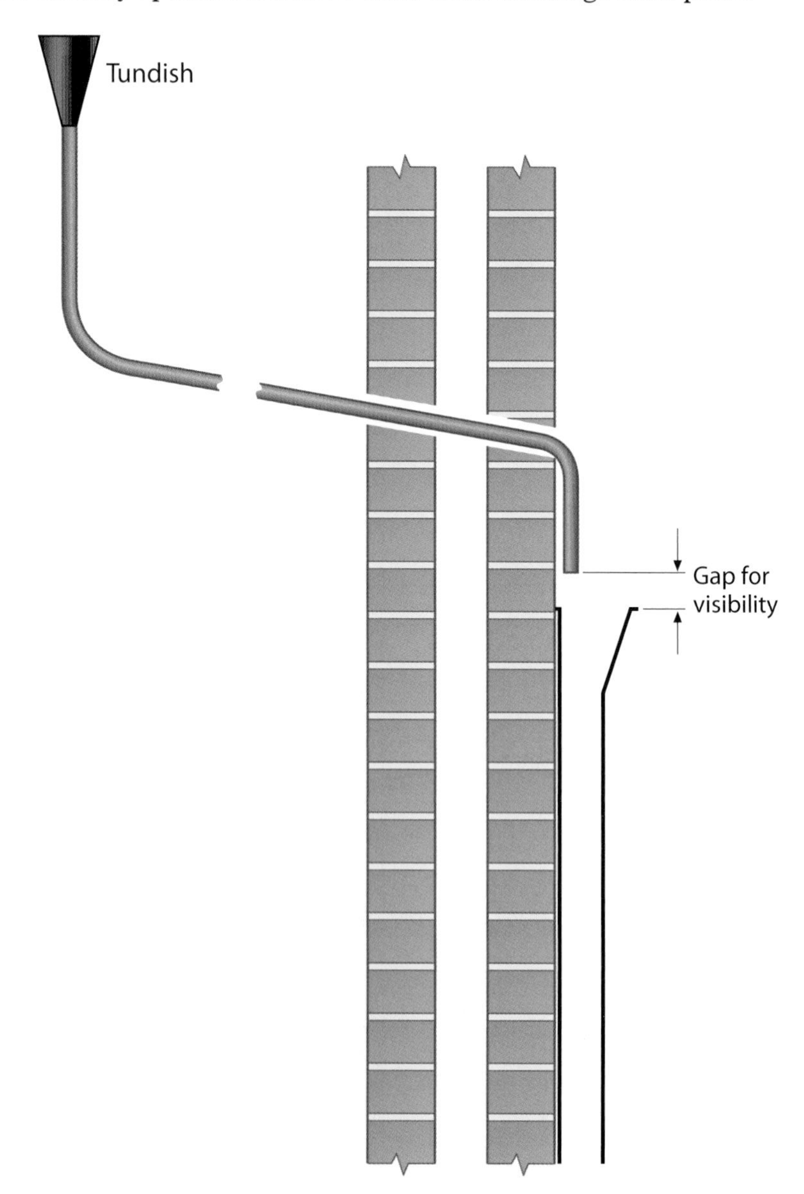

Figure 4.18: Discharge at high level

Professional Practice

On the installation Tomek is doing, he cannot use straight pipes, and elbows and bends create resistance in the pipe. He has to cater for pipework resistance when pipe sizing, or the discharge pipe will not meet requirements. Essentially, the length of straight pipe run is reduced for every bend or elbow used, so to simplify the process of sizing the pipe he uses a table like this.

Valve outlet size	Minimum size of discharge pipe D1	Minimum size of discharge pipe D2 from tundish	Maximum resistance allowed, expressed as a length of straight pipe (i.e. no elbows or bends)	Resistance created by each elbow or bend
G 1/2	15 mm	22 mm	Up to 9 m	0.8 m
		28 mm	Up to 18 m	1.0 m
		35 mm	Up to 27 m	1.4 m
G 3/4	22 mm	28 mm	Up to 9 m	1.0 m
		35 mm	Up to 18 m	1.4 m
		42 mm	Up to 27 m	1.7 m
G 1	28 mm	35 mm	Up to 9 m	1.4 m
		42 mm	Up to 18 m	1.7 m
		54 mm	Up to 27 m	2.3 m

The D2 pipework from an unvented cylinder is run for 13 metres and changes direction five times. What should the size of the D2 be if the size of the temperature-/pressure-relief valve is ½"?

Application of composite valves

Unvented hot water systems now tend to be supplied with composite control valves. These valves contain two or more of the control functions mentioned earlier.

The composite valve contains a pressure-reducing valve, line strainer, expansion valve and check valve (see Figure 4.10 on page 177).

A balanced composite valve contains a pressure-reducing valve, line strainer, expansion valve, check valve and balance cold water supply connection.

A composite valve contains a pressure-reducing valve, line strainer, expansion valve and check valve

Temperature and expansion relief pipework in unvented hot water systems

This is covered under Tundish arrangements on page 183.

Pipework systems incorporating secondary circulation

When designing a hot water system, you need to take into account the length of pipe run from the cylinder to the individual outlets; where possible, you need to avoid excessive lengths of dead legs. If this is not possible, you will need to install a secondary circulation system.

The Water Regulations Guide recommends that uncirculated hot water distribution pipes should be kept as short as possible and, if uninsulated, not exceed the maximum length stated.

Pipe OD	Length	(Seconds)
<12 mm	20 m	(11)
<22 mm	12 m	(25)
< 28mm	8 m	(26)
>28mm	3 m	(>15)

Table 4.4: Maximum lengths of uninsulated pipes

In Table 4.4, the right-hand column gives the approximate length of time it would take to draw off the cool water based on the draw-off rate of a wash basin tap with an 0.15 l/s flow rate. The Health & Safety Legionella Code L8 states that a maximum draw-off period for hot water to reach its correct temperature shall be 60 seconds.

Secondary pipework system

As well as being pumped, secondary hot water systems can be gravity circulation systems. The flow must rise away from the cylinder and the return must fall back to the cylinder as in Figure 4.19. This is essential for circulation to take place.

Pump

As the water is being replenished by fresh water, the pump must be constructed of a non-corrosive material. The most common of these for a secondary circulation pump is bronze. The pump is positioned on the return pipe with full-bore isolation valves either side of the pump to aid servicing and removing the pump.

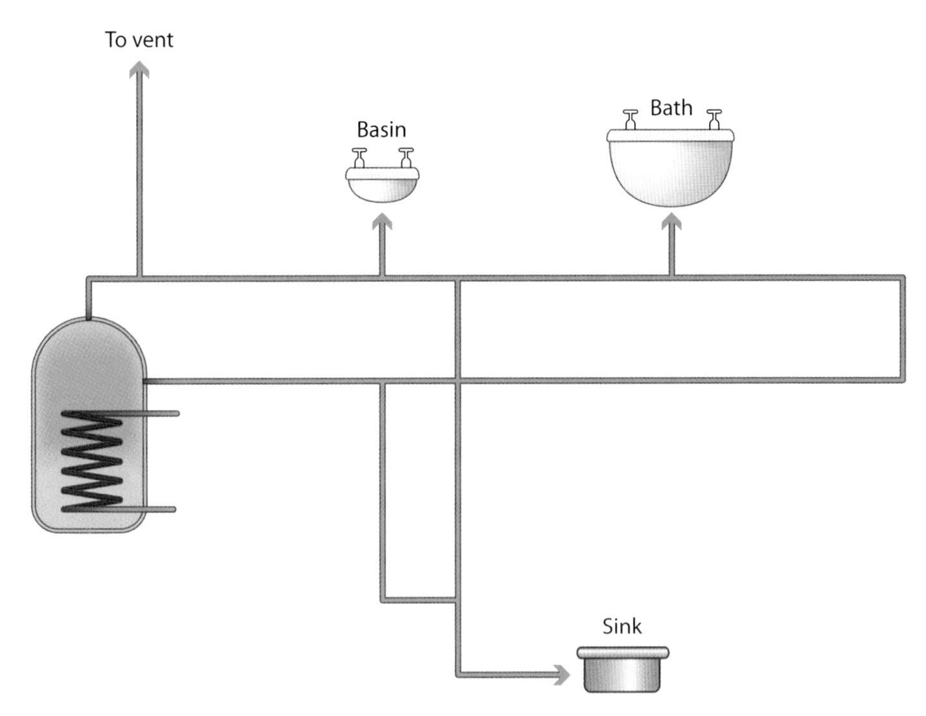

Figure 4.19: Gravity secondary hot water system

Sizing of secondary return pipes

There are two methods of sizing secondary pipework for hot water systems.

- **Formal method:** you will need to calculate the heat loss from all of the flow and return pipe circuits throughout the system. This allows you to establish comparable flow rate, the head loss throughout the system and the duty of the circulating pump.
- **Rule of thumb method:** with this method of sizing you first select a return pipe size that is two sizes lower than the flow. As a guide, select smaller sizes over larger pipe sizes and maintain a check on the hot water supply return pipe velocities.

Balancing the system

On larger systems where the hot water supply return has a number of branches and loops to serve the various parts of the system, you will need to balance the circuit by installing a balancing valve. These valves restrict the flow to the circuits nearest to the pump where the pressure is greatest. These valves force the hot water supply return to circulate to the furthest part of the system. The valves are commonly a double regulating pattern: they permit an accurate 'low flow' setting which, when set, is retained by the valves even after maintenance of the system.

You can use ordinary isolation valves to achieve a crude form of restricting the flow for balancing purposes, but these rarely remain effective or return to their original setting after being shut off or opened for maintenance purposes.

Activity

Find out more about legionella and why this is such a risk by reading the appropriate sections of *The Health & Safety Approved Code of Practice Guidance L8.*

The main purpose of the balancing valve is to maintain the correct temperature, 50°C, within the whole pipework distribution system, to minimise the potential growth of bacteria, in particular legionella.

Preventing reverse circulation

There is always a possibility of reverse circulation in secondary hot water systems, particularly when pumped systems are turned off. This can lead to heat loss. The methods of preventing this are as follows:

- single check valve
- night valve
- anti-gravity valve.

Figure 4.20 shows a system where reverse circulation would cause heat loss.

Key to numbering

1. Stop valve
2. Strainer
3. Pressure-reducing valve
4. Single check valve
5. Expansion vessel
6. Expansion relief valve
7. Single check valve (to prevent reverse return)
8. Pump, fitted to the return

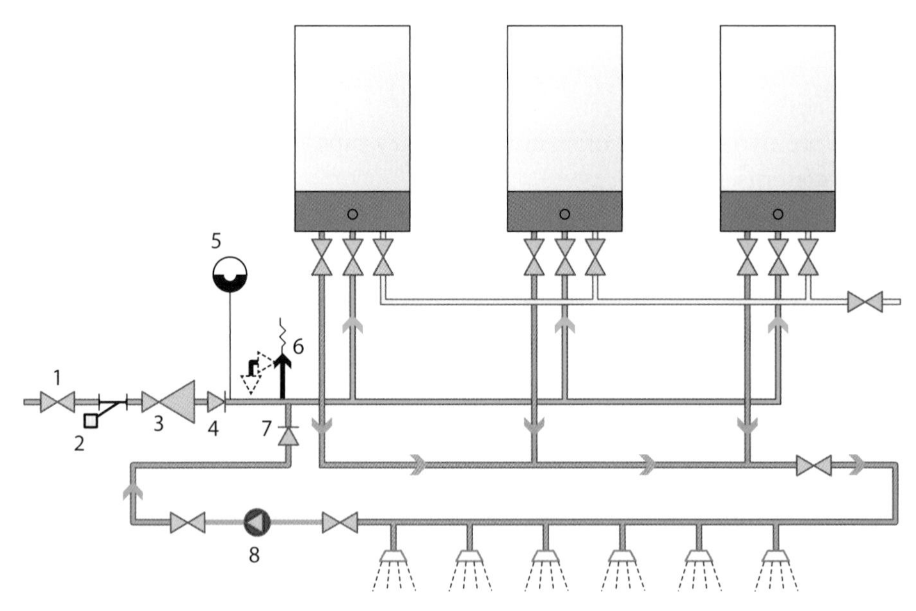

Figure 4.20: Secondary return connected to multiple water heaters

Timing devices

Secondary circuits which are pumped will, in most cases, need to be on a timer or integrated into a system that uses pipe thermostats and timer controls. At peak use the pump may be constantly on, but in places such as schools and nurseries, the secondary circuit would only need to be on during school opening times.

Expansion vessel

Where a secondary circulation system has been installed to an unvented hot water system, an extra expansion vessel may have to be incorporated to allow for the extra water that would expand in the system. This would be directly connected to the secondary pipework circuit.

Trace heating

As an alternative to pumped secondary circulation, the piping can be kept warm by the use of trace heating cabling. The combination of trace heating and the correct thermal insulation for the operating ambient temperature maintains a thermal balance, where the heat output from the trace heating matches the heat loss from the pipe. Self-limiting or regulating heating tapes have been developed and are very successful in this application. When fitted to the pipe they not only monitor and control the temperature but also eliminate the need for return pipework, pumps and valves, as the water can be stored and moved through the pipes at a controlled temperature. These systems offer an intelligent way to instantly supply hot water and are ideal for use in large buildings such as hotels and offices.

PROGRESS CHECK

1. What is the maximum number of systems that can be discharged into a single D2 pipe?
2. Taking the answer from Question 1, what size should the D2 be where more than one system is discharged into it?
3. Which individual components does a composite valve house?
4. What is the purpose of a balancing valve on hot water systems?
5. What are the three levels of safety control on an unvented hot water storage system?

2. Know the uses of specialist components in hot water systems

Hot water system components

For this section you need to know about a range of components:

- infrared operated taps
- non-concussive taps
- combination bath tap and shower head
- flow-limiting valves
- spray taps
- shower pumps – single and twin impellor
- pressure-reducing valves
- shock arrestors and mini expansion vessels.

All these components were covered in Unit 3. Look back at pages 105–116 to remind yourself of the key features of each component.

Components to overcome temperature and pressure effects

For this section, look through the Professional Practice below and answer the questions.

Professional Practice

Lucas is going to install a 10-litre capacity unvented displacement heater. He knows there are a number of different installation scenarios, and different component kits that can be supplied by the manufacturer to suit the installation.

Firstly, he checks the temperature-/pressure-relief valve. As the storage capacity is less than 15 litres he has no obligation to fit this, although the manufacturer recommends it as an additional safety feature. Lucas thinks the best policy, though, is to go for safety at all times!

Lucas is aware that the functional controls can be different. He looks at the manufacturer's diagram, which shows the minimum level of functional control that may be provided.

The only real functional control Lucas has to consider is the expansion valve. This minimum level of control can only be used where the supply inlet water pressure is below 4.1 bar (this may be different for other manufacturers) and where it is possible to accommodate expanded water in the supply pipe as an alternative to providing an expansion space in the storage vessel or an expansion vessel.

For these small heaters it is possible to eliminate the need for an expansion device. The expanded water is taken up in the supply pipe. As the water in the storage vessel is heated, the volume of water increases and is pushed back down the cold supply pipe and ultimately into the mains. However, Lucas also has to make sure that expanded hot water cannot be drawn off through the supply pipework, so a minimum distance is specified between the last cold water draw-off and the point of cold water connection on the heater. Here it is indicated on the diagram as:

- 2.8 metres for 10-litre capacity
- 4.2 metres for 15-litre capacity.

Also, the manufacturer requires that the supply pipe right back to the mains must not contain check valves, stop valves with loose jumpers or fittings that prevent reverse flow.

How will Lucas install the heater if he cannot expand back into the mains?

Lucas uses the expansion vessel and single check valve provided. He knows that he is dealing with supply pressures below 4.1 bar. But what would he do if the supply pressure was above 4.1 bar?

Did you know?

You need to be able to understand the manufacturer's instructions, but most provide a simple flow chart to identify which control kits must be fitted to suit the particular installation.

Activity

Why do manufacturers require that the supply pipe must not contain check valves, stop valves with loose jumpers or fittings that prevent reverse flow?

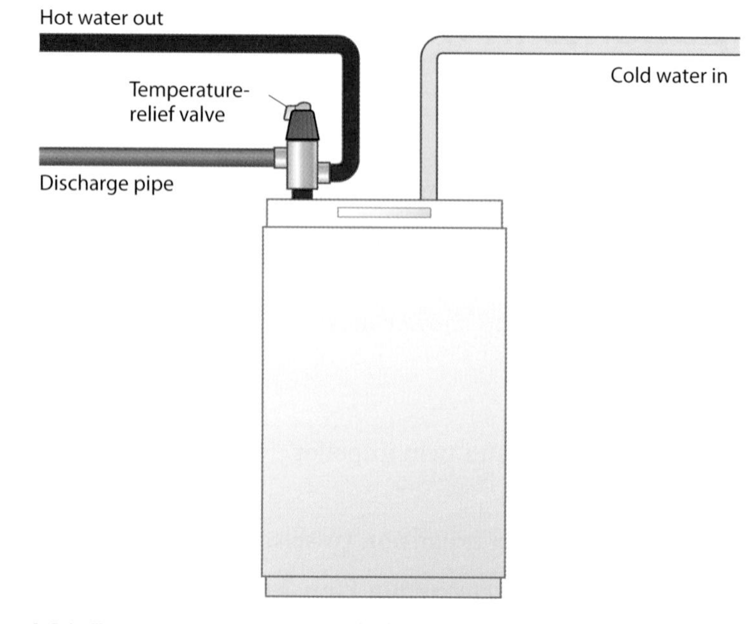

Figure 4.21: Temperature-/pressure-relief valve

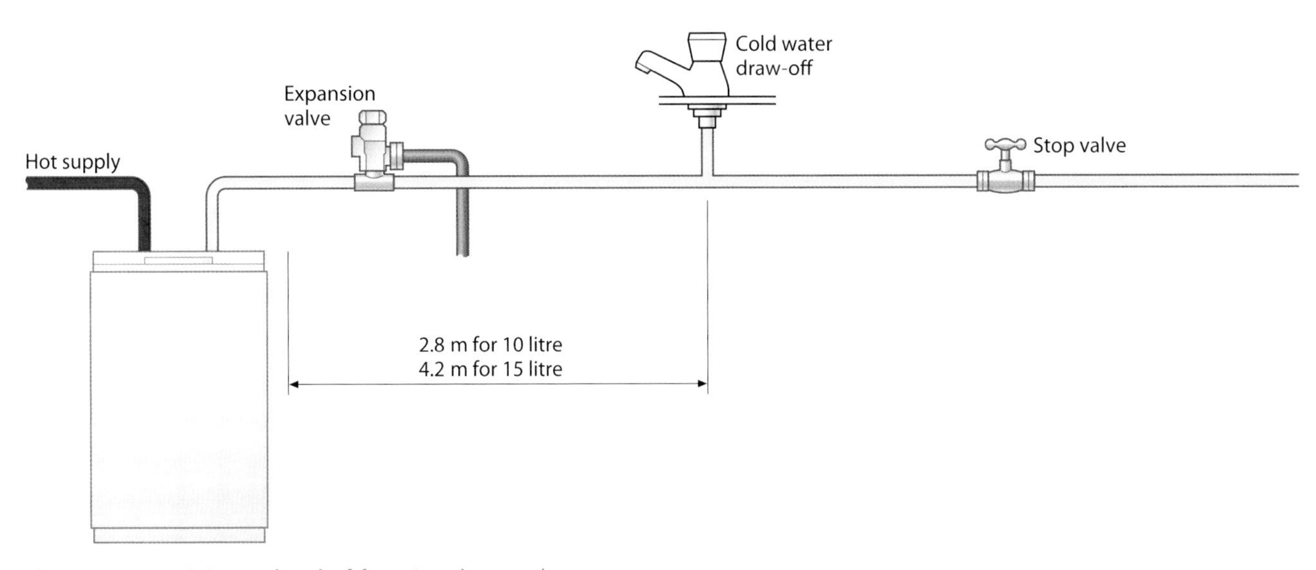

Figure 4.22: Minimum level of functional controls

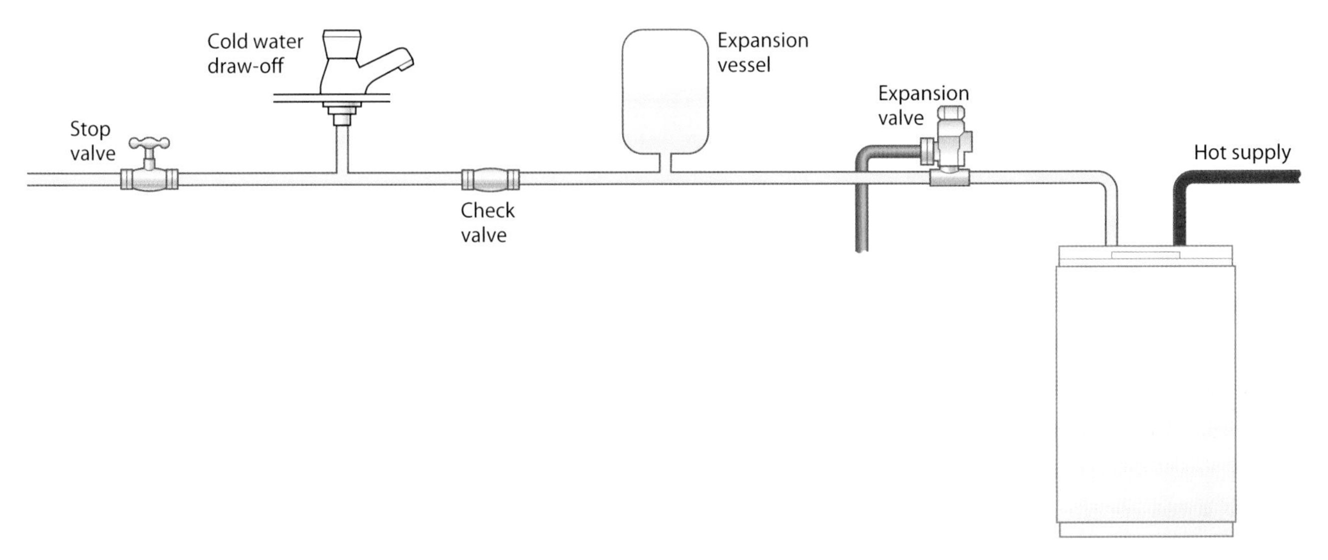

Figure 4.23: Heater installation where it is not possible to expand back to the mains

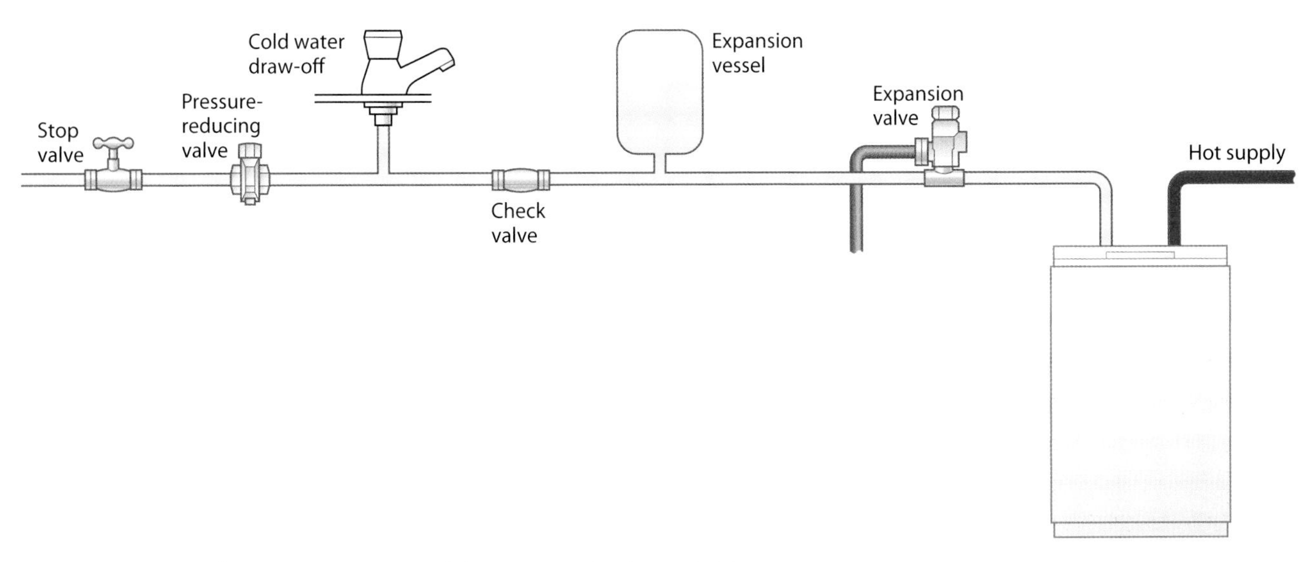

Figure 4.24: Installation arrangement if the pressure is above 4.1 bar

PROGRESS CHECK

1. If the outlet from a shower head was higher than the level of water in the cold water storage system, what methods could be used to ensure that water will flow from the shower head?
2. When a combination bath tap with a shower hose is fitted and fed from a combination boiler, which type of backflow-prevention device must be fitted?
3. When would a shock arrestor be used?
4. What is the purpose of installing a non-concussive tap?
5. What is the purpose of an expansion valve?

3. Know and be able to apply the design techniques for hot water systems

Selecting a hot water system

In some cases, the decision about the choice of system in a dwelling may be taken by someone else. Typically, this would include:

- where the customer (or their representative) has specified the system for a one-off dwelling
- on a large multi-dwelling housing development, where the systems have been specified by an architect or services design engineer.

There may be other circumstances, but as a plumber operating at Level 3, you may be required to advise a customer on the best systems to suit their needs. This will be influenced by:

- customer requirements
- budgets
- technical requirements and limitations.

Customer needs

BS 6700

On a new build the customer needs would be decided during the design stage of the building. Many of these needs would be the basic requirements specified in BS 6700. On an existing dwelling where the customer wants to update the current system, you will need to take into account:

- how many people live in the dwelling
- whether they use baths or showers more regularly
- the quantity and type of other appliances.

A number of factors influence the selection and design of a hot water system:

- amount of hot water required
- temperature of stored water at outlets
- cost of installation and maintenance
- fuel energy requirements and running costs
- economy of water and energy use
- user safety.

Building layout and features

Planning work activities for plumbing installations requires you to have a working knowledge of how a typical domestic dwelling is 'put together'. This will help you to determine whether you need to make provision for any specific fixing materials or specialist tools. It will also make you aware of the nature of the construction into which you intend to carry out the installation work, so that you can inspect buildings to confirm that provision for the system or components is suitable. You need to be aware of the legislation that governs construction work.

Suitability of system

When choosing a system, the options range from a simple single-point arrangement for one outlet to the more complex, centralised boiler systems supplying hot water to a number of outlets. BS 6700 sets out a number of ways of supplying hot water, as summarised in Figure 4.25.

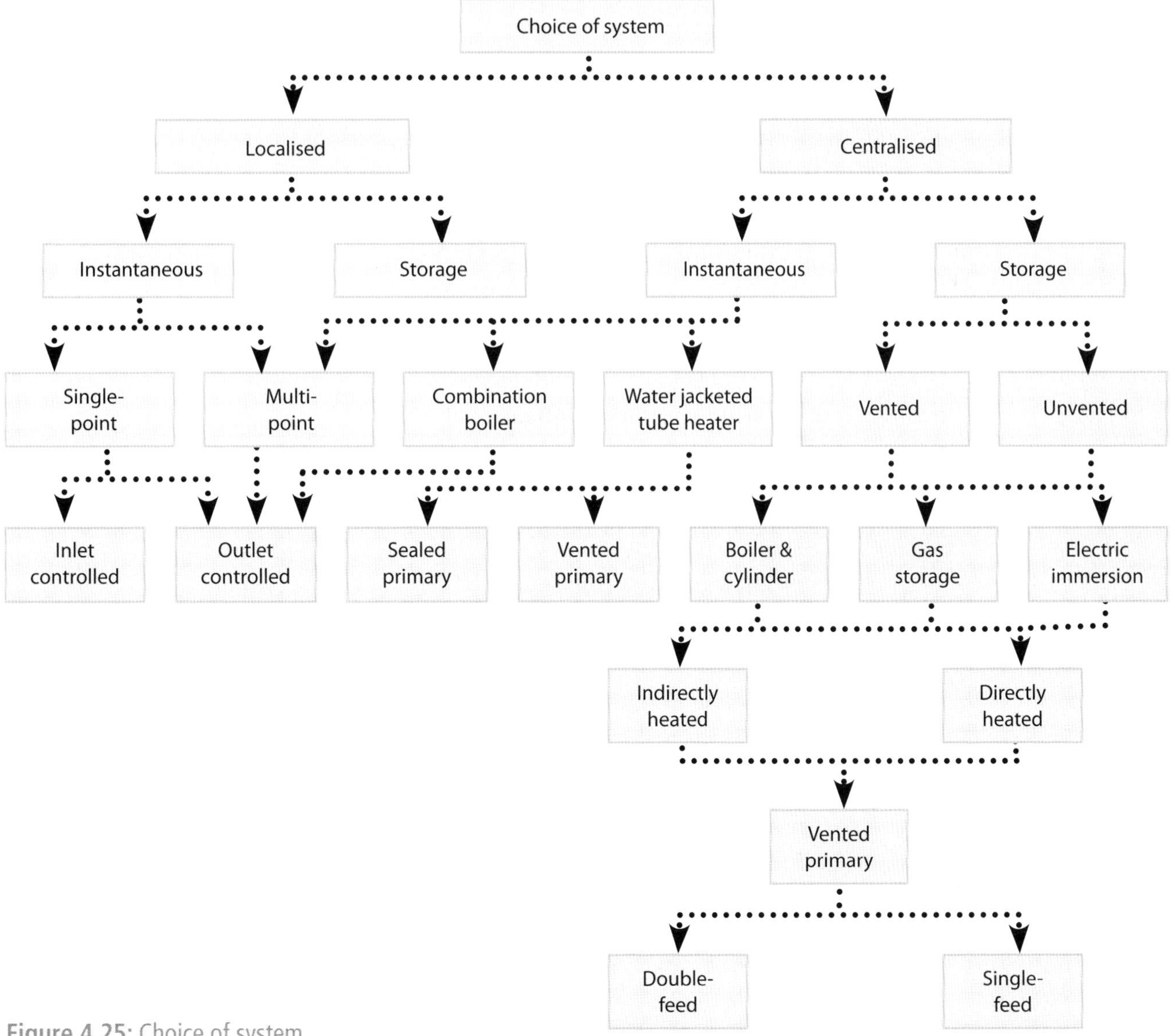

Figure 4.25: Choice of system

Energy efficiency

In this age of sustainability, energy efficiency should be foremost when designing a new hot water system. You must comply with the requirements of the domestic heating compliance guide (available as a free download from the government: www.planningportal.gov.uk).

The minimum provision for new systems in new and existing dwellings is:

- vented copper hot water storage vessels should comply with the heat loss and heat exchanger requirements of BS 1566:2002
- unvented hot water storage systems should:
  - comply with BS EN 12897:2006 (formally BS 7206:1990) or
  - be certified by the British Board of Agrément, the Water Research Council; or
  - be certified by another accredited body as complying with Building Regulations
- primary storage systems should meet the insulation requirements of sections 4.3.1 or 4.3.2 of the Water Heater Manufacturers Association performance specifications for thermal stores
- all hot water storage vessels should carry a label with the following information:
  - type of vessel
  - nominal capacity in litres
  - standing heat loss in kWh/day
  - heat exchanger performance in kW
  - a label on the product such as BSI Kitemark or an equivalent quality control scheme
- vented cylinders which are not of copper construction should be labelled as complying with the heat loss and heat exchanger requirements of BS 1566.

BS EN 12897:2006

Selection criteria

Environmental impact

Being able to reduce the amount of energy used to heat the system is of the utmost importance when designing a hot water system. This will help improve the ability to comply with Part L of the Building Regulations for heating and hot water systems(s), which states:

- Reasonable provision for the performance of heating and hot water system(s) would be:

- the use of an appliance with an efficiency not less than recommended for its type in the Domestic Heating Compliance Guide
- the provision of control requirements as given in the Domestic Heating Compliance Guide for the particular type of appliance and heat distribution system.

Information sources

This topic was covered in Unit 3. Look back at pages 118–121 now to refresh your memory.

Construction legislation

Just like the plumbing industry, the construction industry has to meet the requirements of Health and Safety legislation, including:

- Health and Safety at Work etc. Act 1974
- Electricity at Work Regulations 1989
- Construction (Design and Management) Regulations 2007
- Control of Substances Hazardous to Health Regulations 2002 (COSHH)
- Confined Spaces Regulations 1997.

There are a number of other Regulations, but here you need to know about those that affect the actual design and construction of buildings, namely the Building Regulations.

Building Regulations

Building Regulations approval is required if someone wants to:

- build a house
- carry out alterations to a property, extend it, replace or add new window frames or carry out internal alterations
- build a commercial building, shop, office, or carry out alterations or extend it
- convert (or change the use of) a building or part of a building
- alter, replace or install a controlled service or fitting (i.e. sanitation, drainage, new or replacement heating, a new or replacement roof, new wall surfaces (external or internal) or energy-using equipment)
- create or alter a dwelling or flat in an existing building.

The Regulations exist to provide minimum standards for building work in order to safeguard the health and safety of people in or around buildings. They also include minimum standards for easy access and facilities for disabled people, for the conservation of energy, and water use and disposal. Until 1965, the construction of buildings was controlled by local building byelaws; these were superseded by national Building Regulations, which form the basis of the ones used today.

The current format (simple Regulations plus Approved Documents giving technical guidance) of the Building Regulations was created in 1985. The Regulations are divided up into 'Parts', each Part dealing with a specific technical, construction or design topic.

Several major amendments were made in April 2002, affecting Parts H (Drainage and waste disposal), J (Combustion appliances and fuel storage systems) and L (Conservation of fuel and power).

On 1 July 2003 a new Part E (Resistance to the passage of sound) was introduced. This extends the range of (new and extended) buildings that must offer users relief from excessive noise, both from noise sources within a building and from adjoining buildings. All new (and extended) educational buildings must also design out excessive noise, to give acceptable acoustics in teaching areas within schools.

The current Building Regulations were issued in December 2000 and they have recently been amended. The Building Regulations extended the range of the building work that is now controlled, with effect from 1 April 2002.

The Building Regulations do not offer general advice on how to design 'green' buildings.

However, green drainage design and water disposal are encouraged and illustrated in Part H (2002 edition). The Building Regulations do not prevent the building of 'eco homes' and 'sustainable design solutions', so long as these designs meet or exceed the minimum requirements of the current Building Regulations.

For the installation requirements for hot water systems and components, see Unit 3 pages 105–116.

4. Know and be able to apply the fault diagnosis and rectification procedures for hot water systems and components

Service and maintenance of unvented systems

Periodic maintenance and inspection should usually be carried out on an annual basis. Experience of local water conditions may indicate that more frequent inspection is desirable: for example, when water is particularly hard or scale forming, or where the water supply contains a high proportion of solids such as sand.

The user should, however, be encouraged to report faults such as discharges from relief valves and have them rectified as soon as they occur.

The following steps show a typical servicing procedure for unvented hot water systems.

1) Check that all approved components are still fitted and are unobstructed. Check to see if all valves are still in position and all thermostats are properly wired. Be careful with the electrics: it can be common for immersion heaters to fail and to be replaced with a standard immersion heater rather than one using the proper thermostats.
2) Check for evidence of recent water discharge from the relief valves, visually and by questioning the customer if possible.
3) Manually check the temperature-relief valve – lift the gear or twist the top on the integral test device on the relief valve for about 30 seconds to remove any residue that may have collected on the valve seat. Check that it reseats and reseals.
4) Manually check the expansion valve – lift gear or twist top on the integral test device on the relief valve for about 30 seconds to remove any residue that may have collected on the valve seat. Check that it reseats and reseals.
5) Check discharge pipes from both expansion and temperature-relief valves for obstruction and that their termination points have not been obstructed or had building work carried out around them.
6) Check the cylinder operating thermostat setting (e.g. 60°C–65°C).
7) Isolate gas and electrical supplies to the heating appliance, turn off water supply and relieve water pressure by opening taps:
   a) Drain inlet pipework where necessary to check, clean or replace the line strainer filter.
   b) Remove strainer. These can be located in several places:
      - within a composite valve
      - in a strainer valve
      - where the pressure-reducing valve is of an older model.

Safety tip

Always use manufacturer-approved replacement parts that meet the original system specification and never make temporary repairs – this is an important safety issue.

Cleaning a strainer on an unvented system

Step 1: Isolate water supply.

Step 2: Release pressure from the hot and cold water system.

Step 3: Use the correct tool to loosen the fitting.

Step 4: Remove head from body of combination valve.

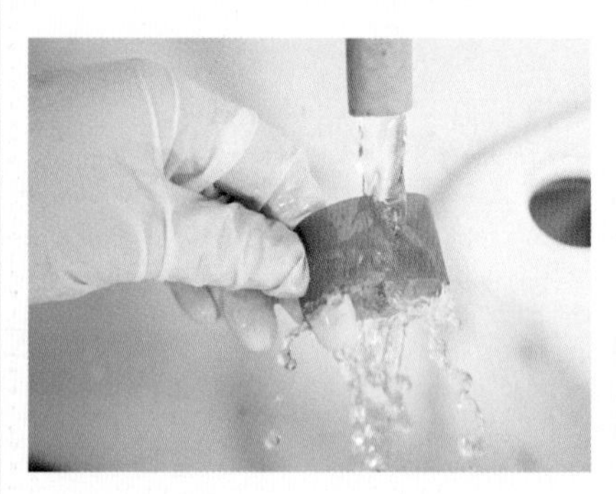

Step 5: Clean strainer using water and check for corrosion and damage.

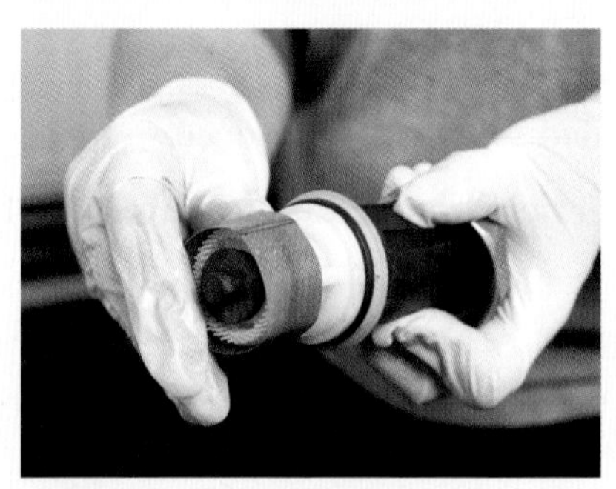

Step 6: Replace strainer onto combination valve head.

Step 7: Reinsert into valve body and tighten.

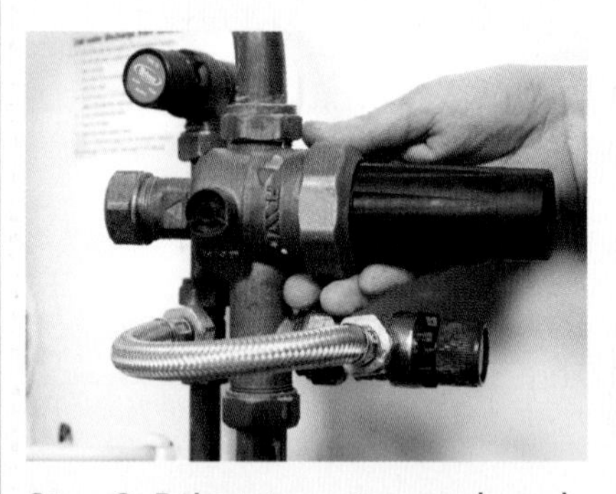

Step 8: Reinstate water supply and check for leaks.

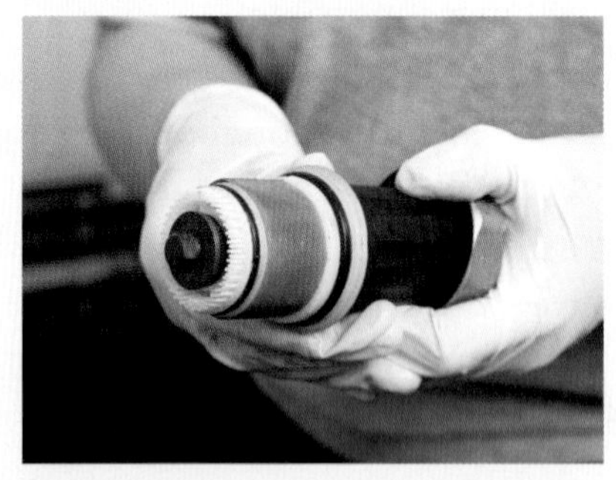

Pressure-reducing valve location of strainer.

Step 9: Check pressure in expansion vessel and top up as necessary while the system is empty or uncharged. Use a tyre gauge and a pump to recharge to operating pressure. If the cylinder includes an air gap as the expansion device, reinstate the air gap as part of the procedure in line with the manufacturer's instructions.

Step 10: Reinstate water, electricity and gas supply. Run the system up to temperature and ensure that control thermostats are working effectively and relief valves are not discharging water.

Step 11: Complete any maintenance records for the system.

Fault finding on unvented systems

Most manufacturers provide good guidance on fault finding in systems, usually in the form of a flow chart to help diagnose the fault. Figure 4.26 gives the chart for a fault where water is discharging from the temperature-relief valve.

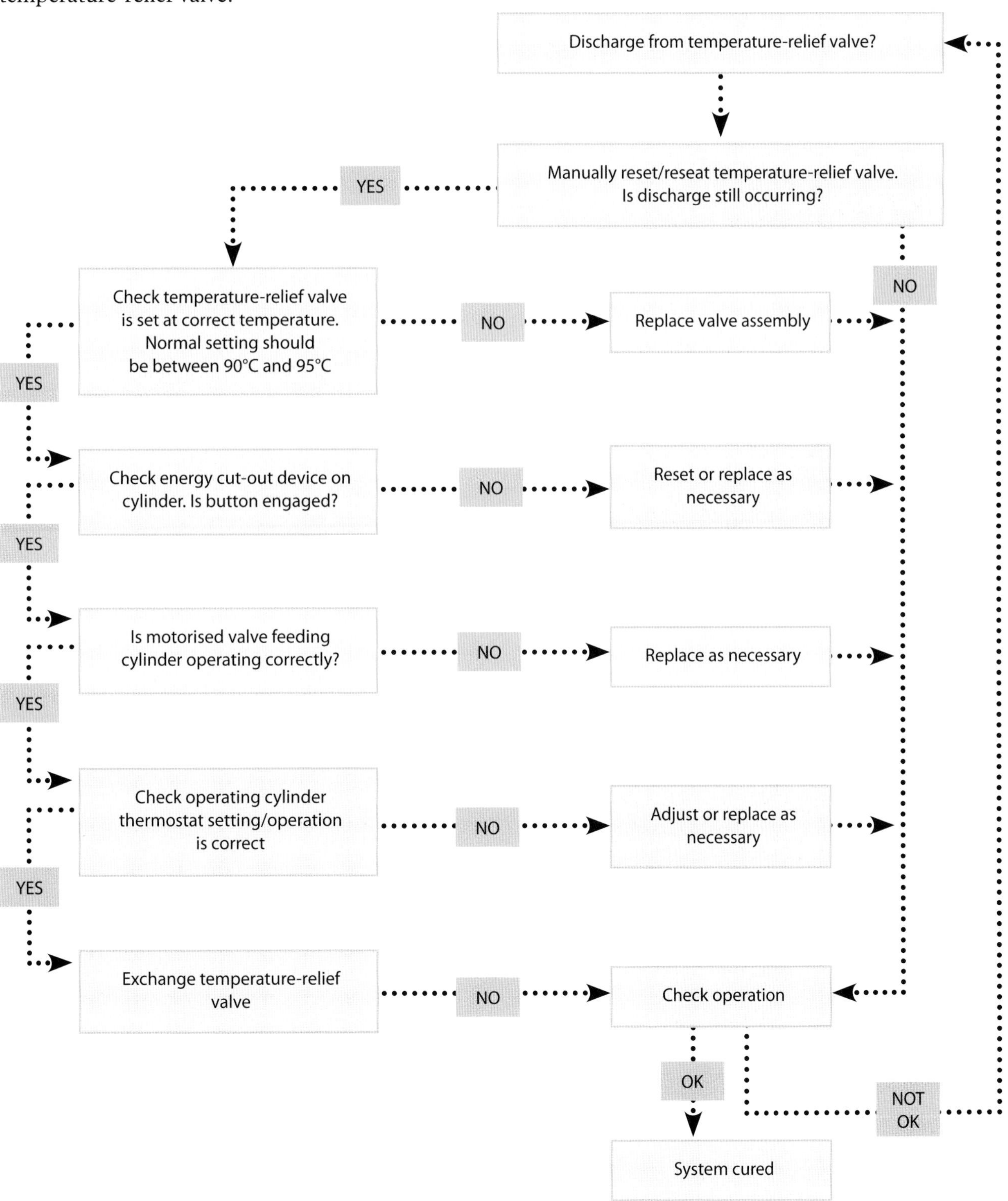

Figure 4.26: Fault finding flow chart for water discharging from the temperature-relief valve

Figure 4.27 shows a fault finding flow chart to be used in situations where there is poor flow rate at taps or outlets.

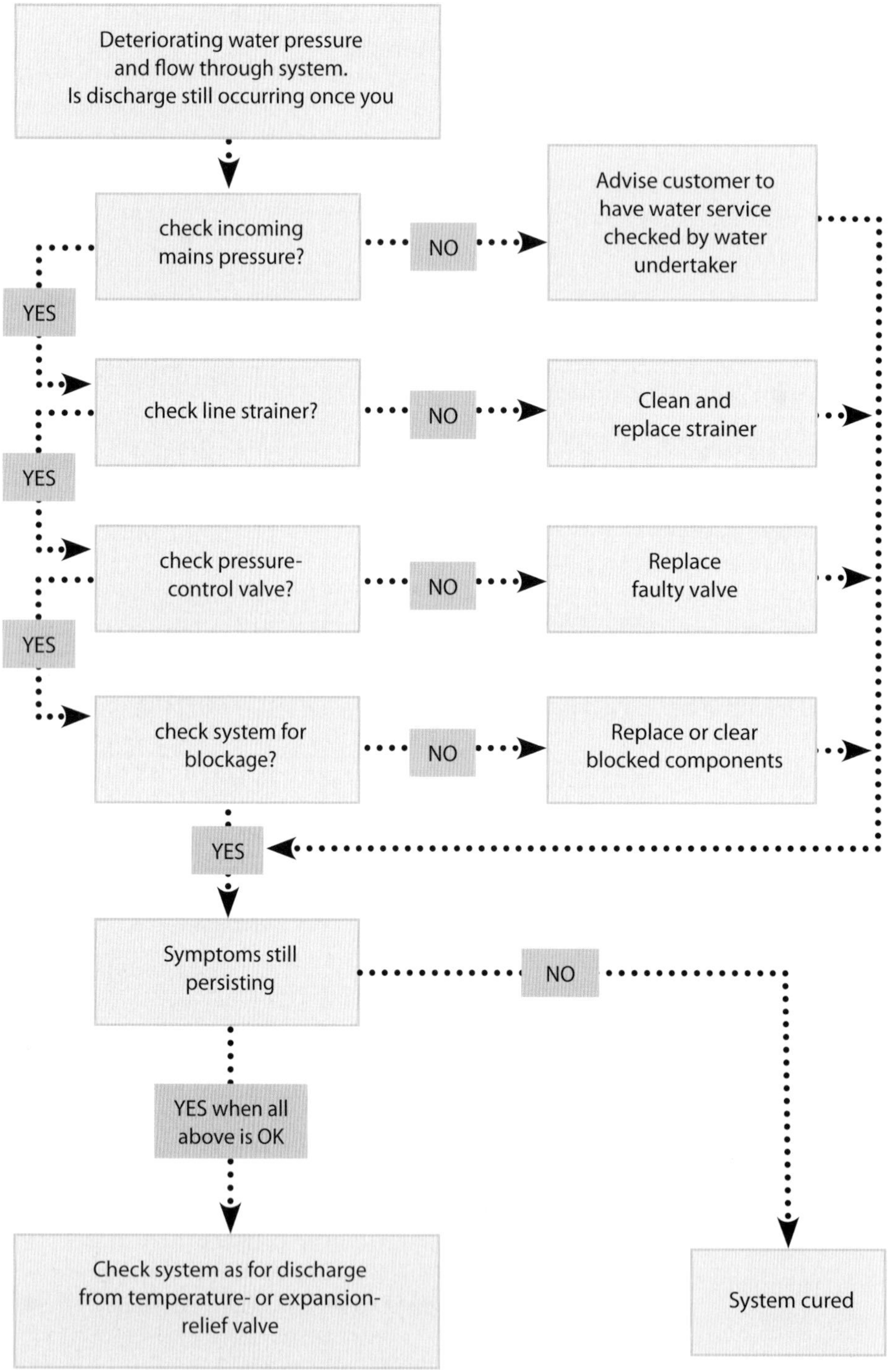

Figure 4.27: Fault finding flow chart for poor flow rate at taps or outlets

Remember

With composite valves, individual parts of the valve can be replaced, rather than the full valve, such as the line strainer part or expansion valve. You can usually order these parts, each with their own part number.

Another fault is water discharging from the expansion valve. Figure 4.28 shows the fault finding flow chart for a system that has an expansion vessel.

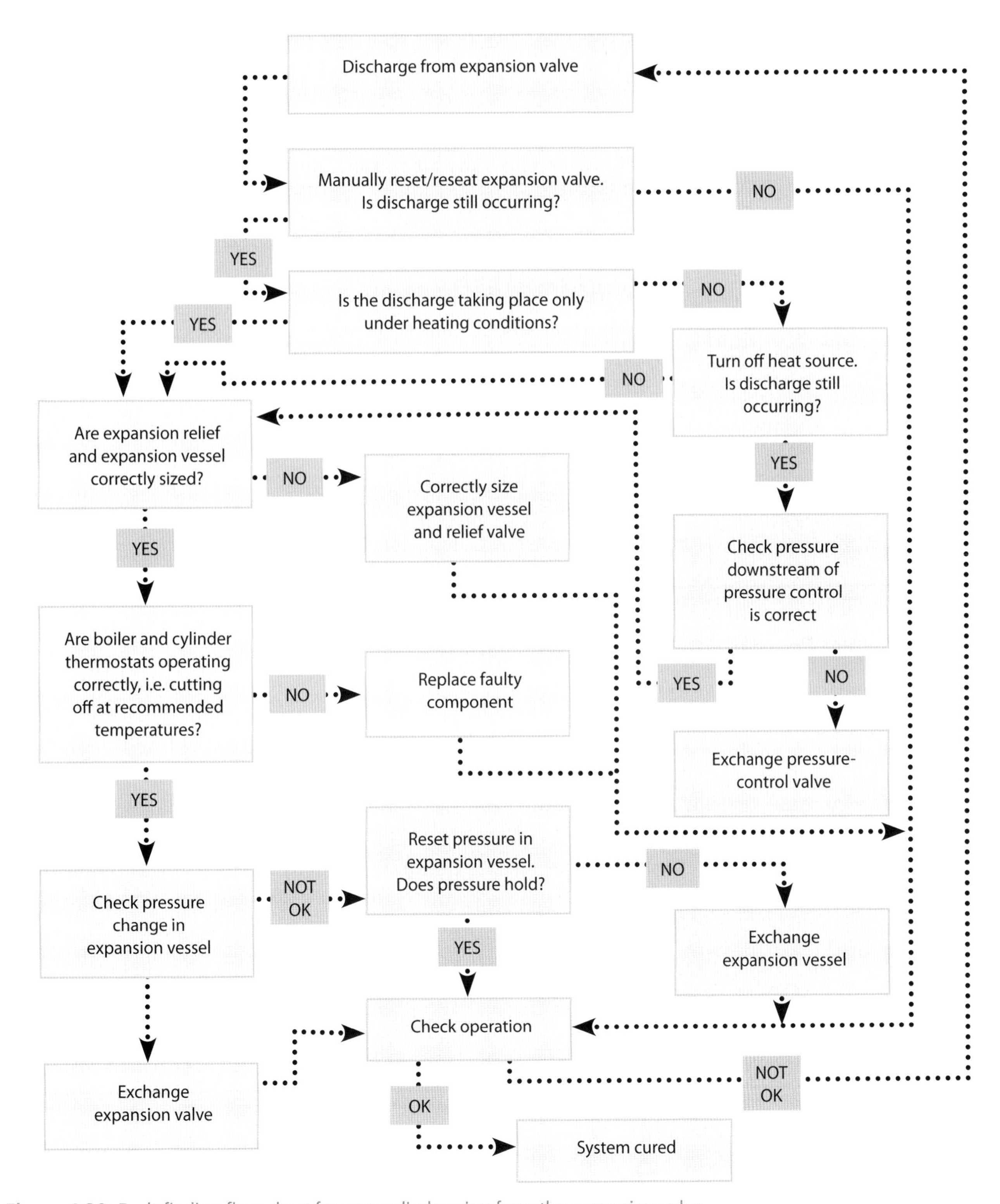

Figure 4.28: Fault finding flow chart for water discharging from the expansion valve.

With cylinder air gaps as the expansion device, intermittent discharge from the expansion valve can often mean that the air gap or bubble has reduced in volume. The solution is to:

- turn off the water supply to the cylinder
- open the lowest hot water tap in the system
- hold open the temperature-relief valve until water ceases to run from the cylinder
- refill the system – the air volume will automatically be recharged as the unit refills.

Safe isolation

It is important when working on any appliance or component that is supplied with electricity, gas or water that they are isolated until the work is completed. With electrical supplies to appliances you must follow the safe isolation procedure – lives could depend on it, including yours.

Safe isolation is covered in detail in Unit 7, on pages 407–415. Read this section now. This is such an important issue that you must make sure you study Unit 7 thoroughly when you come to it.

Locating faults

An important aspect of working with vented and unvented hot water systems and components is the ability to fault-find. When you go to the job the customer is the most important source of information to guide you to the cause of the fault.

Shower booster pump unit

With a shower booster pump, as with any appliance or component, make sure you have the manufacturer's instructions. In most instructions you will find a troubleshooting guide.

Before going through any other processes check that the pump has been installed correctly, especially if the pump is a fairly new unit. If the installation is incorrect, tell the customer and explain why.

Table 4.5 shows some common faults that can occur with boosted shower pumps.

Symptom	Likely cause	Action/remedy
Pump will not start	No electricity supply	Check fuse and electrical supply
	Flow switch not working	• Check water flow from shower head • May require negative head switch
Pump noisy – 'squealing'	Possible aeration	Check pump installed in accordance with instruction regarding siting, system connections and temperature
Pump noisy – 'rumbling'	Worn motor brushes	• Replace pump unit • Check correct installation
Poor water flow pressure from shower mixer valve	Possible aeration	Check pump installed in accordance with instruction regarding siting, system connections and temperature
	Water supply to/from pump	• Check inlet filters • Check for kinked flexible hoses
Fluctuating shower temperature (manual shower mixers)	Possible aeration	Check pump installed in accordance with instruction regarding siting, system connections and temperature
Poor performance	Air in system	To completely remove all air from the plumbing system, stop and start the pump consecutively 5 times, 30 seconds on, 30 seconds off

Table 4.5: Common faults with boosted shower pumps

Safety devices and thermostats

Within an unvented system there are two tiers of safety devices, which are thermostats. This section covers both unvented and vented cylinders; both types of system incorporate the use of thermostats.

The customer may complain that the hot water is extremely hot, or has not heated when timed to do so. Your actions will depend on the type of system the thermostat is used with.

To check a thermostat that is not heating the water in a cylinder

- Isolate the electricity, using the safe isolation procedure (see Unit 7, page 415).
- Using a multimeter set to ohms (Ω), test for continuity across a circuit.
- Place the red probe on the live (1 on Figure 4.29) in to the thermostat and the black probe onto the outlet side of the thermostat (2).
- Twist the thermostat temperature until you hear it 'click'.
- Take the reading from the multimeter.
- If the ohms are greater than 1.00, you have a faulty thermostat.

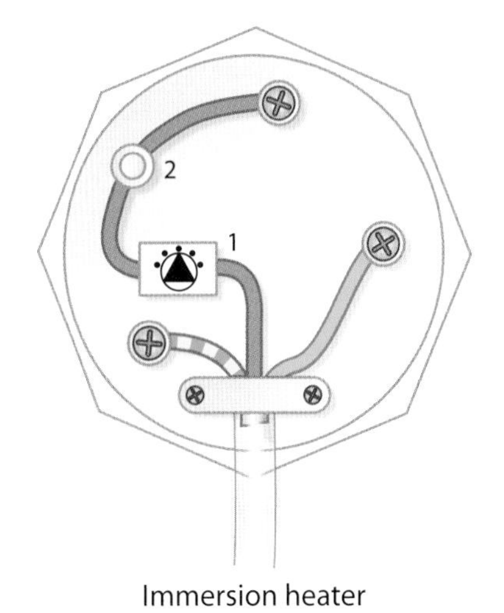

Figure 4.29: Immersion heater with thermostat

Remember

Never bypass the thermostat as a temporary measure while waiting for a replacement.

To check a thermostat that is not 'switching off' when the temperature is reached

- Isolate the electricity, using the safe isolation procedure (see Unit 7, page 415).
- Use a multimeter and set to ohms (Ω) again. Test for continuity across a circuit.
- Place the red probe on the live (1 on Figure 4.29) in to the thermostat and the black probe onto the outlet side of the thermostat (2).
- Check the reading on the multimeter.
- Twist the thermostat temperature until you hear it 'click' off.
- Take the reading from the multimeter.
- If the ohms do not change, you have a faulty thermostat.

Expansion devices

If the expansion valve is discharging water during the heating up of the hot water vessel, you need to carry out tests on the expansion vessel. If the expansion is external, check that there is sufficient pressure in the vessel (normally specified on the data plate). To check the pressure, use a tyre pressure gauge connected to the Schrader valve on the top of the vessel. Read the pressure and, if it is under the required level, use a foot pump to pump up to the required pressure. If the vessel still does not hold the pressure, the diaphragm could be faulty. You should also check that the Schrader valve is not leaking.

For a bubble top unvented hot water vessel, check that the bubble is set.

- First check the manufacturer's instructions.
- Run the hot water tap on the same floor level as the water heater.
- Close the tap.
- Shut off the cold water supply.
- Open the hot water tap.
- Measure the amount of water discharged.
- If no water is discharged, there is no bubble left.
- If the unit is filled to atmospheric pressure, it should discharge approximately 10 per cent of the total water heater.

If you find that the bubble is not in place, reinstate it according to the manufacturer's instructions (normally by draining and refilling the vessel). After reinstatement, check that the bubble is set as before; if not, contact the manufacturer and/or the customer. If the vessel is out of guarantee, installation of a new vessel may be necessary.

Checking and charging an expansion vessel

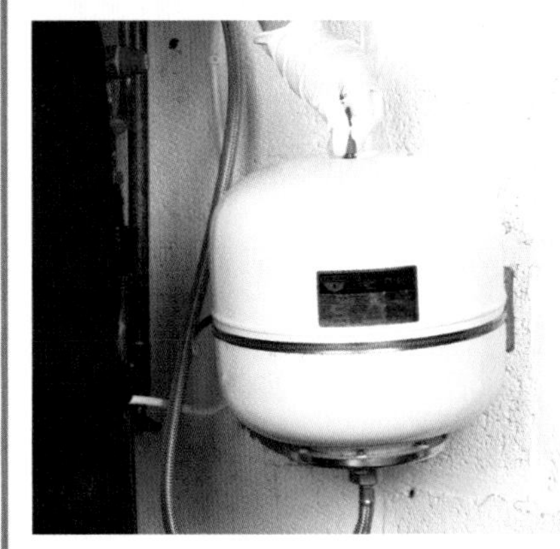

Step 1: Remove cap to Schrader valve.

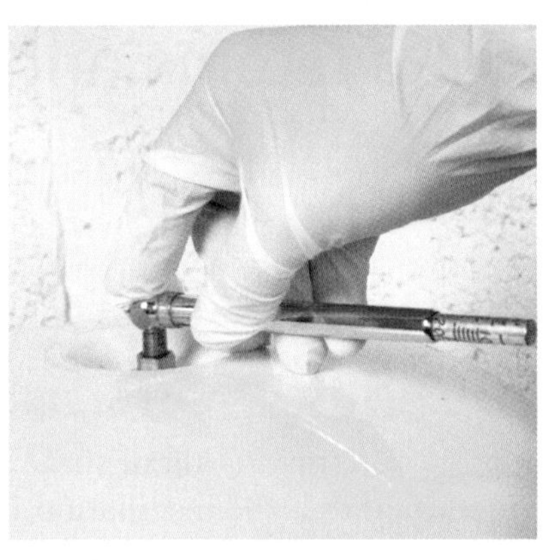

Step 2: Press tyre pressure gauge onto Schrader valve and take pressure.

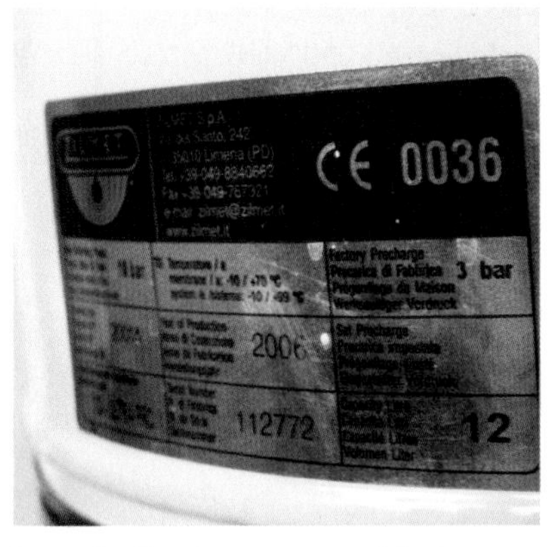

Step 3: Check pressure with data plate on expansion vessel. If less, will need recharging.

Step 4: Isolate water supplies and release pressure from the hot and cold pipework.

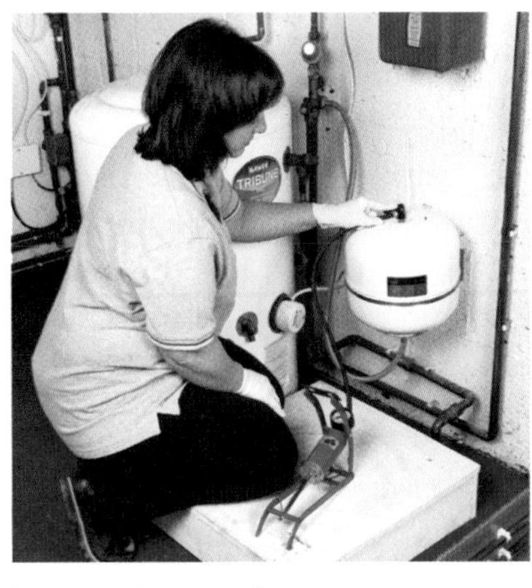

Step 5: Connect foot pump to Schrader valve.

Step 6: Pump up to pressure that is shown on data plate.

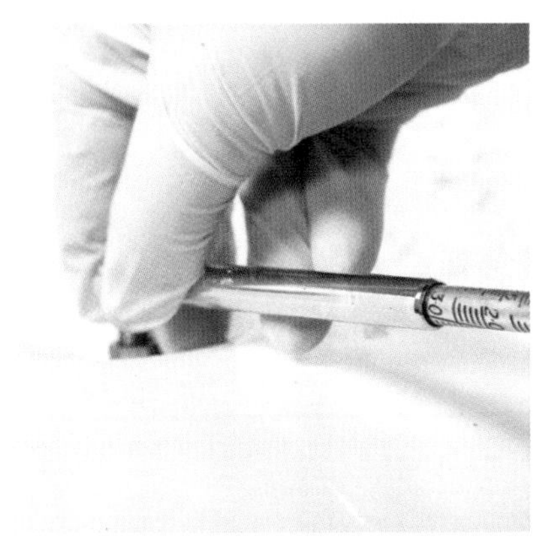

Step 7: Check again with tyre pressure gauge.

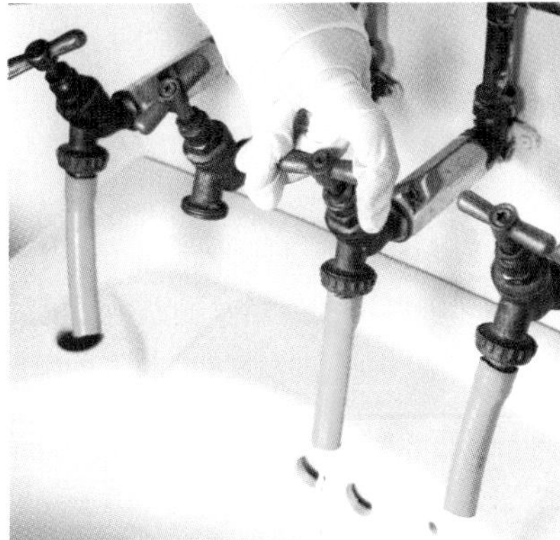

Step 8: Turn off outlets and re-energise system. Make sure that the hot water supply flows from taps.

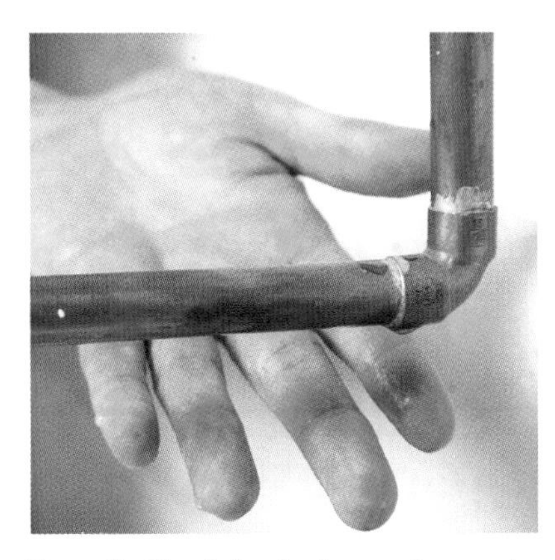

Step 9: Check for leaks to pipework.

Step 10: Fill out service sheet on completion.

Remember

When commissioning, you will need to consult a range of documents and legislation, including:

- job specification
- British Standards
- Building Regulations
- manufacturer's instructions
- building services drawings.

5. Know and be able to apply the commissioning requirements of hot water systems and components

This outcome has largely been covered in Unit 3. Look back at pages 145–156 before reading on.

Defects found during commissioning

During the commissioning of some systems or components you may have identified that the installation is not correct or you are unable to achieve the correct design output for a given appliance. This could be due to poor design and calculations and/or pre-system checks not being carried out by the installer, particularly with unvented hot water systems.

Defective components

If you find a defective component the first thing to check is, has it been damaged during the installation stage? If so, it will be the responsibility of your company to replace the defective part. If you can prove that it is the manufacturers' responsibility, the manufacturer or supplier must be contacted. As the units are new they will either:

- send out one of their own engineers
- pay your company to replace the part and the manufacturer supplies a replacement part.

System defects

Professional Practice

Lucas has been given the task of commissioning the unvented hot water system that the company he works for had previously installed. During the commissioning he finds that the water supply is insufficient for the system to function correctly. Discuss with your classmates:

- What action should Lucas take regarding the water supply?
- Who should he inform regarding the fault?
- What remedial action should Lucas take to overcome the problem?

Sample commissioning record for an unvented hot water cylinder

Commissioning Form

Form completed by: ____________________

Date: ____________________ Signature: ____________________

Installation address: ____________________

System details: ____________________
(type, manufacturer, model, location in property, etc.)

Cold water supply pressure: ____________ Hot water supply pressure: ____________

Cold water flow rate: ____________ Hot water flow rate: ____________

Manufacturer's recommended minimum supply pressure: ________ Manufacturer's min flow rate: ________

Item	Fitted/connected/completed?	Fitted correctly?
Servicing valve	❑ Yes ❑ No	❑ Yes ❑ No
Line strainer	❑ Yes ❑ No	❑ Yes ❑ No
Pressure-reducing valve	❑ Yes ❑ No	❑ Yes ❑ No
Tee for balanced cold water draw offs	❑ Yes ❑ No	❑ Yes ❑ No
Single check valve	❑ Yes ❑ No	❑ Yes ❑ No
Expansion relief valve	❑ Yes ❑ No	❑ Yes ❑ No
Provision for expansion	❑ Yes ❑ No	❑ Yes ❑ No
Temperature- and pressure-relief valve	❑ Yes ❑ No	❑ Yes ❑ No
Direct-fired thermostat and energy cut-out	❑ Yes ❑ No ❑ N/A	❑ Yes ❑ No ❑ N/A
Indirectly-fired thermostat, energy cut-out and zone valve	❑ Yes ❑ No ❑ N/A	❑ Yes ❑ No ❑ N/A
'Competent' installer's details displayed on vessel	❑ Yes ❑ No	
Is unit or package certified, e.g. BBA, WRC or equivalent?	❑ Yes ❑ No	
Are all hot and cold supplies balanced?	❑ Yes ❑ No	

Discharge pipework	Correct?
Metal	❑ Yes ❑ No
Correctly sized upstream of tundish?	❑ Yes ❑ No
Correctly sized downstream of tundish?	❑ Yes ❑ No
Tundish within 600 mm of temperature-relief valve	❑ Yes ❑ No
No bend within 300 mm of tundish	❑ Yes ❑ No
Correctly terminated	❑ Yes ❑ No

Item	Defects/comments

Figure 4.30: Example of a commissioning record

PROGRESS CHECK

1. What would be used to take the pressure and flow readings of a hot water system?
2. When disinfecting a system in a small hotel, who should be notified?
3. What does commissioning of a hot water system include?
4. State the procedure when testing for leaks.
5. State the information that should be sent to the water undertaker.

You will carry out practical tasks to learn how to commission hot water systems and components in your workplace. Find out more on *Level 3 NVQ Diploma Plumbing Tutor Resource Disk.*

Check your knowledge

1. By how much does water usually expand when heated under normal operating conditions?
   - a 4%
   - b 10%
   - c 15%
   - d 20%

2. Unvented hot water systems can be purchased in both ____________ and in ____________ form. Which of the following are the two matching words that complete the statement?
   - a sealed, vented
   - b assembled, self assembly
   - c packaged, unit
   - d pressurised, unpressurised

3. On an unvented domestic hot water system a pressure-reducing valve should be fitted:
   - a on the hot distributing pipe outlet
   - b on the incoming cold water supply
   - c on the heating coil primary flow
   - d on the expansion vessel

4. Which of the following would indicate that the air gap has been lost on an unvented domestic hot water storage system with a bubble top storage vessel?
   - a Discharge of high pressure from the temperature relief valve
   - b Discharge from the expansion relief valve is intermittent
   - c Discharge from the temperature relief valve is intermittent
   - d Discharge of high pressure from the pressure-relief valve

5. In an unvented hot water system, which one of the following system components is installed after the stop valve but before the pressure reducing valve?
   - a Expansion vessel
   - b Expansion valve
   - c Line strainer
   - d Check valve

6. Building Regulations Part G3 and The Water Supply (Water Fittings) Regulations 1999 with unvented domestic hot water storage systems does not apply to:
   - a instantaneous water heaters
   - b instantaneous water heaters with a storage capacity of 25 litres
   - c systems with a storage capacity of more than 50 litres
   - d systems with a storage capacity of 15 litres or less

7. What component prevents potential backflow into the supply pipe in an unvented system?
   - a Service valve
   - b Double check valve
   - c Single check valve
   - d Expansion valve

8. What is the maximum length of the D1 discharge pipe?
   - a 300 mm
   - b 500 mm
   - c 600 mm
   - d 700 mm

9. The high limit stat on a storage vessel of an indirectly heated unvented domestic hot water storage system activates the ____________ to shut down the ____________ heat supply to the storage vessel. Which of the following are the two matching words that complete the statement?
   - a motorised valve, primary
   - b temperature-relief valve, discharge water
   - c tundish, temperature-relief valve
   - d expansion valve, discharge valve

10. A high limit thermostat should operate at which one of the following temperatures?
    - a 65°C
    - b 70°C
    - c 85°C
    - d 95°C

11. The Water Regulations stipulate that the discharge pipe from an expansion valve should pass through a visible tundish with a type:
    - a AUK 1 air gap
    - b AUK 2 air gap
    - c AUK 3 air gap
    - d AA air gap

12. When ordering an unvented unit, how would it be assembled prior to delivery?
    - a Only the pressure control valve is assembled by the manufacturer
    - b All control devices are factory assembled by manufacturer
    - c Only the expansion vessel is assembled by the manufacturer
    - d None of the control devices is assembled by the manufacturer

Preparation for assessment

Now that you have completed all of your learning outcomes it is time to prepare for assessment.

Use the following revision strategy for the underpinning knowledge test:

- Do not try to take in all of the information in one go, take your time and do a little each day.
- Reread Unit 4 *Understand and apply domestic hot water system installation, commissioning, service and maintenance.*
- Go through the notes you have taken over the weeks of teaching.
- Go over any handouts from your tutor.
- Research more information about anything which you are not sure of.
- Check any queries with your tutor.
- Do the practice questions in your book.
- Check your answers with your tutor.

You will also have to take practical assessments. In these assessments you will have to:

- install an unvented hot water cylinder, to current regulations and standards (this also includes the installation of primary flow and return and the installation of a two-port valve)
- commission an unvented hot water cylinder
- fault find on an unvented hot water system
- fault find on a boosted shower pump

Follow the practical sheets which are on the Training Resource Disk.

As part of the qualification you will also be undertaking a design assignment for hot water in a single occupancy dwelling. This will include being able to:

- use information sources when designing hot water systems
- calculate the size of hot water system components:
  - cistern
  - cylinder
  - pipework
  - secondary circulation pump
  - booster pump (Shower and full system)
- present the design calculations in an acceptable format:
  - a single line drawing, not to scale
  - details for insertion into a quotation or tender for work.

This assessment is in assignment format and must be completed as part of your unit award. On completion of this your course tutor will mark your work.

Good luck and we hope you do well.

5 Understand and apply domestic sanitation system installation, commissioning, service and maintenance techniques

This unit builds on the knowledge and understanding of sanitation systems covered at Level 2. It covers the advanced knowledge you will need to work on sanitation systems and layouts and design techniques. It will help you to apply this to systems, installation requirements and fault diagnostics and rectification. You will also cover the commissioning requirements of pipework systems and components.

This unit covers the following learning outcomes:

- know the types of sanitation system and their layout requirements
- know and be able to apply the design techniques for sanitation and rainwater systems
- understand the installation requirements of sanitation system components
- know and be able to apply the fault diagnosis and rectification procedures for sanitary pipework systems and components
- know and be able to apply the commissioning requirements of sanitary pipework systems and components.

1. Know the types of sanitation system and their layout requirements

Legislation and system components

There are four types of stack system, with specific legal requirements, for their installation:

- primary ventilated stack system
- ventilated discharge branch system
- secondary modified ventilated stack system
- stub stacks.

Primary ventilated stack system

The primary ventilated stack system is commonly used in domestic properties. Mistakes are often made in its design, with pipe runs being too long or pipe falls being too shallow or too steep.

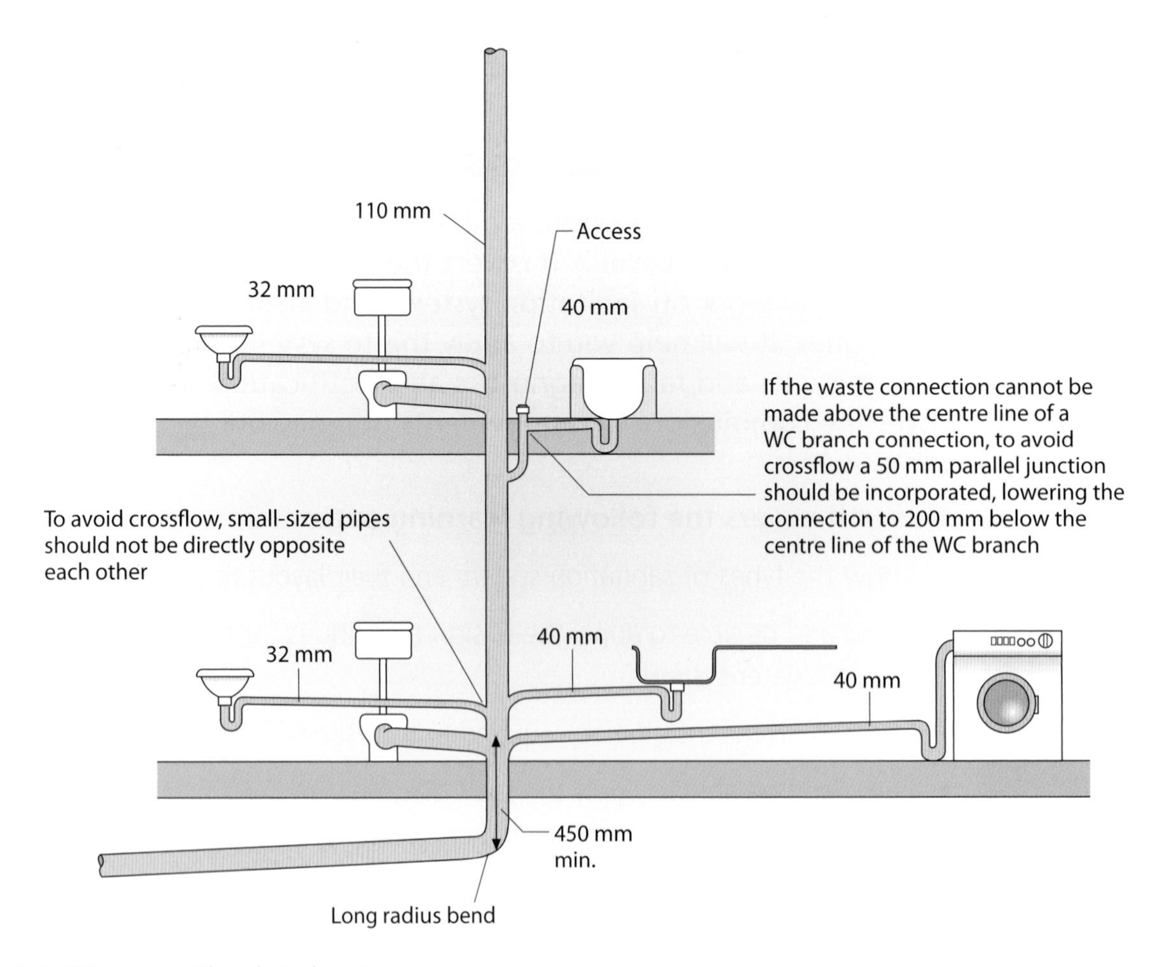

Figure 5.1: Primary ventilated stack system

The Regulations lay down specific requirements for the connection of branches into soil stacks.

Pipe size	Maximum length	Slope
32 mm	1.7 m	See design curve (Figure 5.2)
40 mm	3 m	18–90 mm/metre
50 mm	4 m	18–90 mm/metre
WC	6 m	18 mm/metre minimum

Table 5.1: Design specifications for single branch connections

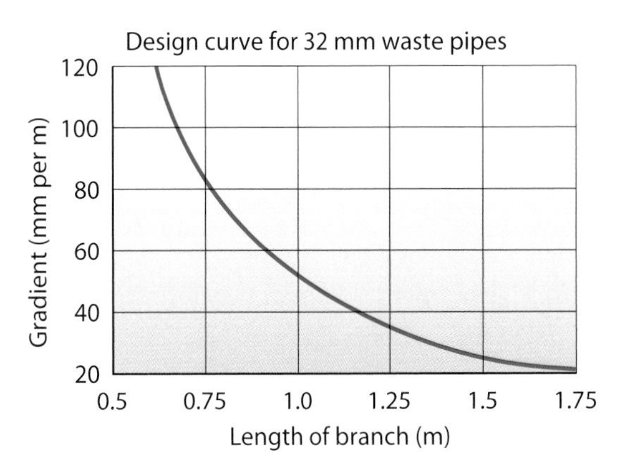

Figure 5.2: Design curve for 32 mm waste pipes

Activity

What is the likely outcome of pipework that is laid with a fall that is too shallow?

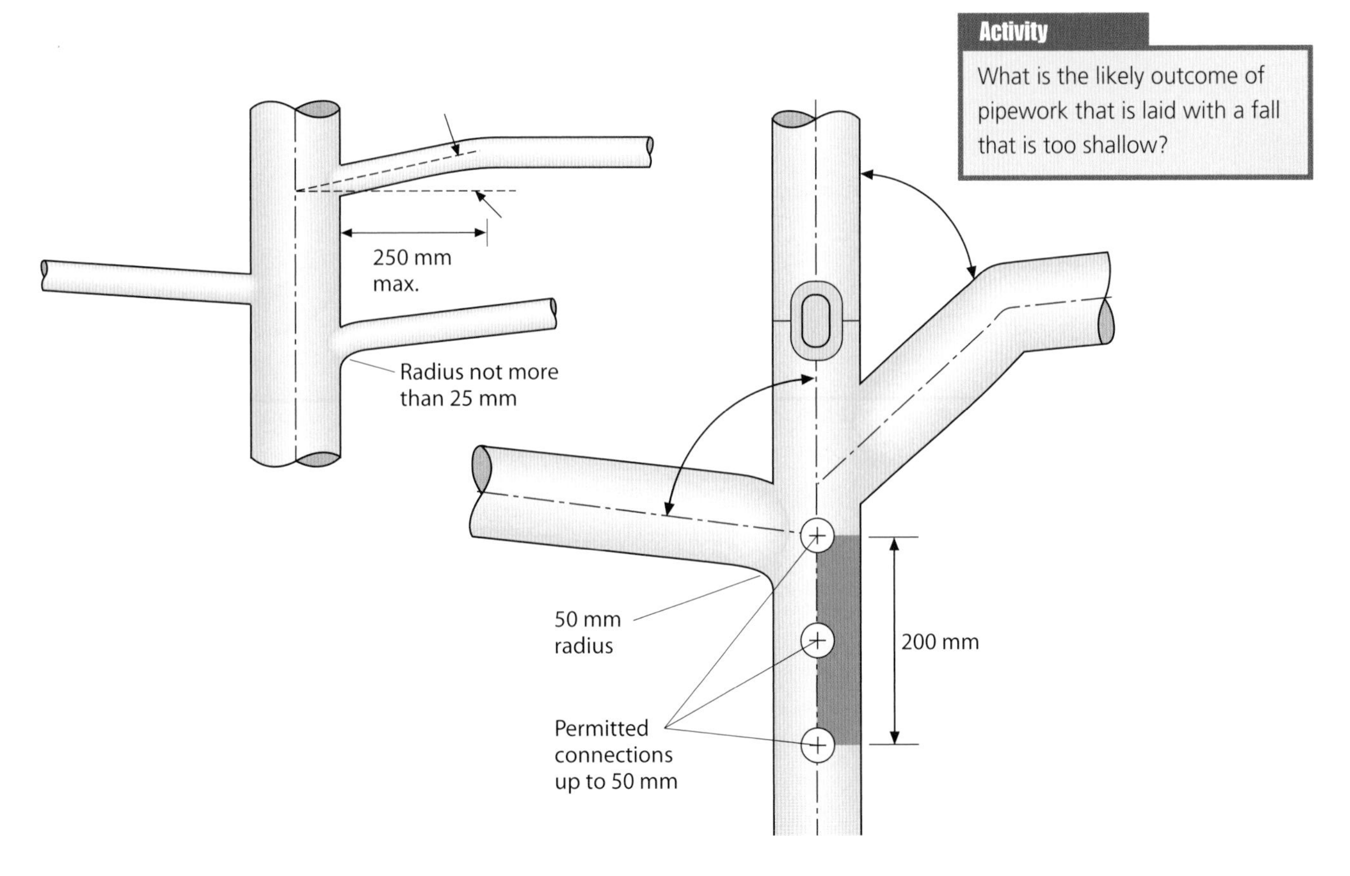

Figure 5.3: Branch connections for waste pipes. For the top diagram, junctions, including branch pipe connections of less than 75 mm, should be made at a 45° angle or with a 25 mm bend radius. The bottom diagram shows the prohibited zone distance (opposite the WC connection) in which a branch pipe may not be connected into within a distance of 200 mm. Branch connection pipes of over 75 mm diameter must either connect to the stack at a 45° angle or with a minimum bend radius of 50 mm.

Foot of the stack

The Regulations state that the lowest point of connection into the stack should not be within 450 mm of the invert of the drain, and the bend at the foot of the stack should have as large a radius as possible, but no less than 200 mm.

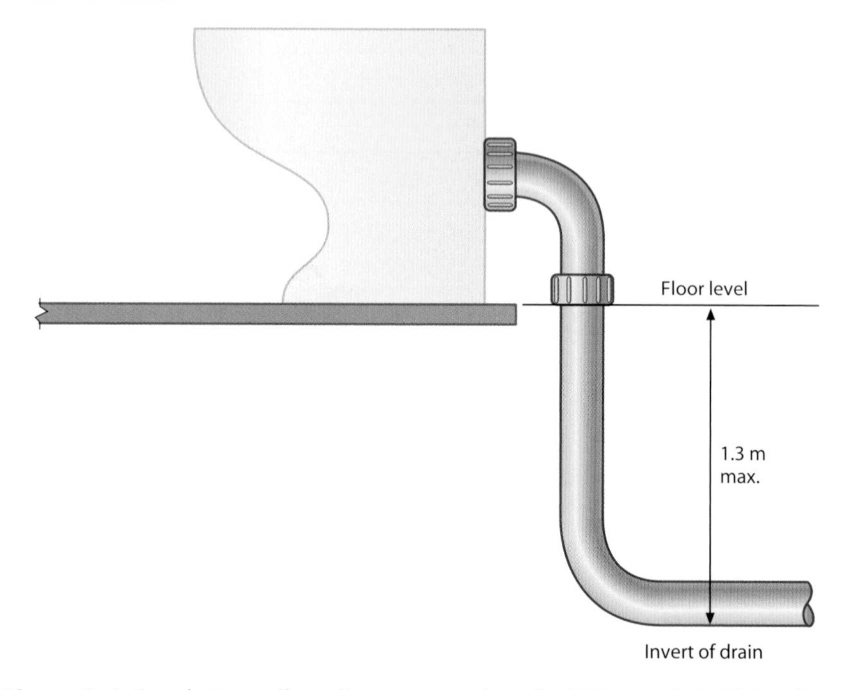

Figure 5.4: Regulations allow direct connection of a WC to a drain if the distance of the drain invert to floor level is 1.3 m or less

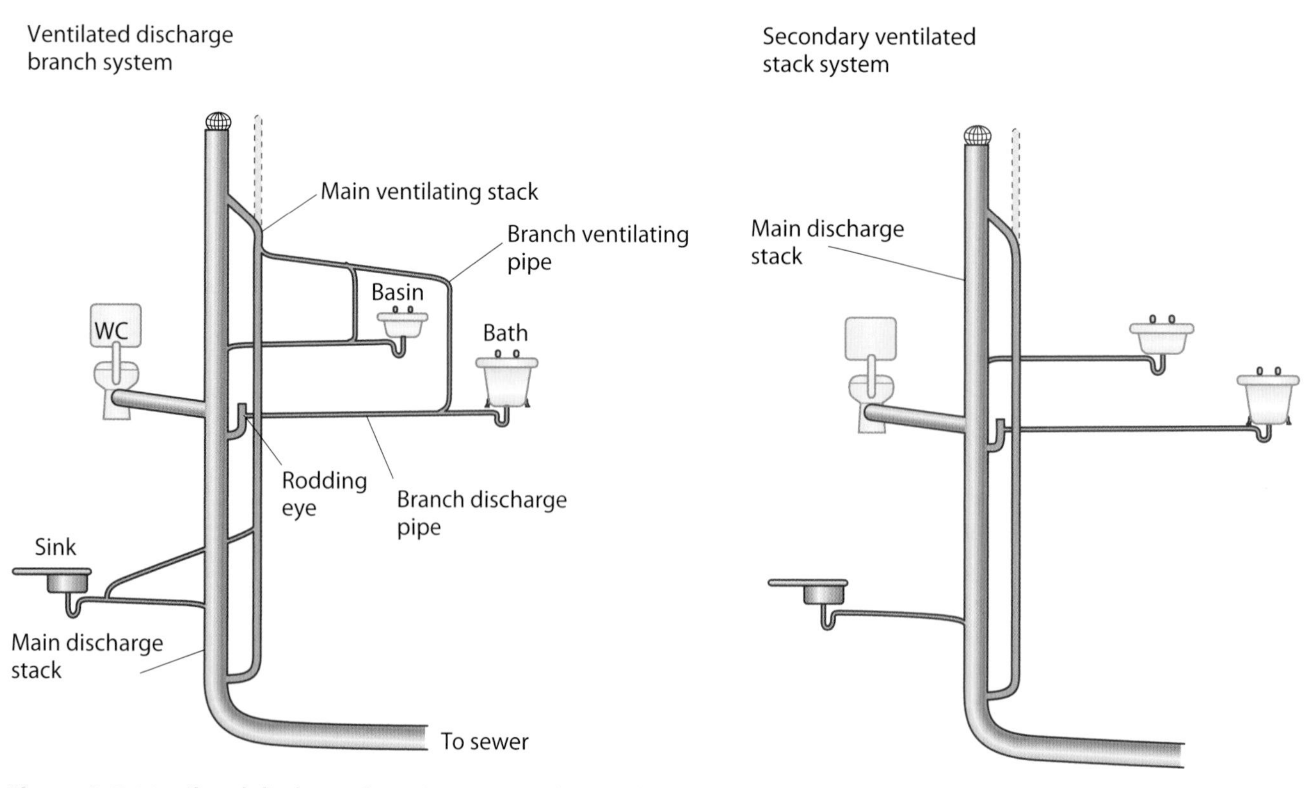

Figure 5.5: Ventilated discharge branch system and secondary stack system

Ventilated discharge branch system and secondary system

These two stack systems are mostly found in older, larger buildings. For more detail, see *Level 2 NVQ Diploma Plumbing*.

Stub stack

With stub stacks you usually see a short length of pipe rising to above floor level in the room, which is capped. There is no ventilating pipework to the system, but the drain to which it is connected must be ventilated. Stub stacks are usually found in a second bathroom where the other bathroom is at the head of the drain and has a properly ventilated stack. There are specific requirements for the connections into the stack given in Figure 5.6 below.

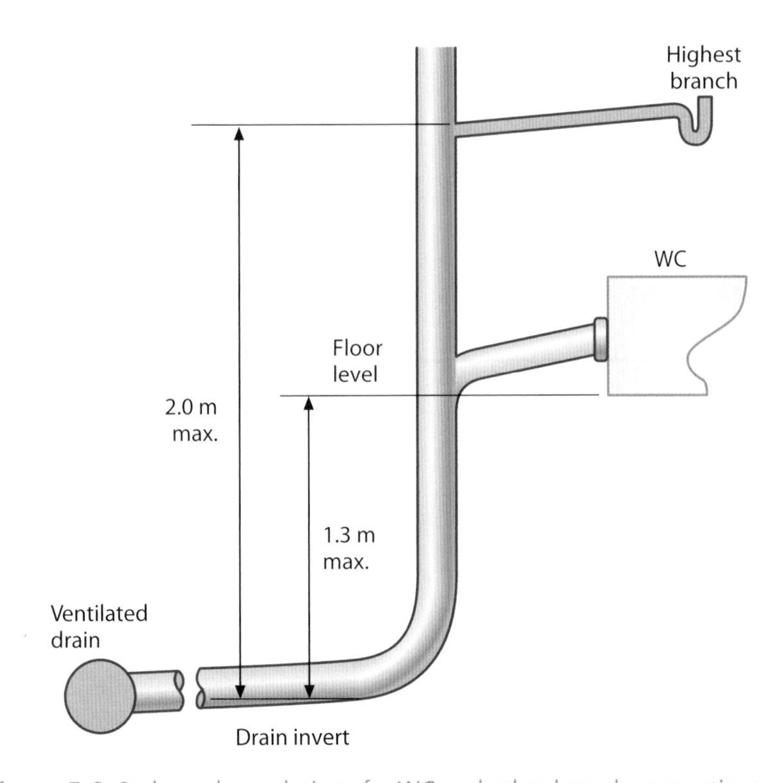

Figure 5.6: Stub stack regulations for WC and other branch connections

Air-admittance valves

Using an air-admittance valve on a primary ventilated stack system removes the need for a stack ventilating pipe. It gives greater flexibility because there are no restrictions on the height of connections above drain invert level.

An air-admittance valve removes the need for a stack ventilating pipe

Remember

The head of the drain (the highest point) should always be ventilated.

There are a number of key requirements for installing an air-admittance valve:

- it can only be used inside a building
- it should not adversely affect the amount of air needed for the below-ground drainage system to work
- it should terminate above the highest water level in the system
- it should be placed in a position where air is easily available at the inlet.

Air-admittance valves are now commonly used because they save costs: you do not have to take the ventilating pipe through the roof. In an unheated area, they may require insulating to prevent them from freezing and malfunctioning. However, you cannot fit them to every property on a site, as the below-ground drainage system needs to ventilate itself to the atmosphere to work correctly.

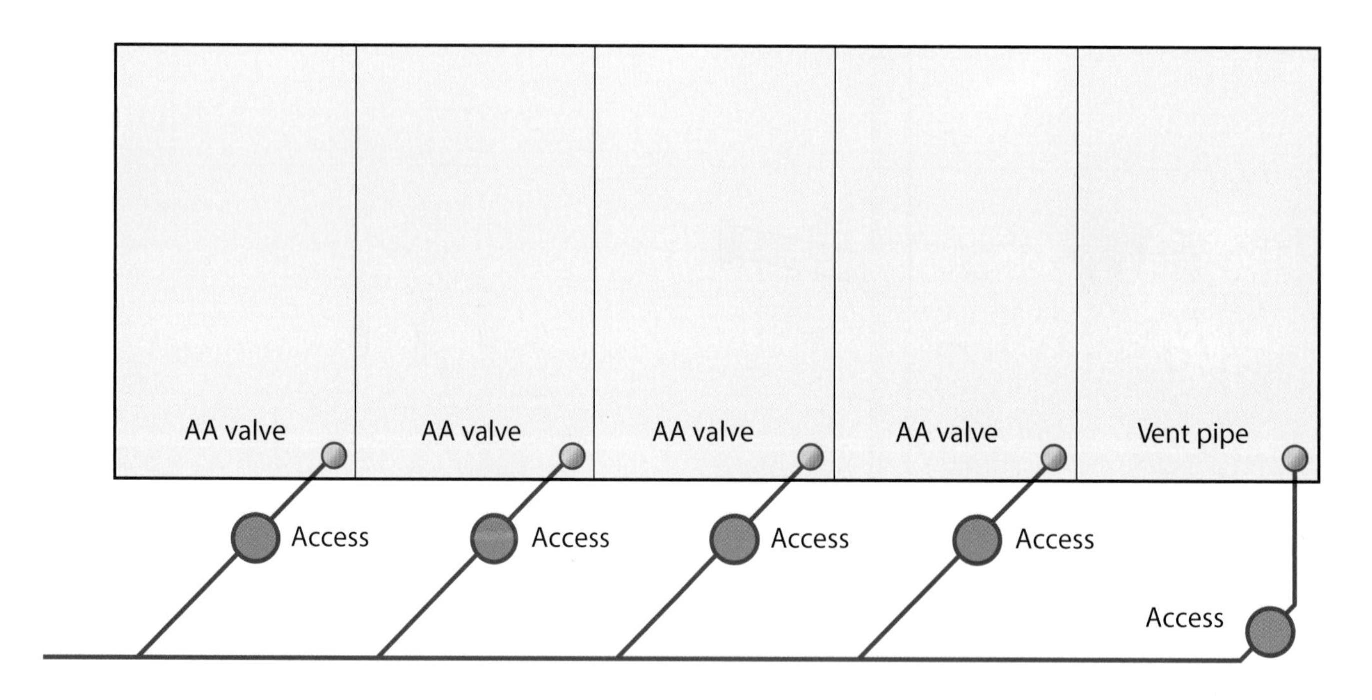

Figure 5.7: A typical layout for ventilating pipes to properties. In this example, every fifth property on a drainage run has a ventilating pipe; other properties are served by air-admittance valves

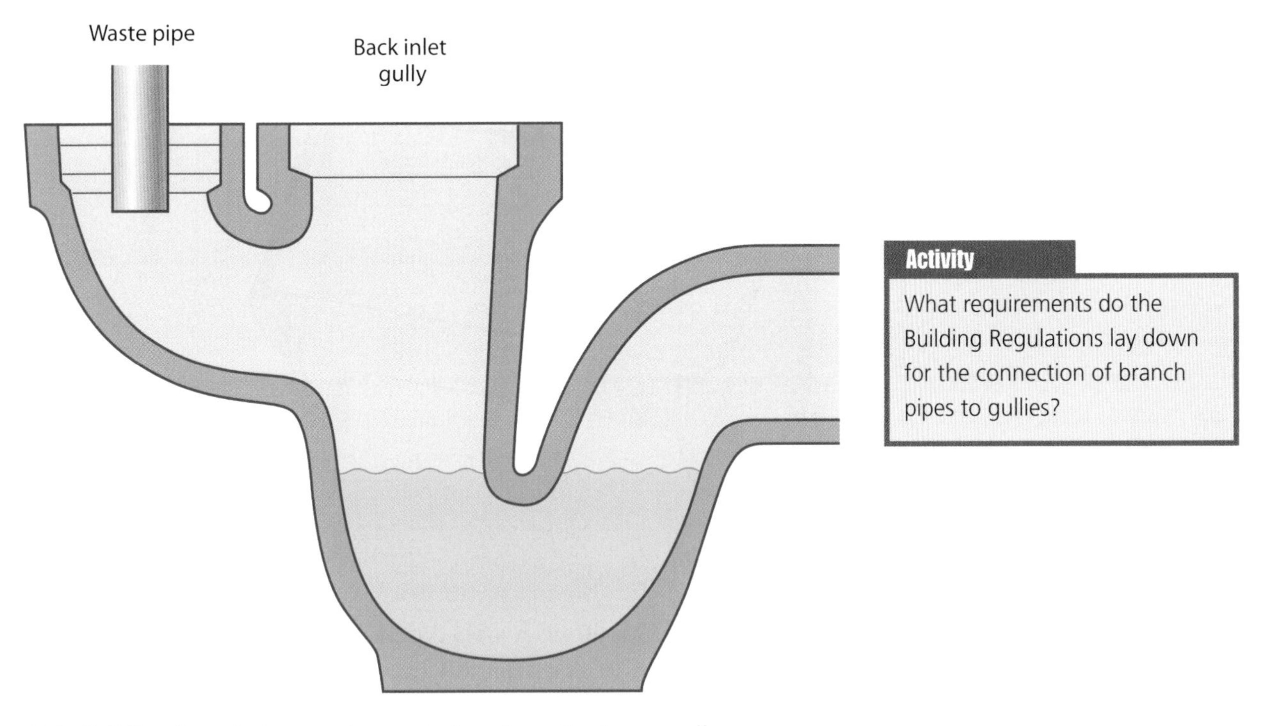

Figure 5.8: Building Regulation requirements for a branch pipe to gully

Activity

What requirements do the Building Regulations lay down for the connection of branch pipes to gullies?

Urinals

Although urinals are not used in domestic properties, you are likely to install them for small businesses. Figure 5.9 shows a bowl-type urinal range, which is most commonly used. It is directly connected by a branch pipe system to a discharge stack. The outlets are trapped in much the same way as for a basin (32 mm is the minimum size). For adults, the front lip of the bowl is placed about 600 mm above floor level; for children, it will be lower and will depend on the children's age.

Figure 5.10 on page 220 shows a trough-type urinal. It is used in toilets where there may be a high risk of vandalism. The trough is sized for the maximum number of people that are going to use it and can be made in various lengths. The outlet is put at one end of the trough, which has a slight fall across the base. It then connects to a discharge stack, usually via branch pipework. The size of the branch pipework and trap are determined by the size of the trough and the distance of the branch pipe to the discharge stack.

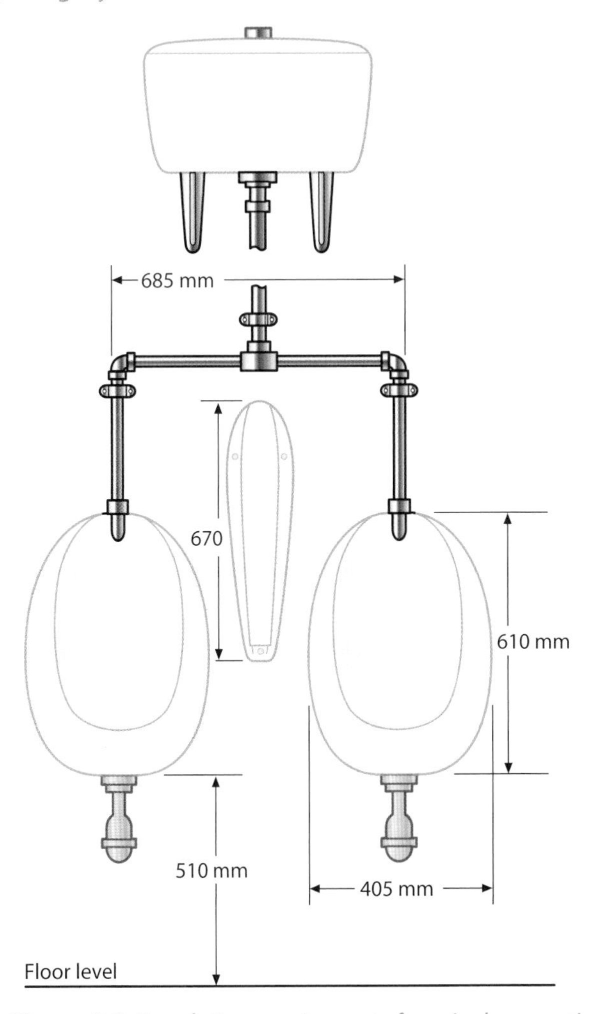

Figure 5.9: Regulation requirements for urinal connections

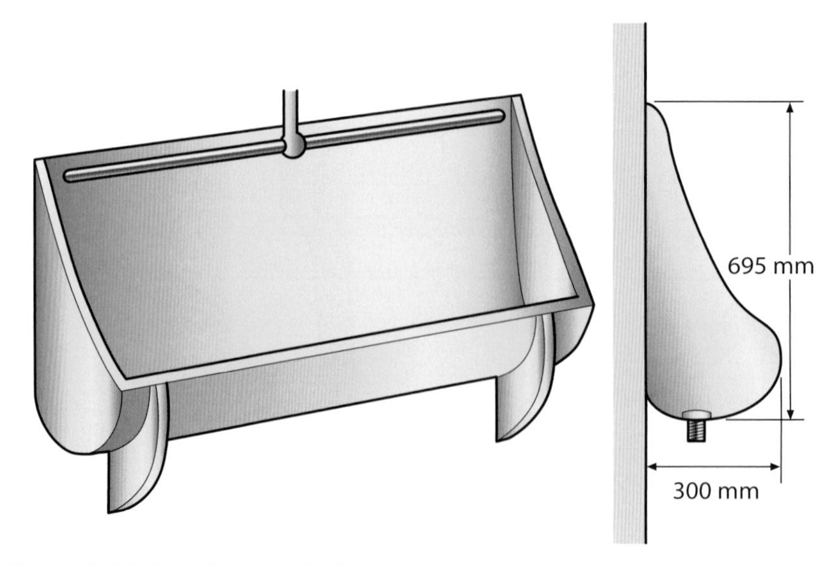

Figure 5.10: Trough-type urinal

Figure 5.11 below shows a slab-type urinal. This urinal is supplied in a number of pieces to assemble on site. If it is on the ground floor of the property, the waste connection is made directly to the drainage system via a trapped gully. If it is on the first floor or above, the connection is made to a discharge pipe system (also trapped).

Figure 5.12 opposite shows an example of a slab urinal manufactured in one piece (and usually not able to accommodate more than two people).

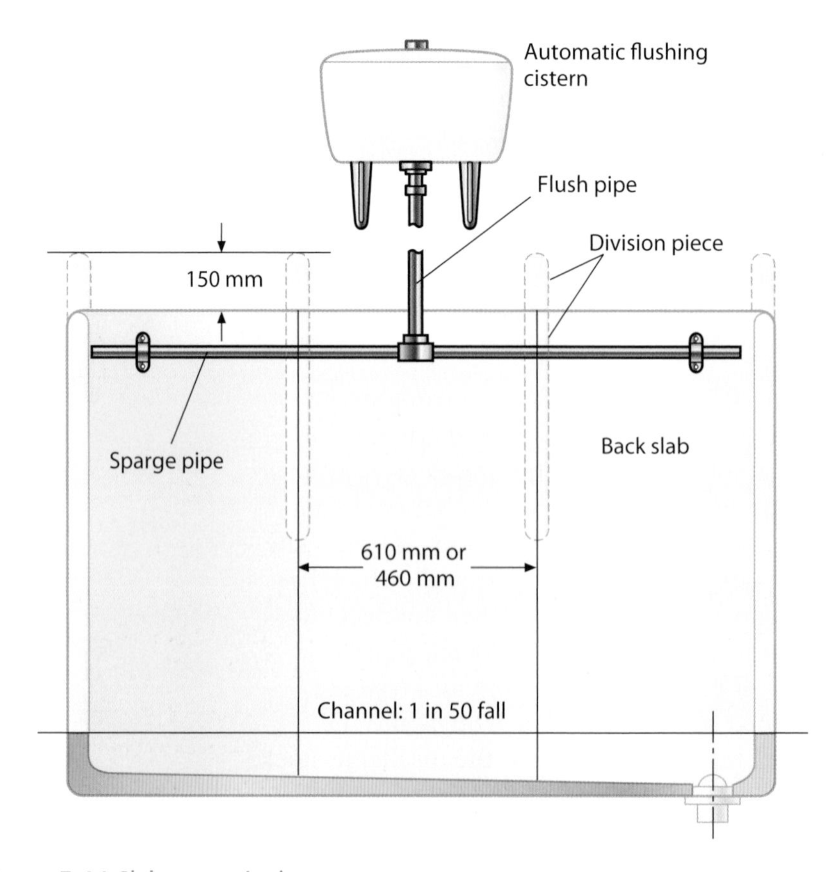

Figure 5.11 Slab-type urinal

It is connected in a similar manner to the slab mentioned previously.

Waterless urinals are now becoming more common. They use a self-cleansing surface coupled with a cartridge where the trap would be, so there is no need for a water supply. The cartridge prevents smells and crystal formation in the waste pipe and has to be replaced at regular intervals.

Figure 5.13 shows an example of a clay trap provided to connect a urinal straight to the drainage system on the ground floor of a property.

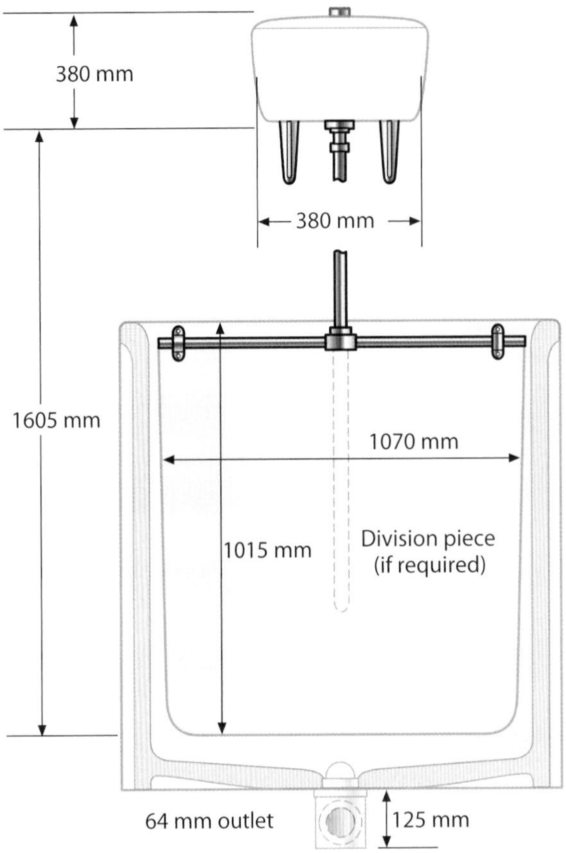

Figure 5.12: One-piece slab urinal

Did you know?

A waterless urinal saves on average 90 000 litres of water a year.

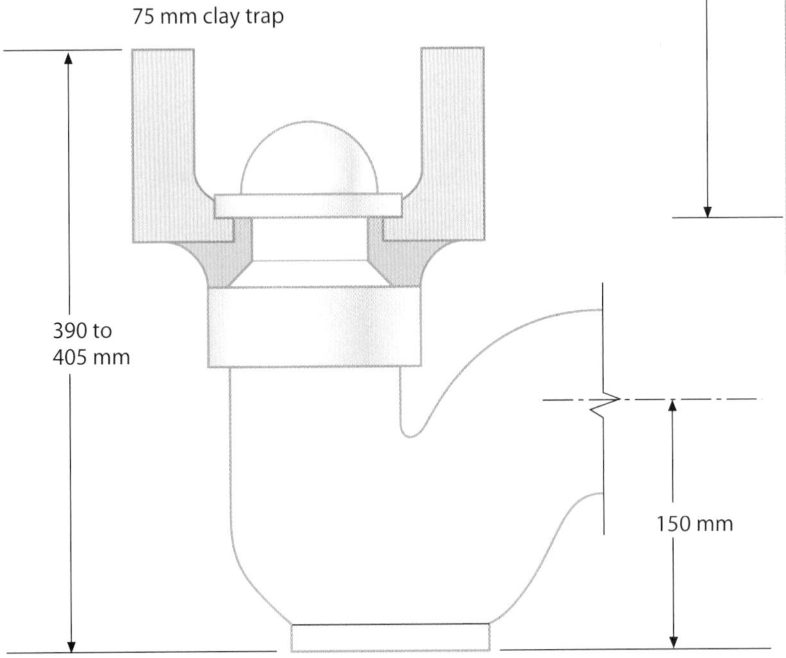

Figure 5.13: Clay trap (illustration purposes only, not to scale)

Valves as an alternative to traps

The Hep$_v$O valve (made by Hepworth Building Products) can be used as an alternative to the trap. Rather than using a water seal, the Hep$_v$O valve uses a tough, collapsible membrane to make the seal.

When water is discharged down the valve, the membrane is in the open position. Once the water has discharged, the membrane returns to its normal state, making an airtight seal. Any back pressure on the system forces the membrane into a closed state and no water can be discharged back into the appliance. The valve therefore overcomes the effects of any form of siphonage. The Hepworth valve has no trap seal to be lost, so problems from back pressure do not exist.

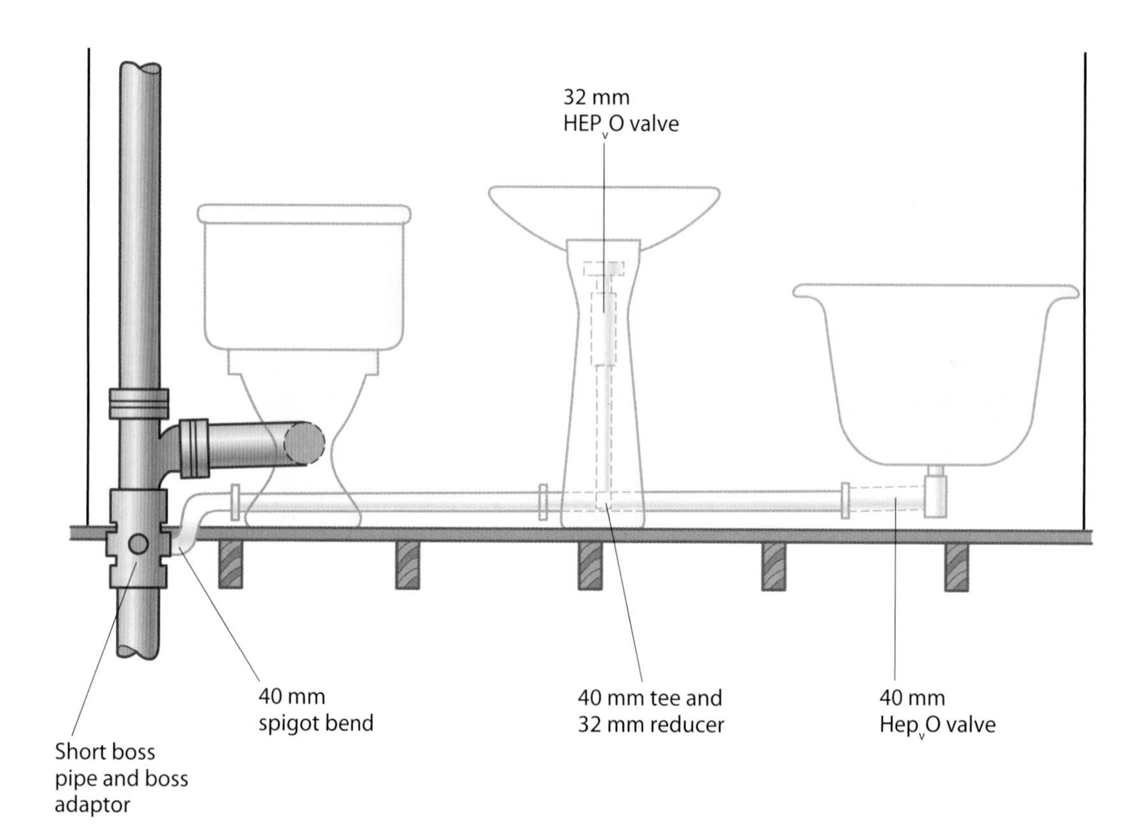

Figure 5.14: A typical example of a Hep$_v$O system with combined waste to bath and basin. The basin discharges vertically to branch into the waste pipe. The collection of two appliances (or more) would normally be unacceptable with a standard trapped system, due to the possibility of induced siphonage

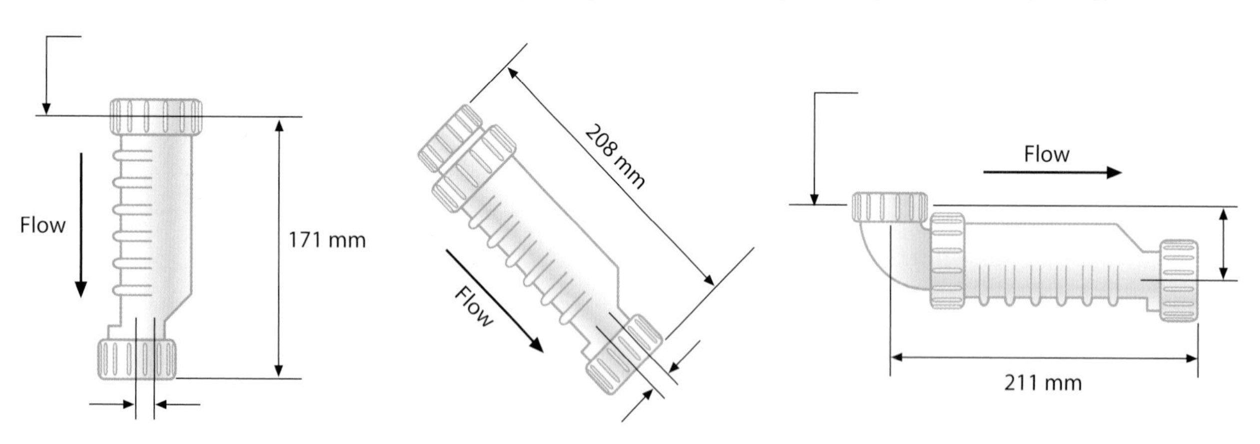

Figure 5.15: Hep$_v$O valve dimensions. The valve can be positioned at different angles and comes with adaptors so that it can be put horizontally, under the bath, or fixed in-line

Did you know?

The Hep$_v$O valve can fix problems that occur with standard traps, such as excessively long pipe runs, incorrect pipe falls, incorrect trap seal depth and problems with combined wastes.

Sanitary facilities for people with disabilities

When looking at requirements for sanitation for people with disabilities, you need to consider what is needed in terms of personal space and in terms of the appliances that you use.

It is important to think about the spacing requirement if the householder is to use the facilities properly. Be aware that body size will have an impact on the use of the sanitary equipment.

The Building Regulations Approved Document M lays down the requirements for sanitary accommodation for people with disabilities. It details the size of the accommodation and the layout and positioning of the WC. The sanitary accommodation provided will depend on whether it is needed for people who can walk, or for people who are wheelchair users. Wheelchair access requires larger areas for turning.

Disabled toilets are designed for specific access requirements

In accordance with Part M, all new domestic builds must have a WC facility on the ground floor. The residents should be able to gain access to the toilet from any habitable room on that floor, without having to go upstairs. The entrance to the property and the toilet must provide access for wheelchair users without any difficulties. The toilet door must open outwards and there should be clear access to the WC. Recommended spacing requirements for sanitary equipment are illustrated in Figure 5.16. Further details of disabled access and disabled toilet facilities can be found in the Building Regulations Part M.

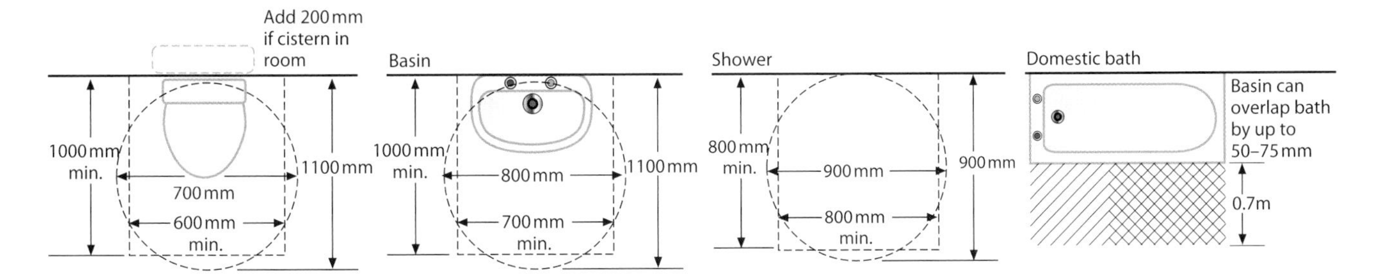

Figure 5.16: Recommended spacing requirements of sanitary equipment for average-sized people

Professional Practice

Lukasz is working on a new extension to a local restaurant. He is asked to install a new toilet facility for disabled people/wheelchair users. Check through Part M of the Building Regulations to see what equipment the new facilities would require.

1 Draw a sketch to show what the new facilities would look like.
2 Include the size of the facility required.

Did you know?

Part F of the Building Regulations details the requirements for ventilation in bathrooms. Installing ventilation will not usually be part of a plumber's job, but you may be asked if ventilation meets Building Regulations requirements.

Key term

Leaching – The movement of contaminants, such as water, soluble pesticides or fertilisers, carried by water downward through permeable soils.

Foul tanks in sanitation systems

Foul tanks come in one of two forms: cesspits or septic tanks. Foul tanks must work effectively and safely to prevent any risk to users from contaminated water, directly or via any **leaching** into the environment. To ensure this, for each type you need to know how to size it correctly, and how to maintain it and keep it clean.

Cesspits

A cesspit is an underground storage tank for effluent when the building is too far away from the main drain to connect to it. The effluent will need to be treated at the nearest sewage works and has to be emptied at regular intervals using specially designed effluent waste tankers.

Cesspits are designed in the form of a sealed tank, which takes all of the property's liquid waste, including waste from toilets, sinks, baths, showers and washing machines. Cesspits are completely sealed apart from the manhole and vent. They need to be emptied regularly, as calculated by the design engineer, to avoid problems.

Cesspits are usually installed in preference to septic tanks because of ground conditions or to remove the potential of pollution associated with septic tanks. Cesspits need to have a capacity below the inlet pipe level of at least 18,000 litres (18 m^3) for two users; this is increased by 6800 litres (6.8 m^3) per additional user.

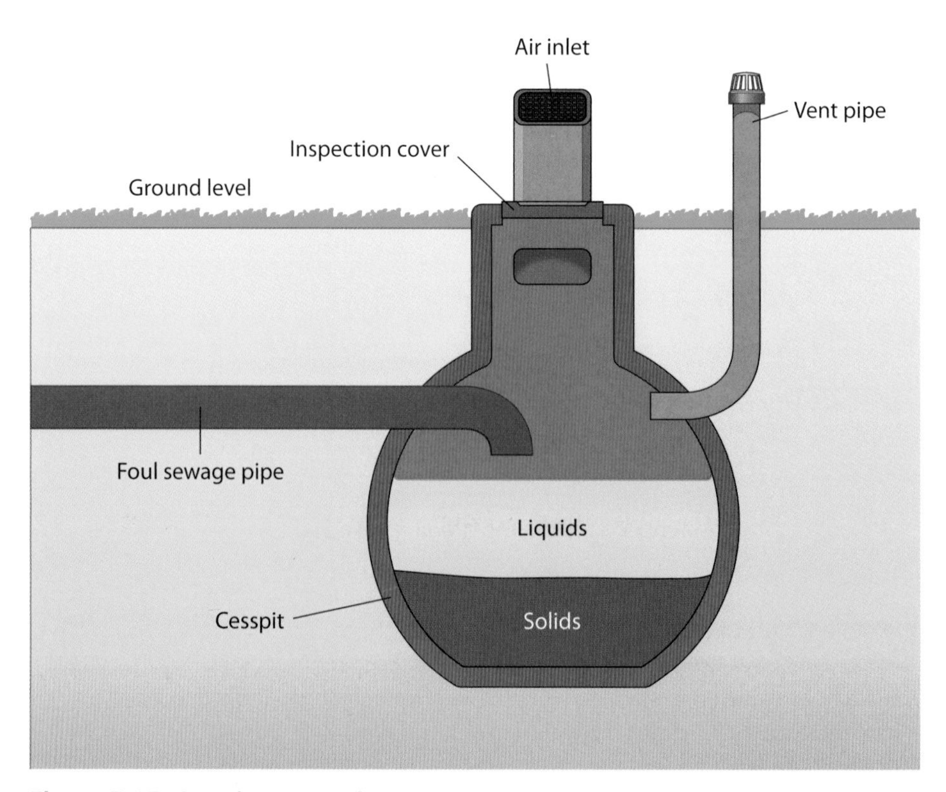

Figure 5.17: Cesspit construction

Worked example

A family of four people would require a capacity of:

$18\ m^3 + 2 \times 6.8\ m^3 = 31.6\ m^3$ tank capacity

The cesspit construction should not allow any escape of contents into the surrounding ground, and should be built to comply with BS EN 12566-1.

BS EN 12566-1

Maintenance and cleaning

The tank access cover should be durable and lockable to prevent accidental access to the chamber, so that it does not overflow and pollute the surrounding ground.

The cesspit should have a notice identifying it and the necessary maintenance required; this should be positioned within the building it serves. The cesspit should be inspected fortnightly for overflow. The owner is legally responsible for ensuring the cesspit does not pollute or create a health hazard, and could be fined if they break the law.

Septic tanks

Septic tanks are a method of sewage disposal that only needs emptying once a year. However their design is more complicated than that of a cesspit. Septic tanks have to be used in conjunction with a secondary treatment, such as a drainage mound (see page 227).

The positioning of the tank is critical to the overall success of the treatment of the effluent. The tank should be at least 7 m from the building it serves but should be within 30 m of a vehicle access point; the invert level of the tank needs to be 3 m below the vehicle access point. The hose from the tanker to the septic tank should take a route outside of any buildings or place of work and without any hazards being created. The septic tank capacity must be at least 2700 litres ($2.7m^3$) for up to four users, and an additional 180 litres for each additional user.

Worked example

A family of five people would require a capacity of:

2700 litres + 180 litres = 2880 litre capacity

Did you know?

Lightweight tanks can float above ground level if the water table is high, so they may have to be anchored down.

This is a lot less than for a cesspit: the effluent is not stored for long because it is treated and released into the environment. A septic tank should be built to comply with BS EN 12566-1. Particular care has to be taken to ensure that any septic tank remains stable throughout its life. It can be made from glass-reinforced plastic, polyethylene, steel or brick.

BS EN 12566-1

The drain inlet should be constructed to prevent disturbance of the surface scum or settled sludge, and should be made of two chambers (see Figure 5.18 on page 226). On entry to the tank, the effluent settles so that solid heavy material falls to the bottom: this is called the sludge, which will have to be removed once a year. The lighter particles of matter float in the liquid above the sludge and are broken

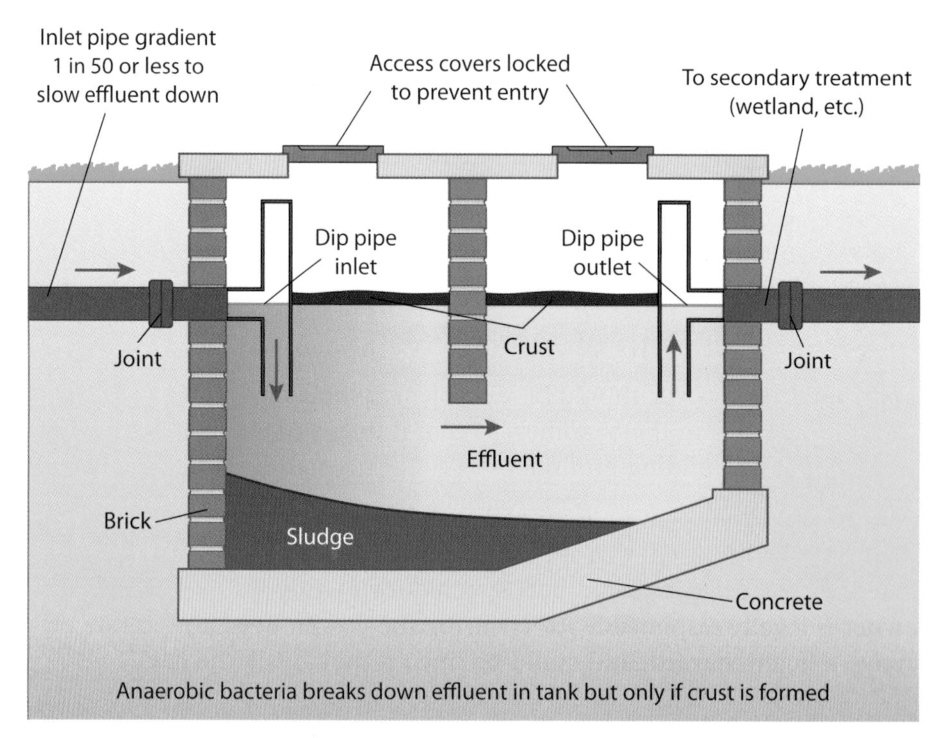

Figure 5.18: Brick-built septic tank

down by bacteria below the crust that forms on the top of the effluent. The final stages produce a clear effluent that can be discharged to the environment where it can act as nutrient for plants.

Maintenance and cleaning

The domestic septic tank requires monthly inspections of the effluent to check that it is free-flowing and clear, and it should be emptied every 6-12 months by a specialist contractor. A commercial septic tank requires emptying four times a year. Always refer to the manufacturer's instructions. As with cesspits, a notice should be provided within the building it serves. The tank is normally jet-washed to remove any stubborn material and pumped out to a tanker. The contractor should be registered and licensed to remove waste from effluent tanks. The customer should receive a transfer note from the contractor as proof of correct disposal of waste; they should keep this for two years.

> **Did you know?**
>
> Dishwasher products could affect the bacterial process in the septic tank, so you should follow the manufacturer's instructions closely.

Secondary treatment of foul waste water

The secondary treatment of foul waste water is to filter it through the ground or vegetation. There are three main types of treatment: drainage fields, drainage mounds and wetlands.

Drainage fields

Drainage fields have to be well drained and away from any water courses, to prevent contamination of the local water networks. A drainage field is basically a series of perforated pipes positioned across the field to allow aerobic contact between the liquid effluent and the subsoil, so that further breakdown of the liquid can take place. A typical drainage field can be seen in Figure 5.19.

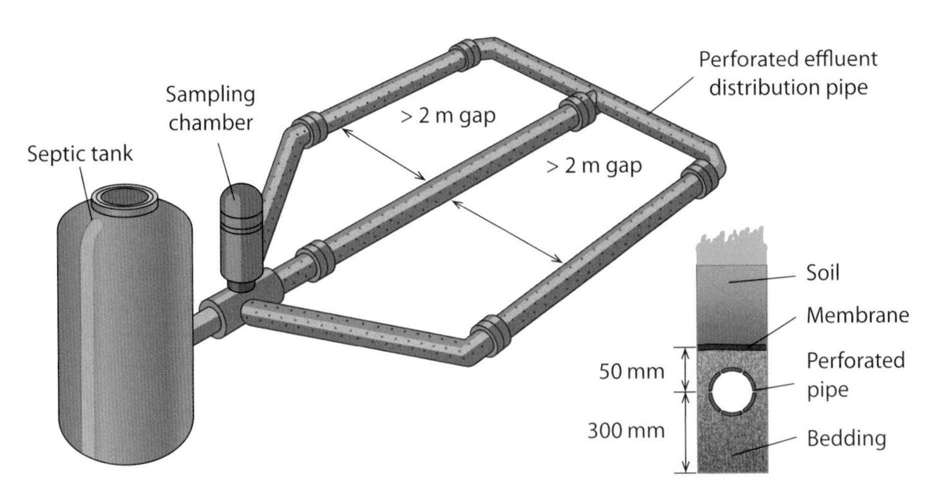

Figure 5.19: Drainage field

Drainage mounds

If the ground is too solid or the water table too high, a mound can be constructed to filter the foul liquid through. This will come at some cost, and the mound will need covering with grass so that it blends into the local environment (see Figure 5.20).

Wetlands

There are two main designs of wetlands or reed beds: horizontal flow and vertical flow systems. Creating a wetland system is a specialist

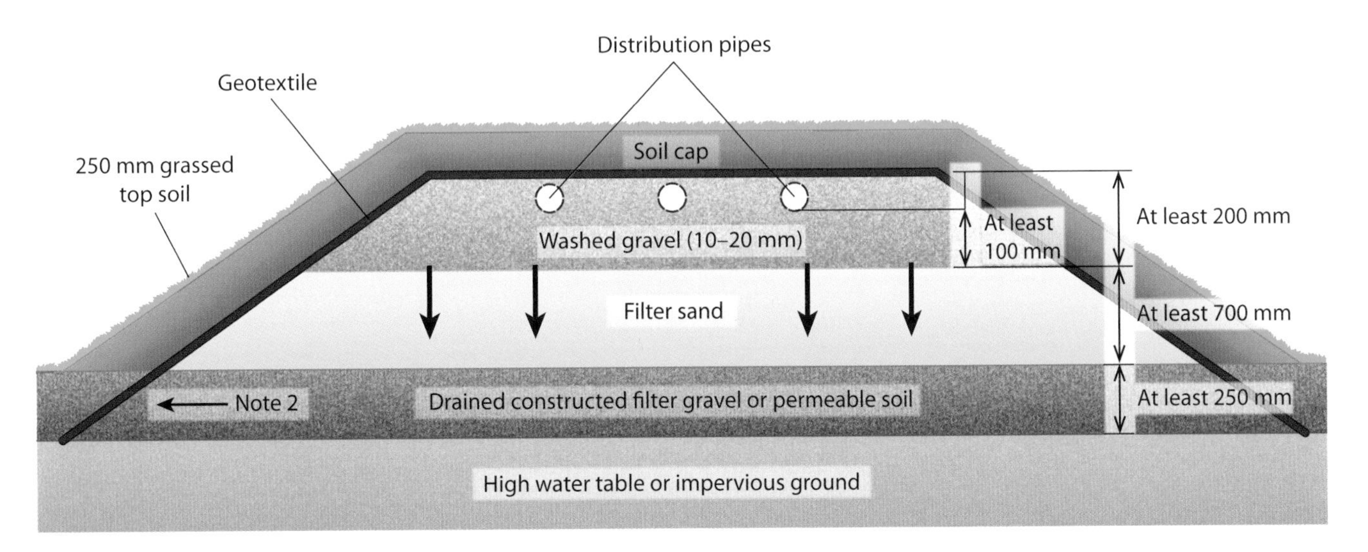

Notes

1. To provide venting of the filter, the upstream ends of the distribution pipes may be extended vertically above mound level and capped with a cowl or grille.
2. Surface water run-off and uncontaminated seepage from the surrounding soil may be cut off by shallow interceptor drains and diverted away from the mound. There must be no seepage of waste water to such an inceptor drain.
3. Where the permeable soil is slow draining and overlaid on an impervious layer, the mound filter system should be constructed on a gently sloping site.

Figure 5.20: Drainage mound

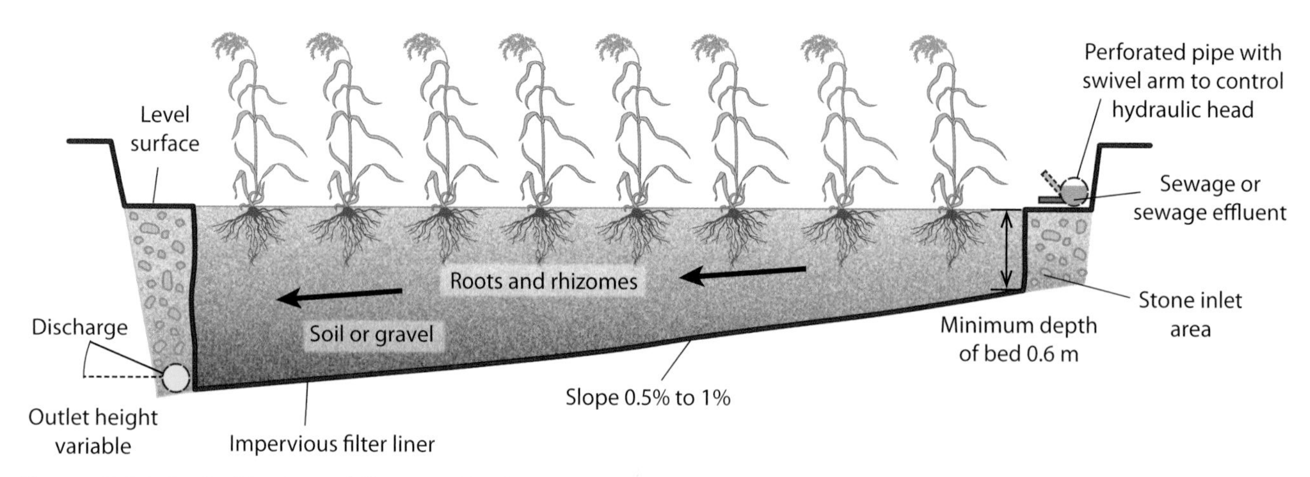

Figure 5.21: Typical horizontal flow reed bed treatment system

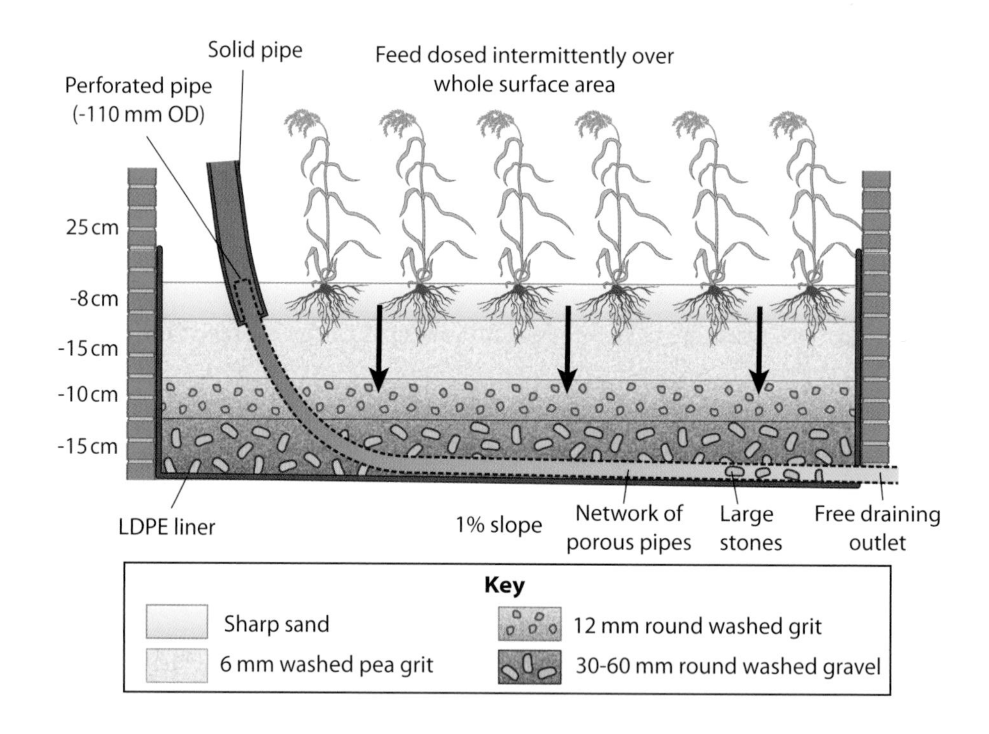

Figure 5.22: Typical vertical flow reed bed treatment system

task, as the correct type of plants and reeds have to be chosen carefully. Figures 5.21 and 5.22 illustrate the differences in function between the two designs. The vertical flow system gives the best treatment of the two, allowing greater oxygenation of the effluent and so breaking down the solids further.

Specialist sanitary components

A range of specialist components are associated with sanitation. Technological innovations, new customer requirements, changing aesthetic tastes and environmental concerns all play their part in creating a demand for new and more specialised components, so you need to keep yourself up to date.

WC macerators

The most widely used WC macerators are manufactured by Saniflo. The key features of a WC macerator are:

- it can only be used in a property where there is access to a WC discharging directly to a gravity system (Building Regulations G4 Approved Document)
- wiring must be carried out to BS 7671. An unswitched fused electrical supply point should be provided by a qualified electrician near the point of connection to the unit
- the unit will usually be connected to 22 mm discharge pipework (depending on the length of pipe run), giving flexibility as to where the WC can be positioned. Some models require 32 mm discharge pipework.

Safety tip

You should only work on the electrical supply if you are trained and competent to do so. Make sure you follow the manufacturer's instructions.

Figure 5.23 shows that the basin is connected to the unit, so a single pipe connection is used (this gives greater location options for a cloakroom). The pump can lift the discharge contents vertically, as well as moving them horizontally.

BS 7671

Key requirements for a macerator unit installation are:

- horizontal pipe runs must have a minimum fall of 10 mm/metre
- maximum horizontal pumping distance of 100 m
- on horizontal runs, the 22 mm pipe should be increased ideally to 32 mm after about 12 metres to enhance drainage
- any vertical lifting must occur at the beginning of the pipe run, not at the end
- 90° elbows must not be used – use 2 x 45° bends instead or 90° bends
- if the horizontal pipe run is significantly below the height of the unit, an air admittance valve should be fitted to the high point on the pipework
- a macerator discharge pipe should have its own dedicated run to the gravity drain (i.e. do not connect into sink, basin, shower, etc. waste pipe)
- to avoid noise transmittance, do not jam or fit macerator tight against the wall or WC pan. If on bare floorboards, fit anti-vibration material under the unit
- any outside pipework should be insulated.

Remember

Always check manufacturer's instructions for pipe sizes and vertical/horizontal pumping performance.

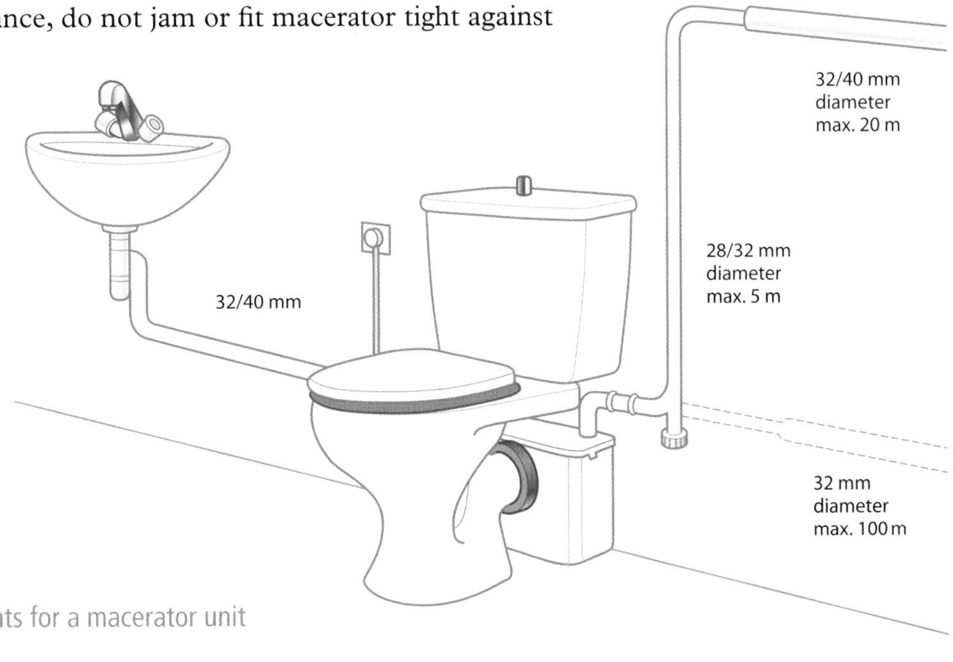

Figure 5.23: Design components for a macerator unit

Remember

Always protect the property before removing a Saniflow unit and ensure no spillage occurs from the open ends.

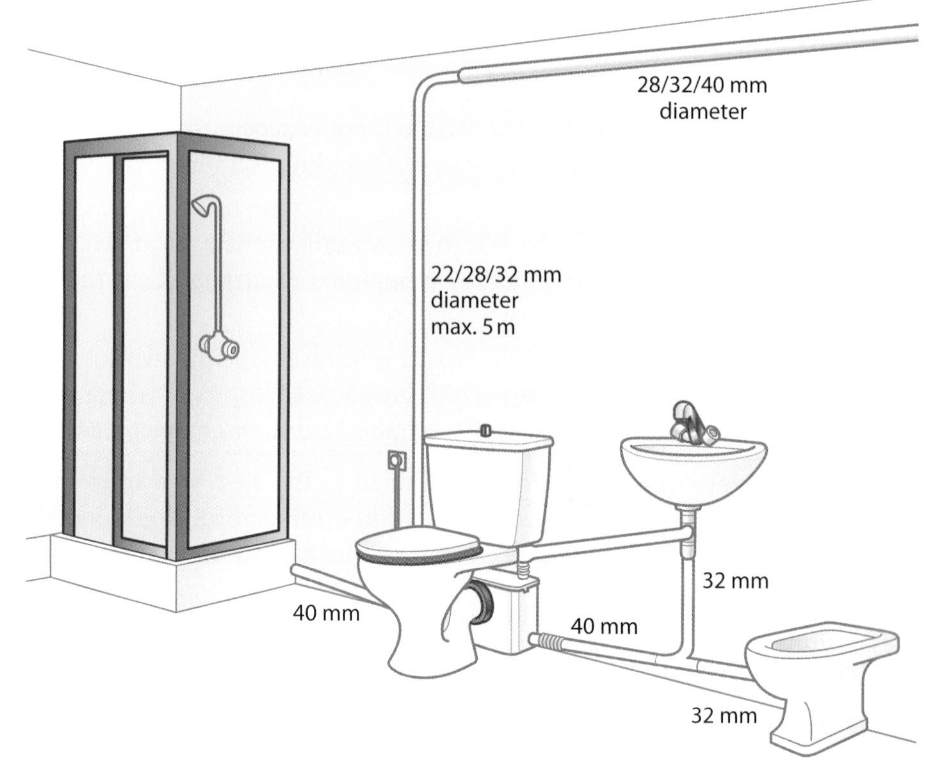

Figure 5.24: Design components of another type of macerator unit. This unit is capable of dealing with the outlets from all the appliances in a shower room-type installation. Always check the manufacturer's instructions for required height of shower tray above floor level.

Sink waste-disposal units

A sink waste-disposal unit is designed to remove waste food and cooking products from the kitchen and discharge them to the drainage system. The waste disposal unit can deal with all food matter, including bone, using a number of cutters to turn the matter into a thin paste. Water is flushed down the unit when it is working and the products are taken out to the drainage system. Cutter blades are driven by an electric motor. In Figure 5.25 you can see that there a rubber splashguard at the inlet, which is common to prevent food splashes and debris being thrown back into the room.

Figure 5.26 shows a typical sink mounting arrangement, an example of the rubber splashguard and an insert fitted into the waste, known as a cutlery saver.

This outline of the unit shows the swivel elbow connection, which can be adjusted to suit the required position. The outlet of the unit is a standard 38 mm or 1½". A trap should be fitted on the unit to stop smells. The discharge pipe should be laid to a fall of 1 in 12 for horizontal runs (to remove the discharge products effectively). The unit can discharge either to a gully or directly to the soil stack. The unit must not have a bottle trap or a grease trap connected to it as these block easily.

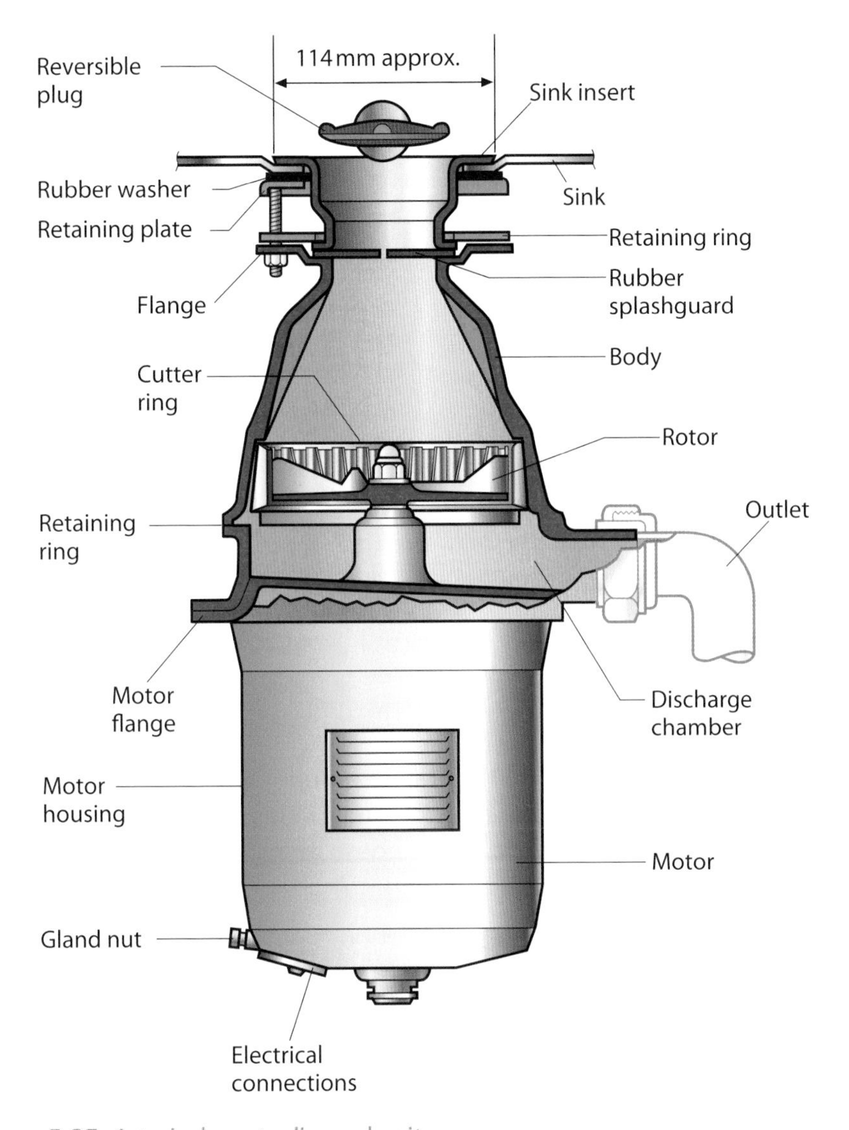

Figure 5.25: A typical waste-disposal unit

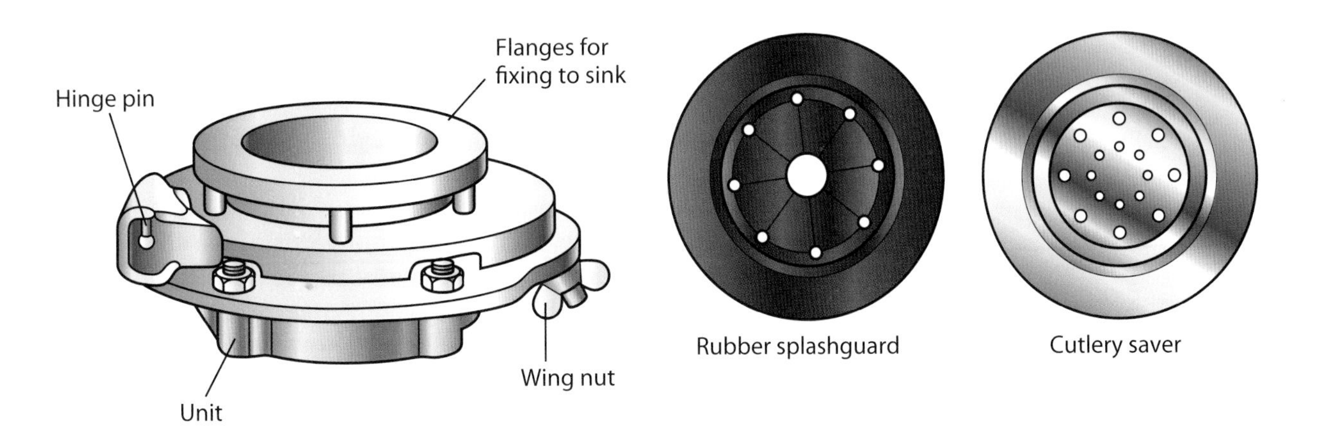

Figure 5.26: Waste-disposal mountings

Safety tip

Always ensure that the system is electrically isolated before starting any decommissioning operation.

Professional Practice

A client asks you to replace a waste-disposal unit in a kitchen with a larger unit because it is more powerful, but there is not enough space in the kitchen cupboard. Explain to the customer how it is possible to install the new waste disposal unit without damaging the existing cupboard unit.

Remember

The manufacturer's instructions are the best documents to refer to for installation requirements.

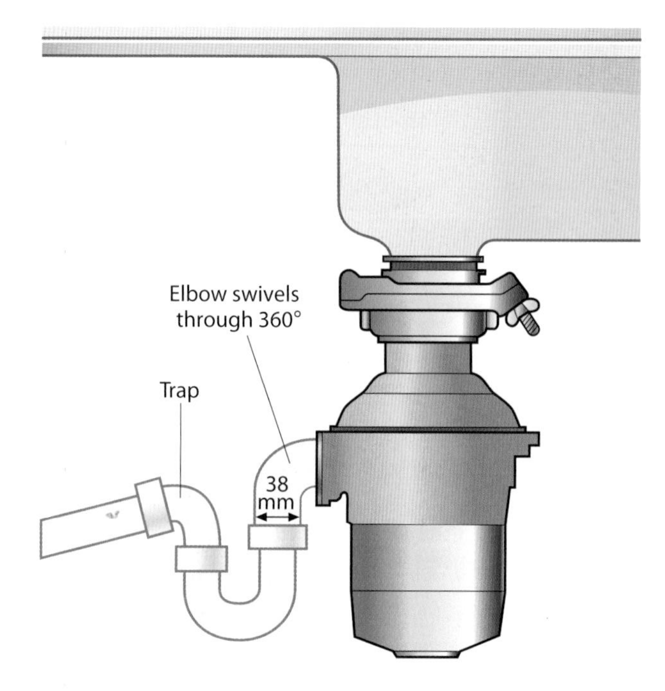

Figure 5.27: Waste-disposal discharge trap and elbow connection

Waste water lifters

The purpose of a waste water lifter is to pump waste into a sewer that may be above the level of the house drainage system. It also ensures that the sewage cannot return to the house in back-surges caused, for example, by flooding of the main drain.

Environmental damage

The damage caused by sewage flooding into a house can be extensive and costly. In no more than 30 minutes a house or basement can flood with untreated sewage to more than 1 m deep. Damaged plaster walls and timber flooring will need to be removed, together with electrical wiring and carpets before replacing can begin. Any furniture, electrical equipment, personal effects will be beyond recovery due to foul waste contamination.

The process of the property drying out can take months, during which time the property is often uninhabitable and the household will have to find alternative accommodation.

Contaminated waste material has to be removed to a tipping site where it adds approximately 10 m^3 to landfill for each affected property.

Changing climate

The threat of global warming brings with it forecasts of raised sea and river levels and more flooding. Coupled with this the need to build more houses on often unsuitable and low-lying areas where the risk of flooding is greater can only lead to more incidents of property damage. For these reasons, the use of waste water lifters is going to increase.

The public highway is usually used as the flood level unless other information is available, such as local river flooding records. If the sanitary appliances of the property are below that level, it can be presumed that the property may be at risk of backflow from the sewer. The customer has two possible solutions:

- install a pump with a backflow loop fitted (see Figure 5.28), or
- have an anti-flood valve installation, if there is a natural fall to sewer (see Figure 5.29).

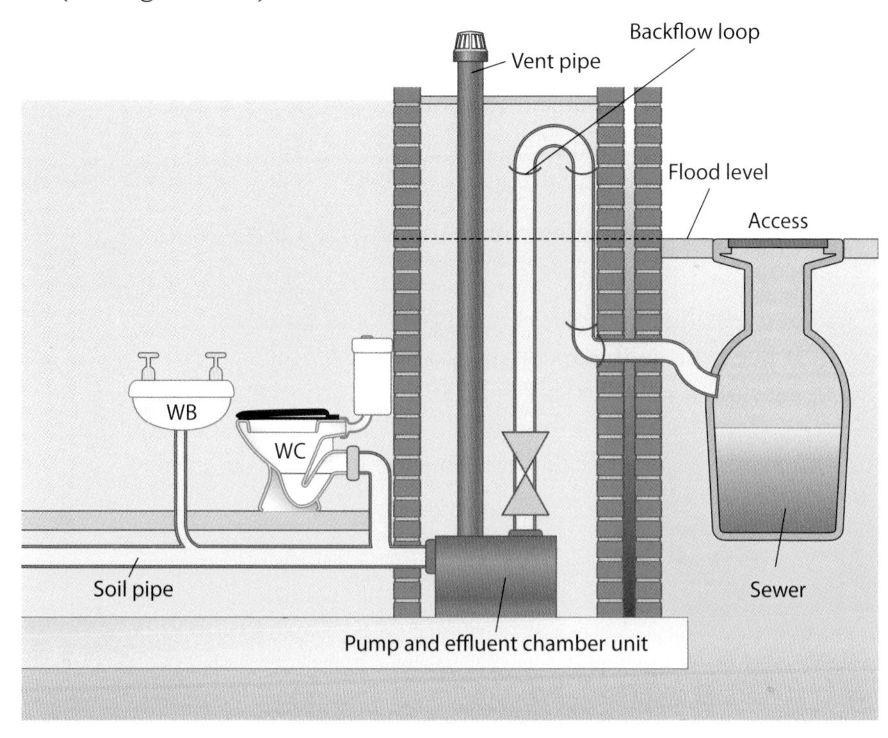

Figure 5.28: Layout of backflow loop

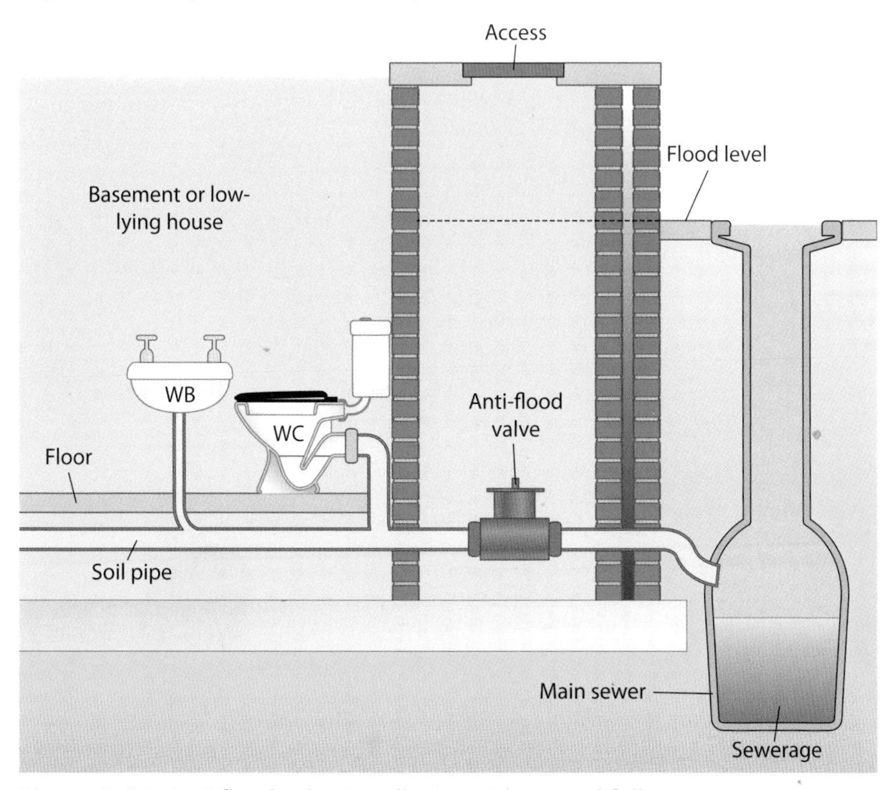

Figure 5.29: Anti-flood valve installation with natural fall to sewer

As you can see from Figure 5.28 (page 233) the backflow loop is just a pipe that goes above the flood level before returning to the sewer level, thus preventing flood sewage going into the building. The anti-flood valve is a flap attached to a float; if a surcharge occurs, the float lifts to swing the flap down to shut off the route back into the property.

Note that this type of device can only be installed if:

- there is a fall on the drain to the sewer
- the number of users is small
- at least one toilet is above the flood level
- the health of the occupancy will not be affected if the room does flood.

The size of the effluent receiver chamber should be taken into account. It needs enough capacity to contain 24 hours-worth of inflow in case of storm overflow conditions or other disruption to the sewer system. The minimum discharge of foul drainage should be 150 litres per person per day in domestic premises.

PROGRESS CHECK

1. Where are stub stacks usually found?
2. What does an air-admittance valve remove the need for?
3. Waterless urinals have a cartridge. What does this do?
4. The Hepworth valve uses a collapsible membrane. How does this work?
5. Which Approved Document lays down the sanitary requirements for the disabled?
6. What is a cesspit?
7. How often should a cesspit be inspected?
8. What is a septic tank?
9. How often should a commercial septic tank be emptied?
10. What must the minimum fall be for macerator horizontal pipework?
11. What is the purpose of a water lifter?

2. Know and be able to apply the design techniques for sanitation and rainwater systems

Selecting sanitation systems

The factors that affect the choice of sanitation system have been dealt with at Level 2, so this section just gives a short recap. As you have seen, most systems come in four types:

- primary ventilated stack system
- ventilated discharge branch system
- secondary modified ventilated stack system
- stub stack system.

Whichever system you use, it must comply with the Building Regulations Part H1.

The main aim of the design of the system is to prevent the loss of the trap seal, which prevents smells entering the building. Discharge piping systems should be designed to use the minimum of pipework necessary to carry water and effluent away from sanitary appliances as quickly and silently as possible, without risk to health.

The main factors to consider are:

- cost
- environmental impact
- minimising the risk of blockage
- distance of appliances from main stack
- number of appliances
- type of appliance
- number of bends in stack
- number of bends in branches
- height of building
- purpose of building
- number of people using the building
- disability
- gender.

Client's needs

With sanitation systems, as with all plumbing work at this level, it is important that you establish the customer's needs at the beginning of the design process. Look back to the section in Unit 2 on pages 16–17 where the best ways to communicate with clients in a clear and effective way were covered.

Building layout and features

It is a requirement to identify the suitability and location of existing plumbing services and take note of the site's overall condition, and the room sizes and type of construction used for the building. The checks on services should cover checking the pressure and flow rate of water services and checking for the type of any existing sanitation systems.

The best way to record your notes in a professional way is to use a survey sheet as in Figure 5.30.

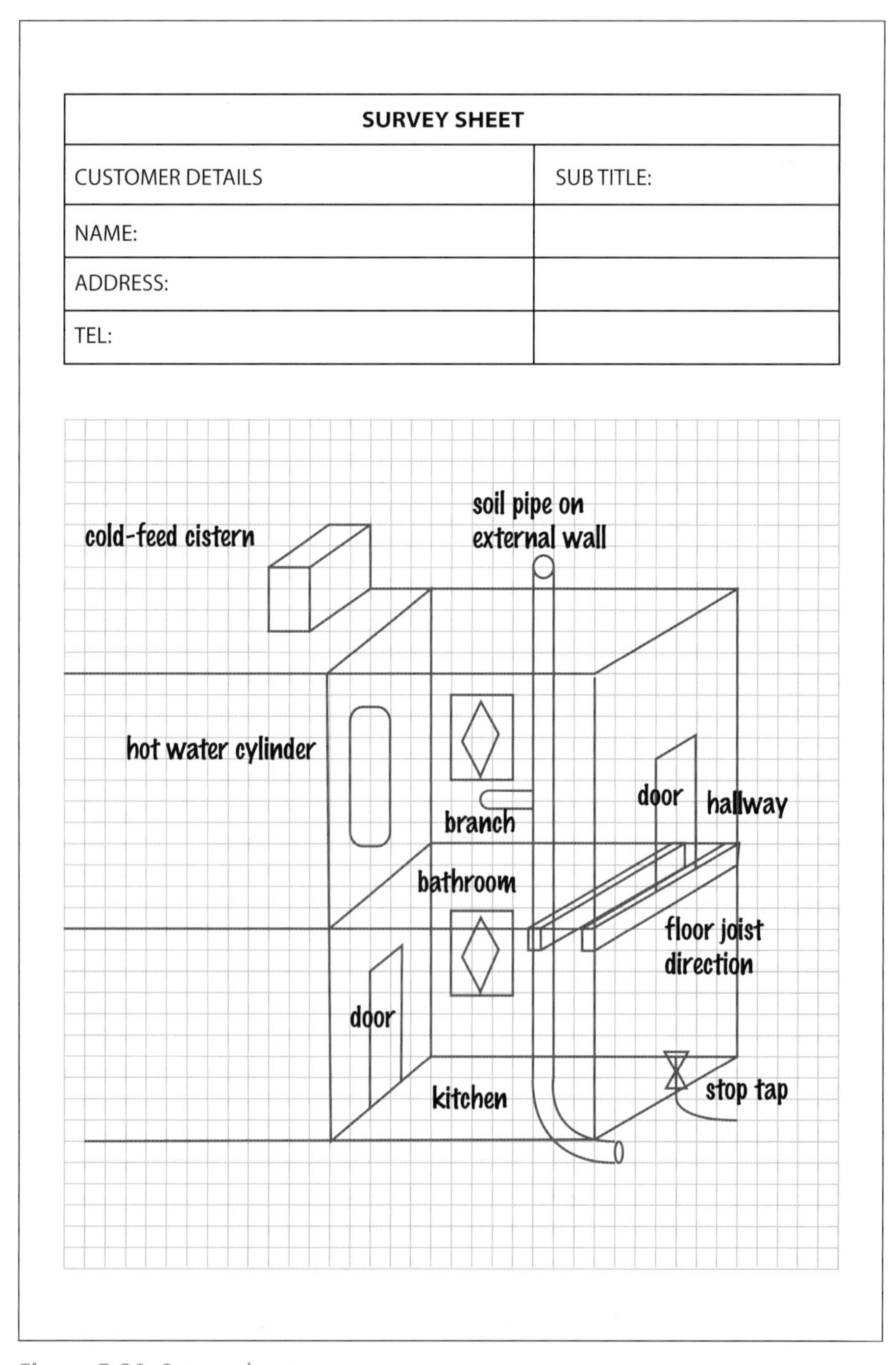

SURVEY SHEET	
CUSTOMER DETAILS	SUB TITLE:
NAME:	
ADDRESS:	
TEL:	

Figure 5.30: Survey sheet

When planning soil pipe runs, it is important to take into consideration the building's construction, the wall and floor types, the direction of joists and walls and the position of any cupboards or pipe ducts that may be available to conceal pipes. Pipes should be run vertically and horizontally in relation to the walls and floors, although this is not always practicable. With sanitation pipework, the need for a slight gradient towards the drain and large-diameter sanitation pipes requires careful consideration to prevent difficulties in the installation phase.

Ventilation

You should always read Building Regulations Approved Document Part F when designing any bathroom or toilet accommodation. Adequate ventilation is a necessity for all rooms containing sanitary appliances. It may be natural ventilation through windows or skylights with openings direct to the outside air, or it can be mechanical ventilation. A window provides a view into toilets or bathrooms, so you should use obscured glass. Any window openings in toilets should have a total free opening area of at least 1/20th (1/30th in Scotland) of the floor area of the room, with some part of the ventilation opening at least 1.75 m above the level of the floor. Windows should also have the facility to provide background ventilation, such as trickle ventilators.

Did you know?

Natural ventilation can be supplemented by mechanical ventilation to help reduce moisture levels in a building.

Passive stack ventilation or mechanical ventilation should be provided in kitchens, utility rooms, and rooms containing baths or showers, in addition to any window provided, to reduce the possibility of condensation. It should be installed in accordance with BS 5720, BS 5250 and BS 5925.

BS 5720
BS 5250
BS 5925

Energy efficiency and environmental impact

Along with hygiene, low water and energy consumption should be of prime consideration with any sanitation design. Conventional sanitation limits itself to sanitising the home, preventing disease and promoting health, by preventing the population's contact with pathogenic germs. In conventional households water conservation is the main argument put forward to the public. The advice usually concerns the water saving of appliances connected to the sanitation system. The public is advised to:

- reduce the water volume of WC cistern flushes
- repair any water leaks
- use low-flow shower heads and taps
- shut the tap while brushing teeth
- do laundry with full loads only
- recycle rainwater and greywater.

These are all measures designed to reduce water usage from the main water supply (see Unit 3 for further information). The designer should always be concerned about water-saving features when selecting appliances.

Interpreting information

When designing a sanitary pipework system, consultation between clients, architects and engineers is essential at every stage of a building's design. This enables efficient and economic planning of the sanitary installations and the discharge system, and the provision and positioning of ducts in relation to the building design as a whole.

Details of drains, sewers and any precautions necessary to ensure satisfactory working of the discharge systems should be obtained from the bodies responsible for the systems: for example, information on the possibility of drains and sewer surcharging and statutory regulations. You should also find out any specific requirements made by the sewerage undertaker.

Alterations or extensions to existing work may need a survey, which should include:

- the type of drainage system in use
- drain and sewer loading
- details and positions of appliances connected to the system
- a description of the existing pipework and its condition
- details of the ventilation of the system
- the results of system testing.

Statutory regulations

Building Regulations and industry standards

As a specialist area, sanitation is covered by its own section of the Building Regulations, Part H. This is a statutory regulation that gives powers to the local planning and Building Control. Building Control Officers and Approved Inspectors are responsible for the enforcement of the relevant regulations. The information they may require includes:

- information on the number, position and types of appliances to be installed and details of the proposed use of the premises
- notification on the appropriate forms and particulars of the proposed work
- drawings and specifications.

Remember

The manufacturer's instructions should always be used when designing or installing a system.

BS EN 12056
BS 6465-1

Before starting work, the installer should be in possession of drawings as approved by the appropriate authorities, together with the specification and any further working drawings and information necessary to carry out the work.

The main standard for sanitation design is BS EN 12056; all the design elements in this section are based on this standard. BS 6465-1 covers commercial and industrial appliance provision.

See Unit 2 for details of the information sources you will need to be familiar with and able to interpret.

Fire protection

Figure 5.31 shows enclosure arrangements for drainage and water supplies.

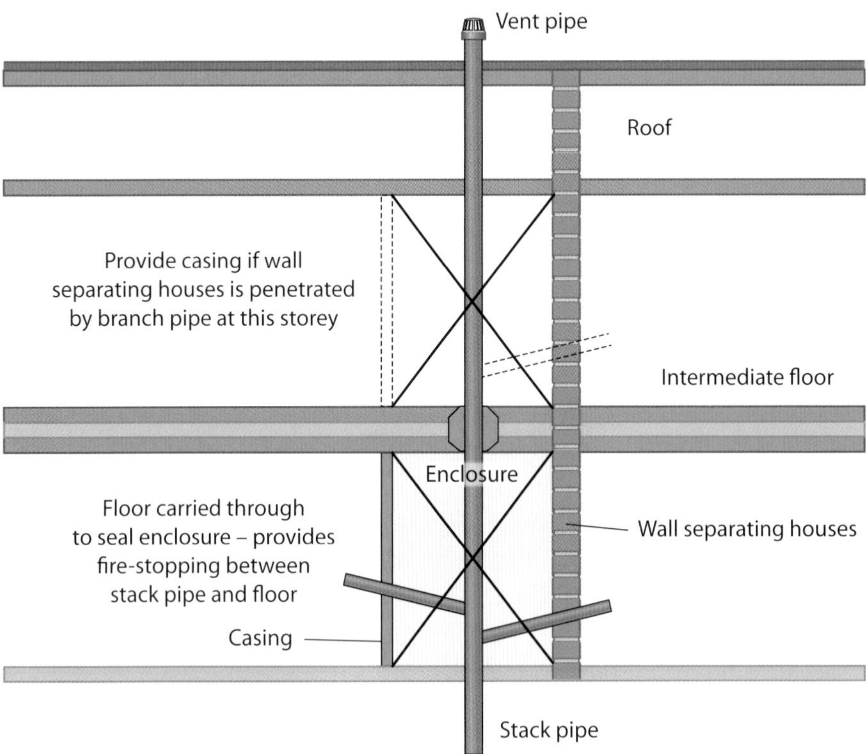

1. The enclosure should:
   a. be bounded by a compartment wall or floor, an outside wall, an intermediate floor or a casing (see specification below);
   b. have internal surfaces (except framing members) of class 0 (National class) or Class B-a3, d2 or better (European class) Note: When a classification includes a3, d2, this means that there is no limit set for smoke production and/or flaming droplets/particles);
   c. not have an access panel which opens into a circulation space or bedroom;
   d. be used only for drainage, or water supply, or vent pipes for a drainage system.
2. The casing should:
   a. be imperforate except for an opening for a pipe or an access panel;
   b. not be of sheet metal;
   c. have (including any access panel) not less than 30 minutes fire resistance.
3 The opening for a pipe, either in the structure or the casing, should be as small as possible and fire-stopped around the pipe.

Figure 5.31: Enclosure for drainage or water supply pipes

Calculating sanitary provision requirements

Private dwellings

The design of sanitary facilities in private dwellings should be in accordance with the recommendations shown in Table 5.2 on page 240.

Sanitary appliance	Number of sanitary appliances per dwelling	Remarks
WC	1 for up to 4 persons; 2 for 5 persons or more	
Wash basin	1	There should be a wash basin in or adjacent to every toilet
Bath or shower	1 per 4 persons	
Kitchen sink	1	

(taken from BS 6465-1)

BS 6465-1

Table 5.2: Minimum provision of sanitary appliances for private dwellings

Notes:

- A WC with a wash basin should be provided on the entrance storey of every private dwelling.
- A room containing a WC should not be able to be entered directly from a bedroom unless it is intended for the sole use of the bedroom occupants such as an en-suite, and a second WC is provided elsewhere in the private dwelling for visitors.
- In blocks of flats, toilets should be provided for any non-residential staff working in the building.
- A cleaner's sink must be provided for the cleaning of any communal areas in blocks of flats.

Did you know?

For houses in multiple occupancy (where occupants do not live as a family but share facilities), the local authority licensing department needs to be consulted as to their minimum requirements.

Worked example

A customer lives with their partner and four children. The customer asks you to design a bathroom. What sanitary appliances should they have in their house?

Looking at Table 5.2 you can see that, as there are six occupants, you will need to ensure that they have: 2 WCs, 2 wash basins, 2 baths or showers, 1 kitchen sink.

Did you know?

Calculations for commercial and industrial settings can also be found in BS 6465-1.

Sizing and selecting gradient for branch pipework

Domestic systems' pipe branches are usually on a primary ventilated stack system. Because of the need to maintain close grouping of the appliances, the branch should be no less than the trap size serving the appliance. If the pipe serves more than one appliance on a primary ventilated stack system, Table 5.3 can be used.

For larger domestic and industrial installations, a discharge unit method is used. This is derived from statistical data analysis. A numerical value is given to different types of sanitary appliance, which have different flow rates and frequency of use.

A low flow limit of one-quarter capacity for the discharge stack and one-half capacity for the branch discharge pipe is adopted; this is to prevent plugs of water developing, which would pull the trap seal out from the trap.

Appliance	Diameter DN	Min. trap seal depth mm	Max. length (*L*) of pipe from trap outlet to stack m	Pipe gradient %	Max. number of bends No.	Max. drop (*H*) m
Wash basin, bidet	30	75	1.7	2.21[1]	0	0
Wash basin, bidet	40	75	3.0	1.8 to 4.4	2	0
Shower, bath	40	50	No limit[2]	1.8 to 9.0	No limit	1.5
Kitchen sink	40	75	No limit[2]	1.8 to 9.0	No limit	1.5
WC with outlet greater than 80 mm	100	50	No limit	1.8 min.	No limit[4]	1.5
Floor drain	50	50	No limit[3]	1.8 min.	No limit	1.5

1 Steeper gradient permitted if pipe is less than maximum permitted length.
2 If length is greater than 3 m noisy discharge may result with an increased risk of blockage.
3 Should be as short as possible to limit problems with deposition.
4 Sharp throated bend should be avoided.

Table 5.3: Limitations for unventilated branch discharge pipes, system III BS EN 12056

Appliance	Discharge units
Wash basin, bidet	0.5
Shower without plug	0.6
Bath	0.8
Kitchen sink	0.8
WC with 6–7.5 l cistern	2.0
Floor gully DN 50	0.8
Used for unventilated stack only	–

Table 5.4: Discharge units table

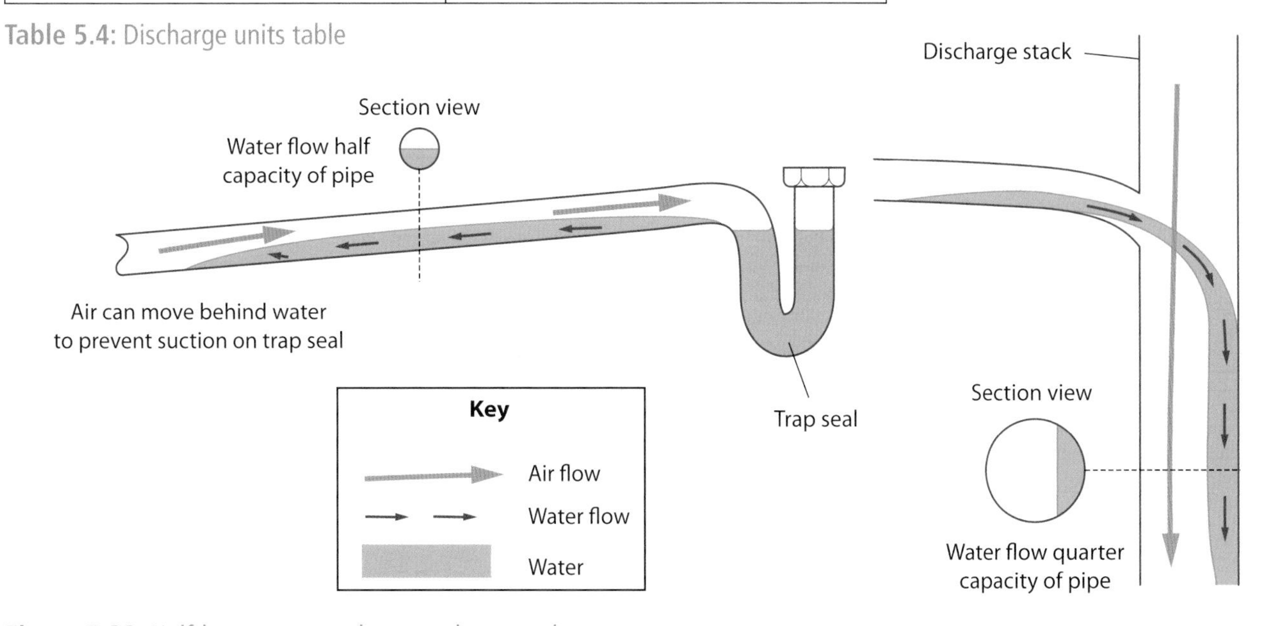

Figure 5.32: Half-bore water and quarter-bore stack

Designing and sizing sanitary pipework system

Stack sizing

Most domestic stack sizing is done using the following table.

Note: Hydraulic capacity (*Q*max) and nominal diameter (DN)

Stack and stack vent	System I, II, III, IV *Q*max (l/s)	
DN	Square entries	Swept entries
60	0.5	0.7
70	1.5	2.0
80*	2.0	2.6
90	2.7	3.5
100**	4.0	5.2
125	5.8	7.6
150	9.5	12.4
200	16.0	21.0

* Minimum size where WCs are connected in system II.
** Minimum size where WCs are connected in system I, III, IV.

Table 5.5: Stack and stack vent sizing

All you need to do is to work out the total discharge volume going into the stack. Use the discharge unit table to work out the discharge rate from each appliance. Add them together, then multiply by the frequency factor (how often the appliance is likely to be used) taken from Table 5.6.

Usage of appliances	K
Intermittent use, e.g. in dwelling, guesthouse, office	0.5
Frequent use, e.g. in hospital, school, restaurant, hotel	0.7
Congested use, e.g. in toilets and/or showers open to public	1.0
Special use, e.g. laboratory	1.2

Table 5.6: Typical frequency factors (K)

The formula is total√discharge units (square root) x frequency factor = total discharge volume. Then look at Table 5.5 on hydraulic capacities to find your stack size.

Worked example

A house with 2 WCs, 2 wash basins, 1 bath, 1 kitchen sink.

- WCs 2 x 2 = 4
- Wash basin 2 x 0.5 = 1
- Sink 1 x 0.8 = 0.8
- Bath 1 x 0.8 = 0.8

Total discharge units = 6.6

Times frequency factor of 0.5 for dwellings

$$\sqrt{6.6} \times 0.5 = 1.3 \text{ l/s}$$

Check this against Table 5.5 and you see that, with swept entries, you can use a 70 mm stack and, for square entries, you need to use a 70 mm stack. However as 70 mm stacks are uncommon and would leave no room for system alteration at a later date, a designer may opt for 100 mm, even on the swept system. This is because 100 mm pipe is the most common pipe size sold, so it may be easier to obtain at a lower cost.

The same method is used for non-domestic systems. Here is an example of a small school:

WCs	20 x 2	= 40
Wash basins	23 x 0.5	= 11.5
Sinks	10 x 0.8	= 8
Shower	10 x 0.8	= 8
Total discharge units		= 67.5

Frequency factor 0.7 for schools
Discharge $\sqrt{67.5} \times 0.7 = 5.75$ l/s

Look at Table 5.5 and you can see that a 125 mm stack is required.

Ventilation requirements

Air-admittance valves

Where air-admittance valves are used to vent branches or appliances, they must comply with BS EN 12380 and be sized in accordance with Table 5.7.

System	*Qa* l/s
I	1 x *Qtot*
II	2 x *Qtot*
III	2 x *Qtot*
IV	1 x *Qtot*

Q = Flow rates in litres per second
Qa = Minimum air flow rate in litres per second (l/s).
Qtot = Total flow rate in litres per second (l/s).

Table 5.7: Minimum air flow rates for air-admittance valves in branches

Worked example

A branch pipe on a type III system with a total flow rate of 2 l/s will require an air flow rate of 2 x 2l /s = 4 litres of air per second. An air-admittance valve should be selected with 4 litres per second as the minimum size. Information can be obtained from the valve manufacturer.

Air-admittance valves for stacks

Where air-admittance valves are used to ventilate stacks, they must all comply with EN 12380 and be sized with *Q*a not less than 8 x *Q*tot. So the total flow rate multiplied by 8 will give you the air flow rate required. To select the air-admittance valve, you can check the manufacturer's specification.

Ventilating pipes

The size of ventilating pipes to branches from individual appliances can be 25 mm; however, if they are longer than 15 m or have more than five bends, a 30 mm pipe should be used. If the connection of the ventilating pipe is liable to become blocked due to repeated splashing or submergence on a WC branch, it should be larger, but it can be reduced in size when it gets above the spill-over level of the appliance. Ventilating pipes should be connected to the stack above the spill-over level of the highest appliance to prevent blockages. Connections to the appliance's discharge pipe should normally be as close to the trap as practicable, but within 750 mm to ensure effectiveness. Ventilating pipe connections to the end of branch runs should be at the top of the branch pipe, away from any likely backflow, which could cause blockage.

Designing and sizing a rainwater system

Sizing gutters

There are two factors to take into account when specifying the size of a gutter:

- the effective maximum roof area to be drained
- gutter flow capacity.

Effective roof area

This is calculated using the formula:

- length of roof x width x table 1= effective roof area in m^2

The information contained in Table 5.8 calculates the effective roof area.

If a roof's dimensions are length = 10 m, width = 4 m and 30^0 pitch the effective roof area would be:

$$10 \times 4 \times 1.15 = 46 \text{ m}^2$$

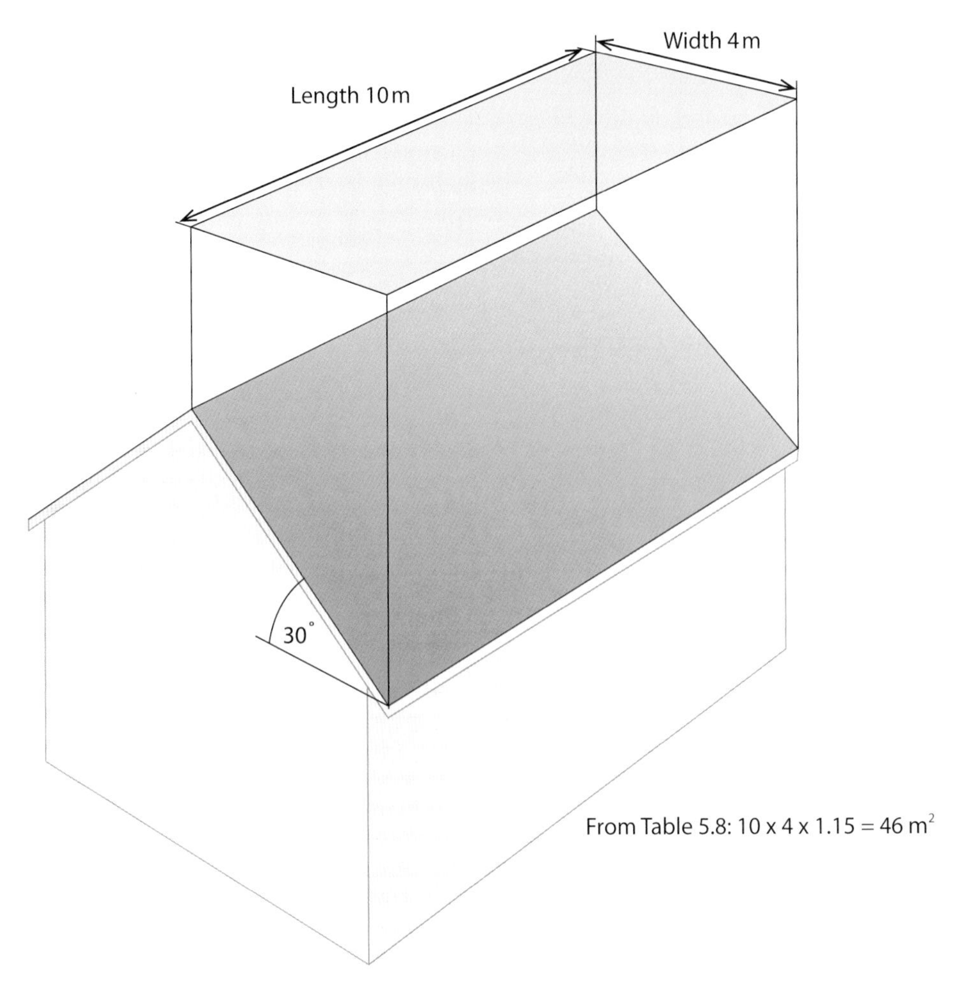

Figure 5.33: Roof area calculation

Type of surface	Affective design area m²
1 Flat roof	Plan area
2 Pitched roof at 30° 3 Pitched roof at 45° 4 Pitched roof at 60°	Plan area x 1.15 Plan area x 1.4 Plan area x 2.0
5 Pitched roof over 70°	Elevation area x 0.5

Table 5.8: Calculation of roof area (extract from Building Regulation H)

Gutter flow capacity

This depends on three factors:

- system type
- outlet position
- fall of the gutter.

Use Table 5.9 to specify the gutter for a 1:600 fall.

Max effective area m^2	Gutter size mm	Outlet size mm	Flow capacity l/s
18.0	75	50	0.38
37.0	100	63	0.78
53.0	115	63	1.11
65.0	125	75	1.37
103.0	150	89	2.16
Refers to half-round guttering			

Table 5.9: Gutter and outlet size (extract from Building Regulation H)

Remember

Make sure you present all your calculations in a way that is clear and easy to understand.

A 115 mm half-round gutter with 63 mm outlet with a 1.11 l/s flow capacity would be selected.

The changing climate means it is not possible to rule out flooding. Usually, it is satisfactory to design to a rainfall intensity of 75 mm per hour where overflow will not cause damage within the building.

For other rainfall intensity levels, calculate as follows:

effective roof area x rainfall intensity (mm per hour) ÷ 3600 = l/s

Worked example

Effective roof area 46 m^2 x 90 mm per hour (say a very exposed site) ÷ 3600 = 1.15 l/s.

From Table 5.9 you would select a 125 mm gutter.

Did you know?

Roof outlets positioned in the centre of a flat roof can drain larger areas than outlets located at the edge or corner of the roof.

Professional Practice

The client has asked you to calculate the gutter size for a domestic property. What information are you going to need before you can carry out the calculations? Think about how you are going to present the calculations to the client.

Gutter type selection and outlet positioning

You will have covered this at Level 2. See Unit 9 pages 374–378.

PROGRESS CHECK

1 What are the main factors to consider when designing a sanitation system?

2 When planning soil pipe runs, what must you consider?

3 A bathroom has an openable window. What should the total free openable area be?

4 How should a permanent record of any agreed alterations to a plan be recorded?

5 When is it alright to be able to enter a room containing a WC directly from a bedroom?

6 What should the minimum trap seal depth be on a wash basin?

7 What is the frequency factor for dwellings when calculating stack size?

8 Gutter flow capacity depends on which three factors?

9 If a gutter has a flow capacity of 38 litres/second, what minimum size should it be?

10 What is the rainfall intensity generally taken as?

3. Understand the installation requirements of sanitation system components

Walk-in wet rooms

A wet room is a fully watertight bathroom with no separate shower tray; instead, it is designed so that water produced through showering is drained through a plug or drain hole. The bath is removed, leaving a much more spacious and luxurious open room.

The 'walk-in' shower area is usually level with the surrounding floor but with a slight slope to the drain, which is fitted directly into the floor. The shower unit and controls are fitted to one of the walls within the wet room.

The floor can be laid by a specialist or a plumber with training. To ensure a watertight seal, careful planning and installation of the drain hole is essential. If the height or position of the drain hole is incorrect, the whole floor will have to be relaid, at considerable expense.

The finished look of a wet room can enhance a property, as well as providing access to wheelchair users.

Installation and fixing for walk-in wet rooms

Shower trays

A wet room shower tray is a level, preformed waterproof tray and fully load-bearing floor, which is prelaid to ensure correct drainage. A shower tray can be laid directly onto existing wooden or concrete floors. If the shower area is to be level with the floor, the shower tray can either be cut into the floor and reinforced underneath, or the surrounding area can be raised up to the correct level using tile backing boards. Alternatively, the shower tray base could be above the rest of the floor and a raised step could be added around the tray.

Activity

The Bathroom Manufacturers Association website has useful information on walk-in wet rooms. www.bathroom-association.org. Make some notes on your findings.

Initial planning

Both timber and concrete are suitable substrates for wet rooms. A wet room does not need to extend over the entire bathroom or bedroom area; it can be restricted to a corner of a bathroom or a corner of a bedroom for an en-suite. A key point is the need to waterproof only the 'wet' area, which can reduce costs to one third of normal levels. However, many wet rooms are completely sealed across the floor and turned up 100 mm at the edges to collect any accidental spillage and to protect the building's structure.

While a concrete floor needs only sharp projections removing, timber requires floorboards or chipboard lifting and discarding around the wet area. Lay a perimeter strip of plywood around the underlay (as shown in Figure 5.35), all screwed, or reinforce the floor by overlaying with plywood.

The finished floor height will be determined by:

- type of adhesive
- type of tile
- inclusion and type of underfloor heating system
- inclusion and size of the tiled backboard for heat insulation and waterproofing purposes.

In new-builds, the finished floor height should be considered at the beginning and incorporated into the design, so that the finished floor is flush with the external floor area. In renovation work, however, the extent to which the floor level is altered can be a problem: for example, large steps in the room may not be suitable. Check to see if the door swings inwards; if so, it may be possible to alter the door. Otherwise, a new one may be required.

Service pipes are pre-planned to run outside of the floor's wet zone. However, when penetrating the vertical water barrier in the shower area,

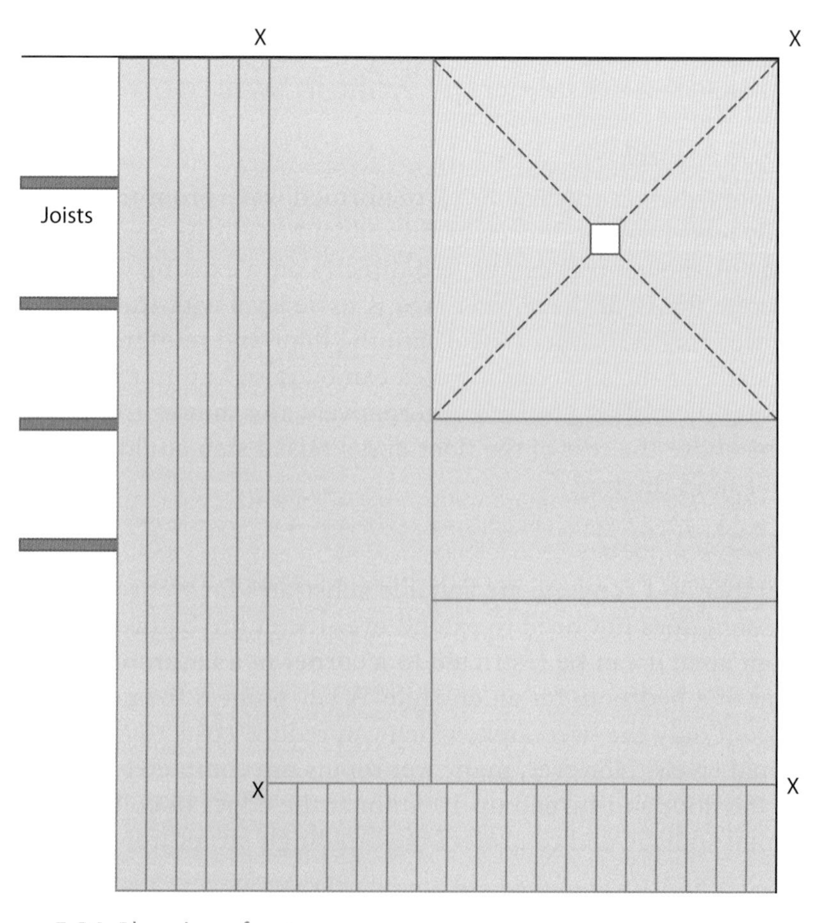

Figure 5.34: Plan view of wet area

you must use the sealants provided by the manufacturer. The exception to this is for the waste pipe and trap, which are built into the floor.

Drainage pipework

You must consider the floor depth of your room in order to determine what type of floor drain is suitable for the bathroom. There are several types of drain design, including a horizontal outlet design and a vertical outlet design. The horizontal outlet has a slim profile, at just a few millimetres deep below the shower tray. This maximises space usage to accommodate buildings with a smaller floor depth, and is ideal for renovation work. The vertical design is better when there is plenty of space under the floor. Access to drains may not always be possible, so specialised pump units may have to be used to remove the waste water.

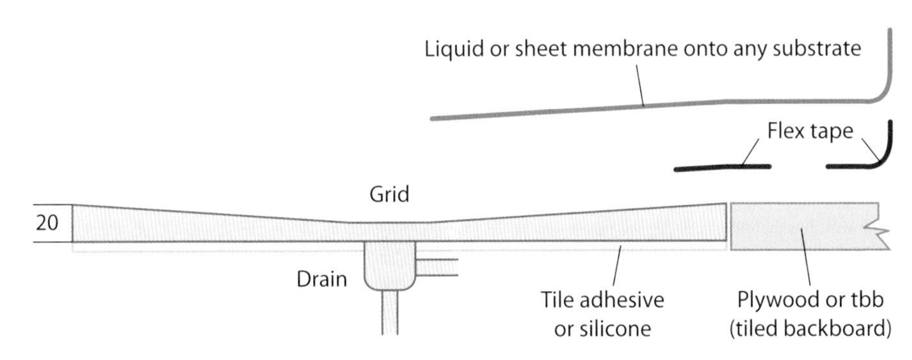

Figure 5.35: Side view of tray installation

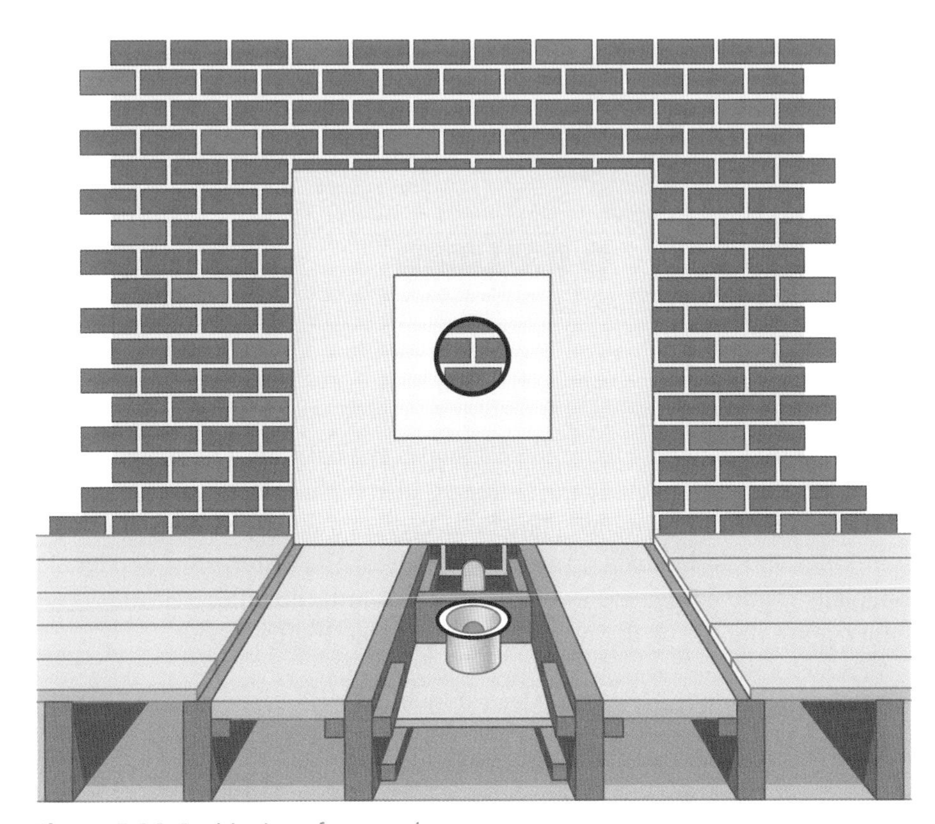

Figure 5.36: Positioning of trap and waste

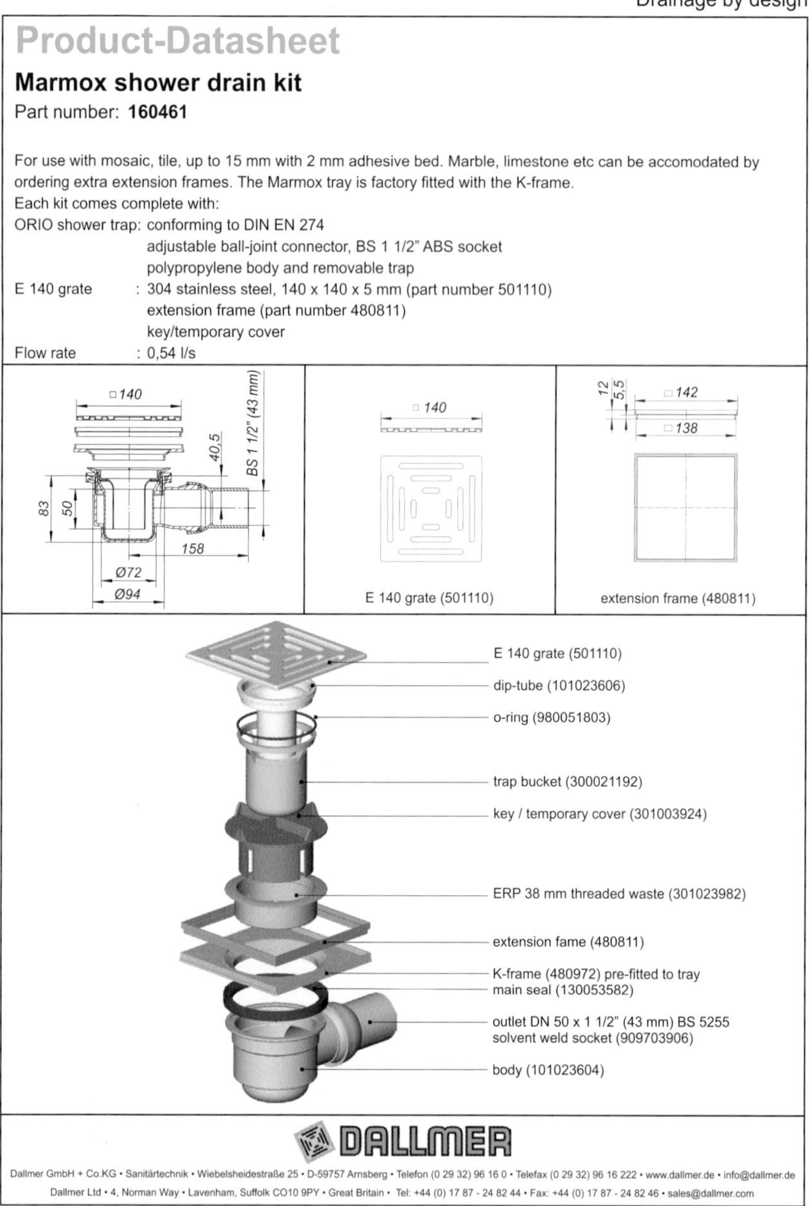

Product-Datasheet

Marmox shower drain kit

Part number: **160461**

For use with mosaic, tile, up to 15 mm with 2 mm adhesive bed. Marble, limestone etc can be accomodated by ordering extra extension frames. The Marmox tray is factory fitted with the K-frame.
Each kit comes complete with:
ORIO shower trap: conforming to DIN EN 274
adjustable ball-joint connector, BS 1 1/2" ABS socket
polypropylene body and removable trap
E 140 grate : 304 stainless steel, 140 x 140 x 5 mm (part number 501110)
extension frame (part number 480811)
key/temporary cover
Flow rate : 0,54 l/s

Dallmer GmbH + Co.KG • Sanitärtechnik • Wiebelsheidestraße 25 • D-59757 Arnsberg • Telefon (0 29 32) 96 16 0 • Telefax (0 29 32) 96 16 222 • www.dallmer.de • info@dallmer.de
Dallmer Ltd • 4, Norman Way • Lavenham, Suffolk CO10 9PY • Great Britain • Tel: +44 (0) 17 87 - 24 82 44 • Fax: +44 (0) 17 87 - 24 82 46 • sales@dallmer.com

Figure 5.37: Exploded view of trap

Tile
Lug
Waterproof membrain
Hole drilled into tile

Figure 5.38: Fixing depth in wet room

The exploded view in Figure 5.37 shows a typical trap. The installation engineer must not lose any parts during installation, but must keep the parts together until installed. The floor creates a waterproof barrier, so no fixings should be drilled through the tile or the waterproof barrier. Specially made fixings are available that slot into tile holes, without penetrating the tile or the waterproof layer (see Figure 5.38).

Important requirements of a wet room

- Stable (non-flexing) flooring
- Correctly sloped flooring and floor drainage to ensure no water build-up
- Effective waterproofing by 'tanking' the room

- Good ventilation to remove moisture
- Protection of neighbouring rooms from water migration (shower area)

Table 5.10 shows the main features and benefits of a walk-in wet room.

Feature	Benefit
Flat and level access to the shower area	Easy and safe access Suitable for someone with limited mobility or a wheelchair user
Fully watertight	Protects the building from moisture damage and leaks Prolongs the life of the tiles and grout
No shower enclosure or screen required	Allows the installation of a shower where a traditional shower tray and enclosure may not have been viable Can be installed on wooden and concrete floors with no predetermined shower sizes or shapes
Fully tiled or panelled	Opens up the room and creates a more attractive space. May be contemporary minimalist style or more traditional
Flexible layout and design	No space or size limitations to the shower area
Slip resistant floor surface	Safety
No shower tray required	Compatible with underfloor heating

Table 5.10: Features and benefits of a walk-in wet room

It is usually recommended that the whole floor is 'tanked' with a turn-up of 100 mm onto the walls. Fully tanking the walls in the shower area is essential to prevent water penetrating the building structure.

Professional Practice

A customer has asked you to look into whether a wet room would be suitable to install in his house. Discuss the advantages and disadvantages of wet rooms and if it is suitable for the needs of the customer, who has an infirm mother living with him.

Positioning and fixing designer sanitary appliances

Glass sanitary appliances

Glass wash basins are becoming increasingly common installations for the plumber and require a more delicate approach than some other material types. The two types – above-counter and recessed bowls – need to be fitted according to the manufacturer's instructions.

For the recessed type you need to cut the top of the counter to shape, so that the basin will sit in snugly. The tap is usually close to the bowl, so you should install this first before fixing the bowl into place. Apply silicon around the cut shape and then place the bowl into it. Remove

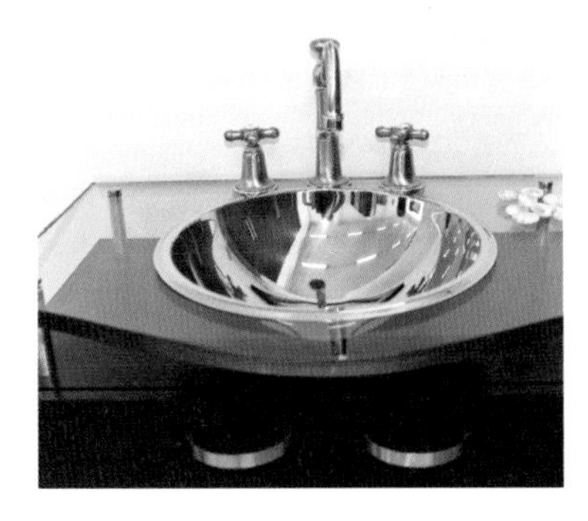

Recessed bowl type basin

Above-counter type basin

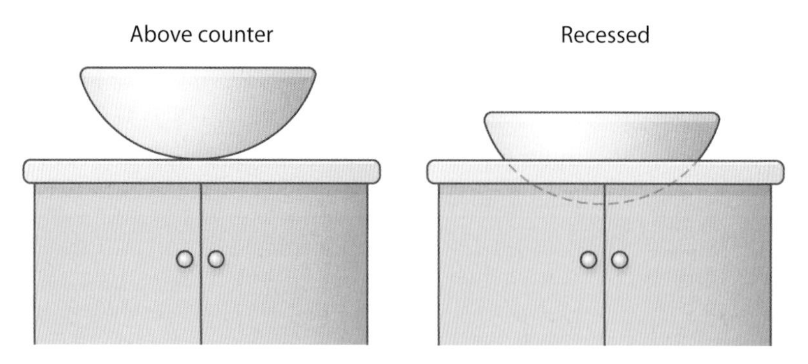

Figure 5.39: Recessed bowl and above counter types

any excess silicon and leave to cure. You can then fit the trap to the bowl, taking care not to overtighten or the bowl may crack.

The above-counter type looks like it is balancing on the top, although it is securely fixed down by the basin waste. Once you have fitted the taps, you need to cut the waste hole as tightly as possible, then place clear silicon around the hole to bed the basin on. Fix a bracket via the waste pipe under the counter top to give rigidity to the installation, then fit the trap and waste (see Figure 5.40).

Cutting through the top is usually done using a jigsaw with a special blade that cuts downward to prevent chipping of the work surface.

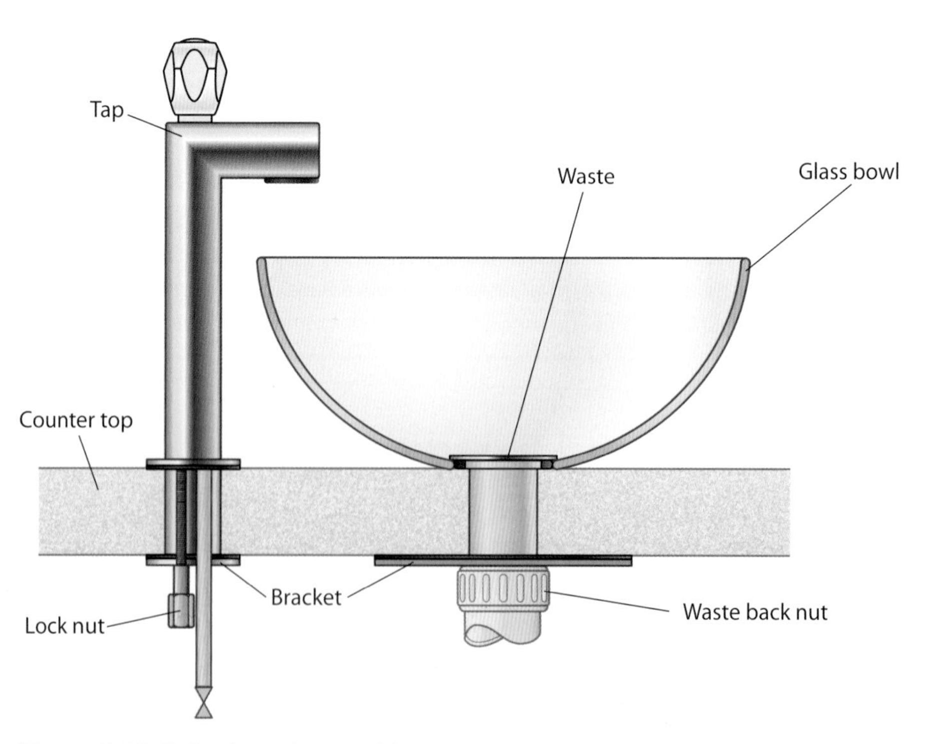

Figure 5.40: Fixing bowl into position

Antique-style sanitary appliances

Antique styling is often a way to give customers the distinctive design they are looking for. The choice of bathroom appliances is large and the methods of installation are equally varied, so you must always read the manufacturer's instructions during the planning stage to prevent problems occurring.

As you can see from the picture, the slipper bath looks elegant but the pipework – hot and cold water and the waste pipe – has to come through the floor. In this example, the bath has tap holes so careful alignment is required to ensure the holes in the floor match those of the bath.

A slipper bath

This exposed waste overflow pipe fitted to the bath is going to be on show, so you need to avoid any damage such as tool marks.

The chrome trap and pipework need careful positioning to the waste pipe under the floor, which in turn must be positioned and secured to stop movement or sagging creating future problems.

The shrouds shown above are designed to fit over the hot and cold pipework to give them a polished effect. On some installations, the copper pipe is left exposed and lacquered to maintain a copper finish. However, free-standing taps are becoming more popular.

Victorian basins are fully supported by their legs

A Victorian basin such as the one pictured left is fully supported by its legs, which give it a classical look. You should follow the manufacturer's instructions carefully as the legs require fixing to the floor, so make sure no pipework or cables are underneath. The fixing screws should be at least 50 mm in length to give the correct anchorage into the wall to support the load.

The classical WC looks the part, but of course has a modern flushing mechanism. The height of the cistern is critical to its safe operation and, because of the leverage that can be applied to the chain, it has to be well secured to the wall. Always check with the manufacturer's instructions before installation. The arrangement for high-level WC suites with wall-mounted cistern should be approximately 2 m above the floor; this provides the greatest amount of potential energy for flushing.

The height of a Victorian WC's cistern is critical

Remember

Chrome should be cleaned with non-abrasive specialist cleaner.

The flush pipe connecting the cistern to the pan can be supplied in one, two or more sections. The typical diameter of a high-level flush pipe is 35 mm. It is essential that pans for use with high-level cisterns have flushing rims designed to utilise the high velocity flush water, without undue splashing of the toilet area. This makes it difficult to change just the pan, so a whole new WC sometimes has to be purchased.

Sanitary appliances with floor-mounted taps

Floor-mounted taps are popular with the modern free-standing bath. Fixing these taps requires a good, solid flooring of wood or concrete; the bracketing firmly fixes the tap to the floor. You must plan the pipework carefully to ensure a good connection to the base.

The standpipes shown below will look good in an antique style bathroom. The floor must be firm enough that there is no movement when the taps are being used; otherwise the connections could loosen, causing a leak.

PROGRESS CHECK

1. What is a wet room shower tray?
2. What four factors will decide the finished floor height of a wet room?
3. What should the fixings not penetrate?
4. Name one of the features of a wet room.
5. What are the two common types of glass wash basins?
6. When working on chrome waste pipework, what must you take care not to do?
7. Why must fixing screws for Victorian basins be at least 50 mm long?
8. What height must a high-level cistern be above the pan?
9. What must the floor be like when using stand taps?
10. What must you not use when cleaning chrome?

4. Know and be able to apply the fault diagnosis and rectification procedures for sanitary pipework systems and components

Getting information when repairing faults

The user of the system will be the first one to be able to tell you if there is something wrong – but you need to be able to work out just what they know. For more about dealing with end-users, see Unit 3 page 136.

Activity

Before you read on, make a note of some of the keys to good communication with clients, then look back at Unit 2. How much did you remember?

The details of required maintenance and testing will be set out in the appropriate scheme document and manufacturer's instructions, and in BS 6465 Part 3. Maintenance schedules will depend on the amount of use and the degree of soiling that occurs. In domestic dwellings, weekly cleaning is usually sufficient; however, in public facilities, such as a college, a constant programme of cleaning and maintenance is needed.

BS 6465 Part 3

- Cast iron soil pipes will accumulate rust in bends and will require cleaning from time to time.
- Access cleaning eyes and caps may require new sealing rings after use.
- Any chemical cleanser should be checked for its suitability before use.
- Care should be taken when using high-pressure cleaning devices to ensure they do not damage the system.
- Any equipment used to rod the pipework should not damage the internal pipework.
- Mechanised equipment should only be used by trained operatives.
- Any damage to paintwork identifying the use of the system should be renewed in accordance with BS 1710.

BS 1710

Routine checks and diagnostics

The following checks can be made as part of routine maintenance.

Sanitary system diagnosis is usually straightforward. Leakages or blockages are the most common pipe system fault, but require a logical diagnostic procedure to locate the blockage or leak.

Appliance faults are common and can be diagnosed by operating the appliance and observing its function. Water backing up into the appliance means there is a blockage. If a blockage is on a branch with several appliances:

- if all are blocked the blockage is between the last appliance and the drain
- if the first two appliances are blocked but the last two are not, the blockage is in the pipework between the two sets of appliances (see Figure 5.41).

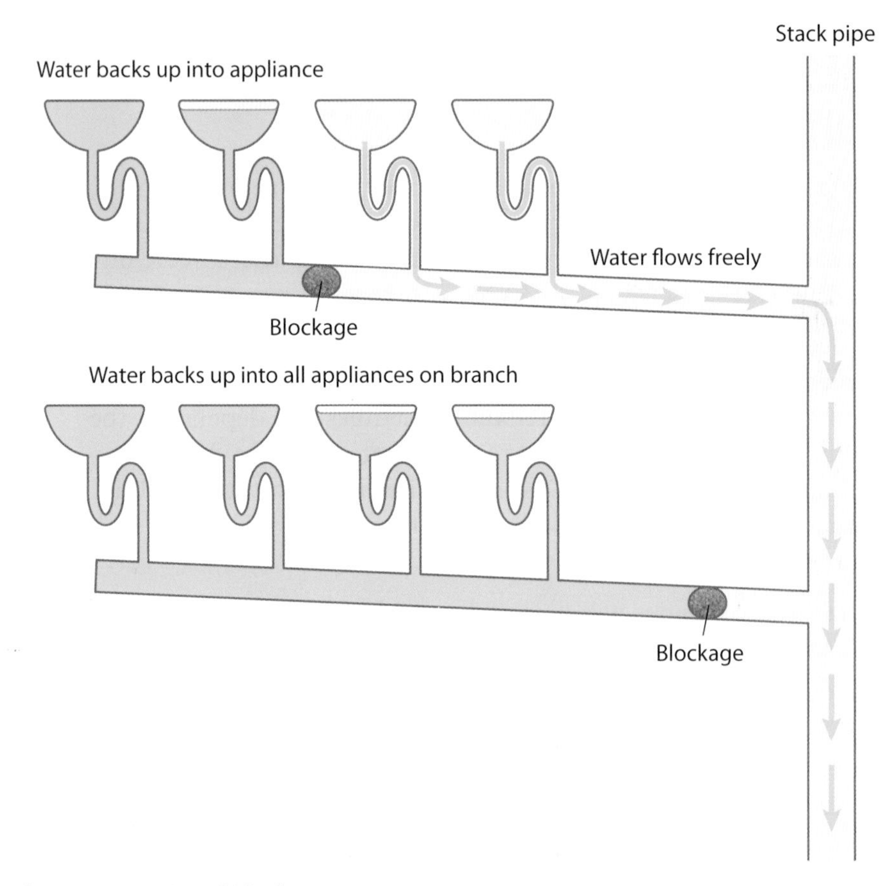

Figure 5.41: Site of blockage

Cisterns

Cisterns with outlet valves

Cisterns that appear to be filling when not being used should be suspected of having leaking outlet valves. The leak from a WC flush valve is not always easy to spot. By sprinkling a suitable dusty powder (such as talcum powder) on the surface of the WC pan, any moving water will be seen.

Cisterns with siphons

The diaphragm inside a siphon may break or tear as it reaches the end of its life. An indication of this is difficulty in priming the siphon so that flushing takes place. If a demountable siphon has been fitted, it may be simple to replace it; otherwise replace the complete siphon assembly.

Cleaning

Cleaning materials incorporating corrosives, abrasives or acids should be avoided, as they can damage the sanitary appliance or supply and discharge pipework. Always follow the manufacturer's instructions for sanitary cleaning or descaling materials.

Did you know?

BS 1710 gives all the service pipe colour codes for identification purposes.

Removing scale and limescale

To remove scale, encrustation and deposits, cleaning materials containing acid and alkalis should be used carefully to prevent damage to sanitary appliances and water supply fittings or injury to those doing the cleaning. To avoid damage, use descaling materials containing corrosion inhibitors. See Table 5.11.

Application	Method	Notes
The removal of limescale accumulations in discharge stacks and branch pipes The removal of grease and soap residues from the discharge pipes from wash basins and sinks	Apply diluted, inhibited, acid-based descaling fluid directly to scale. Apply these measured quantities of fluid into the pipes at predetermined points on the pipeline, or by using a drip feed method (acid strength approximately 15% inhibited hydrochloric acid, 20% ortho phosphoric acid). For heavy limescale encrustations, undiluted descaling fluid can be used (30% inhibited hydrochloric acid, 40% ortho phosphoric acid). The softening scale can be removed by thorough flushing and where practicable by the use of drain rods and scrapers. On completion of the work, the system should be thoroughly flushed with clean water. Particular care should be given to the traps of appliances to ensure that all traces of acid are removed from the trap water seals when the work is finished.	Acid-based descaling fluid will attack linseed oil bound putty. Care should be taken to avoid unnecessary or prolonged contact of descaling fluid with the jointing material used in the jointing of the outlet fittings and wash basins and urinals. Drip feed method: The acid-based descaling fluid is allowed to drip slowly into the discharge pipe at a rate of about 4 litres over a period of 20 min. Repeat, after flushing with clean water, if necessary for very heavy deposits.
NOTE Acid-based cleaners in contact with chlorine bleach will produce chlorine gas. It is essential that discharge systems be thoroughly flushed before acid-based cleaners are used, to remove as far as possible all traces of chlorine bleach residues. All windows should be opened in the areas where acid-based cleaners are being used.		

Source: Table BS 12056-2 (Table g2 (extract))

Table 5.11: The use of descaling materials

Urinals

Depending on the urinal, the maintenance requirements may be daily, weekly, monthly or even less frequent. Rubbish blocking the urinals may have to be removed every day. Blockages due to deposits may require specialist contractors to remove the problem. In any case, it is important that blockages in flushed urinals are cleared immediately or there is a risk of flooding with automatic-flushing apparatus. If the blockage cannot be removed quickly, it is essential that the water supply for flushing is turned off and the facilities closed until the fault is rectified.

Did you know?

Urinals flushed with hard water are likely to develop hard waste pipe deposits that may need to be drilled or chiselled out. The pipework may even need to be completely replaced.

Checking component operation

WC macerators

A macerator WC has many parts, as you can see from Figure 5.42. Most parts are available as spares, including the motor. These units are fairly reliable, but some common faults are listed in Table 5.12.

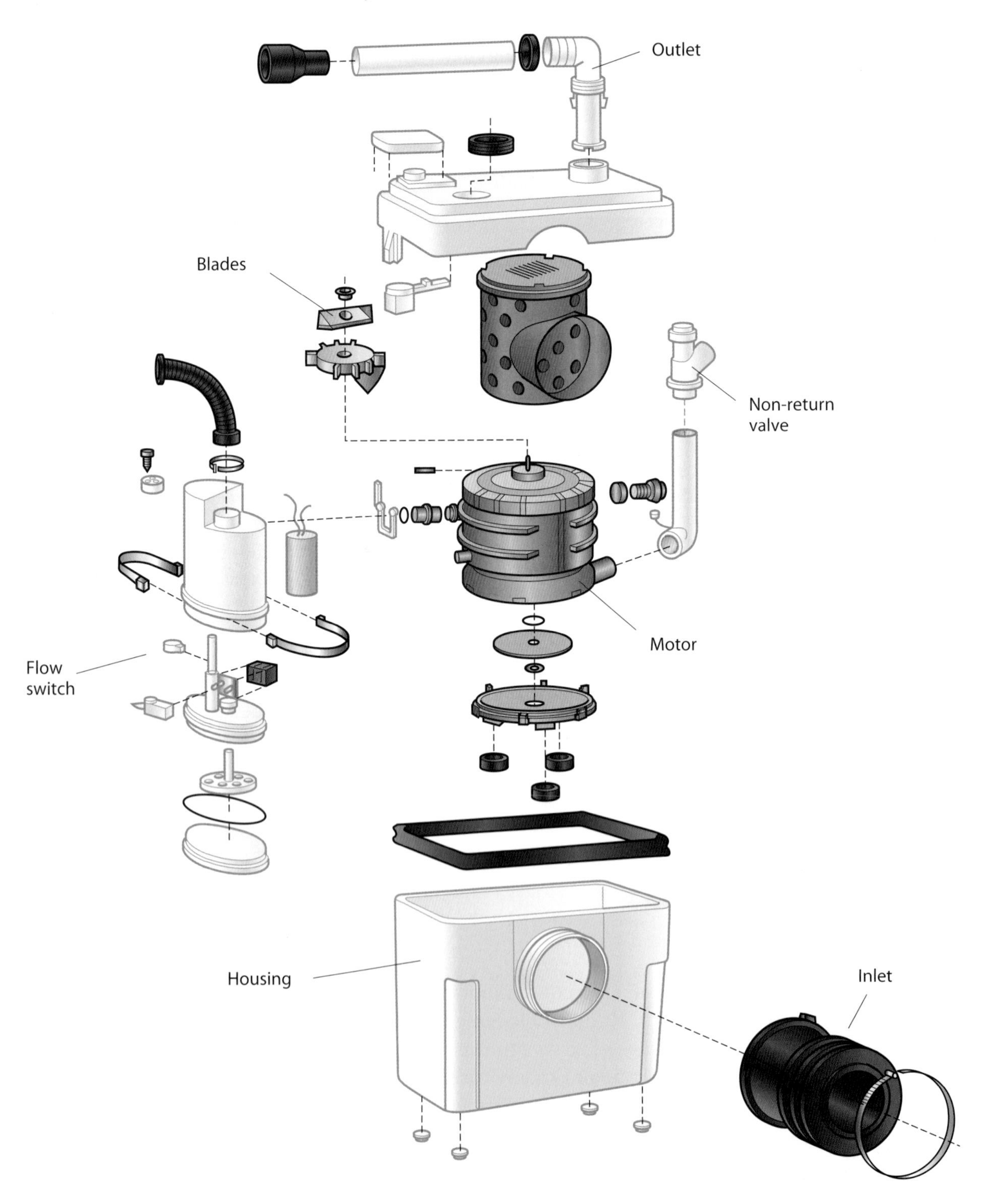

Figure 5.42: Working parts of a macerator WC

Fault	Possible cause of fault
The motor operates with an intermittent on/off action	There is a dripping tap or float-operated valve. Macerator non-return valve faulty
The water in the pan only discharges slowly	The inner grille is blocked up and needs cleaning. Partial blockage of entry to macerator
The motor operates but runs for a long time	Pipework could be partially blocked or the activating pressure membrane is coated with limescale so it stays activated
The motor does not activate	The electricity may be off, or there may be a defective motor or micro switch
A rattling or crunching sound is heard	A foreign object, such as a toilet block holder, has made its way to the grille
The motor hums but does not run	The capacitor or the motor is defective or blades jammed with a foreign object

Table 5.12: Common faults found in macerator WCs

The procedure is first to do a visual inspection to see if anything is obvious. Then turn the macerator on, check what is going wrong and identify the fault. Isolate the appliance and decommission, remove and clean. Carry out the repair, then reinstall and recommission.

Sink waste-disposal units

The major problem with a sink-waste disposal unit is usually a foreign body in the unit (which tends to be cutlery). This can jam the blades, causing a thermal cut-out device on the electrical supply to stop the motor. Sink waste-disposal units usually come with a de-jamming tool to help you remove the offending article. Some models have a reverse option to assist with blockages, sending the problem item back into the sink. Bottle traps should not be used with the disposal unit.

Remember

Always wear protective gloves when working on existing sanitary systems.

Air-admittance valves

The most common problem with air-admittance valves is when moisture has formed in the valve and frozen. This will be due to the valve being situated in an unheated area, where it needs to be insulated to prevent it freezing. As appliances are flushed, the air admittance valve should be checked to see if the valve opens and closes correctly.

Safe isolation

You should always put in the appliance last to prevent it being used before it is connected.

You should always seal open ends of waste and soil pipes to stop vermin getting into the building or debris entering the pipework and causing

BS 6700

Safety tip

Always ensure that the system is isolated before starting any maintenance operation, and inform everyone in the property.

blockages. Water services should be isolated and capped in accordance with BS 6700. A notice should be displayed to inform people of the situation. Discussion with the user is vital; always communicate with them and other trades.

Safe isolation is vital, for your own safety and the safety of others. For more about safe isolation, look at Unit 7 pages 407–415 and read carefully.

Professional Practice

Malcolm is an experienced plumber who has an apprentice called Ian. Their employer has asked Malcolm to show Ian how to carry out routine maintenance on a domestic sanitation system. Discuss a checklist of maintenance tasks that will be required during the routine maintenance.

PROGRESS CHECK

1. How should you gain information from a customer regarding a system fault?
2. Where can details of appliance testing procedures be found?
3. Which BS gives details of pipe colour system identification?
4. How could a blockage in the pipework of a branch connecting four basins be found?
5. How should limescale on sanitary appliances be removed?
6. If a urinal bowel is blocked and cannot be repaired immediately, what should be done?
7. What tends to be the most common type of blockage for a waste-disposal unit?
8. What is the most common problem with an air-admittance valve?
9. When isolating an appliance, what should be done to the water supply?
10. Draw a picture of a waste-disposal unit de-jamming tool.

5. Know and be able to apply the commissioning requirements of sanitary pipework systems and components

Getting information

BS 5572/6465/12056

BS 5572/6465/12056 should be complied with when commissioning a sanitation system along with the Building Regulations Parts G and H. Manufacturers' instructions should always be complied with at all times; failure to do so will lead to warranties and guarantees becoming void and the cost of any rectification and fines would be expensive.

Visual inspection

You should carry out a visual inspection during the installation phase, correcting any defects as the work progresses. Large installations or areas that are to be concealed before final tests can be tested in sections. The final tests – usually an air test and performance tests – have already been covered in *Level 2 NVQ Diploma Plumbing*. You should record the test findings and any making good in accordance with the project specification and any statutory requirements and notices to be given for inspection.

All sanitary appliances should be checked for signs of damage and correct operation, and any faults should be repaired before handover to the client. All appliances should drain quickly and quietly through the installation. Performance tests are required to leave a minimum of 25 mm in every trap in peak working conditions.

Remember

At Level 2, you will have learned how to carry out an air test on a sanitary pipework system, and how to carry out a performance test for trap seal retention. Look back at *Level 2 NVQ Diploma Plumbing* page 430 to remind yourself what is involved.

Testing

Testing branch discharge pipework

Test for self-siphonage

Self-siphonage is when an appliance siphons out its own trap seal. To test for self-siphonage, make sure the trap is full using a dipstick then fill the sanitary appliance to its highest level, such as the overflow level in a wash hand basin. Remove the plug and let out the water at full bore, simulating the worst condition scenario for self-siphonage. Measure the trap again: the seal depth should be at least 25 mm. Perform this test three times on each appliance and record the results. If the appliance fails again, you will need to investigate further and carry out repairs or alterations.

To measure the trap seal using a dipstick, take the trap tube depth away from the overall depth, leaving the actual seal depth (see Figure 5.43).

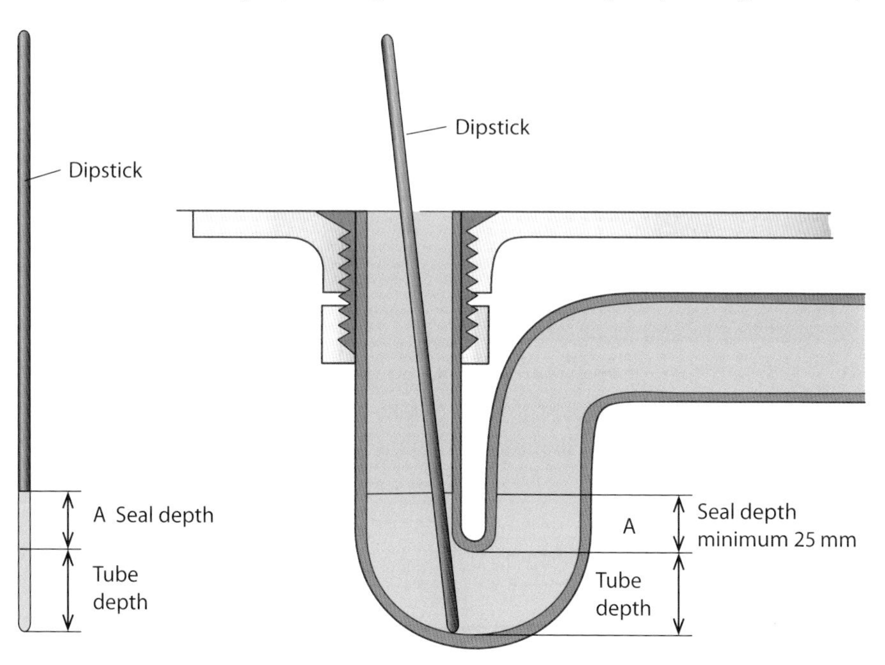

Figure 5.43: Trap seal depth measurement

Test for induced siphonage

Induced siphonage is when the operation of other connected appliances causes the trap seal to be lost on another appliance on the same branch.

To check for induced siphonage, first check that the trap seal depth is full on all the appliances on the same branch. Then fill each appliance to its overflow level, and discharge all the appliances at once. Measure the trap seal loss on all appliances to check they are all 25 mm or more deep; do this three times, recharging the trap or traps before each test. Take remedial action if any faults are found. The maximum loss of seal in any one test, measured by a dipstick or small diameter transparent tube, should be taken as the significant result and all test results recorded.

Testing for induced siphonage

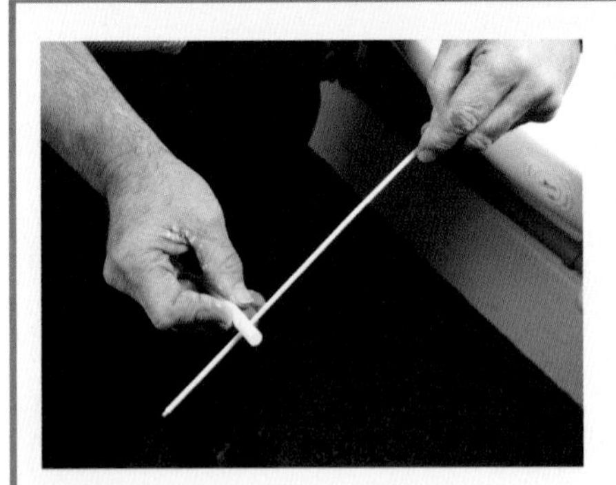

Step 1 Chalk up the dipstick.

Step 2 Fill traps with water. Measure initial water level with dipstick.

Step 3 Put in plugs and fill all basins with water.

Step 4 Empty basins simultaneously or as close as possible after one another.

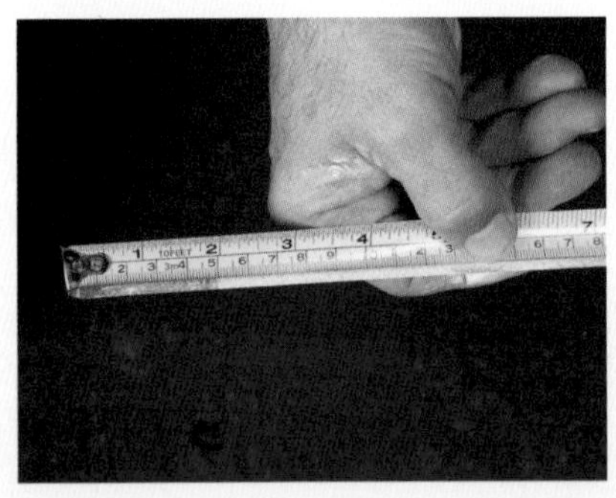

Step 5 Take a second measurement. Repeat the process three times and take the lowest reading. Take remedial action if there are signs of induced siphonage.

Main discharge stack

Test for induced siphonage and compression

One test is used to identify any induced siphonage or compression within the stack. This is done by selecting a range of appliances from Table 5.13 and discharging them simultaneously from the highest floor. The results should show no loss of seals greater than 25 mm across all appliances or traps discharging back into the appliances. If the system fails remedial work needs to be carried out and the system needs to be retested. The table does not include baths as they are considered to be used over too long a period to add to the peak flow of the system.

Type of use	Number of appliances of each kind on the stack	Number of appliances to be discharged simultaneously		
		WC	Wash basin	Kitchen sink
Domestic	1 to 9	1	1	1
	10 to 24	1	1	2
	25 to 35	1	2	3
	36 to 50	2	2	3
	51 to 65	2	2	4
Commercial or public	1 to 9	1	1	–
	10 to 18	1	2	
	19 to 26	2	2	
	27 to 52	2	3	
	53 to 78	3	4	
	79 to 100	3	5	
NOTE These figures are based on a criterion of satisfactory service of 99%. In practice, for systems serving mixed appliances, this slightly overestimates the probable hydraulic loading. The flow load from urinals, spray tap basins and showers is usually small in most mixed systems, hence these appliances need not normally be discharged.				

Source: Table G abstracted from British Standard 12056 – 2

Table 5.13: Number of sanitary appliances to be discharged for performance testing

Worked example

For a domestic property with two toilets, two wash hand basins, one shower and one sink you need to check Table 5.13. This shows that you need to discharge simultaneously from the highest floor one WC, one wash hand basin and one kitchen sink.

Commissioning macerator WCs

- Do a visual inspection. Checks should include the soundness of the water supply and the discharge pipework; also ensure that the electrical supply connection is correct and has been properly tested.
- The electricity supply should be fused at 5 amp and protected by a residual current device (RCD) with a maximum rating of 30 mA (see page 392).
- Turn on the water and fill the cistern. Check for leaks at the cistern and the supply or distributing pipework. Then set and adjust the float-operated valve to the correct level.
- Turn on the electrical supply and flush the WC once.
- The motor on the unit should run for 3 to 10 seconds (depending on the length and height of the pipe run).

- If the motor runs for longer than this, check whether:
  - there is blockage to the pipe run, joint blocked by solvent cement
  - there are kinks in the flexible hose connections
  - the unit's non-return valve is working properly (this should be sited at its inlet).
- Flush the WC several times to check all the water seals. The discharge pipe should be fully checked for leaks, as should the pipework from other appliances.
- Check the float-operated valve and appliance taps for dripping (annoying short-term activation of the pump can occur).

Dealing with defects when commissioning

Blockages

WCs

Blockages occur every now and then. They are usually due to too much toilet paper being used or a sanitary towel being thrown down the pan. Typically, the material will lodge on the outlet side of the trap, out of view. The blockage can often be removed by discharging a full bucket of water into the pan at the same time that the WC is flushing (creating increased pressure). Alternatively, a disc plunger can be used. Another cause of poorly performing WCs can be that a detergent-block holder has come loose from the pan. These can make their way to the pan outlet and wedge themselves there, causing a build-up or blockage. The only solution is to remove the WC – then you can get at the item causing the blockage.

Traps

Blockages occur in traps because deposits such as hair and soap build up. Traps may need cleaning from time to time. Sink traps have a greater blockage risk due to grease and fat deposits. Because of this, bottle traps should not be used on a sink. Blocked bottle traps will need replacing rather than just cleaning. Smells from the appliance tend to be a common fault.

Pipework

Blockage in a pipework system should not be a common occurrence. If pipework does block, it is likely to be in a horizontal run that is not laid to the correct fall. This means you will be faced with a pipework modification job.

Creating a commissioning record for a sanitation system

A commissioning record will need to include: the date the appliances were commissioned; any alterations or repairs required; performance test results; soundness tests of pipework; the name of the commissioning engineer and their signature; any inspection from building control.

SANITATION COMMISSIONING SHEET			
Address **Engineer's name**	**Visual inspection report**		
Soundness test 38 mbar for 3 min No pressure drop	Yes	No	Fault
Performance test self-siphonage			
Appliance 1			
Appliance 2			
Appliance 3			
Appliance 4			
Appliance 5			
Appliance 6			
Appliance 7			
Performance test induced siphonage			
Appliance 1			
Appliance 2			
Appliance 3			
Appliance 4			
Appliance 5			
Appliance 6			
Stack induced siphonage and compression test			
Report overall condition of system			
Engineer's signature **Date / /**			
Note: tick box in relevant field If more appliances than in table add to report area			

Figure 5.44: Sample of a sanitation commissioning sheet

Notifying works to the relevant authority

The Building Control Office at the local county council sometimes has to be informed. The customer does not usually need to apply for planning permission for repairs or maintenance on drainpipes, drains and sewers. However, if the bathroom or kitchen is part of an extension, planning permission may be required. Occasionally, your customer may need to apply for planning permission for some of these works because your council has made an Article 4 Direction withdrawing permitted development rights.

Although the work itself may not require planning permission, you should clarify ownership and responsibility before modifying or carrying out maintenance. Drains, sewers and manholes may be shared with neighbours or owned by the relevant water authority. Failure to confirm these details or to comply with relevant standards and legislation could lead to legal and remedial action at your own cost.

If you are working in a listed building, you will need listed building consent for any significant works, whether internal or external, and you should make sure the home owner is aware of this. The work should always comply with the Building Regulations Parts G and H, particularly in relation to sanitation systems.

The self-certification scheme allows plumbers to register their own work through different organisations on installation of new bathrooms and kitchen appliances. The plumber should check in Annex 2 of the self-certificating scheme found in the Building Regulations Approved Document G.

Remember

Very rarely you will get a rogue customer who will damage appliances then blame you or co-workers, so they do not have to pay for the work. A signature on a customer satisfaction form on completion can save problems later.

Handing over to the end-user

Instructions on operation or maintenance of the system and appliances should be provided to the user and, where applicable, attached to the installation. If possible, you should show the customer or user how to operate any appliances in the system. You should also give a copy of the self-certification scheme paperwork to the customer for future reference, along with a customer satisfaction form for the customer to sign as part of the handover procedure.

PROGRESS CHECK

1. What source documents should be used when commissioning sanitation systems?
2. When should the inspection of an installation take place?
3. How many times should an appliance be tested for self-siphonage?
4. Before carrying out any performance testing, what should be done to the traps?
5. To test a stack for induced siphonage, how should the appliances be discharged?
6. What should the electrical supply for a macerator be fused at?
7. What usually causes a WC blockage?
8. What must a commissioning record include along with the engineers' signatures?
9. What does the self-certification scheme enable plumbers to do?
10. What should the customer always be shown?

Check your knowledge

1. The bend at the foot of the stack should have as large a radius as possible but not less than:
   - **a** 400 mm
   - **b** 150 mm
   - **c** 200 mm
   - **d** 1000 mm.
2. Air-admittance valves are now commonly used because they:
   - **a** let air out of the system
   - **b** save costs
   - **c** are non-mechanical so do not break down
   - **d** conform to BS 6700.
3. A macerator can be used only in a property where there is:
   - **a** no soil pipe
   - **b** disabled access
   - **c** soft water supply
   - **d** another bathroom with a traditional soil pipe.
4. The purpose of a waste water lifter is to pump waste into a sewer which may be:
   - **a** above the level of the house drainage system
   - **b** below the level of the house drainage system
   - **c** blocked upstream of the house
   - **d** contaminated with water.
5. The frequency factor indicates how often the appliance is:
   - **a** maintained
   - **b** cleaned
   - **c** broken
   - **d** used.
6. The changing climate means it is not possible to rule out flooding. Usually, it is satisfactory to design to what hourly rainfall intensity?
   - **a** 25 l
   - **b** 50 mm
   - **c** 75 mm
   - **d** 100 mm
7. What is one of the important requirements of a wet room?
   - **a** Stable (non-flexing) flooring
   - **b** Able to withstand rain
   - **c** It is permeable
   - **d** It has a large area
8. What is a shroud designed to fit?
   - **a** A basin
   - **b** A cistern
   - **c** Pipes
   - **d** Mirrors
9. What do sink waste-disposal units usually come with?
   - **a** De-jamming tool
   - **b** Method statement
   - **c** Steel fork
   - **d** Bottle trap
10. Sink traps have a greater blockage risk due to?
    - **a** The trap being only 32 mm
    - **b** More detergent being used
    - **c** Grease and fat deposits
    - **d** Large pipe diameters

Preparation for assessment

The information in this unit of the book, as well as the continued practical assignments that you will carry out in your college or training centre, will help you with preparing for both your end-of-unit test and the diploma multiple-choice test. It will also support you in preparing for the practical assignments you will need to do to demonstrate your understanding of and ability to apply above ground drainage systems installation, commissioning, service and maintenance techniques.

There are opportunities throughout the unit for you to test your progress in and understanding of the required underpinning knowledge; this will enhance your preparation for the forthcoming assessments, so make good use of them.

The unit will be assessed by the following assessment method:

- external set knowledge assessment
- externally set assignments.

With regards to the mechanical services industry, you will need to know:

- the types of sanitation system and their layout requirements
- the design techniques for sanitation and rainwater systems
- the installation requirements of sanitation system components
- the fault diagnosis and rectification procedures for sanitary pipe work systems and components
- the commissioning requirements of sanitary pipe work systems and components.

Using your knowledge from this unit practise by testing yourself on the following key points: Design curve, Prohibited zone, Air-admittance valves, Slab-type urinal, a self-cleansing surface, Approved Document M, Foul tanks, Cesspits, Septic tanks, BS EN 12566-1, Maintenance and cleaning, Drainage mound, Invert level, Effluent, Wetlands, Drainage fields, Oxygenation, WC macerators, Sink waste-disposal units, Waste water lifters, Backflow loop, Effluent receiver chamber, Customers' needs, Building layout, Pathogenic germs, Sewer surcharging, Fire protection, Minimum provision, Discharge unit, Stack sizing, Hydraulic capacity, Intermittent, Frequency factor, Air flow rates, Sizing gutters, Effective roof area, Gutter flow capacity, Walk-in wet rooms, Initial planning, Finished floor height, Vertical water barrier, 'Tanked', Glass sanitary appliances, Antique-style sanitary appliances, Chrome trap, High-level WC suites, Flushing rims, Maintenance schedules, Diagnostics, De-jamming tool, Visual inspections, Performance tests, Self-siphonage, Dipstick, Induced siphonage, Compression, Disc plunger, Commissioning record, Notifying works, Self-certification scheme.

6 Understand and apply domestic central heating system installation, commissioning, service and maintenance techniques

Working on complex central heating systems requires detailed knowledge and understanding to do your job well. This unit provides advanced information about installation, maintenance, commissioning and design of a range of central heating systems and components in dwellings and large properties. This unit is a combination unit: it contains both performance- and knowledge-based contents and assessment. Within this unit, there will be knowledge and skills acquired from Unit 7 Electrical principles. Make sure you read these cross-referenced pages carefully.

This unit covers the following learning outcomes:

- know the types of central heating systems and their layout requirements
- know and be able to apply the design techniques for central heating systems
- know the commissioning requirements and be able to commission central heating systems and components
- know and be able to apply the fault diagnosis and rectification procedures for central heating systems and components
- know the installation requirements and be able to install underfloor central heating systems and components.

1. Know the types of central heating systems and their layout requirements

Space heating zoning requirements for large single-occupancy dwellings

The Building Regulations give minimum requirements for space heating zoning in Part L, and further guidance is given in the Domestic Building Services Compliance Guide. The guide divides heating systems into new systems and existing systems.

New systems

Key term

SEDBUK – Seasonal Efficiency of Domestic Boilers in the UK.

The boiler efficiency should be no less than 86 per cent. To check a boiler's efficiency, you can access the Boiler Efficiency database online at www.boilers.org.uk. This is updated regularly to give installers the latest **SEDBUK** value.

Developed under the Government's Energy Efficiency Best Practice Programme with the co-operation of boiler manufacturers, SEDBUK provides a basis for fair comparison of the energy performance of different boilers.

SEDBUK is the average annual efficiency achieved in typical domestic conditions, making reasonable assumptions about pattern of usage, climate, control and other influences. It is calculated from the results of standard laboratory tests together with other important factors such as boiler type, ignition arrangement, internal store size, fuel used, and knowledge of the UK climate and typical domestic usage patterns.

In existing dwellings, in exceptional circumstances the boiler efficiency may be no less than 78 per cent for a non-condensing boiler, but this will require an Approved Assessment Procedure outlined in Appendix A of the Domestic Heating Appliance Guide.

The Standard Assessment Procedure (SAP) forms a key part of heating design. It produces a Target CO_2 Emission Rate (TER) for new houses to achieve, and generates Dwelling CO_2 Emission Rate (DER) as built. The Building Regulations specify that the DER must be as good as, if not better than, the original TER. Existing properties that change ownership have to comply with a Reduced Data Standard Assessment Procedure (RdSAP), which is commonly referred to as the Energy Performance Certificate.

For all new systems, system circulation must be fully pumped for the space heating and also the hot water primary circuits. Any bypass valves should be installed to meet the requirements of the boiler manufacturer's instructions.

Replacing systems in existing dwellings

Replacement not involving a fuel or energy change

The seasonal efficiency of the new system should be the same as that of a new dwelling, and no worse than 2 per cent lower than the system it replaces. This 2 per cent allowance is due to changes that might be made to the type of boiler or controls used from one manufacturer to another.

Did you know?

If you do not know the existing efficiency of the system to be replaced, you can take the values from SAP Table 4a or 4b.

Replacement involving fuel or energy switch

If a new heating system or heat-generating appliance uses a different fuel, the efficiency of the new service should be multiplied by the ratio of CO_2 emission factors of the fuel used in the service being replaced. You should take the CO_2 emissions factors from SAP Table 12.

As in the new installation category, existing systems must be converted to a fully pumped system.

Space heating zones

A dwelling with a total usable floor area up to 150 m^2 should be divided into at least two space heating zones (see Figure 6.1) with independent temperature control, one of which has to be in the living room.

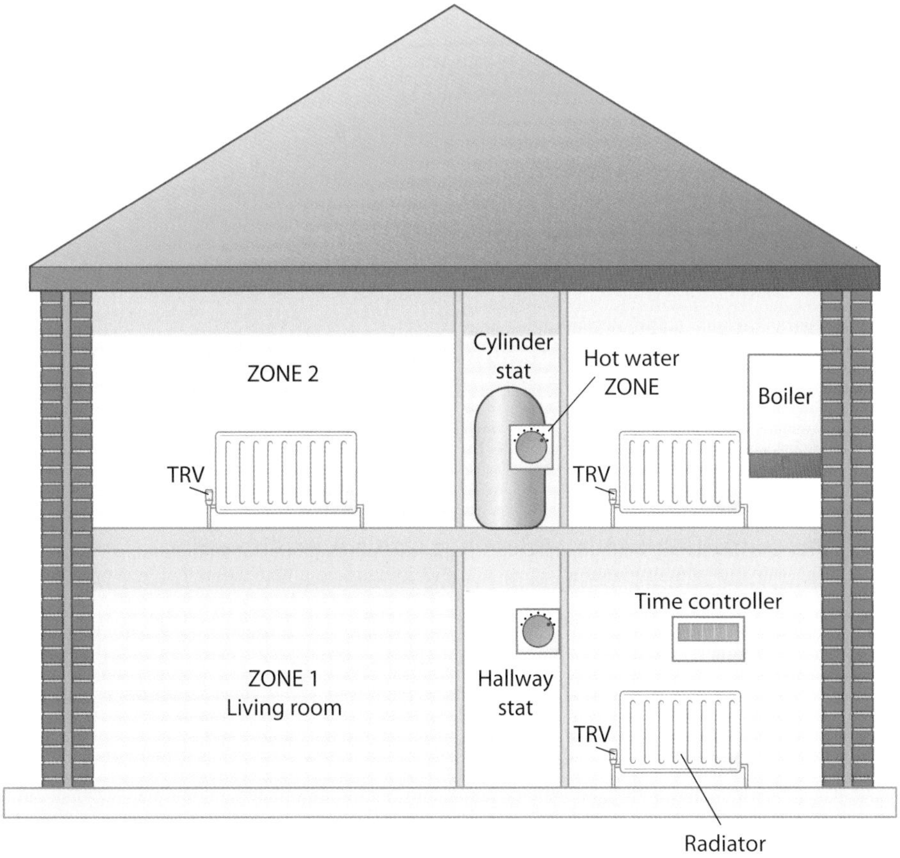

Figure 6.1: Space heating zones for floor area up to 150 m^2

Dwellings with a total usable floor area greater than 150 m² should be provided with at least two space heating zones, each one having a separate timing and temperature control (see Figure 6.2).

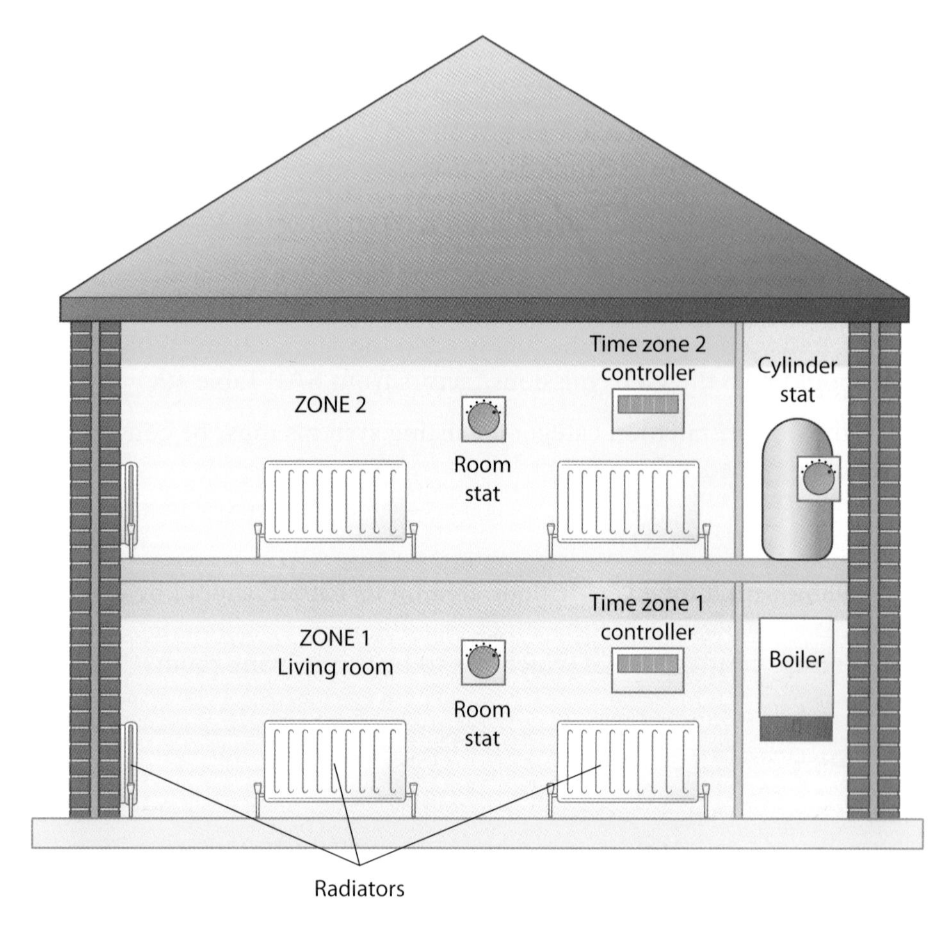

Figure 6.2: Space heating zones greater than 150 m²

Sub-zoning is not appropriate for single-storey open-plan dwellings in which the living area is greater than 70 per cent of the total floor area.

Water heating zones

All dwellings are required to have a separate hot water zone in addition to heating zones, unless hot water is produced by instantaneous supply, such as a combination or 'combi' boiler. In dwellings with a total floor area greater than 150 m², it is reasonable to provide more than one hot water circuit with separate timing and temperature control.

Components used in central heating systems

A range of special components enable the end-user to control, programme and make the best use of their central heating system.

This is not only for the end-user's convenience; it also means that systems can be made to work as efficiently as possible, minimising the impact on the environment.

Zone control

Systems with regular boilers must have separately controlled circuits via either three-port or two-port motorised valves for hot water cylinders and radiators; both circuits must have pumped circulation. Large properties must be divided into zones not exceeding 150 m^2 in floor area. This allows both the timings and temperatures to be controlled separately. Two-port valves can also be used to provide zone control: for example, giving lower temperatures in a sleeping area or different heating times. There are a few exceptions to this requirement, for which you should consult the Building Regulations Part L.

Time control

The time control of space and water heating up to a floor area 150 m^2 should be provided by one of the following options:

- a full programmer with separate timing to each circuit
- two or more separate timers providing timing control to each circuit
- programmable room thermostat(s) to the heating circuit(s), with separate timing of the hot water circuit.

The time control of space and water heating for a floor area greater than 150 m^2 should be provided by one of the following options:

- multiple heating-zone programmers
- a single multi-channel programmer
- programmable room thermostats
- a separate timer to each circuit
- a combination of the previous two options.

There are some exemptions from the above: see the Domestic Heating Compliance Guide for further information.

Controllers

With the need for more efficient heating systems, it is now more important than ever to use controls that will save energy. As every installation is different, it is becoming more important to identify the best controls for the system being used.

Full programmer

A full programmer allows the time settings for one or two space-heating zones and hot water to be fully independent. There is a range of programmers; some have a variety of options and multi-zones.

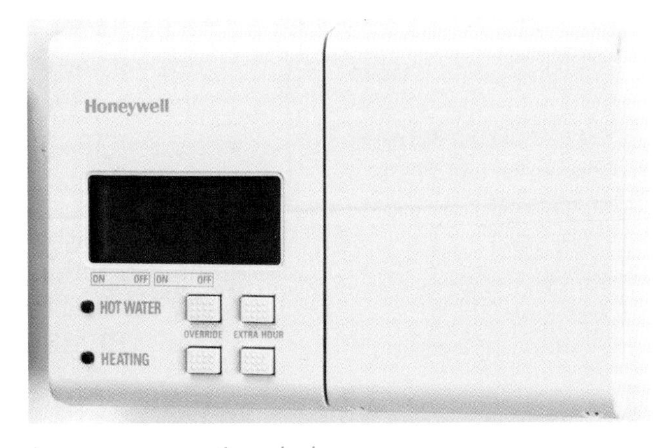

A programmer – time clock

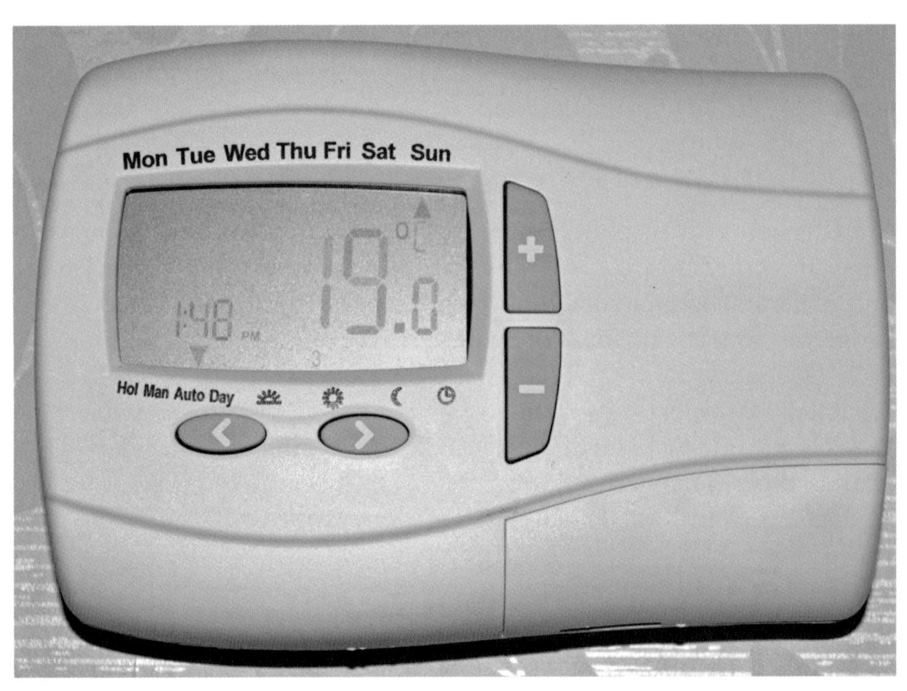

An optimiser can be used to vary the start times of domestic heating systems

Optimiser

This is an electronic device that can be used to vary the start times of domestic heating systems. Using sensors, the optimiser measures and compares the outside and inside temperatures. This measurement is then used to adjust the programmed start time of the heating system. This can either bring on or hold off the programmed start time.

The warm-up time needed for building heating systems depends on the outside air temperature and the residual temperature in the building. Optimum start controllers continually measure these temperatures and change the heating start times accordingly. The customer programmes in the times when the building is occupied and the inside temperature they want to maintain. The controller automatically starts the heating to give the required temperature by the time people start to arrive at the building.

Activity

Some optimisers are 'self-learning'. Do some research to find out what this means. Where would self-learning optimisers be useful?

Most optimum start controllers can also be configured to:

- turn off the heating during the day if the outside temperature rises above a set level (known as day economisation)
- turn off the heating early if the inside temperature will still be comfortable until people leave for the day (known as optimum off).

Optimum start controllers can be stand-alone devices or may be combined with other control functions, such as weather compensation. Many can be connected into site-wide building energy management systems (BEMS).

Common problems

The location of the sensors is very important. The external sensor needs to be on a north-facing wall, so that it does not get sun at any point of the day.

The internal sensor needs to be in a typical location in a colder (but heated) area of the building. Avoid direct sunlight, heat from process equipment, office equipment or draughts. Do not put the sensor in an office where people like to open the windows or switch off their heating in the winter.

When programming the start time, it is important not to build in any pre-heat time. The programmed start time should be the start of occupancy – the controller will bring the heating on in advance of this, as needed.

Professional Practice

A customer complains that a newly fitted heating system turns itself off every time the rooms get to temperature, but does not bring itself back on when the rooms cool. Discuss what may be going wrong with the controls in the new system to cause this problem.

Weather compensation

Weather compensation takes into account the outside temperature and reduces the boiler heat output accordingly. As with the optimiser, the electronic weather compensator measures outside and inside temperatures, plus the heating flow temperature. The resulting calculation is used to position a three-port mixing valve to give the correct flow temperature for the prevailing conditions inside and out.

Weather compensators are best suited for the larger systems; they are generally also used with an optimiser for greatest efficiency. The external sensor should be located on a north-facing wall so that direct solar heat does not affect the **ambient temperature** readings.

Key term

Ambient temperature – the air temperature surrounding a body.

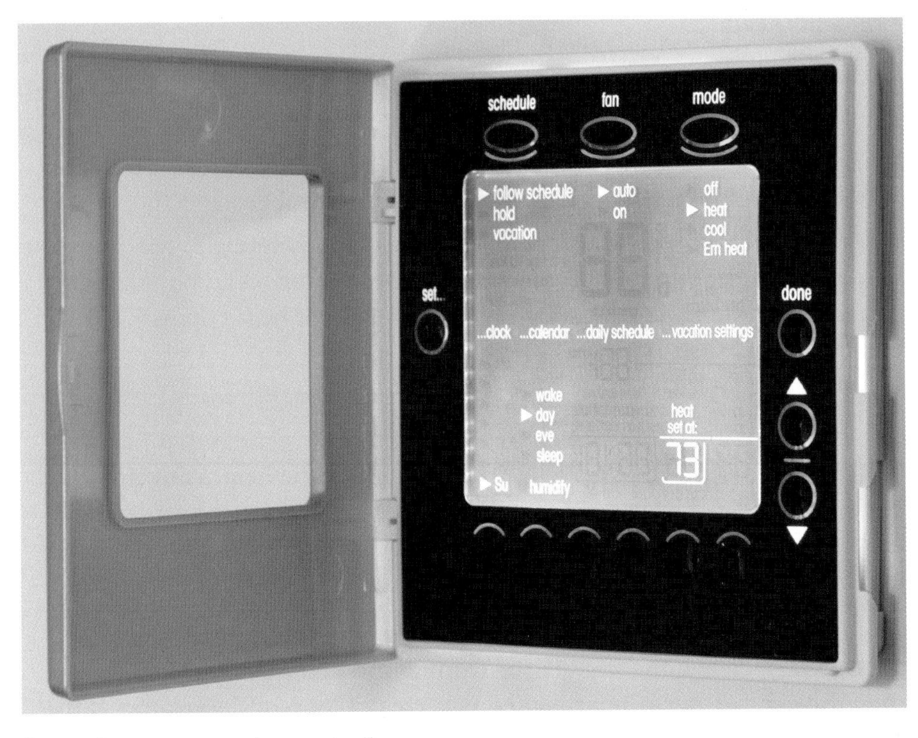

A weather compensation controller

Home automation system

Home automation systems are systems that enable homeowners to integrate features such as security, lighting, appliances and heating for maximum operational control by the user. Modern systems use secure and reliable radio frequency (RF) links throughout the home, to save time and money compared with wired systems. At a touch of a button, heating can be operated even when you are a long way from home. Voice and web interfaces enable users to use a phone to switch between various lifestyle profiles, which they can preset to suit their needs and preferences. A profile can regulate the temperature to unique values in each room and adjust these values at times chosen by the user (see Figure 7.22 on page 363).

PROGRESS CHECK

1. Where can you get further information and guidance on Part L of the Building Regulations?
2. What is the lowest boiler efficiency allowed according to SEDBUK?
3. What does TER stand for?
4. What do existing properties that change ownership have to comply with?
5. Regarding upgrades to existing systems, what does the system have to be made into?
6. What should dwellings with a total usable floor area greater than 150 m^2 be provided with?
7. What type of valve should be used to provide control over two or more heating zones?
8. If you are required to have a separate timer to each circuit, what is the floor area of the dwelling?
9. What can most optimum start controllers be configured to?
10. When placing external sensors, in which direction should the wall face?

Heat sources for central heating systems

Central heating systems can draw their heat from a number of different sources, which come under three main categories.

- High-carbon or non-renewable energy sources are natural resources that cannot be regenerated, such as coal, natural gas and fuel oils.
- Low-carbon or renewable energy sources, such as sunlight, wind or wave power, can be replenished. Power from these sources can be channelled through systems such as solar panels, fuel cells or heat pumps.

Ground-source heat pumps

Ground-source heat pumps (GSHPs) take low-level heat that occurs naturally underground and convert it to high-grade heat using an electrically driven or gas-powered heat pump. This heat can then be used to provide space heating for a building. GSHPs can also be driven in reverse to provide comfort cooling.

Did you know?

Heat pumps cover a wide range of capacities, from a few kW to many hundreds of kW.

The heat is collected through a series of underground pipes laid about 1.5 m below the surface, or from a borehole system. In both of these options, water is recirculated in a closed loop underground and delivered to the heat pump, which is usually located inside the building.

The installation of GSHPs requires a large amount of civil engineering works, such as sinking boreholes (50 m+) or digging 1–2 m deep trenches to house the collector pipe. The feasibility of doing this will depend on the geological conditions at the site.

Connecting a GSHP into an existing heating system is often constrained by the requirement of the system to operate at temperatures above that delivered by the GSHP. This can often be overcome, but at an extra cost. GSHPs are best suited to new-build projects, where they can be included in the building design.

Air-source heat pumps

Air-source heat pumps (ASHPs) take low-level heat that occurs naturally in the air and convert it to high-grade heat using an electrically driven or gas-powered heat pump. Such systems typically use an air-source collector, which is located outside the building. The heat generated can be used to provide space heating for a building. ASHPs can also be driven in reverse to provide comfort cooling.

Installation of an ASHP involves siting an external unit and drilling holes through the building wall (this may require planning permission). Some degree of additional pipework may also be required. ASHPs are a good alternative to GSHPs where lack of space is a problem. The performance of an ASHP varies dramatically with the external air temperature and this should be taken into account when considering the use of an ASHP system. In mild climates, such as the UK, frost will accumulate on the system's evaporator in the temperature range 0–6°C. This can lead to reduced capacity and performance of the heat pump. You should check with manufacturers, as some units have anti-frost cycling, which will reheat the evaporator to prevent this from happening.

How do air-source heat pumps work?

Heat from the air is absorbed at low temperature into a fluid that is a refrigerant. This fluid then passes through a compressor, increases in temperature, and gives off higher temperature heat to the heating and water circuits of the house.

There are two main types of air-source heat pump system: air-to-water and air-to-air.

- An air-to-water system distributes heat via the wet central heating system. Heat pumps work much more efficiently at a lower temperature than a standard boiler system would, so they are more suitable for underfloor heating systems or larger radiators, which give out heat at lower temperatures over longer periods of time. Solar panels for hot water will improve efficiency of the heat pump in the summer months.
- An air-to-air system produces warm air, which is circulated by fans to heat the home. These systems are unlikely to provide hot water as well.

Planning permission

In Wales and Northern Ireland, air-source heat pump installations require planning permission. In Scotland, they may be considered as 'permitted development', in which case you will not need planning permission. However, the criteria are complex so it is always a good idea to check with your local planning office. In England, domestic air-source heat pump systems are now classed as permitted development provided that they comply with certain criteria, including:

- there is no wind turbine at the property
- the external unit is less than 0.6 m^3 in size
- the unit is more than one metre from the edge of the householder's property
- it is not on a pitched roof, or near the edge of a flat roof
- it meets additional criteria if in a special area, such as a conservation area or World Heritage Site.

This list is not comprehensive, so you should contact your local planning office for full details.

Did you know?

Heat pump technology is well established on the continent, especially in Scandinavia.

Professional Practice

A customer has asked you to look into whether an air-to-air heat source pump would be suitable to install in his house, for heating and hot water needs. Discuss the advantages and disadvantages of heat pumps and whether this sort of pump would be suitable for the hot water and central heating needs of the customer.

Micro combined heat and power

Micro combined heat and power, or micro-CHP for short, is the process of generating both electricity and heat from the same source, close to where it is to be used. It provides efficient gas central heating and hot water like any other boiler, but also produces up to 1 kW of electricity per hour from most units. Any electricity that is not used can be sold to the electricity supplier for an agreed tax-free and inflation-linked price. Electricity suppliers will pay householders a generation and export tariff for every kilowatt hour (kWh) of electricity generated and for every kWh of electricity exported back.

There are three main micro-CHP technologies, each with a different way of generating electricity.

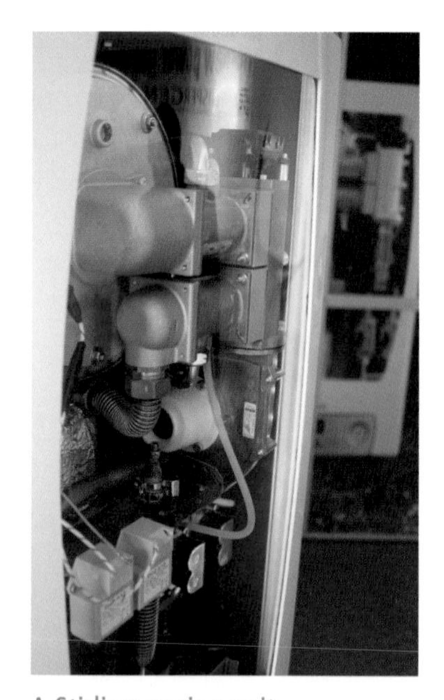

A Stirling engine unit

Stirling engine micro-CHP

This is new to the market, although the principle of the Stirling engine is well established. The electrical output is small relative to the heat output (about 6:1), but this is not necessarily a problem for micro-CHP. The only system currently available is a Stirling engine unit, and these are now being installed across the UK.

Internal combustion engine CHP

This is the most proven technology. These are essentially truck diesel engines modified to run on natural gas or heating oil, connected directly to an electrical generator. Heat is then taken from the engine's cooling water and exhaust manifold. They can have a higher electrical efficiency than a Stirling engine, but are larger and noisier. They are not currently available for the normal domestic market.

Fuel-cell CHP technology

This is also new to the market, in the UK and globally. Fuel cells work by taking energy from fuel at a chemical level, rather than by burning it. The technology is still at developmental stage and not widely available to consumers, but hydrogen fuel cells are looking promising.

Installation

For the installer, there is very little difference between a micro-CHP installation and a standard boiler. If the house already has a conventional boiler, a micro-CHP unit should be able to replace it as it is roughly the same size. However, the installer must be approved under the Micro-generation Certification Scheme.

Did you know?

Heat can be extracted from boreholes in granite rock, and water can be pumped in, to be used for heating.

Multiple boiler installations incorporating low loss headers

A boiler heat exchanger will only function at its peak efficiency when the water velocity passing through it is maintained within design limits. Boiler manufacturers should tell you what the limits are for each make and model of boiler.

In a few cases, the flow rate through the system will increase above the recommended maximum flow rate through the boiler heat exchanger, or it may be that the system flow rates are simply unknown. In other cases the reverse is true: the boiler flow rate exceeds the maximum system flow rate (particularly true in some multi-boiler installations). Installing a low loss header allows the creation of a primary circuit, within which water velocity can be maintained at the required flow rate, regardless of changes or requirements in the secondary circuits.

The water velocity is important, but so is the water temperature. There are two main problems here: thermal shock and temperature of the heat exchanger.

Thermal shock

If the temperature difference between the flow and return is too great, it puts a strain on the boiler, through thermal expansion and contraction, on the heat exchanger.

The temperature of the water passing through the heat exchanger is important, particularly with condensing boilers: these have their own nominal requirements to operate at maximum efficiency. For a boiler to enter into 'condensing mode', the return temperature should be no greater than approximately 55°C, depending on boiler manufacturer. In some cases, temperature sensors are installed on the header to allow adjustment of the primary circuit temperature.

The primary circuit is acting like a circulating reservoir of hot water into which the secondary circuits can dip as needed and increase the speed of warm-up periods of the radiators. Secondary circuits are therefore normally pumped to control the flow. Flow connections are usually at the top and the return at the bottom of the header.

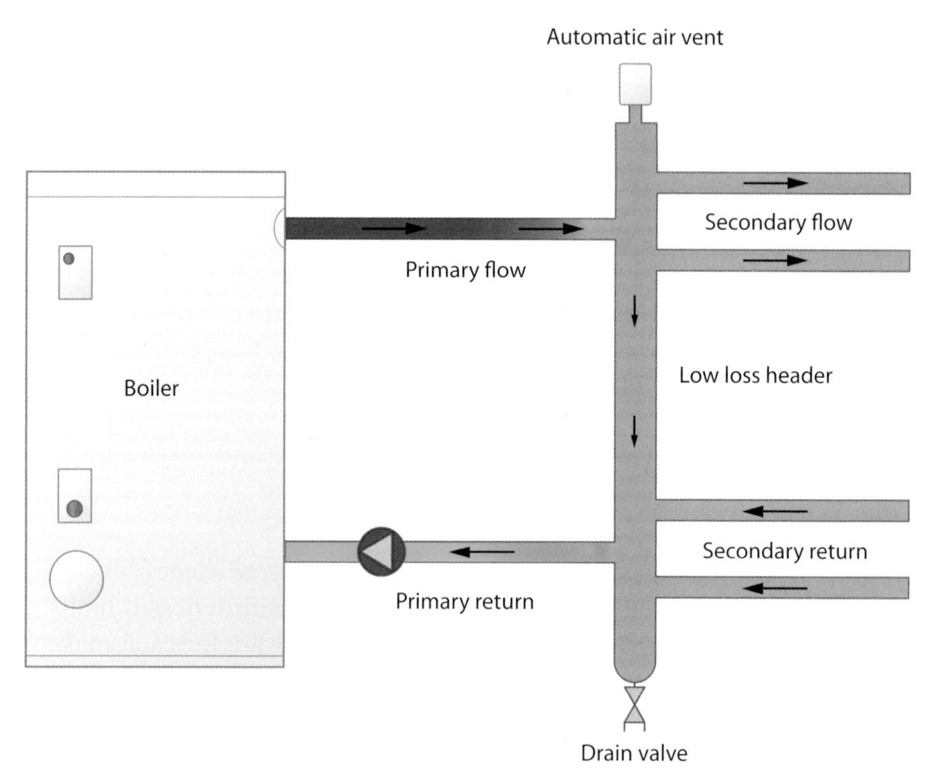

Figure 6.3: Low loss headers

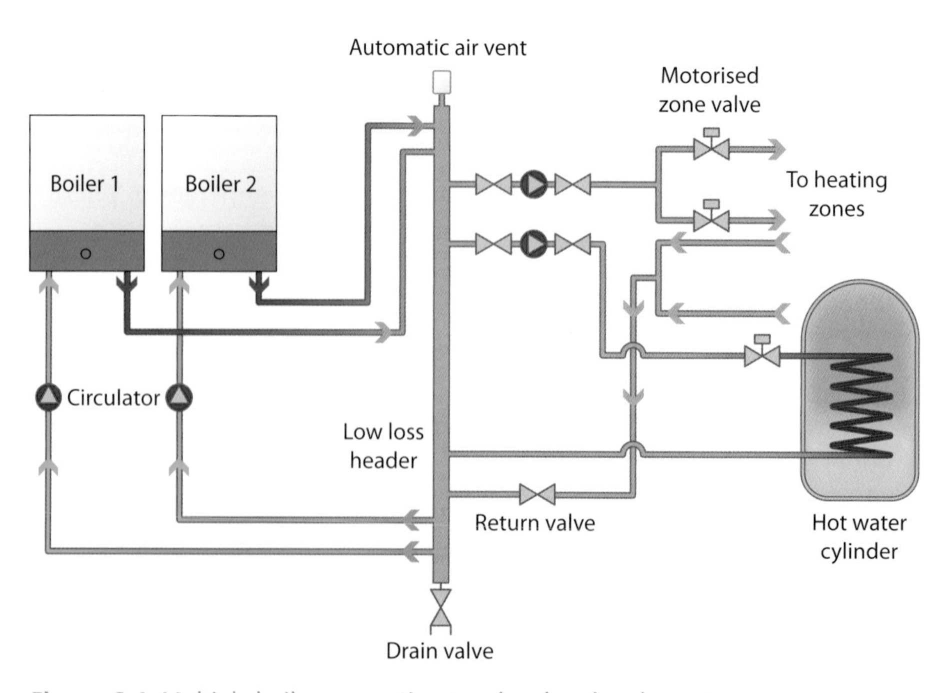

Figure 6.4: Multiple boiler connection to a low loss header system

The low loss header is designed to allow the boiler flow rate to be maintained, keeping it within the manufacturer's design. When coupling two or more boilers to a larger than normal heating system, this is best done through a low loss header allowing a more balanced pressure throughout the system. The low loss header acts as a neutral point in the system, so all circulators must pump away from the header, ensuring that the pipework and boilers are all under positive pressure. As with all installations, you should always follow the manufacturer's instructions.

Poor boiler connection into a low loss header

The incorrect positioning of flow and return connections to a low loss header can cause reverse circulation and poor circulation: water that has not been cooled by passing through the heating load (radiators, and so on) will go back to the boilers through the return pipe. This will cause the boilers to shut down as the circuit is too small, and the rest of the system will be running below design temperature.

Did you know?

Low loss headers are available from manufacturers who can design each one to suit the individual system's needs.

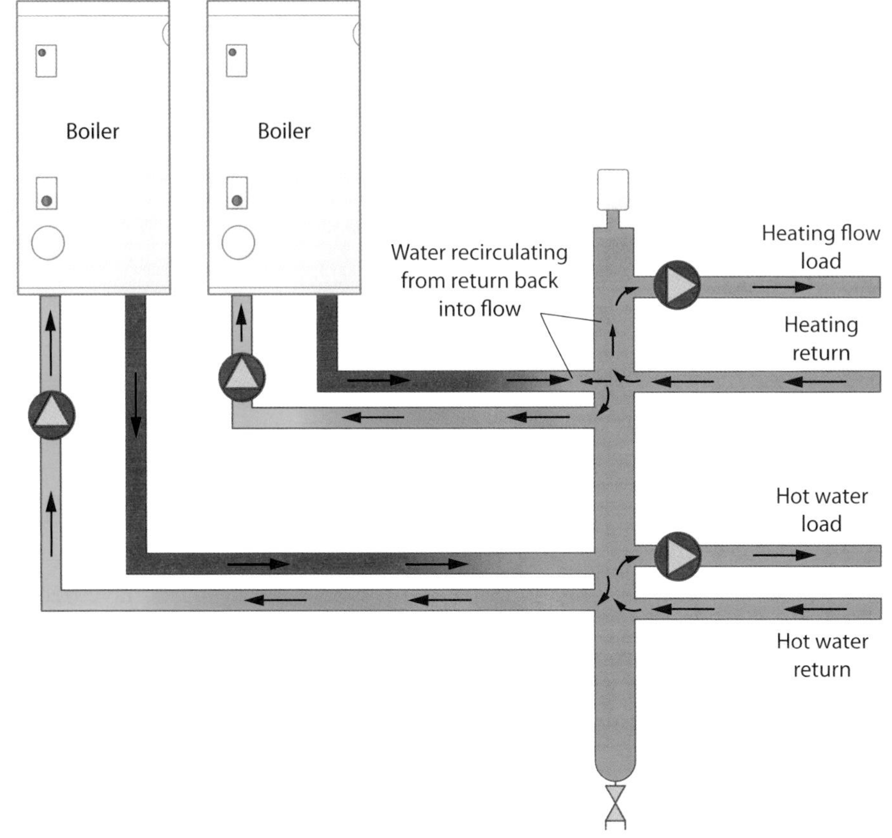

Figure 6.5: Incorrect pipework low loss header

Electrics and wiring for controlling central heating

Central heating is one of the areas of plumbing where you will need a solid understanding of the electrics involved, even if you do not have to wire the system yourself. If you fully understand how central heating system controls work, mechanically and electrically, you will be in a far better position to maintain them properly (see pages 360–368).

PROGRESS CHECK

1. What is the name for a system that requires a borehole as its energy source?
2. What does ASHP stand for?
3. What is the liquid called that absorbs the heat in the heat pump?
4. A heat pump must be installed within the boundary of the property by no less than what measurement?
5. What are the three main types of micro-CHP technology?
6. What is the peak efficiency at which a boiler will function?
7. What does a low loss header allow the creation of?
8. What characteristic do secondary circuits in a low loss header usually have?
9. What can a low loss header connect two or more of?
10. Under what pressure do you need to make sure low loss headers are?

2. Know and be able to apply the design techniques for central heating systems

Selecting central heating systems

In some cases, the decision about the choice of system in a dwelling may be taken by someone else. This could be:

- where the customer (or their representative) has specified the system for a one-off dwelling
- on a large, multi-dwelling development, where the systems have been specified by an architect or services design engineer.

However, as a plumber operating at Level 3, you may be required to advise a customer on the best system to suit their needs. The suitability of any system will be influenced by:

- customer needs
- budget
- technical requirements and limitations.

Customer's needs

Communication with clients was covered in Unit 2. Look back at page 16 before reading on.

For central heating, your discussion with the customer about design should include:

- thermal design of the building
- opportunities to save energy, such as the use of insulation
- siting of boiler
- flueing route and plumbing from terminal
- position of feed and expansion (f&e) cistern if required
- filling and draining points
- hot water requirements
- position of cylinder, if required
- type of heating system (radiator, underfloor heating)
- route of system pipework
- system controls
- frost protection
- night set-back (depending on system type).

Budget

Budgets can be difficult to deal with, especially for systems as complex and with as many options as central heating. Often the customer will not be aware of the likely cost of materials or installation. Your employing organisation/merchants will provide cost options for the materials, in terms of quality of appliances and components, and at the planning stage should be able to provide a general idea of labour costs. Once the customer is comfortable with an outlined cost, you or your organisation can prepare a more detailed estimate.

Technical requirements and limitations

Technical requirements and limitations are probably the key planning considerations. Some customer aspirations may be technically impossible, or technical difficulties may make the cost prohibitive. You will need to consider:

- building layout and features
- suitability
- energy efficiency
- environmental impact.

These points have already been covered these in Unit 4 on pages 195–197. Look back now to remind yourself of what is involved. See also the section in Unit 3 on pages 118–121 about the information sources you will need to draw on while assessing the technical limitations.

Regulations, standards and instructions

It is important that you know how to deal with all appropriate industry standards and the Building Regulations. If a person carrying out building work contravenes the Building Regulations, the local authority or another person may decide to take them to the magistrates' court where they could be fined up to £5000 for the contravention, and up to £50 for each day the contravention continues after conviction.

Industry standards were covered in Unit 3 on page 145, and the Building Regulations in Unit 4 on pages 197–198. Look back now to refresh your memory.

The Appendix covers some of the details of how to deal with the Building Regulations, including checklists for before and after inspection.

Remember

Double-check all your details against the Water Regulations, Building Regulations or design information.

If you are working on an existing appliance on a maintenance job (particularly boilers), you must have access to the manufacturer's instructions. You should be able to obtain copies of instructions from most manufacturers.

Evaluating heating requirements of rooms

To create the most efficient and suitable system, you must be able to evaluate each room's heating requirements and use this as a basis for your work.

Taking measurements of building features

Unit 3 provided information about the different range of sources that provide plumbers with the information they need for different tasks. Before reading on, go back to pages 118–121 and remind yourself of the range of plans and drawings you need to be familiar with.

Calculations

As a plumber, you can take the measurements from a dwelling using a tape measure or laser measuring equipment. You will need to measure the room size and building features, such as windows, to get an accurate measurement for the heat loss calculations. You will also have to identify the type of building fabric in order to calculate the heat loss through it. You should record all this information on a survey sheet, along with the measurements of each room and wall positions, noting any fixed appliances and the types of roof and flooring. This will make it easier to ensure that the design matches the physical requirements of the building.

Heat loss and gain in dwellings

Heat loss from a building occurs when the temperature outside the building is lower than the temperature inside; heat migrates to the colder outside area. The reverse is true in summer as heat moves from the outside air into the building; even on sunny winter days, this effect can be seen in the temperature rise inside the dwelling. All building fabric allows some heat loss, but this happens at different rates (measured in Wm^2). When you calculate the heat requirements of buildings, you have to take this loss into consideration.

Remember

How you present your design calculations is important. Clear details that are easy to understand will save any misunderstandings, and can save money, time and disappointment.

Sizing heat emitters (panel radiators)

Heat emitters must be capable of providing sufficient heat output to maintain a comfortable temperature in the room in which they are installed. Installing undersized heat emitters is uneconomical, as it would result in higher fuel costs and poor system efficiency, as well as complaints from the customer. However, oversized heat emitters also create efficiency problems – so you need to size them correctly.

There are several different ways to establish the correct size of heat emitter for a room:

- using a central heating calculator, such as a Mears wheel
- using a manufacturer's heat emitter computer program, such as the Myson heat loss calculator
- by mathematical calculation.

With any type of heat loss calculator, you need to know how to use it correctly. If you can understand the mathematical calculations involved, you will be able to make good use of any type of calculator.

When sizing a radiator in a property, the calculation centres on the rate of heat that will be lost from that room. You need to know that heat is lost in two ways:

- through the building fabric (this is primarily affected by the level of insulation in the building structure: high-insulation qualities lower the heat loss. A key measure of the insulation properties of a building component is the U-value)
- due to ventilation through air changes and natural ventilation – air circulation in the room due to the number of air changes that occur in the room per hour.

So, when you are starting to size a radiator in a room, you must first work out the heat loss using two different calculations.

Rate of heat loss due to ventilation

The formula to use here is:

heat loss (W) = room volume (m^3) × temperature difference (°C) × number of air changes per hour × a constant figure of 0.33

The number of air changes in a room are identified from tables in BS 5449. The air change rate table will be part of the same table that quotes room temperatures (see page 286).

Did you know?

In new properties, room air changes per hour have been significantly reduced to prevent heat loss.

Rate of heat loss due to the building fabric

The formula to use here is:

heat loss (W) = surface area (m^2) × temperature difference (°C) × U-value

(W/m^2/°C) room size

BS EN 6946

Part L of the Building Regulations (England and Wales) requires that 'U-values are calculated to BS EN 6946'. Before looking at U-values in depth, it is necessary to look at temperature difference.

- In the case of a component that makes up the building fabric, the difference is measured across that fabric. This could be between inside a lounge and the outside (an external wall), between the lounge and the kitchen (an internal wall), or between the lounge and the next-door property (a party wall).
- In the case of ventilation, the difference is always taken as the difference between the required temperature in the room and the outside fresh-air temperature.

The external (outside the building) temperature is usually identified as –1°C, although a lower figure of –3°C is often used for more exposed locations.

Recommended internal air temperature and air change rates		
Room	**Temperature (°C)**	**Air changes (per hour)**
Living room	21	1.5
Dining room	21	1.5
Bedsitting room	21	1.5
Kitchen	18	2.0
Bedroom	18	1.0
Hall/landing	16	2.0
Bathroom	22	3.0
Toilet	18	3.0

Table 6.1: Internal design temperatures in accordance with BS 5449 Section 3 1990

Remember

The U-values for different walls will depend on the resistant factors for the different materials the wall is made from.

Be careful when working out the temperature difference. For example: a lounge is at a temperature of 21°C and the outside temperature is –1°C. The temperature difference = 21 – (–1) = 22°C (minus a minus and it becomes a positive). Internal design temperatures are generally in accordance with BS 5449 Section 3 1990. These are shown in Table 6.1.

BS 5449

Be careful when looking at the number of air changes required in rooms with solid-fuel open fires, as they need to increase greatly; refer to BS 5449 for details. Where a building or dwelling adjoins another – as in a semi-detached house – assume a 10°C temperature difference.

The U-value

A U-value is the thermal transmittance rate from the inside to the outside of a building, through the intermediate elements of constructions. The U-value is defined as the energy in watts per square metre (W/m^2) of construction for each degree of Kelvin temperature difference between the inside and outside of the building (W/m^2K). Note that various textbooks and the table below refer to this as $W/m^2/°C$. U-value tables can be found in a variety of system design guides; if you are going to do design calculations, you will need to access these.

Table 6.2 gives approximate U-values through building fabrics, but to do the job accurately you will need to use the proper tables.

Construction	$W/m^2/°C$	Construction	$W/m^2/°C$
External solid wall	2.0	Ground floor – solid	0.45
External cavity wall	1.0	Ground floor – wood	0.62
External cavity wall (filled)	0.5	Intermediate floor – heat flow up	1.7
External timber wall	0.6	Intermediate floor – heat flow down	1.5
Internal wall	2.2	Flat roof	1.5
Window – single glazed	5.7	Pitched roof (100 mm insulation)	0.34
Window – double glazed	3.0	Pitched roof (no insulation)	2.2
Internal wall – solid block	2.1		

Table 6.2: Approximate U-values through building fabric

Remember

The rate at which heat is lost by conduction through the building elements (parts of the structure) is affected by:

- temperature difference
- the area
- the building elements' ability to conduct heat.

Professional Practice

The client has asked for the radiator sizes for each room so that they can select a suitable radiator for each room. Discuss how you would present the information to the customer so that it was clear and easy to understand.

PROGRESS CHECK

1. To advise a customer on the best system, which three factors do you need to take into account?
2. If you are discussing energy-efficient systems with a client, what must you include?
3. What do local authorities administer and enforce?
4. What are the two reasons for making sure that an installation meets industrial standards?
5. What do most plumbers use manufacturers' instructions for?
6. What can you use to measure a dwelling?
7. What does a block plan provide?
8. What do assembly drawings show?
9. What is a specification?
10. In what units is heat loss through the building fabric measured?

Worked example

Imagine you have to provide the answer to this question:

Using the floor plan of the detached bungalow in Figure 6.6, work out the heat emitter requirement of the lounge.

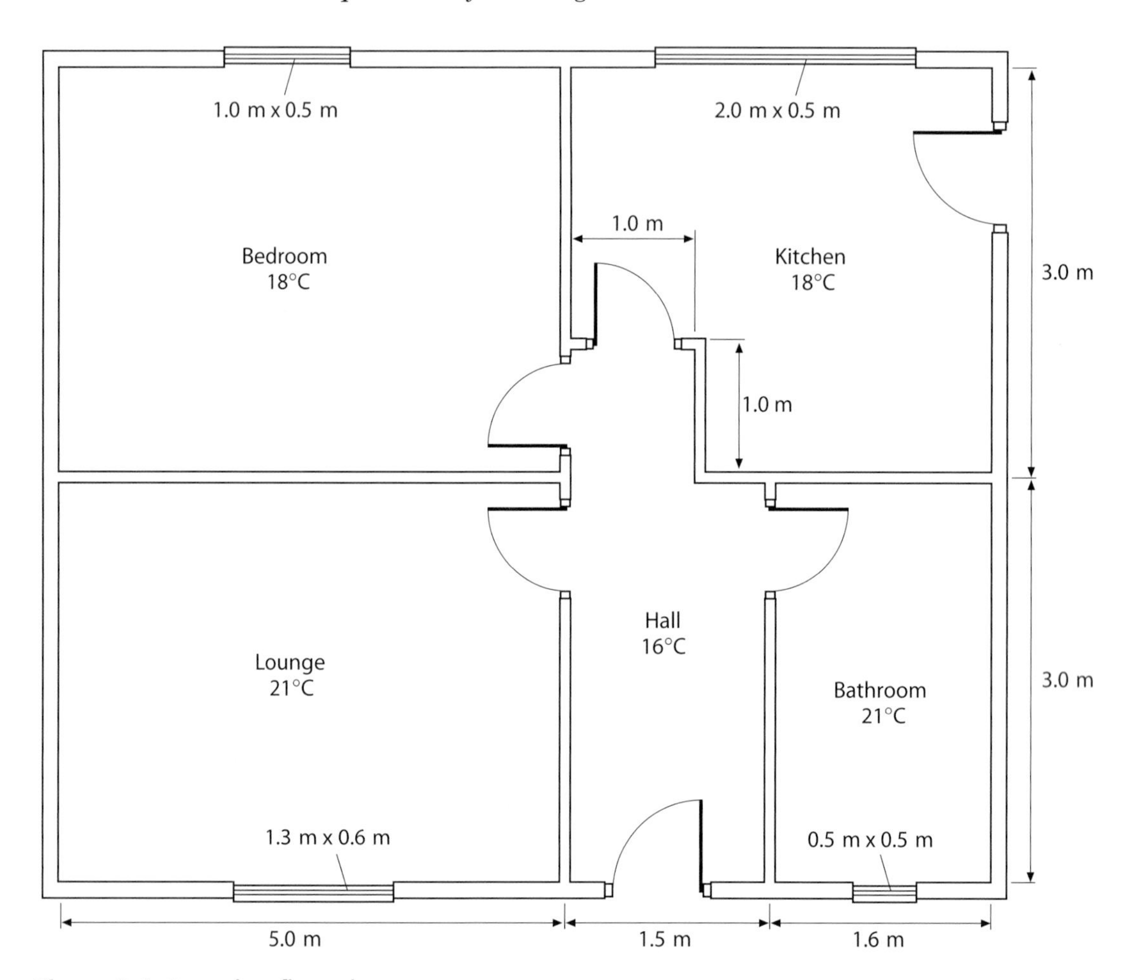

Figure 6.6: Bungalow floor plan

The question comes with some notes for you to use:

- *all dimensions to the bungalow are in metres*
- *the bungalow has solid brick external walls and a solid floor*
- *the windows are single-glazed, with double-glazed doors*
- *the roof insulation is 100 mm; the bungalow has a pitched roof*
- *the height of the rooms is 2.4 metres*
- *the internal walls are solid block.*

This is the calculation procedure you would follow to work out the lounge heat loss. The key points are:

- the window or door heat loss is done first
- the total area of glazing must be deducted from the wall area for the external heat loss calculation
- internal doors are treated as wall surface.

Fabric-loss element	Area: length x breadth (m²)	Temperature difference	U-value W/m²/°C	Heat loss (W)
Window	1.3 x 0.6 = 0.78	x 22	x 5.7	97.8
External walls	8 x 2.4 = 19.2 –0.76 = 18.44	x 22	x 2.0	811.4
Internal wall – bedroom	5.0 x 2.4 = 12	x 3	x 2.1	75.6
Internal wall – hall	3.0 x 2.4 = 7.2	x 5	x 2.1	75.6
Floor	5.0 x 3.0 = 15.0	x 22	x 0.45	148.5
Roof	5.0 x 3.0 = 15.0	x 22	x 0.34	112.2
			Total	1321.1
Ventilation loss				
Volume	**Air change**	**Temperature difference**	**Factor**	
5 x 3 x 2.4 x	1.5 x	22 x	0.33 = 392.0	

Table 6.3: Heat loss calculation

The total heat loss for the room is:

$$1321.1 + 392.0 = 1713.1 \text{ watts.}$$

However, so far you have only calculated the amount of heat that will be lost from the room. In cold weather conditions, the amount of heat shown would not be sufficient to raise the temperature in a room in a reasonable timescale, so you need to add a percentage margin to the total room heat loss for intermittent heating of between 10 per cent and 20 per cent, depending on the system controls and the size of the property. With good controls and a small property, the percentage will be average here, so you can assume 15 per cent. So the heat loss is: 1713.1 watts × 1.15 (15 per cent add on) = 1971.0 watts.

Having done these calculations, you can begin to select a radiator, for which you will need a radiator catalogue.

Radiator selection

The figures quoted in the manufacturer's catalogue are to a test standard. The pipework to them is connected flow at the top and return at the bottom – opposite ends of the radiator; if bottom opposite-end connections are used (the norm in domestic properties) a correction factor needs to be applied to the figures in the catalogue. There is a table you can use to find this correction factor, which is known as 'f2'.

Connection	Factor
Top and bottom opposite-end connections	1.00
Bottom opposite-end connections with blind nipple	0.97
Bottom opposite-end connections	0.90

Table 6.4: Radiator correction factor – f2

You will also need to use a second correction factor, known as 'f1'. The radiator is tested in a room with a difference between the mean water temperature in the radiator and the air temperature in the room of 60°C.

You will probably notice that, in this bungalow, different temperatures are used, so you will need to apply the further correction factor.

The mean water temperature is the average of the flow and return water temperatures. In most systems, the flow will be 80°C and the return will be 70°C; so the mean water temperature will be 80 + 70 divided by 2 = 75°C.

You then need to deduct the room temperature from the mean water temperature, to find the difference between mean water and air temperature. In the case of the bungalow lounge, this is 75° – 21° = 54°C; you then apply this to Table 6.5 to determine the f1 correction factor.

Remember

The U-values for different walls will depend on the resistant factors for the different materials the wall is made from.

Professional Practice

Your employer has asked you to calculate a room's radiator size. Discuss and make a list of what information you will require to do this.

Did you know?

To operate at maximum efficiency, a condensing boiler requires a flow and return temperature difference of 20°C.

Temp diff. °C	f1	Temp diff. °C	f1
40	0.605	56	0.918
41	0.624	57	0.938
42	0.643	58	0.958
43	0.662	59	0.979
44	0.681	60	1.000
45	0.700	61	1.020
46	0.719	62	1.041
47	0.738	63	1.062
48	0.758	64	1.062
49	0.778	65	1.104
50	0.798	66	1.125
51	0.818	67	1.146
52	0.838	68	1.168
53	0.858	69	1.189
54	0.878	70	1.211
55	0.898	71	1.232

Table 6.5: Temperature difference factor – f1

To work out the radiator size, multiply f1 × f2 (see Table 6.4) to arrive at an overall correction factor. For the bungalow lounge at 21°C, f1 is 0.878 and f2 is 0.90:

$$0.878 \times 0.90 = 0.79$$

The size of the radiator required is the total room heat loss, including the intermittent use margin, divided by the overall correction factor:

= 1971.0 watts / 0.79 = 2495 watts.

The nearest sized radiator above this figure may be selected from the catalogue to suit space requirements.

Sizing pipework and circulators

Pipe sizing calculations – space heating and hot water circuits

Pipe sizing affects the size of the pump required and the setting it should be placed at during commissioning.

To determine the size of pipework for a system you need to be able to identify the flow rate down the pipe required to get the desired amount of heat from the radiators. This is usually measured in kg/s. In moving that water flow through the system, you have to overcome the frictional resistance of the pipework and fittings that make up the circuit, so you need to apply pressure via a pump. The pressure will be greater at the beginning of a particular circuit than it is at the end; this is because the resistance will reduce as you get further down the circuit.

To undertake pipe sizing, the first thing to know is the flow rate through a particular section of pipework. Take the lounge radiator that you worked out earlier, which totalled (with an allowance for intermittent heating) 1971 watts. The flow rate is calculated as follows:

Heat in kW = flow rate (kg/s) × the specific heat capacity of water (which is a constant figure of 4.2) × temperature difference flow pipe to return pipe (normally 10°C)

So, moving the figures around in the equation:

$$\text{Flow rate} = \frac{\text{heat in kW}}{10 \times 4.2} = \frac{\text{kW}}{42}$$

This will change if the temperature difference between flow and return temperatures is different. So for the lounge radiator, the flow rate to the pipework immediately supplying that radiator is:

$$\frac{1.969 \text{ kW}}{4.2} = 0.047 \text{ kg/s}$$

If there is exposed pipework, remember to add the allowance to the radiator output.

The next thing you need to know is the length of pipework throughout the circuit. For this you will need a simple layout drawing of the system. Draw the flow and return pipes as single lines, but remember that the length of each circuit will be the length of both flow and return pipes together, so the true length will be doubled.

Most plumbers use a chart for pipe sizing to make it easier, but it is important that you understand the principles behind this.

Start with section 1 on Figure 6.7 (page 292). First find the flow rate through the pipe, adding any mains pipe heat losses to the radiator output. So section 1 carries:

3 kW × 1.1 (10 per cent mains pipe heat loss) = 3.3 kW.

Therefore the flow rate in that section for a 3.3 kW load is:

3.3 kW ÷ 42 = 0.08 kg/s

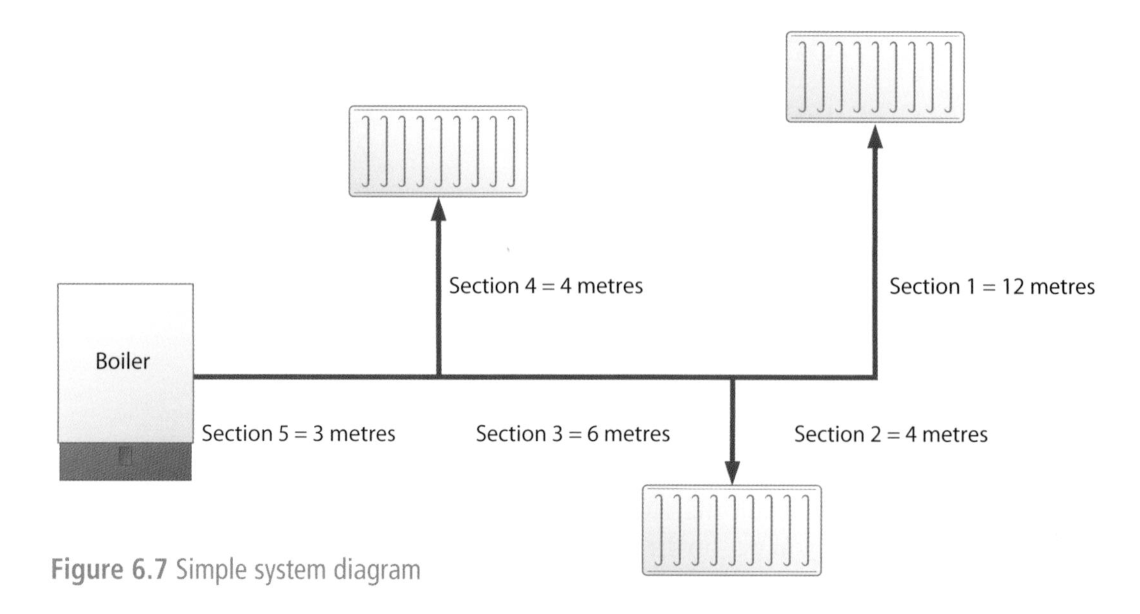

Figure 6.7 Simple system diagram

The length of pipe run is 12 metres, but this does not include any pressure loss due to fittings. Here, use a percentage addition of 33 per cent if the pipework has an average number of changes of direction; if a lot of changes are used, this figure should rise to 50 per cent. This will give an overall effective pipe length of:

12.0 m ×1.33 (adding 33 per cent for fittings) = 16.0 m.

Use the chart in Figure 6.8 to determine the pressure loss per metre run of pipe.

You need to select a pipe size that you think might be able to meet the requirement. However, the pipe selected needs to fit within a maximum velocity reading:

- 1.0 m/second for standard small-bore pipework; and
- 1.5 m/second for microbore pipework.

The velocity readings are the stepped scale from the right-hand side to the left-hand side of the chart. So try an 8 mm pipe size first for the 0.08 kg/s. Look for 0.08 kg/s under the section of the table for 8 mm pipe. You will see that this size is not even on the table, which indicates that it is not suitable.

Now try 15 mm pipe. You can see that 0.080 kg/s is equal to 0.030 loss/metre run of pipe from the right-hand scale, and the velocity is between 0.5 and 0.75 m/s, so this is acceptable. So the resistance in the section of pipe is the head loss/metre run × the effective pipe length. The head loss is:

$$0.030 \times 16.0 \text{ metres of pipe} = 0.48 \text{ metres}$$

Now look at section 3. Section 3 carries the heat loads of sections 1 and 2. You already know the flow rate through section 1 as 0.080 kg/s. You therefore need to know the heat load in section 2:

$$(2.1 \div 42) \times 1.1 \text{ (10\% mains loss)} = 0.055 \text{ kg/s}$$

The total load on the section is therefore sections 1 and 2 = 0.08 + 0.055 = 0.135 kg/s. The effective pipe length for the section is 6 m × 1.33 (fittings resistance) = 7.98 m.

Pressure loss (rn/rn)	8 mm kg/s	10 mm kg/s	15 mm kg/s	22 mm kg/s	28 mm kg/s	35 mm kg/s	Velocity rn/s
0.008		0.0108	.0380	0.109	0.227	0.400	0.50
0.009		0.0114	0.040	0.117	0.235	0.424	
0.010	0.0064	0.0122	0.042	0.124	0 250	0448	
0.011	0 0067	0 0129	0.044	0.131	0.263	0.475	
0.012	0.0071	0.0135	0.047	0.137	0.277	0.499	
0.013	0.0074	0.0141	0.049	0.144	0.289	0.523	
0.014	0.0077	0.0147	0.052	0.150	0.302	0.543	
0.015	0.0081	0.0154	0.054	0.156	0.314	0.564	
0.016	0.0084	0.0159	0.056	0.161	0.325	0.594	
0.017	0.0086	0.0165	0.058	0.167	0.336	0.604	0.75
0.018	0.0089	0.0171	0.060	0.172	0.348	0.623	
0.019	0.0092	0.0176	0.061	0.178	0.359	0.645	
0.020	0.0095	0.0182	0.063	0.183	0.369	0.669	
0.021	0.0098	0.0185	0.065	0.188	0.380	0.686	
0.022	0.0101	0.0192	0.067	0.193	0.390	0.704	
0.024	0.0106	0.0203	0.070	0.203	0.408	0.735	
0.026	0.0111	0.0212	0.073	0.212	0.428	0.773	
0.028	0.0116	0.0221	0.076	0.221	0.446	0.805	1.00
0.030	0.0120	0.0230	0.080	0.230	0.464	0.838	
0.032	0.0125	0.0238	0.082	0.238	0.482	0.869	
0.034	0.0129	0.0245	0.085	0.247	0.500	0.898	
0.036	0.0133	0.0253	0.088	0.255	0.518	0.925	
0.038	0.0138	0.0261	0.091	0.263	0.533	0.952	
0.040	0.0142	0.0268	0.094	0.270	0.548	0.982	
0.042	0.0146	0.0276	0.096	0.278	0.564	1.010	1.25
0.044	0.0150	0.0283	0.099	0.286	0.578	1.035	
0.046	0.0154	0.0290	0.101	0.293	0.592	1.048	
0.048	0.0158	0.0298	0.104	0.300	0.608	1.075	
0.050	0.0162	0.0305	0.106	0.307	0.622	1.100	
0.052	0.0167	0.0312	0.018	0.314	0.637	1.123	
0.054	0.0170	0.0320	0.111	0.321	0.651	1.150	
0.056	0.0173	0.0326	0.113	0.328	0.665	1.178	
0.058	0.0177	0.0332	0.115	0.334	0678	1.194	1.50
0.060	0.0180	0.0339	0.117	0.340	0.691	1.215	
0.062	0.0184	0.0345	0.120	0.347	0.705	1.235	
0.064	0.0187	0.0351	0.122	0.353	0.718	1.253	
0.066	0.0190	0.0358	0.124	0.359	0.724	1.272	
0.068	0.0193	0.0364	0.126	0.364	0.736		
0.070	0.0196	0.0370	0.128	0.370	0.750		
0.072	0.0200	0.0377	0.130	0.375	0.762		
0.074	0.0203	0.0382	0.132	0.381	0.774		
0.076	0.0206	0.0388	0.134	0.386	0.785		
0.078	0.0208	0.0394	0.136	0.391	0.797		
0.080	0.0211	0.0400	0.138	0.397	0.808		
0.082	0.0215	0.0406	0.140	0.402	0.819		
0.084	0.0217	0.0411	0.142	0.407	0.830		
0.086	0.0220	0.0417	0.144	0.412	0.841		
0.088	0.0223	0.0423	0.146	0.417	0.851		
0.090	0.0226	0.0429	0.148	0.422	0.862		
0.092	0.0229	0.0433	0. 149	0.426	0.872		
0.094	0.0231	0.0439	0.151	0.431			
0.096	0.0234	0.0445	0.153	0.435			
0.098	0.0237	0.0450	0.155	0.440			
0.100	0.0240	0.0455	0.156	0.445			
0.102	0.0243	0.0460	0.158	0.449			
0.104	0.0245	0.0465	0.160	0.453			
0.106	0.0247	0.0469	0.162	0.458			
0.108	0.0250	0.0474	0.164	0.462			
0.110	0.0253	0.0479	0.165	0.466			
0.112	0.0256	0.0484	0.167	0.471			
0.114	0.0258	0.0488	0.169	0.475			
0.116	0.0261	0.0493	0.170	0.479			
0.118	0.0264	0.0498	0.172				
0.120	0.0266	0.0502	0.174				
0.130	0.0279	0.0523	0.181				
0.140	0.0291	0.0548	0.189				
0.150	0.0302	0.0568	0.197				
0,160	0.0314	0.0588	0.204				
0. 170	0.0326	0.0608	0.211				
0.180	0.0336	0.0628					
0.190	0.0347	0.0648					
0.200	0.0357	0.0668					

Figure 6.8 Pressure loss chart

Now go back to Figure 6.8. Looking at 15 mm first, 0.135 is between 0.75 and 1.0 m/s and is acceptable; the pressure loss/metre run of pipe to the nearest figure above is 0.078 m/m run of pipe.

The pressure loss across the section of pipe is therefore 7.98m × 0.078 m/m = 0.62 metres. The remaining pipe sections are shown completed in the form of Table 6.6, which makes the calculation process easier.

Section	Section heat requirement (kW)	Mains loss (%)	Total heat loss (kW)	Flow rate (kg/s)	Pipe size (mm)	Actual pipe run (m)	Fittings resistance (%)	Effective pipe length (m)	Pressure loss per metre run of pipe (m/m)
1	3.0	1.1	3.3	0.08	15	12	1.33	16	0.030
2	2.1	1.1	2.31	0.055	15	4	1.33	5.32	0.016
3	-	-	-	0.135	15	6	1.33	7.98	0.078
4	2.5	1.1	2.75	0.065	15	4	1.33	5.32	0.021
5	-	-	-	0.2	22	3	1.33	3.99	0.024

Table 6.6: Pipe sizing calculations

Pump sizing

Now you need to work out the pump size, so that you can specify the right size of pump and also to commission the system. To size the pump you need to know the required flow rate, which is the heat load on the section from the boiler to the first branch. In the worked example, this is the flow rate on section 5; from Figure 6.9, this is – 0.2 kg/s.

Size the pump for the head of pressure to be generated to overcome the pipe losses in the circuit. Do not be tempted to add up all the pipework resistances on all the pipework sections – that would be the wrong thing to do. Instead, the pump requirement is based on the individual pipework circuit with the highest pressure loss. This circuit must have only one radiator on it, and is known as the index circuit.

From the drawing there are three possible circuits:

- Section 5-3-1
- Section 5-3-2
- Section 5-4.

The section with the highest pressure loss is the index circuit.

- Section 5-3-1 = (from Figure 6.9) 0.10 + 0.62 + 0.48 = 1.2 m
- Section 5-3-2 = (from Figure 6.9) 0.10 + 0.62 + 0.08 = 0.8 m
- Section 5-4 = (from Figure 6.9) 0.10 + 0.11 = 0.21 m

You can see that the index circuit is therefore section 5-3-1, which has a pressure loss of 1.2 m head. At this point you need to consult the manufacturer's information to establish whether the boiler generates a significant pressure loss through it (many low-water-content boilers do).

If so, add this to the pressure loss through the pipework. If the head loss through a boiler is 2.0 m and the pipe loss is 1.2 m head, the pump should be sized at 3.2 m head, delivering a flow rate of 0.2 kg/s.

You now need to consult a pump manufacturer's catalogue to select your pump and determine any speed setting that it may be placed on. See the example in Figure 6.9.

To convert 1.2 metres to kPa:

$$= 3.2 \times 10 = 32 \text{ kPa.}$$

Speed setting 2 will meet the requirement.

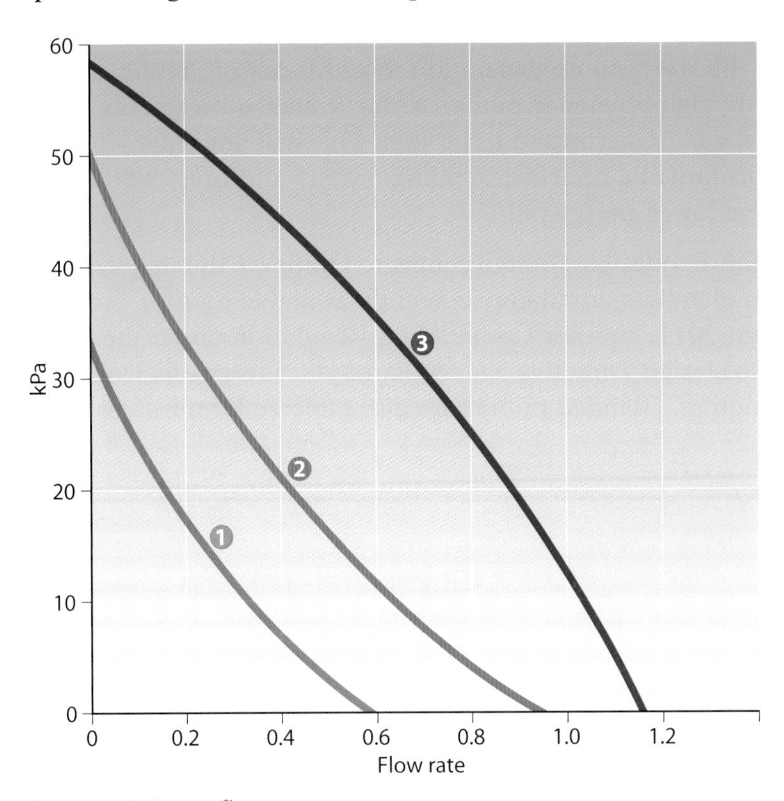

Figure 6.9: Pump flow rate

High flow rate (boilers)

Many newer low-water-content boilers (combis in particular) have a higher flow rate requirement through them than the heating load that the circuit requires. You need to check the manufacturer's minimum water flow rate requirement through the heat exchanger.

If you had used a boiler that required a minimum water flow rate of 0.4 kg/s through the exchanger, it can be seen from the worked example that the system only required 0.2 kg/s, so there is a shortfall. This is where the system bypass comes in. To ensure the minimum flow through the boiler would require 0.2 kg/s to be circulated around a bypass circuit.

Checking back to the pipe size chart, that bypass would have to be sized at 22 mm. The pump would also now need to deliver 0.4 kg/s at 3.2 m head, heading towards speed setting 3.

European Ecodesign Directive

Over 90 per cent of the glandless circulators for heating available on the market today will soon be prohibited for sale. This is due to the enforcement of a Commission Regulation for glandless circulators under the European Ecodesign Directive. Throughout the EU, this directive will introduce increasingly stricter requirements on the energy efficiency of glandless circulators in three stages, starting in 2013.

Currently, many heating systems are equipped with unregulated circulators. As a result, unnecessary amounts of energy are consumed – energy consumption is up to ten times higher than required by the latest pump generation. For this reason, under the directive, only high-efficiency pumps with extremely low energy consumption will be permitted to be sold. This will not only provide environmental benefits; building owners and users will also benefit from a lower energy bill.

The European Ecodesign Directive aims to eliminate the majority of inefficient glandless circulators currently available on the market. From 2011, another Commission Regulation under the European Ecodesign Directive has regulated the energy efficiency of electric motors. Glanded pumps are also affected by this.

Activity

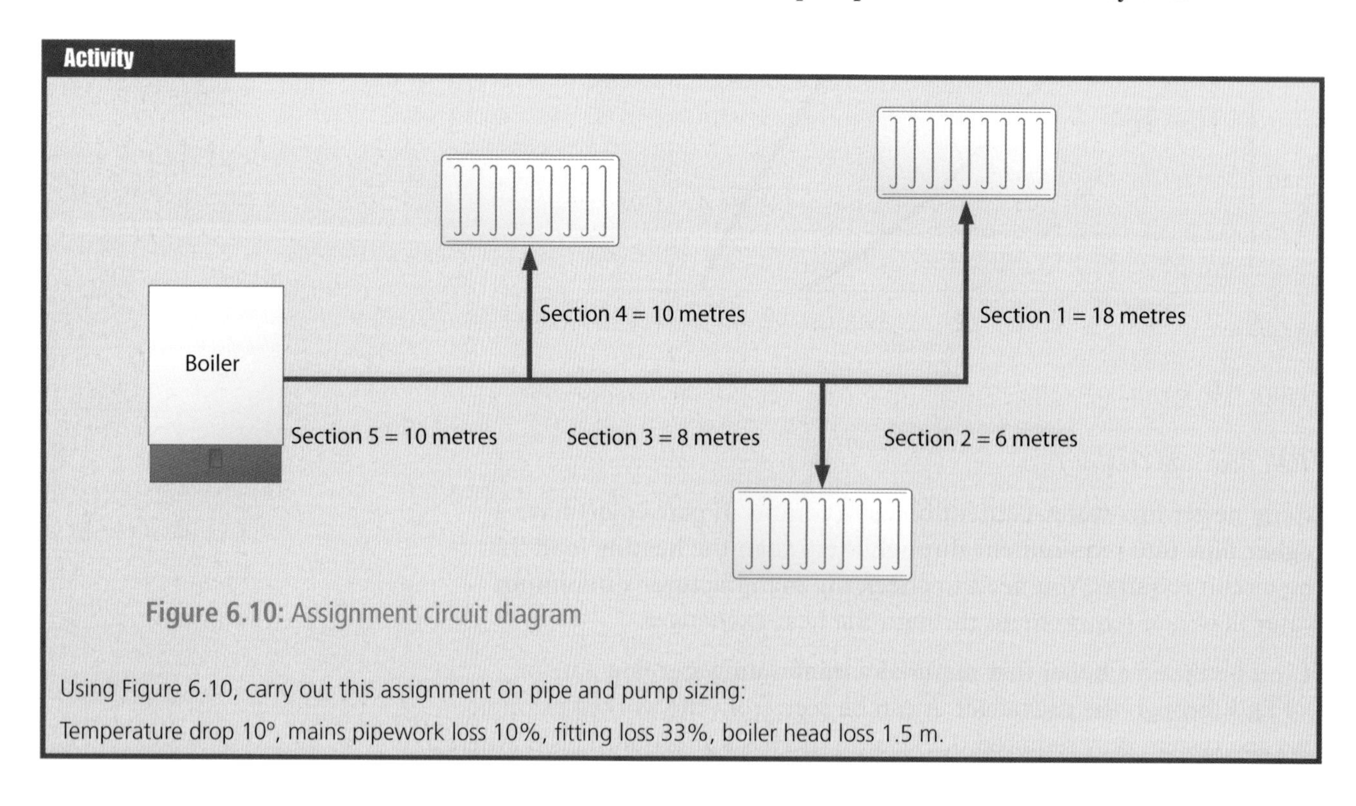

Figure 6.10: Assignment circuit diagram

Using Figure 6.10, carry out this assignment on pipe and pump sizing:

Temperature drop 10°, mains pipework loss 10%, fitting loss 33%, boiler head loss 1.5 m.

Selection criteria for boilers in dwellings

Boiler sizing

Add up all the heat losses from the rooms and add an allowance for pipework heat losses and hot water.

> **Did you know?**
>
> A pipe heat loss allowance will typically be added where a fair proportion of pipework is installed under suspended timber floors or in roof spaces. If it is all surface-mounted in the room, no allowance needs to be added; typically 10 per cent will be added.

Hot water heating load

The hot water heat requirement is usually based on providing 1 kW of boiler heat output for every 50 litres of hot water stored.

Space heating load

When determining the heat requirement for the radiator circuit, use the figure calculated for the room heat loss, with factors for intermittent heating, plus any percentage applied for intermittent use. Do not use the figure used for the radiator size after correction factors have been applied, as this would over-size the boiler.

As an example, the heat loss including intermittent heat margin from four rooms is:

- lounge – 1.7 kW
- bedroom – 0.8 kW
- bathroom – 1.2 kW
- kitchen – 1.1 kW.

This gives a total radiator load of 4.8 kW. The hot water storage capacity of the cylinder is 100 litres, which is 2 kW output.

The total heat load is (4.8 + 2.0) × 1.1 (10% for the pipe losses) = 7.5 kW heat output from the boiler; this can now be selected from the manufacturer's catalogue.

An alternative to this is the whole house method. The whole house method is not as accurate as working out each individual room, but will give good enough accuracy for selecting a boiler replacement only.

The calculation is carried out in eight stages.

1. Identify the shape of the building and, if necessary, divide into sections.
2. Take the measurements of the outside walls, length, breadth, room height and number of floors.
3. Calculate the total wall area.
4. Calculate the heat loss through walls and windows.
5. Calculate the floor and roof area heat loss.
6. Add the total fabric heat loss together and multiply by the location factor.
7. Calculate ventilation loss.
8. Add ventilation loss to fabric loss, plus 2 kW for hot water demand, to give you the boiler size.

> **Did you know?**
>
> Further details of calculation tables and method can be found on the Energy Saving Trust website www.energysavingtrust.org.uk

PROGRESS CHECK

1. What is the formula for ventilation heat loss?
2. What is a U-value?
3. Why do you add a percentage margin to the total room heat loss?
4. What do you need to know to be able to work out the pump size required?
5. What is the index circuit?
6. What does the European Ecodesign Directive regulate?
7. How do you allow for the heat loss from pipework?
8. What is the usual formula for water heating allowance?
9. What is the whole house method?
10. How many stages are in the whole house method?

Sealed central heating systems

Figure 6.11 shows a typical layout for a sealed system. You will need a good understanding of the components that make up this sort of system, so that you can install and maintain them correctly.

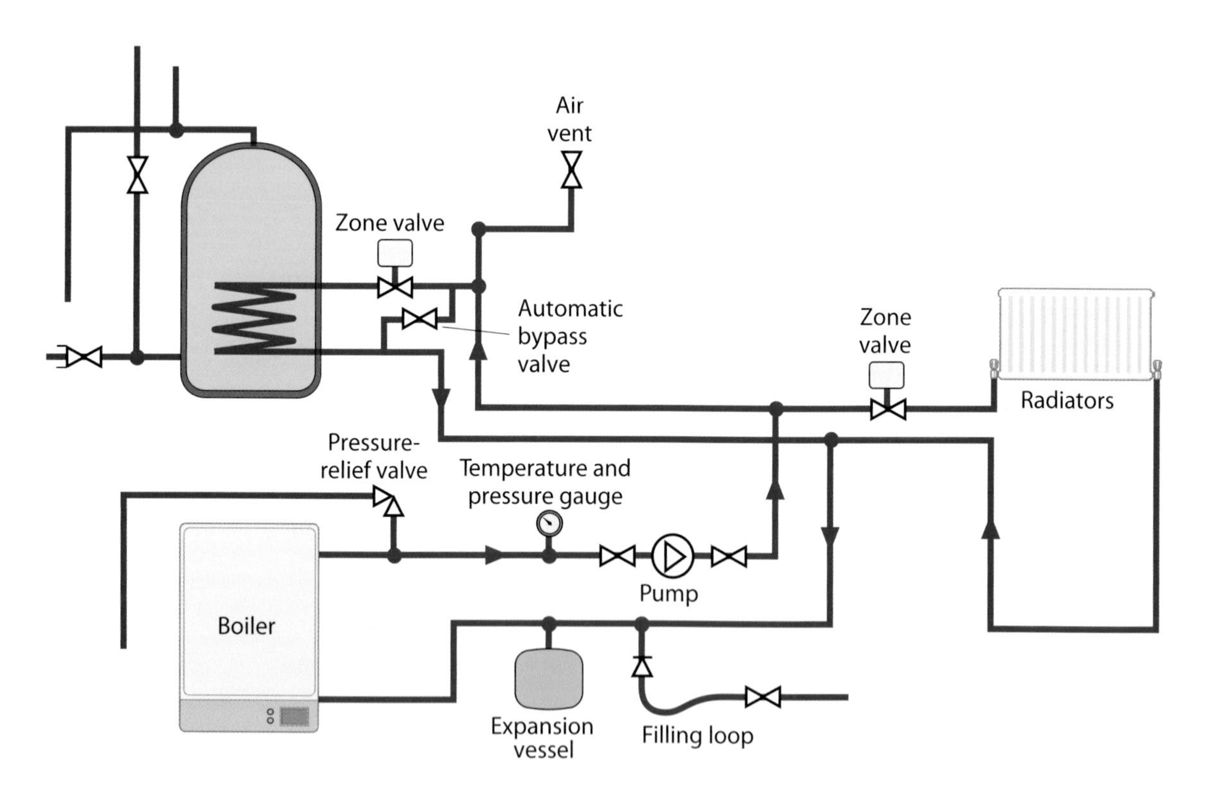

Figure 6.11: A typical layout for a sealed system

Sealed system components

Pressure-relief valve

The pressure-relief valve replaces the open-vent pipe (as seen on open-vented systems). It must therefore be installed on all types of sealed system.

For domestic systems it must:

- be non-adjustable
- be preset to discharge when the system reaches a maximum pressure of 3 bar
- have a manual test device
- have a connection for a full-bore discharge pipe
- be connected either into the boiler or on the flow pipe close to the boiler
- have a metal discharge pipe, which as a minimum should be the same size as the valve outlet, discharging to a tundish which discharges in a safe, visible, low-level location.

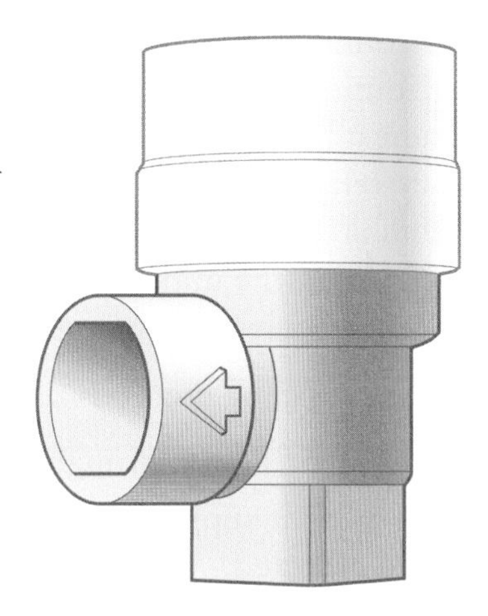

Figure 6.12: Sealed system pressure relief valve

Pressure gauge

A pressure gauge must be provided to assist with system filling and top-up. The pressure gauge should be capable of giving a reading between 0 and 4 bar. It is also preferable that a temperature gauge is provided to measure the boiler flow temperature (this will indicate any potential faults in the system). The temperature gauge should be capable of giving readings up to 100°C. The pressure gauge will usually be sited in the vicinity of the expansion vessel.

Figure 6.13: Sealed-system gauge

Method of filling

There are two main methods of filling the system: filling and pressurising by filling loop, and refilling through a top-up unit.

Filling and pressurising by filling loop

Essentially a temporary connection is made to the system via a flexible filling loop, which includes a flexible hose, stop valve and double check valve. The connection is used:

- to fill the system
- to pressurise the system
- for the purposes of future top-up.

Refilling through a top-up unit

This method is less common, but you should understand it in case you come across it. The unit is essentially a small bottle sited at the highest part of the system and fitted with a double check valve assembly and automatic air eliminator (all provided as part of the package).

The water in the system is topped up manually by the unit, which contains approximately three litres when the water level drops.

The connection to the unit should be made into the return side of the distribution pipework or the domestic return from the cylinder. The system still requires a filling loop for initial filling purposes. However, the system is not pressurised on fill-up; the pressure acting on the system is the head of water exerted by the top-up unit. Take care when determining the initial charge pressure of the expansion vessel so that it functions correctly.

Two other filling methods of filling can be:

- automatic filling through a make-up cistern
- permanent supply connection using a filling loop that can be disconnected.

The use of an automatic filling cistern for domestic applications is very unlikely owing to its cost. The use of a Type CA disconnector came in with the Water Regulations; it is now possible to have a permanent connection to the main supply pipe. However, a well-installed sealed system should not require much in the way of top-up as evaporation is minimal; the need to periodically top up a system is a sign that it contains leaks.

Safety tip

The boiler used with a sealed system must have a high-limit thermostat (energy cut-out device) fitted to it so that, in the event of control-thermostat failure, there is the added protection.

As with any other system, constantly introducing fresh water to the system increases the system's exposure to aerated water and promotes corrosion. Both of these methods of fill connection – pressurisation and top-up – should therefore be avoided for domestic systems.

Expansion vessel

An expansion vessel is used in the sealed system, and it replaces the feed and expansion (F&E) cistern in an open-vented system, taking up the increase in system volume when the system is heated.

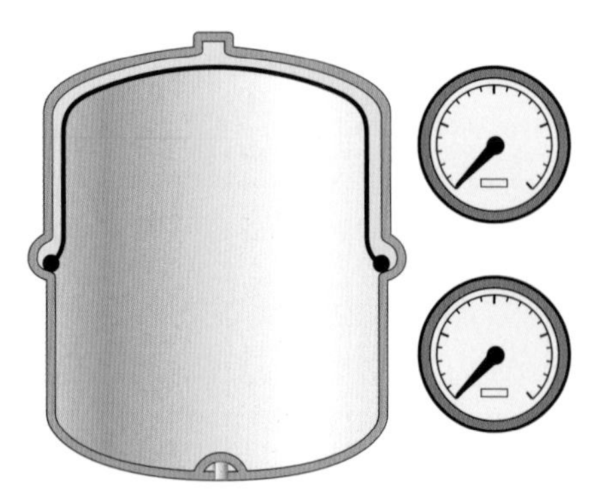

Before filling, the diaphragm is pushed up against the vessel by the preset initial gas charge. The gas charge supports the pressure exerted by the static head of water in the system.

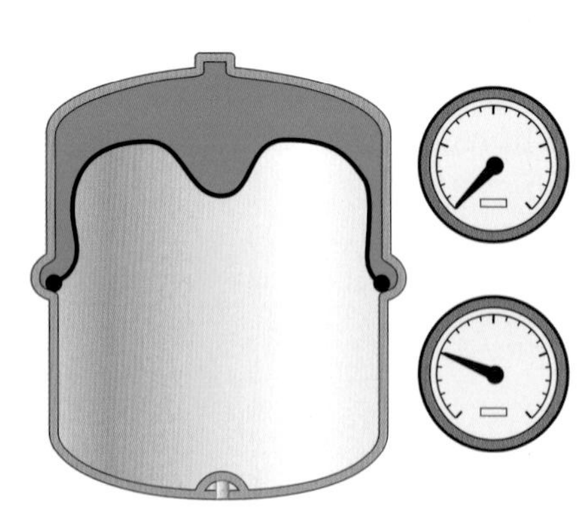

On filling, the vessel contains a small amount of water.

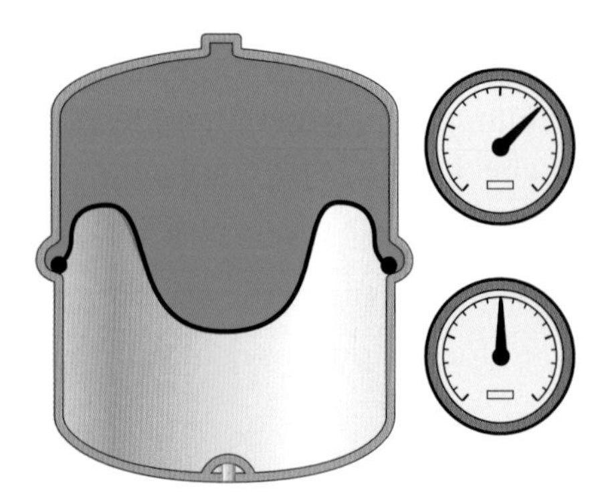

At operating temperature the total mass of expanded water is contained in the vessel. The diaphragm is virtually static with equal pressure on either side.

Figure 6.14: Expansion vessel

The vessel should be located close to the suction side of the pump to ensure that the system operates under positive pressure. The point of connection of the expansion vessel to the system is the neutral point of the system. It is preferable to install the vessel at the coolest part of the system to maximise the lifespan of the flexible diaphragm. The pipe connecting the vessel to the system should be the same size as the vessel outlet, and there must be no isolating valve installed between the vessel and its point of connection to the system.

Different arrangements are available for mounting expansion vessels, including a manufacturer-produced mounting bracket, as shown in Figure 6.15.

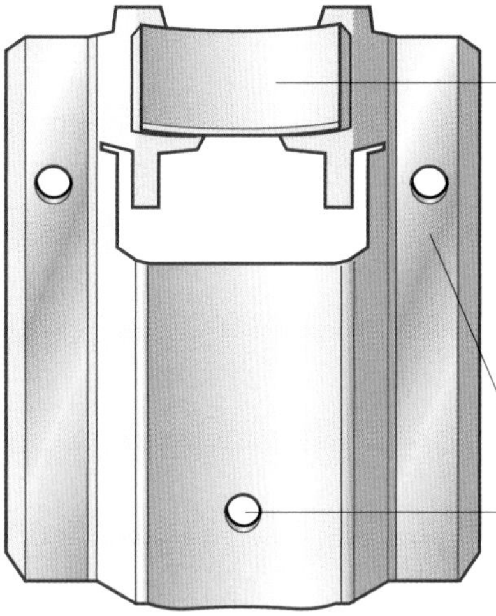

Figure 6.15: Expansion-vessel bracket

Sizing the vessel

When it comes to sizing the vessel there are two points that need to be considered: vessel charge pressure and volume of the vessel.

Vessel charge pressure

Vessels for domestic use tend to be available with an initial charge pressure of 0.5 bar, 1 bar or 1.5 bar. The initial charge pressure should be in accordance with the manufacturer's instructions and must always exceed the static pressure of the heating system at the level of the vessel. For the vessel to work correctly, the air or nitrogen charge pressure on the dry side of the vessel should be slightly higher than the static pressure on the wet side of the system when cold-filled. It follows that, if a vessel was required to be fitted in the cellar of a three-storey property with 8 metres static head of water in the system above it:

1 metre = 0.1 bar, therefore 8 metres = 0.8 bars

We would select a 1 bar vessel and would charge the system to around 0.9 bar (less than the vessel charge of 1 bar).

Volume of the vessel

The volume (size) of vessel required is based on the amount of water contained in the system, which can vary dramatically. The amount of water contained in the major system components can usually be obtained from the manufacturer (usually boiler, cylinder and radiators).

Table 6.7 can be used to determine the amount of water in the system per metre run of pipe.

Pipe OD (mm)	Water content (litres)
8	0.036
10	0.055
15	0.145
22	0.320
28	0.539

Table 6.7: Water in system per metre run of pipe

Worked example

Imagine that you have identified from manufacturers' catalogues that a system you are to install contains the following:

- boiler – 1 litre
- cylinder – 2.5 litres
- radiators – 41 litres
- 15 mm pipe – 78 metres
- 22 mm pipe – 60 metres.

Using Table 6.7:

60 metres of 22 mm pipe contains:

$$60 \times 0.320 = 19.2 \text{ litres}$$

78 metres of 15 mm pipe contains:

$$78 \times 0.145 = 11.3 \text{ litres}$$

Total content = 75 litres.

Apply this information to Table 6.8 (assuming you already know the vessel charge pressure). If the system utilises a 0.5 bar pressure vessel and the pressure valve setting is 3 bar, for a system containing 75 litres, the vessel volume will be 6.3 litres.

Usually the next vessel size up available is 8.3 litres at 0.5 bar pressure; this is the specification for the vessel.

Safety valve setting (bar)	3.0		
Vessel charge and initial system pressure (bar)	0.5	1.0	1.5
Total water content of system (litres)	Vessel volume (litres)		
25	2.1	2.7	3.9
50	4.2	5.4	7.8
75	6.3	8.2	11.7
100	8.3	10.9	15.6
125	10.4	13.6	19.5
150	12.5	16.3	23.4
175	14.6	19.1	27.3
200	16.7	21.8	31.2
225	18.7	24.5	35.1
250	20.8	27.2	39.0
275	22.9	30.0	42.9
300	25.0	32.7	46.8
Multiplying factors for other system volumes	0.0833	0.109	0.156

Table 6.8: Calculating the volume of the vessel

Composite valves

A number of composite valve arrangements are available, such as a combined pressure-relief valve and pressure gauge, as shown in Figure 6.16.

Figure 6.17 shows an example of a combined expansion vessel filling loop, pressure-relief valve and pressure gauge all in one unit. The main point that you need to consider when deciding on a composite valve arrangement is whether placing components together in a composite valve meets the key requirements that must be met when installing the system, such as proximity of the pressure-relief valve to the boiler and the availability of the discharge pipe connection.

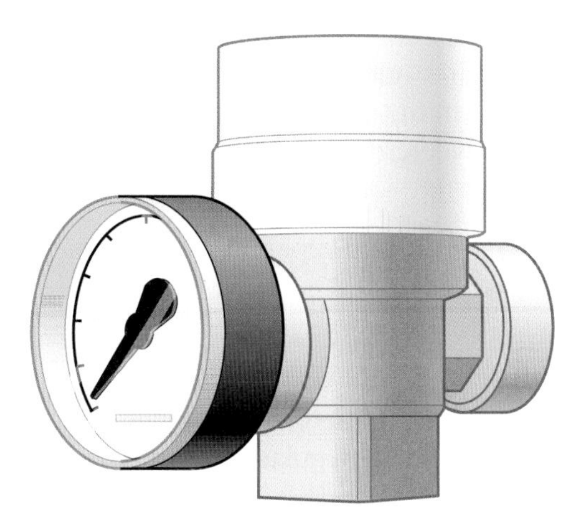

Figure 6.16: Combined pressure-relief valve and pressure gauge

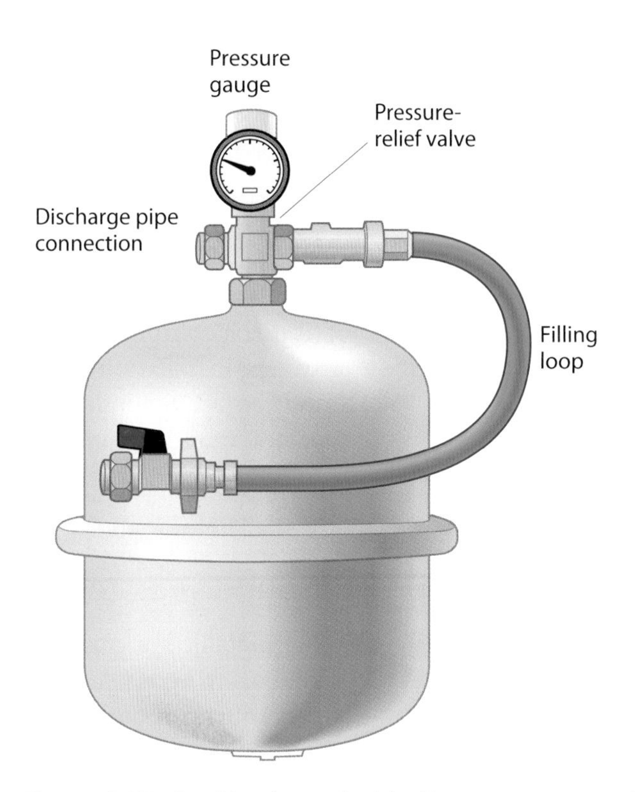

Figure 6.17: Combined vessel with all components attached

Combination boilers and system boilers

These usually contain all the key components that make up a sealed system, and can include the filling loop as well. One key issue to take into account is the size of the expansion vessel, which is based on an average system size as determined by the boiler manufacturer. If the system has a higher water content, an additional vessel may be necessary.

Visual inspection

Just as with hot and cold water pipework, this includes thoroughly inspecting all pipework and fittings, including F&E cisterns, hot-water storage cylinders and expansion vessels. The checklist on page 304 covers the key points.

Checklist

- Pipes and fittings are fully supported.
- Pipes and fittings are free from jointing compound and flux.
- All connections are tight.
- In-line valves and radiator valves are closed to allow stage filling.
- The inside of the F&E cistern is clean.
- All the air vents are closed.
- Before filling, the pump is removed and replaced with a section of pipe, to prevent any system debris entering the pump's workings (it is also good practice to take the expansion vessel out of the system at this stage if possible, or to leave it out until after testing for leaks has been completed, to prevent any dirt or residue getting to the membrane. Remember to plug its connection for the test).
- The customer or other site workers are advised that soundness testing is about to commence.

Checks and tests

Checking for leakage

To fill the central heating pipework with water, turn on the stop tap (if it is a complete cold water, domestic hot water and central heating system installation) or the service valve to the F&E cistern (if it is just the central heating circuit). Allow the system to fill. Turn on the radiator valves, manually open the motorised valve, and open any bypass valves fully, and bleed each radiator.

Visually check all the joints for signs of leaks and repair as necessary.

Safety tip

When testing heating systems, make sure the pressure gauge on test equipment is accurate. Otherwise, the system can be damaged or accidents may occur.

Soundness test

For metallic systems, the Water Regulations do not require closed systems to be tested. However, it is good practice to test equipment and procedures, as outlined in Section 3 of the Water Regulations. On larger jobs, this could also be part of the contract specification. You would usually expect to test the system to 1½ times the normal working pressure; the test pressure would be achieved using hydraulic test equipment.

Hydraulic test equipment

Remember to seal any open ends before testing. Test to 1½ times the working pressure and leave to stand for one hour. Repair any leaks as necessary.

The testing of plastic pipework systems should always be carried out to manufacturer's instructions. British Standards provide a leak testing method for underfloor heating, in BS EN 1264 part 4.

BS EN 1264

The leak test can use compressed air or water, and should be carried out before laying any screed. The pressure test should be between 4 and 6 bar. You should correct any leaks found, then retest.

Flushing requirements

There are three cleansing and flushing options to choose from:

- power flushing
- mains pressure cleanse and flush for sealed systems and open-vented systems with the feed and vent temporarily capped off
- cleanse and flush using gravity, with the assistance of a circulator pump.

Power flushing is most effective as this produces a more thorough clean, but you should check the boiler manufacturer's instructions to establish whether power flushing is acceptable. Power flushing is the most effective method of cleansing existing systems, especially those containing a high level of black magnetite sludge.

With all three methods, reversing the flow will help to remove debris that might otherwise remain trapped. You should choose an appropriate cleanser, according to the manufacturer's instructions, and take the following factors into account:

- the reason for cleaning
- the system materials (for example, aluminium)
- the age and condition of the system (you can use a survey sheet, as shown in Figure 6.19)
- any specific problems identified
- any local restrictions on disposal of the effluent.

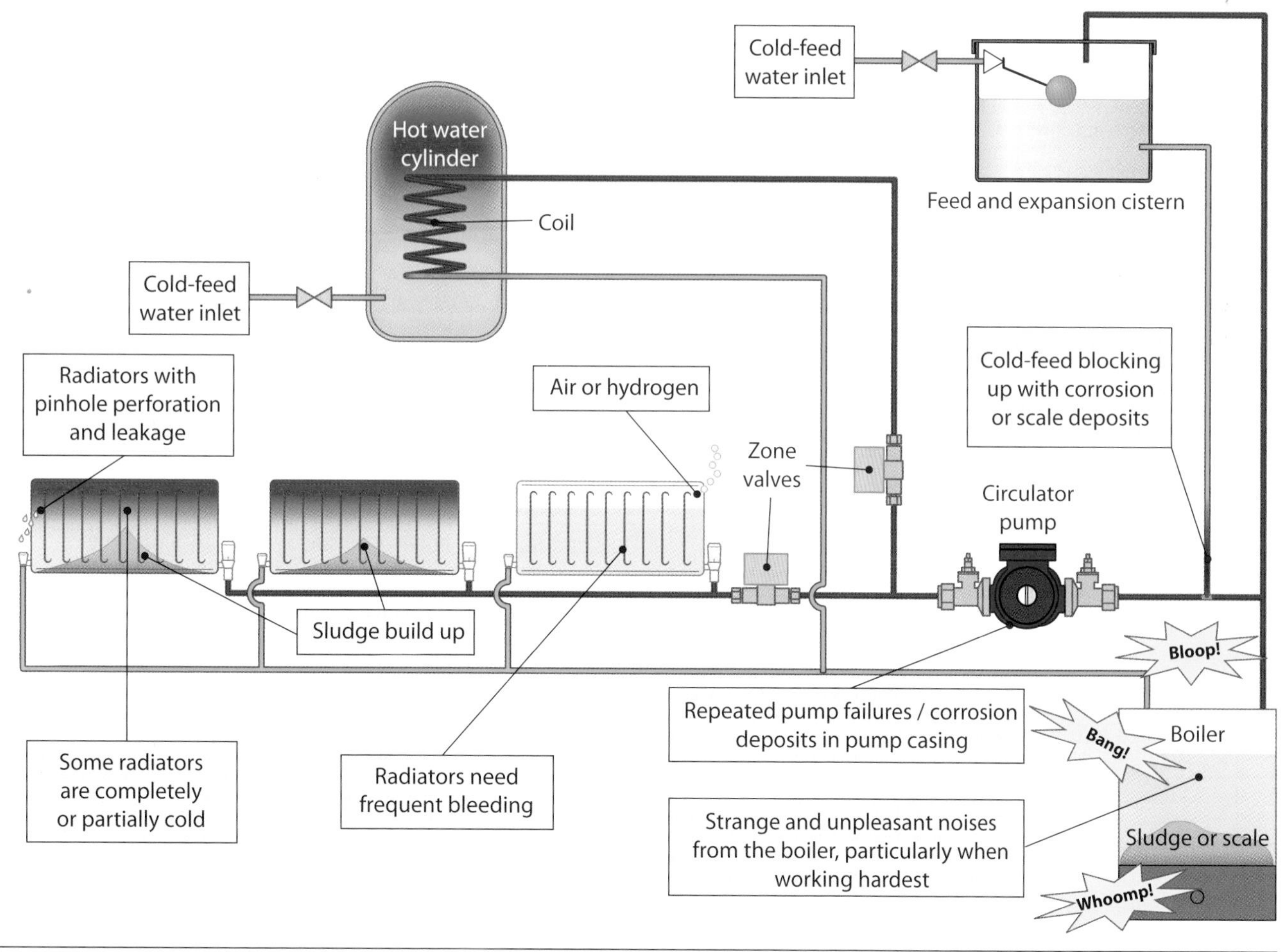

Any of these symptoms could indicate that the system has circulation and flow problems as a result of internal corrosion, scale and sludge formation. Power flushing can remove undesirable corrosion residues and replace aggressive water quickly restoring circulation and efficiency and preventing future problems. The illustration has a great deal of corrosion due to its poor installation. Identify and list as many faults as you can find.

Figure 6.18: Symptoms which indicate that a heating system would benefit from a power flush

If single pipe system, is there circulation (heat) to all radiators? Cold radiators will need removal from a system and individual flushing.
If elderly steel pipework, is system sufficiently sound to power flush? (Or would it be better to re-pipe?)

Location of system circular pump:

In boiler casing	Adjacent to boiler	Airing cupboard	Other

Check location to connect pump:

On to circular pump fittings	On to radiator	Other

Number of radiators?:

Steel	Aluminium	Are they all getting warm?	TRVs fitted?	Any signs of damage / leaks?
			Yes / no	
Do all thermostatic radiator valves (TRVs) operate correctly?				

Zone valves location?

Number of valves	Airing cupboard	Other

F&E tank

Location	Check supports	Condition?

Check place to connect for mains water supply? ________________________________

Check place to locate power flushing pump? ________________________________

Circulator pump fittings	Radiator tails	Flow and return at boiler	Flow and return pipe work from cylinder

Use a drip tray? ________________________________

Check place to run hose to? ________________________________

Toilet	Outside drain	External hopper	Other

Colour of heating system water, as run from lowest point of a radiator:

Clear	Orange	Dark brown	Black

Operative signature ________________________________

Figure 6.19: Survey sheet

Hot flushing is more effective than cold flushing, but you should follow the cleanser manufacturer's instructions.

Power flushing

Preparation

1. Isolate all electrical controls for the system.
2. Isolate the cold water supply to the central heating system.
3. Close all air vents.
4. Open-vented systems, cap-off open ends of vent pipe, etc.
5. Mark the position and settings of all lockshield or other control valves, then open all valves until they are in the full on position.
6. Remove heads of thermostatic radiator valves to enable maximum flush through the valve body.
7. Manually set diverter and zone valves to their on position.
8. Anti-gravity and non-return valves should be bypassed or temporarily removed. If they become damaged they may fail to prevent backflow.
9. Connect the power-flushing equipment to the heating system and follow the manufacturer's instructions.

Procedure

You should follow the manufacturers' recommended operating procedures at all times.

Increased temperatures improve the effectiveness of chemical cleaning, and some power-flushing units allow the boiler to be operated during power-flushing operations.

The power-flushing procedure should include:

- operation of the unit for at least 10 minutes (circulation mode) with all radiator and system valves open, reversing the flow regularly
- dumping the dirty water to a foul drain while mains water is continually added via the power-flushing reservoir tank until the water runs clear
- addition of the chosen cleansing chemical to the reservoir of the power-flushing machine and circulating to disperse throughout the system
- circulating the cleanser through each radiator for at least 5 minutes in turn by isolating the other radiators and the hot water circuit, reversing the flow regularly
- cleansing of the hot water circuit for at least 5 minutes (circulation mode) by isolating the radiators, reversing the flow regularly
- flushing of each radiator in turn for at least 5 minutes by isolating the other radiators and the hot water circuit, and dumping to foul drain until the water runs clear
- flushing of the hot water circuit for at least 5 minutes by isolating the radiators, and dumping to foul drain until the water runs clear

Did you know?

If a new boiler is being fitted, to prevent damage or contamination, the power flushing should either be carried out before the boiler is installed or with the new boiler isolated from the rest of the system.

- flushing of the system with all radiator and system valves open for at least 5 minutes and dumping to foul drain until water runs clear
- continual flushing and dumping to foul drain until all of the cleanser and debris have been removed.

Extract from BS EN 7593

BS EN 7593

Remember

Tapping the radiator with a rubber hammer will help to remove any loose material.

Mains pressure cleansing and flushing

Preparation

Run the boiler until normal operating temperature is throughout the system.

- Turn off all electrical controls and electrically isolate the system (see page 415).
- Isolate the cold water supply to the central heating system.
- Mark the position or note the setting of lockshield or other regulating valves, then fully open all valves. Remove any thermostatic radiator valve (TRV) heads to ensure maximum flow through the valve.
- Dump system water to foul drain. (All radiator and other air vents should be opened to ensure complete removal of system water.)
- Where practical, anti-gravity and non-return valves should be bridged, bypassed or temporarily removed as failure to do so will prevent flow reversal.

Procedure for cleansing

- Dose the system with a cleanser and refill with mains water, bleeding the radiators and any other vent points where necessary.
- Reinstate the electrical supply.
- Circulate cleanser in accordance with the manufacturer's instructions.
- Once the cleanser has circulated within the system for the prescribed period:
  - turn off all electrical controls and electrically isolate the system
  - isolate the cold water supply to the central heating system
  - manually close any automatic air vents
  - for open-vented systems, cap-off or temporarily join together, the open vent and cold feed to the feed and expansion cistern
  - connect a mains pressure hose to an appropriate point in the central heating system and a hose from a drain valve to a foul drain.

Check for any additional open vents, which will also need capping.

Procedure for flushing

- Flush each radiator in turn, dumping to foul drain for at least 5 minutes until the water runs clear, by isolating the other radiators and the hot water circuit.
- Flush the hot water circuit, dumping to foul drain for at least 5 minutes until the water runs clear, by isolating the radiators.

- With system additives, flush the system with all radiator and system valves open, dumping to foul drain, for at least 5 minutes until the water runs clear.
- Continue flushing until all of the cleanser and debris have been removed. Refer to the manufacturer's instructions.

After this procedure, recommissioning should be carried out in accordance with the boiler manufacturer's instructions.

Gravity cleansing (with the assistance of a system circulatory pump)

Note: this method is not appropriate for a system or part of a system that depends on gravity circulation.

Preparation

Where you cannot use the methods of power flushing or mains pressure cleansing described above, this procedure might be appropriate, particularly for open-vented systems. However, you should note the need for adequate drain valves at all low points to ensure debris and cleanser can be effectively removed.

If possible, run the boiler until normal operating temperature is achieved throughout the system.

- Turn off all electrical controls and electrically isolate the system.
- Isolate the cold water supply to the central heating system.
- Mark the position or note the setting of lockshield or other regulating valves, then fully open all valves. Remove any thermostatic radiator valve (TRV) heads to ensure maximum flow through the valve.
- Set any diverter or zone valves to their manual, open position.
- Dump system water to foul drain using all available drain points. (All radiator and other air vents should be opened to ensure complete removal of system water.)
- Where necessary, install drain valves at each low point.

Procedure for cleansing

- Dose the system with a cleanser and refill with mains water, bleeding all radiators and any other vent points where necessary.
- Reinstate the electrical supply.
- Circulate cleanser in accordance with the manufacturer's instructions.
- Once the cleanser has circulated within the system for the prescribed period:
  - turn off all electrical controls and electrically isolate the system
  - isolate the cold water supply to the central heating system.

Procedure for flushing

- Dump system water to foul drain using all available drain points. (All radiator and other air vents should be opened to ensure complete removal of system water.)
- Refill the system, bleeding all radiators and any other vent points where necessary.
- Reinstate the electrical supply.
- Circulate the system water.
- Repeat draining, refilling and circulating a minimum of twice.

Repeat as necessary until the water runs clear at all drain points. Refer to the manufacturer's instructions. After this procedure, recommissioning should be carried out.

Be able to apply design techniques for central heating systems

This a practical outcome, so you will be learning about this in your workplace. Activities and further information can also be found on the *Level 3 NVQ Diploma Plumbing Tutor Resource Disc.*

3. Know the commissioning requirements of central heating systems and their components

Boiler commissioning

Commissioning may be more involved than you expected, but it is vital if the system is to work correctly.

Once the whole system is installed you can think about putting it into service. This is probably the most important phase of the installation: you should be aiming to leave the site with a correctly operating system that meets the design specification. But it is not as simple as just turning it on, both from a safety and an operational perspective.

On completion of a central heating installation, the system should be commissioned in accordance with the manufacturer's instructions.

Interpreting information to complete commissioning

The commissioning of the "fuel supply" will need to be carried out by a "competent person" such as a Gas Safe registered operative in the case of gas fuelled appliances. The following procedure covers the water side of the system and any reference to gas-related activities should be carried out by those who meet the above requirement.

1. Run the boiler with the cleaner in the system. Run it up to temperature to see if the boiler thermostat is working. Set the boiler to its correct operating temperature (about 80°C) and connect a digital pipe thermostat on the flow pipe next to the boiler. It should turn off when the reading is about the right temperature and begin to cycle on and off.
2. Next, run the water up to temperature and establish that the water supplied is at a higher temperature than the cylinder thermostat setting of 60–65°C. It should be higher, as the temperature of the water at the top of the cylinder will be higher than at the point at which the cylinder thermostat is positioned.
3. Now check the room thermostat operation, again using the digital thermometer. Set the room thermostat to just above the temperature currently in the room and run the heat up until it shuts down. The air temperature near to the thermostat should be similar to the point at which the thermostat has turned off.

Balancing a central heating system during commissioning

This is a crucial phase, as you will need to ensure that all the radiators are working uniformly and the components are receiving the correct flow rate from the pump. To understand this fully you need to know something about system design.

A system's design includes a circulation pump that will provide a certain flow rate, to get the required amount of water from the boiler to the component parts of the system's cylinder and radiators. In delivering the required flow to all points of the system, it is necessary to overcome pressure created by the frictional resistance in the pipework. However, this pressure is in relation to the frictional resistance of (usually) just one of the radiator circuits, known as the index circuit (see page 294). In any system this will be the circuit that has a combination of the highest flow rate and highest resistance. So the pump is sized to meet the resistance of just that index circuit. Any other radiator, or indeed cylinder circuit, on the system will have less resistance, so the pump will be able to deliver the flow rate required to those circuits much more easily.

To ensure that all the circuits work uniformly, you need to balance the radiators that are not in the index circuit. This balancing ensures that there is the same frictional resistance at the other circuits as there is in the index circuit, and is achieved by adjusting the lockshield radiator valve, placing a restriction on the other circuit.

The starting point is to put the circulating pump on its correct setting. Now with only the index radiator working (and the valves fully open) and using the digital thermometer (which should be capable of taking readings on both the flow and return pipes), you should have the required temperature drop across the flow and return pipes near the boiler of 80°C and 70°C respectively.

Did you know?

There is a tendency in systems not to balance the hot water circuit but to leave it fully open. It will tend to starve the heating circuit while in operation but will shut off more quickly. This is not particularly problematic if the hot water is brought on with the programmer before the heating circuit.

If the boiler requires a bypass, or the pump is sized much higher than the system flow rate usually requires (for example, a combi boiler with bypass), you should start taking the readings at the flow and return pipes with the bypass fully closed. This will have the effect of raising the return temperature at the boiler. The bypass should be eased open gradually, turn by turn of the valve, until the return temperature achieves the required temperature drop across the boiler – 80°C and 70°C respectively. The bypass should then be left at this setting.

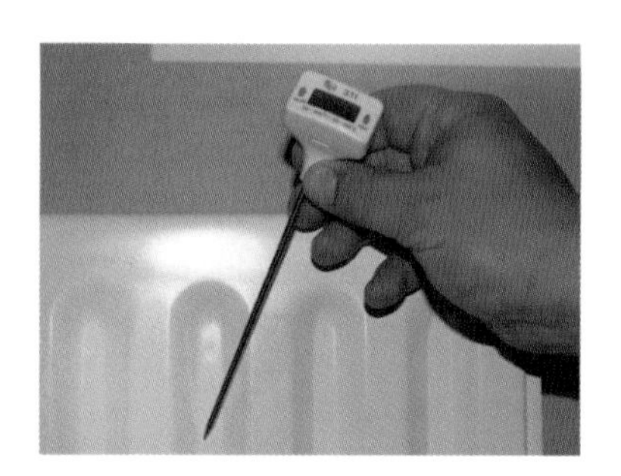

Using a radiator thermometer

The remaining radiator circuits should now be balanced. This means going round the radiators and opening them gradually at the lockshield valve until a point occurs at which the difference in the flow and return pipes at the inlet and outlet to each radiator is 10°C. You might have to go round the system a couple of times to get this right, as opening one valve tends to have an effect on other parts of the system already running.

Once they are all running with the required temperature drop across them, do a final check on the boiler flow and return, which should be 80°C and 70°C, with the 10° drop, and your system is balanced.

The final job is to check to see if there is any discharge from the pressure-relief valve (if it is a sealed system) and that the pressure gauge does not show an excessively high pressure in the system. If it does, and the expansion vessel has been properly charged, it is too small.

Fault diagnosis

If you find a fault during the performance test, you should turn off the system immediately and take remedial action to repair the fault.

Figure 6.20 gives one example of how to identify what is causing the problem if a boiler's burner is blocked.

Once you have rectified the fault, the system should be retested and checked again. If there are no further problems, you can add the inhibitor and hand over.

Inhibitor

The inhibitor you select should be compatible with the system components and water quality and used in accordance with the manufacturer's instructions.

The system inhibitor should be checked at intervals recommended by the product manufacturer to ensure adequate and lasting protection. Unless the manufacturer's instructions state otherwise, you should not mix products from different manufacturers or different products.

Remember

Handing over the system is an important part of the process, and it is your responsibility to do it well. Look at page 155 to remind yourself of how to do this properly.

You should leave a record of the work carried out at the premises, usually close to the boiler. You should also fix a permanent label in a prominent position, usually on the outside of the boiler casing, indicating the make and type of the inhibitor used and the installation date.

Notifying the authorities

If the installer is registered with a competent person registration scheme for the purpose of self-certifying, they should notify their scheme administrator, who will then notify Building Control on their behalf. If not on a registered scheme, the installer or person undertaking the commissioning must inform Building Control directly before commissioning takes place.

The benchmark system can be used to show that commissioning has been carried out correctly. The boiler manufacturer's benchmark commissioning sheet can be found in the literature within the boiler packaging.

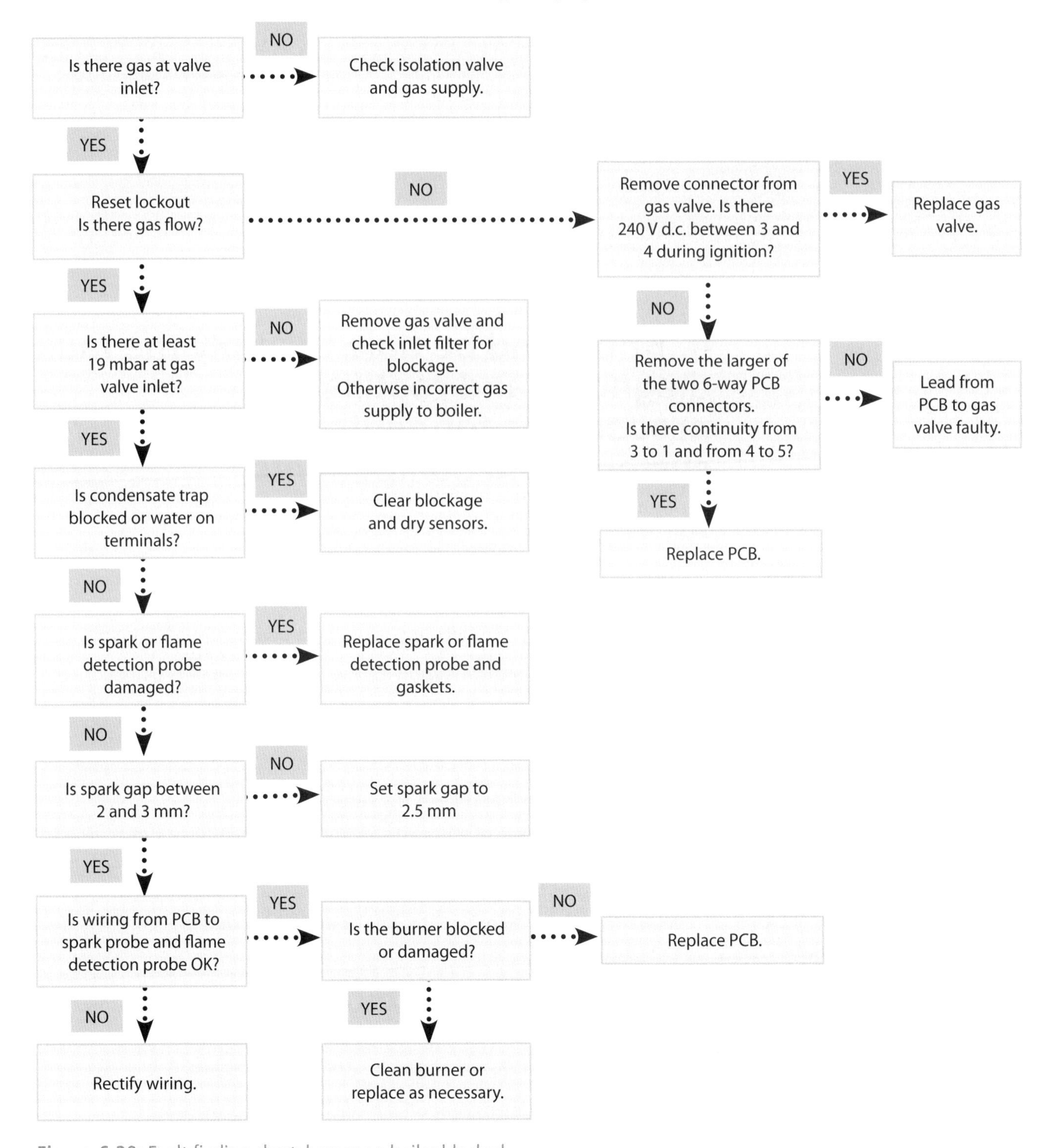

Figure 6.20: Fault finding chart: burner on boiler blocked

Be able to commission central heating systems and components is a practical outcome, so you will be learning about this in your workplace. Activities and further information can also be found on the *Level 3 NVQ Diploma Plumbing Tutor Resource Disk.*

PROGRESS CHECK

1. What does a pressure-relief valve replace?
2. What range must a pressure gauge be capable of reading?
3. What must a boiler which is fitted to a sealed heating system be fitted with?
4. To what pressure should expansion vessels be pre-charged?
5. One metre of static head of water is equivalent to how much in bars?
6. In what circumstances would an extra expansion vessel need to be fitted to a system boiler?
7. What is the first check done when commissioning a system?
8. There are three cleansing and flushing options. Name one of them.
9. When flushing an existing system, what needs to be done to thermostatic valves?
10. What is the first check when commissioning a boiler, excluding the gas system and its controls?

4. Know and be able to apply the fault diagnosis and rectification procedures for central heating systems and components

An important note on safety

It is crucial that you learn how to isolate a system. Safe isolation is covered in detail in Unit 7 (see pages 407–415), so it will be important that you read this section fully, but here are a few short points for the moment.

1. Make sure the electrical supply to the heating system is isolated.
2. Turn off the boiler.
3. Remove the fuse from the spur outlet to the wiring centre or junction box for the controls.
4. Lock off if possible and keep the fuse with you.
5. Advise the customer of what you are doing and tell them not to touch the controls. Only work on the system when the water has cooled.
6. If you speak to the customer beforehand to arrange the job, it is a good idea to ask them to turn off the system so that it is cold when you get there.

Safety tip

Always ensure that the system is electrically isolated before starting any maintenance operation.

Routine checks and diagnostics

A boiler should be serviced once a year and the rest of the heating system should be checked at the same time to ensure the boiler is safe to use.

The best way to check all components and controls is by using a maintenance checklist form like the one opposite, to make sure that you do not miss anything. For individual faults, see pages 316–318 in this unit.

Remember

Always protect the property before removing a radiator and ensure no spillage occurs from the radiator open ends.

Maintenance and service procedure heating system Location address: Service engineer: Date:					
Equipment	**Service task**	**Checked**	**Notes**	**Repair**	**Initials**
Pipes	Check for adequate support Check for signs of corrosion Check for adequate allowance for expansion and contraction Check for correct insulation Check for adequate pipe size Check for soundness Clean pipework				
Control valves	Check they operate correctly Check they are correctly labelled Check for signs of corrosion Check they are readily accessible Check float-operated valves for water level and compliance with Water Regulations				
Cisterns	Check for signs of leakage Check adequately supported Check for sediment Check for stagnation (bio film) Check lid fitting Check warning overflow pipework Check adequate insulated				
Pressure vessel	Check for leaks Check for signs of corrosion Check gas pressure Check adequately supported				
Water	Check chemical inhibitor Check system operating pressure Check filler loop compliance with Water Regulations Check flow and return temperature				
Pumps	Check pump operation Check for signs of corrosion Check for noise				
Pressure-relief valve	Check valve is accessible Check valve discharges when operated and shuts off				
Electrical controls	Check they operate to manufacturer's instructions Time controller/programmer Room thermostat Cylinder thermostat Boiler energy manager				
Earth bonding	Check adequate earth bonding to pipes and equipment				
Heat emitters	Check for adequate temperature Check for signs of corrosion and damage Check for system balance Check for signs of air build-up				
Boiler/heat source	Check heat source operation against manufacturer's instructions				
Note: This is not an exhaustive list – variables will occur from system to system Service engineer report:					

Figure 6.21: Maintenance checklist form

Professional Practice

Sami is an experienced plumber who has an apprentice called David. Their employer has asked Sami to show David how to carry out routine maintenance on a central heating system. Discuss a checklist of maintenance tasks that will need carrying out during the routine maintenance.

Repairing faults and replacing components

System faults

With central heating systems, poorly designed and installed systems account for many call-backs. If you are working on a system that someone else installed, you should carry out a visual inspection for correct compliance with installation standards, as this may reveal a lot.

The system works but there's noise from the pipework

Key questions are:

- Has it just started?
- Has it been there all the time?

If the noise has just started then do the preliminary checks:

- Has the system got water in it?
- Is it topped up?
- Has the float-operated valve stuck?

Then run it up to temperature:

- Is the pump on the right setting?
- If this checks OK, then are all the valves open properly?
- Is the bypass set correctly?

Remember

Expect the unexpected when working out the problem, but always look for the simple points first, as they are usually to blame.

If the problem has been there for a long time, you might be checking on such issues as whether the boiler needs a bypass. Following these checks, one of your findings might be that key pipe sizes in the system are incorrect, particularly for a long-term problem. Here you might have to re-pipe some aspect of the system.

Radiators are not getting hot in some parts of the system

The first thing to identify is which radiators are not heating up.

- Upstairs only and not as hot as they usually are: the pump may have stuck or failed.
- Downstairs only: the system could have air in the upstairs radiators; the system is running dry and may need topping up; the float-operated valve in the cistern may have stuck.
- Individual radiators not as hot as they used to be: have the radiators been off for decorating? Does the system need re-balancing? Is the system getting 'sludged up'? If it is, there will be reasons and these will need investigating.

- Individual radiators not working at all: are the radiator valves open? On thermostatic valves, check the pin that operates the function of the valve: they have a tendency to stick closed in old age and may need greasing or replacing.
- No radiators at all: there could be a component fault or an electrical fault. If hot water is available, you can discount the boiler.

First check the major components to ensure that the fuse has not blown.

If power is available then check the operation in the following order: programmer, to thermostat, to motorised valve (here it could be due to a defective motor or a defective auxiliary switch), to pump.

With a combi, you will need to follow the manufacturer's guidance on checking its components for correct operation.

No hot water

First consider the type of system. If it is fully pumped, you must establish that the system has got water in it. Assuming that the radiators work, you will probably need to look for component faults, in a similar way as for heating-circuit faults.

First check the major components to ensure that the fuse has not blown.

If power is available then check the operation in the following order: programmer, to thermostat, to motorised valve (here it could be due to a defective motor or a defective auxiliary switch), to pump.

With a combi you will need to follow the manufacturer's guidance on checking its components for correct operation. With gravity systems, check whether the circulation pipework is air-locked. If it is, investigate the pipework to make sure it is run properly.

Dirty-coloured hot water

With single-feed indirect cylinders, this is usually a sign that the air bubble has been lost. You may need to investigate further, to see whether the boiler thermostat may have failed, overheated and removed the air bubble, or whether the system may have been extended and the cylinder requires replacement.

Noise at the boiler

The key questions here are: what is the noise like and how long has it been going on?

If the noise has been short term and sounds like a boiling noise, the cause is likely to be a component failure on the boiler.

If it is more long term and like a kettle heating up until it shuts off under temperature, then starts again, this is more likely to be a boiler circulation problem. This could be due to:

- no bypass fitted – the boiler heat exchanger could already be damaged
- pump at the wrong setting (again, the boiler heat exchanger could already be damaged)
- sludge in the boiler causing poor circulation (cleaning may be an option, but an investigation into how the sludge has collected is required).

No power to the system

First check the fuse and try replacing it. If it blows again, it is either a component or a wiring fault. With the electricity off, check the pump. If it sticks, it will probably blow a 3 amp fuse.

After this, you will need to carry out further checks.

- Has water got into any of the electrical components? Badly positioned motorised valves suffer from this problem, quite often with drips from valves entering the electrics and the motor.
- Have any flexes connecting to the boiler strayed too close to the boiler? Badly installed flex to fireback boilers can be a problem, even if it is heat-resistant.
- Have any badly installed cables strayed too close to heating pipes, or has somebody recently been working in the property and damaged a cable?

System keeps filling up with air

This signifies a major problem of some description. The main question to ask here is how often you have to let the air out.

If it is a sealed system, this probably indicates a leak, especially if the system needs ongoing topping-up (normally it should require it only rarely). With an open-vented system, this could indicate a leak, or could be a sign of more serious problems, such as pumping over and sucking down at the F&E cistern, in which case work will need to be done to the system.

Remember

F&E cisterns are sized by the volume of the system, allowing for 4 per cent expansion.

You will also find some unusual faults that experience will help you to deal with.

Be able to diagnose and rectify faults in central heating systems is a practical outcome, so you will be learning about this in your workplace. Activities and further information can also be found on the *Level 3 NVQ Diploma Plumbing Tutor Resource Disk.*

5. Know the installation requirements and be able to install underfloor central heating systems and components

Underfloor heating (UFH) systems have been around since the Roman era. The concept of heating a large surface area at low temperature rather than a small radiator to high temperature is easy to understand.

How underfloor heating works

The basic system works from a heat source (a traditional boiler source or more modern heat pump) piped to a manifold, which branches the pipework to supply each separate room with its own underfloor heating pipework. This is controlled by an electric actuator (motorised valve) on the manifold, controlled by a room thermostat. The circuit has the hot water pumped through a blending valve that controls the temperature of the circuit, stopping the floor getting too hot.

When the room reaches temperature, the circuit is switched off. This gives good control of the temperature in each room. Unlike radiator heating systems, once underfloor heating systems are turned on for the heating season, they are best left on all the time. When sleeping times are reached, a 'set-back' temperature can be brought into operation on a timer control, lowering the average temperature to make it more comfortable to sleep at night.

Did you know?

- Surface heating is based on radiant heat transfer. This interacts directly with the room and occupants in the same comfortable way that the sun warms us.
- Radiator heating is based on air current movement with a hot ceiling and cold floor draughts.
- Convection heating warms the air in the room, which then heats the room and its occupants.

Advantages

- No visible heat panels.
- No cold floors.
- Safe from high temperatures.
- No visible pipes.
- Heat distribution even across room.
- Low maintenance.
- Quiet in operation.
- Does not use up floor space.
- Energy-efficient system.
- Helps with allergy conditions, as it reduces dust movement.

In Figure 6.31, you will find faults found on UFH systems.

Underfloor heating compared to radiators

Much lower water temperatures are required than with radiators as the large surface area of the floor is enough to warm the room efficiently as the heat rises. An underfloor heating system will generally run at around 45° as opposed to 80° used in radiator systems, which means that there are energy and running cost savings, especially when used with renewable heat sources such as heat pumps, condensing boilers, etc.

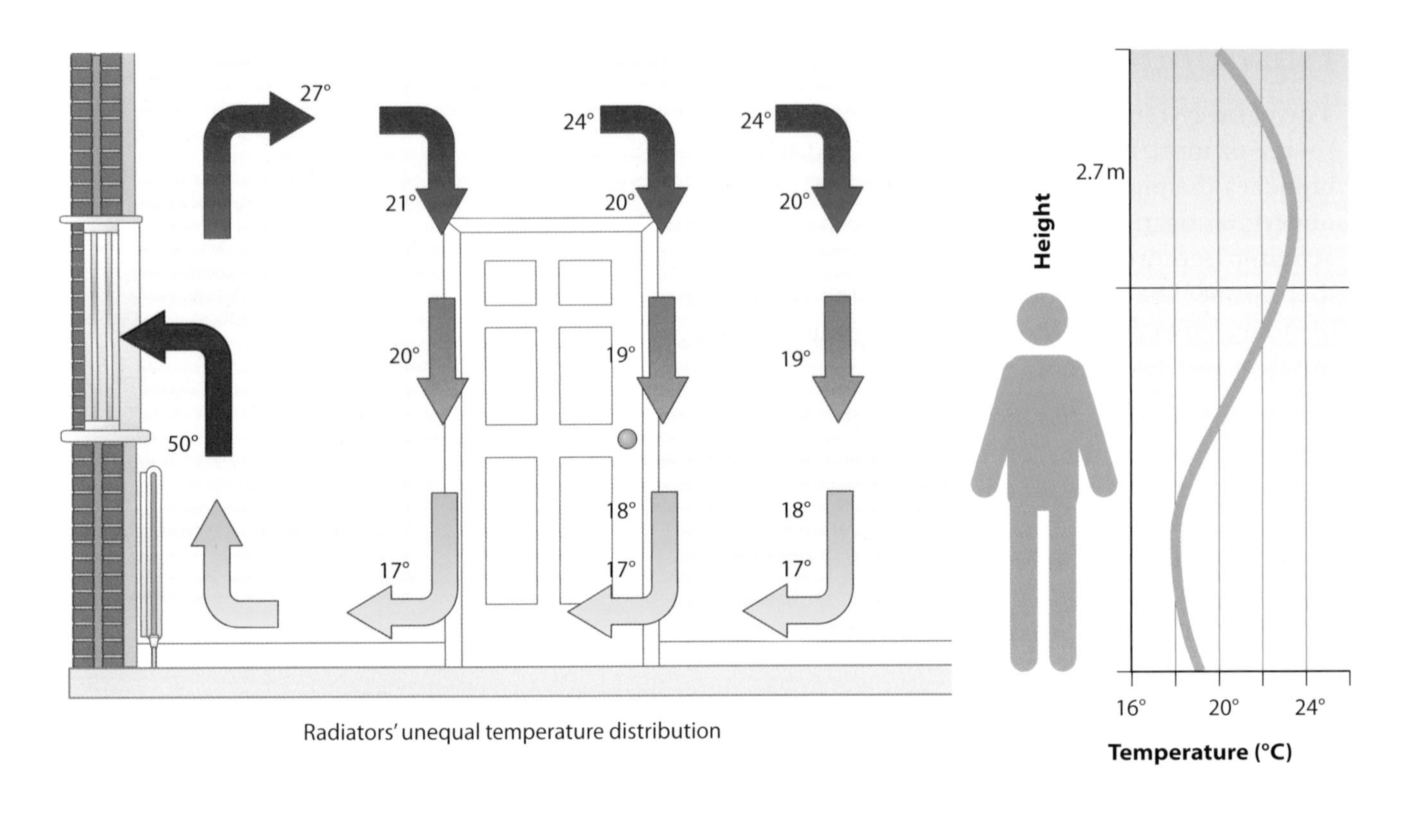

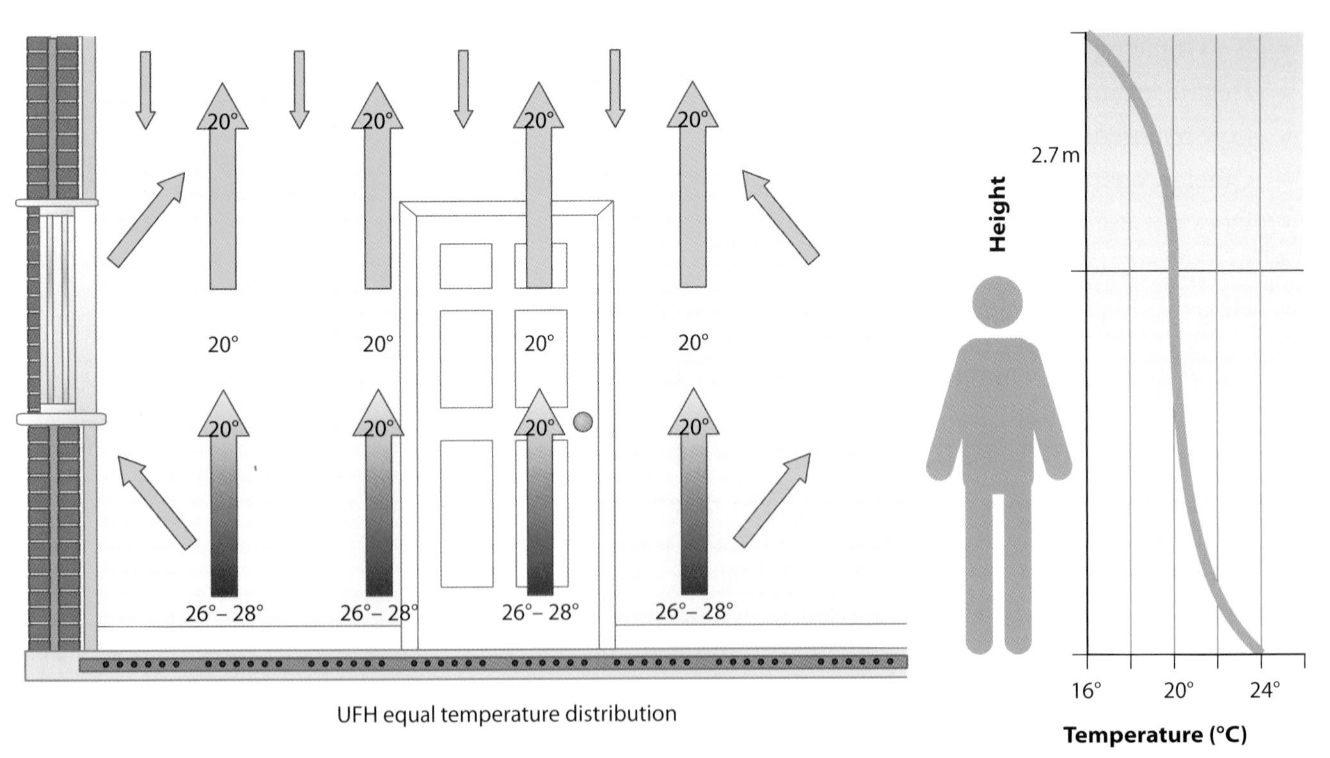

Figure 6.22: Example of temperature distribution between UFH and radiators

Positioning and fixing requirements

The underfloor heating system works on the principle of warm water at 35°–55°C being pumped through plastic or copper pipework buried in the floor. The floor heats up and radiates most of the heat into the room.

The floor temperature should not exceed 29°C, depending on the floor type; lower temperatures may be required to prevent damage to certain types of flooring. The floor can be solid floor (such as concrete and screed) or timber flooring.

No joints should be placed into the floor; the system should be made from a continuous roll of pipework to prevent any leaks occurring (which would be very costly to repair). The pipes should be placed in a pattern specifically to ensure the heat is distributed evenly across the floor.

Remember

A floor that is too hot will be uncomfortable for the customer.

Did you know?

Floor covering can act as insulation and slow down heat gain into a room. It is important to consider this in the design of the system.

Installing underfloor heating

There are three main types of wet underfloor heating design:

- screeded floors
- floating floors
- joisted and battened floors.

Screeded floors

Screeded floor underfloor heating consists of a concrete base plus an insulated layer onto which the pipework is fixed and then screeded over.

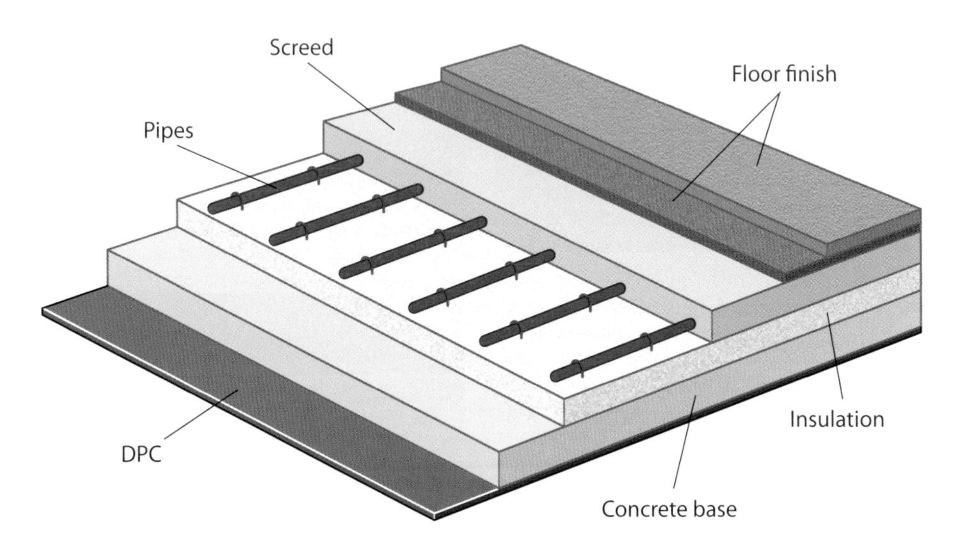

Figure 6.23: Screeded floor

Staple systems

There are two pipe configurations:

- series or serpentine pattern
- bifilar, snail or spiral pattern.

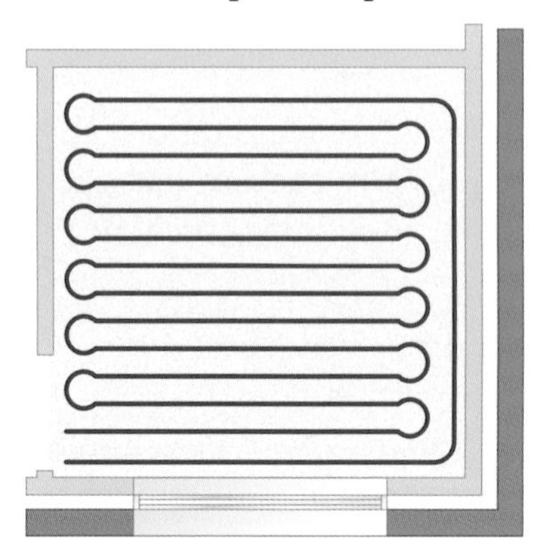

Requirements

- Staple gun to ensure correct fixings of staples
- Staples to anchor the pipe to the insulation

- Pipe designed to withstand the expansion and contraction and spaced to give the correct amount of heat to the floor

Installation steps

Screeded floors

Step 1 Lay edge foam around the room perimeter to allow for expansion and contraction. Place first plate into the corner. Each plate overlaps the next and they lock together as a snug fit.

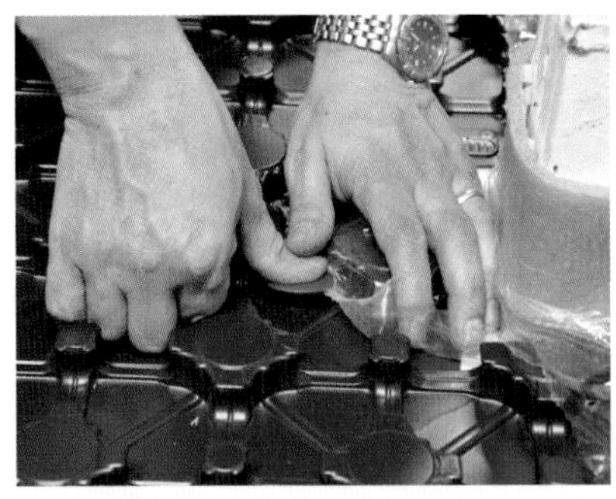

Step 2 Continue until the floor is covered to design specification, trimming with a sharp knife or scissors where necessary.

Step 3 The floor can now be piped. Start at the manifold and run the first loop a minimum of 50 mm from the edge of the room. It is advisable to first mark out where the pipe is to go using a piece of chalk. When making the return in the centre, make a simple 's' shape then bring it back.

Pocketed polystyrene

This system has grooves moulded into the insulation to which the plastic pipe can be fitted, holding it in place.

Installation steps

Pocketed polystyrene

Step 1 Starting with a clear and level floor, lay the return panels at the end of the room. Place the straight panels against the ends, making sure that the end loops match the straight channels.

Step 2 When the floor insulation panels have been laid and the floor is covered, simply walk the pipe into the floor using your feet to press the pipe home.

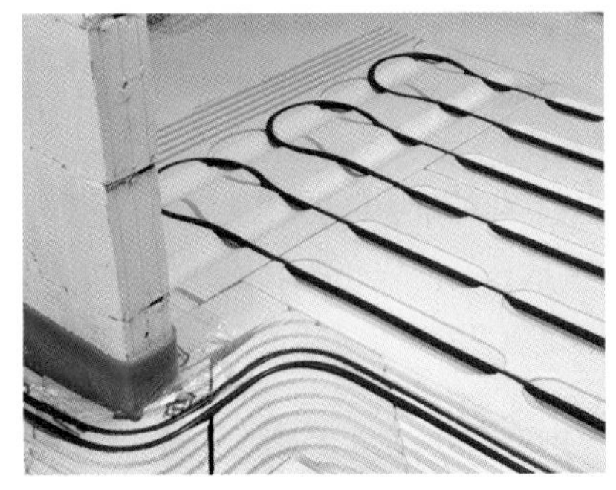

Step 3 Once the pipework has been pressure-tested, it can be screeded over. If you are using liquid screeds, tape the panel joints to stop the screed running under the panels.

Floating floors

As the name implies, the floor (usually chipboard flooring) sits or 'floats' on top of the underfloor heating pipework.

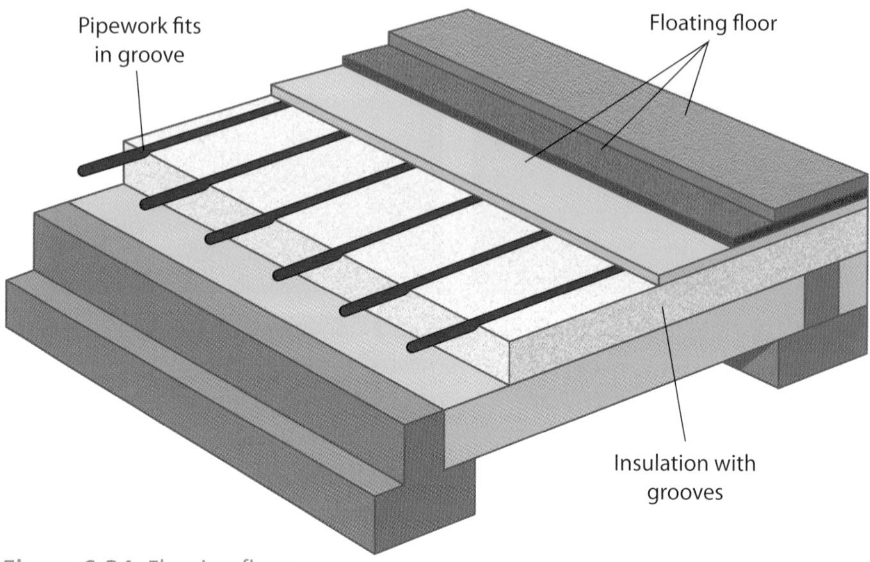

Figure 6.24: Floating floor

Installation steps

Floating floors

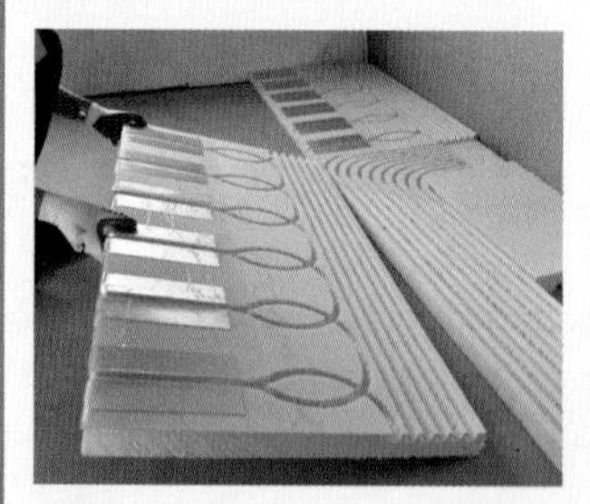

Step 1 Lay the return loop panels at each end of the room up against the wall.

Fill in between the ends with straight heated panels, making sure that the grooves line up with the slots in the return loop panels. Use offcuts in the next row.

Step 2 Place the insulation, then inlay the pipe.

Step 3 Set the floor deck over the top, glueing all tongued and grooved joints.

You can see here a very standard form of floor make up, of a tongued and grooved or planked floor deck over 35 or 50 mm insulation. A high compressive-strength extruded insulation is used because there is no need to fix edge battens around the edge of each room – this avoids the large deflections that can occur with softer expanded polystyrene.

Joisted and battened floors

In the joisted and battened floor the insulation sits between the joist and so it is often used in upstairs underfloor heating flooring.

Installation steps

Joisted and battened floors

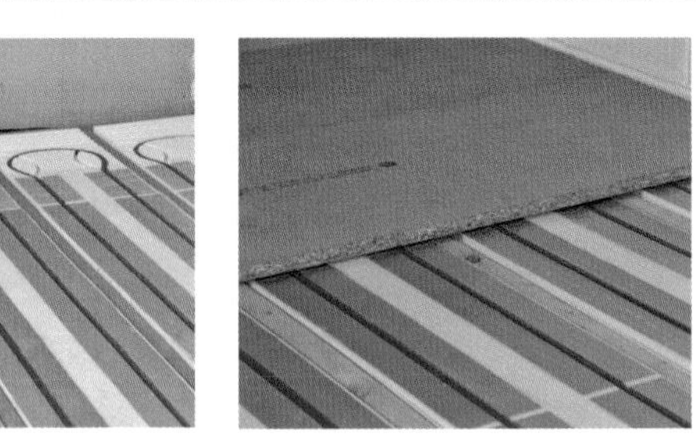

Step 1 Place battens around the edge of the room then, starting in one corner, place the first return loop.

Step 2 Where the pipe must cross the batten/joist, cut a notch.

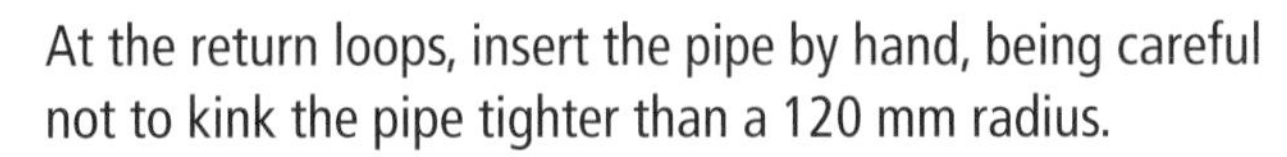

At the return loops, insert the pipe by hand, being careful not to kink the pipe tighter than a 120 mm radius.

Step 3 Once the pipe has been inserted into all of the grooves, lay the floor panels and glue them to the top of the batten/joist.

Chipboard and plywood modules

In this design, the groove is in the chipboard and comes from the factory with the pipework fitted in the groove, ready for assembly.

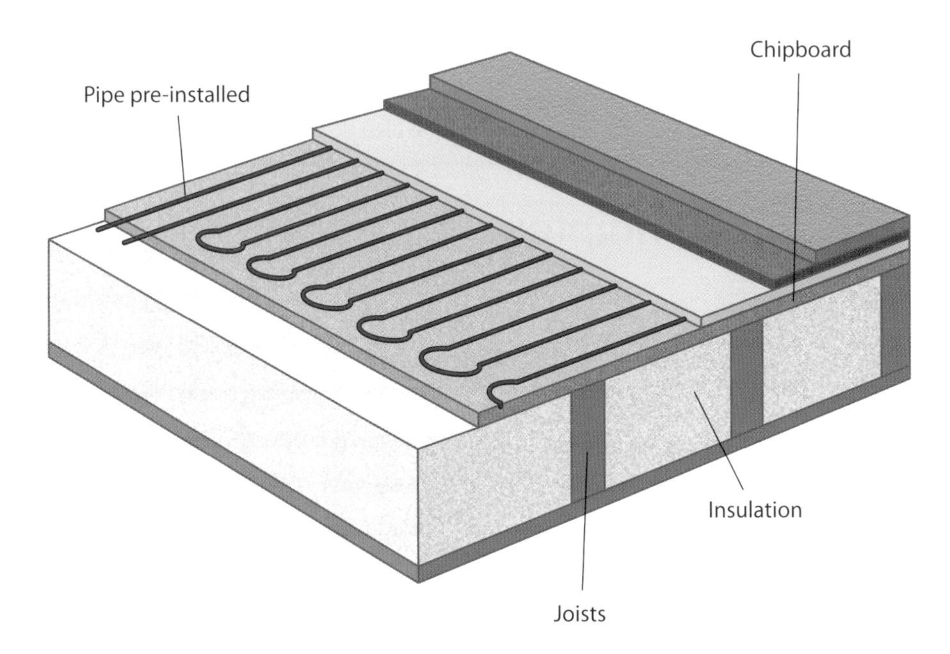

Figure 6.25: Chipboard and plywood modules

The chipboard has a red marking to show the pipe position, to prevent nails or screws being driven into the pipe in later stages by other trades.

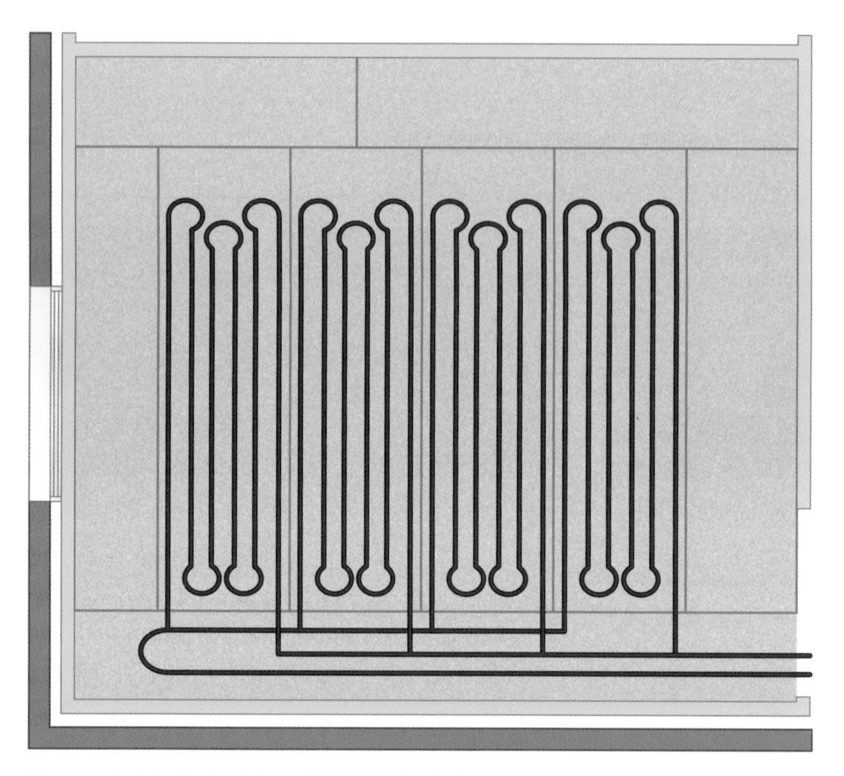

Figure 6.26: Typical floor layout of chipboard system

Professional Practice

A client asks you to remove a radiator in a bathroom and replace it with an underfloor system: there is not enough space in the bathroom with the radiator, but they do not want to alter the tiled floor. Explain to the customer how it is possible to install the underfloor system without damaging or lifting the tiled floor.

Working principles

Energy sources

Underfloor heating gives the flexibility to use any of the modern energy sources.

Traditionally, the primary heat source for heating has been a boiler producing hot water for the system. Modern high-efficiency boilers are ideal for underfloor heating, as the low water temperatures allow the boiler to work in condensing mode.

New technologies, such as heat pumps and solar panels with water storage, will often provide water at 45°C. Often this is sufficient to heat a modern building without the need of additional power. Heat loss and output calculations must be done to ensure sufficient output. Well-insulated buildings will gain the greatest benefit from standard boiler systems.

Combi boiler systems

Combi boilers can easily be incorporated into the underfloor system, as shown below.

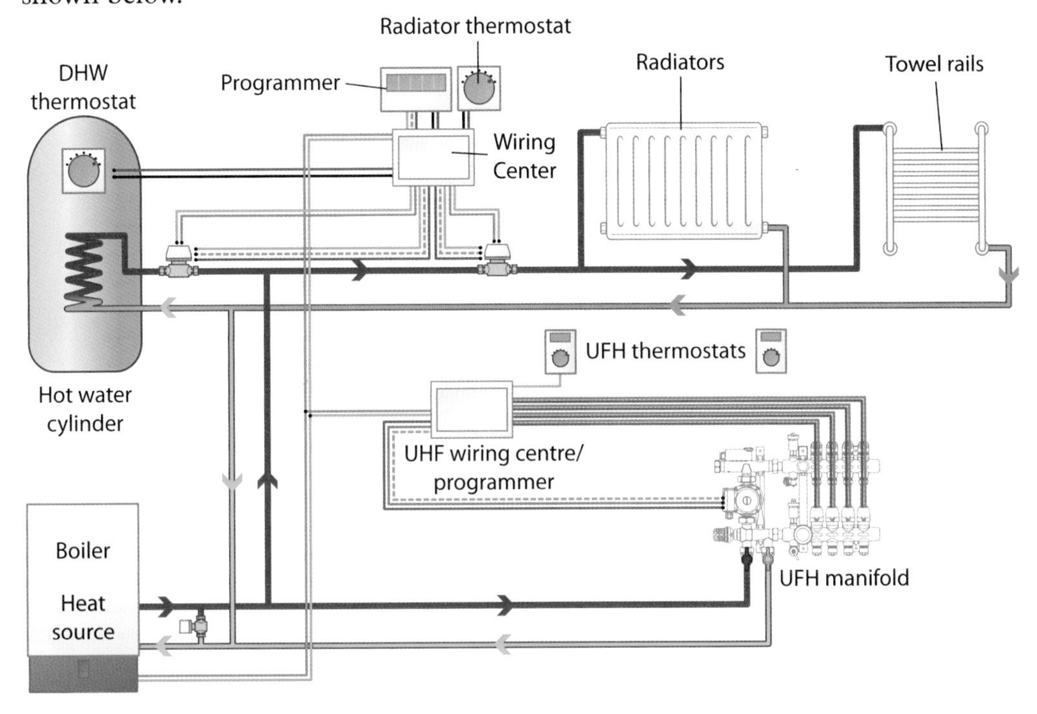

Figure 6.27: Combi boiler systems

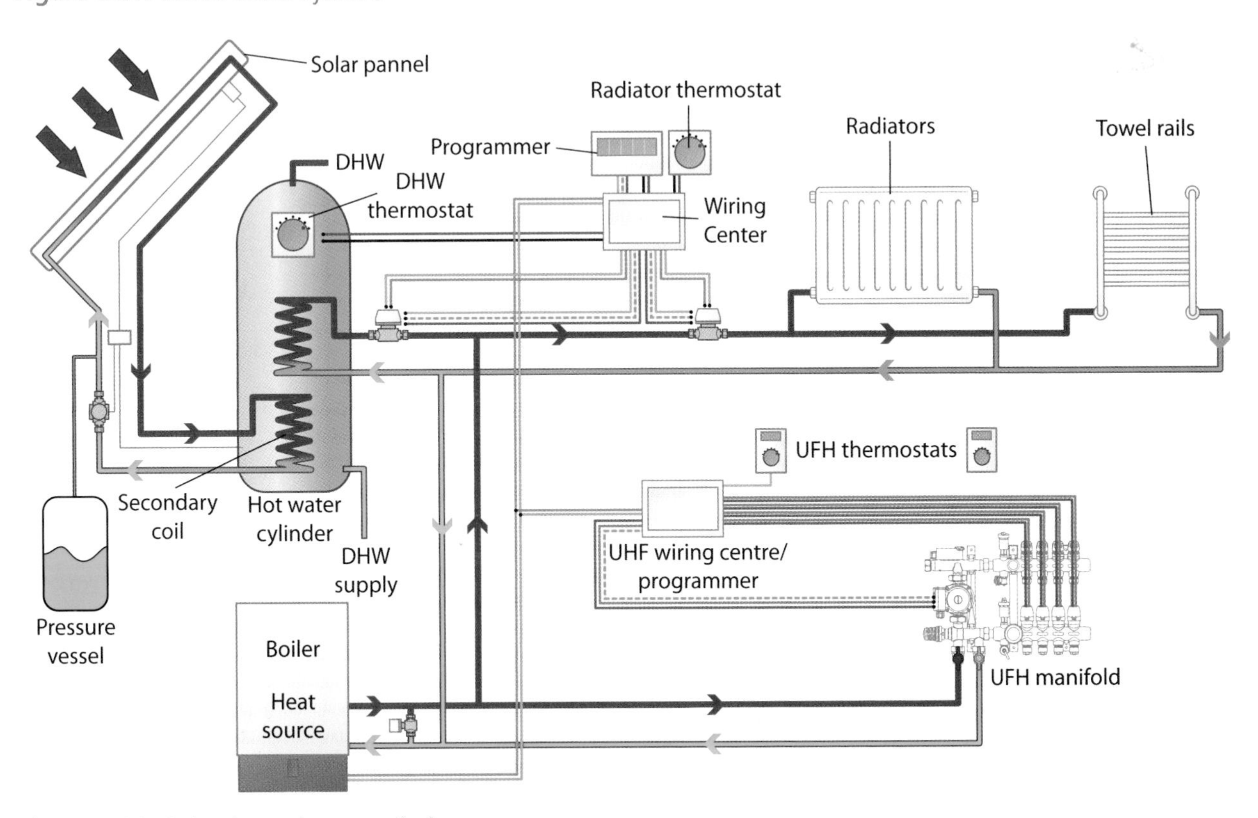

Figure 6.28: Solar thermal store cylinder

Solar thermal heat stores can also be incorporated with little difficulty.

Manifolds

Perhaps the most important component at the heart of the underfloor heating system is the manifold. Manifolds are available from one to twelve ports. They generally mix, regulate and allow effective distribution of low-temperature water to the underfloor heating circuits by drawing off from the primary circuit, which is often at a higher temperature.

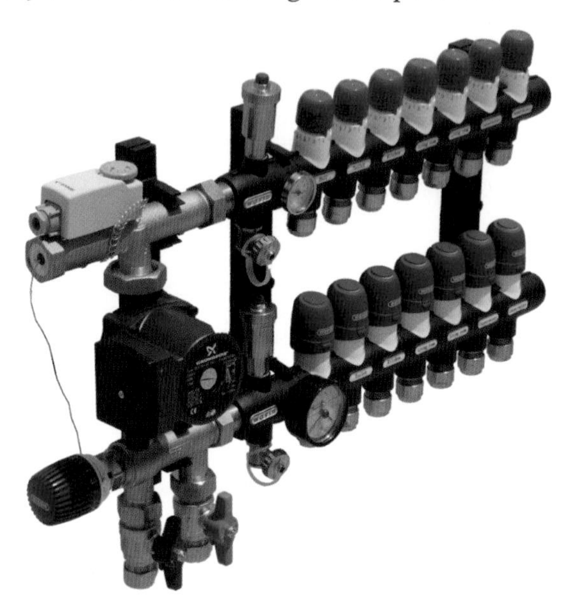

The manifold is perhaps the most important component of the UFH system

How manifolds work

Primary flow from the boiler or heat source passes through a mixing unit. The pump pushes the water into the flow arm and through the balancing valve. Water passes round the circuit and back to the return valve, controlled by the actuator. The cool returning water is then either drawn into the mixing valve to reduce the temperature of the hot primary water, or is returned to the boiler to be reheated.

Control system applications

The basic principles of controlling the underfloor heating system are similar to those of a traditional heating system. However, the following controls have been specifically developed for underfloor heating applications:

- thermostatic mixing valve – used to blend water down to that required for underfloor heating, with a range of 10°C to 70°C; mixes water down by utilising the return from the underfloor heating with hot water from the heat source
- thermal actuator – simple 230 V or 24 V thermoelectric heads which open and close in conjunction with room thermostats and pipe loops to allow water to flow around the floor
- zone control centre – used to simplify the underfloor heating wiring installation by allowing multiple room thermostats and actuators to control the heating pump and boiler interlock

- thermostats – available in various types including wired, wireless, programmable, non-programmable and dial type; used to manage the room temperature during times when room is being used and when it is unoccupied.

The following drawings outline the requirements of zoning underfloor heating under the Building Regulations.

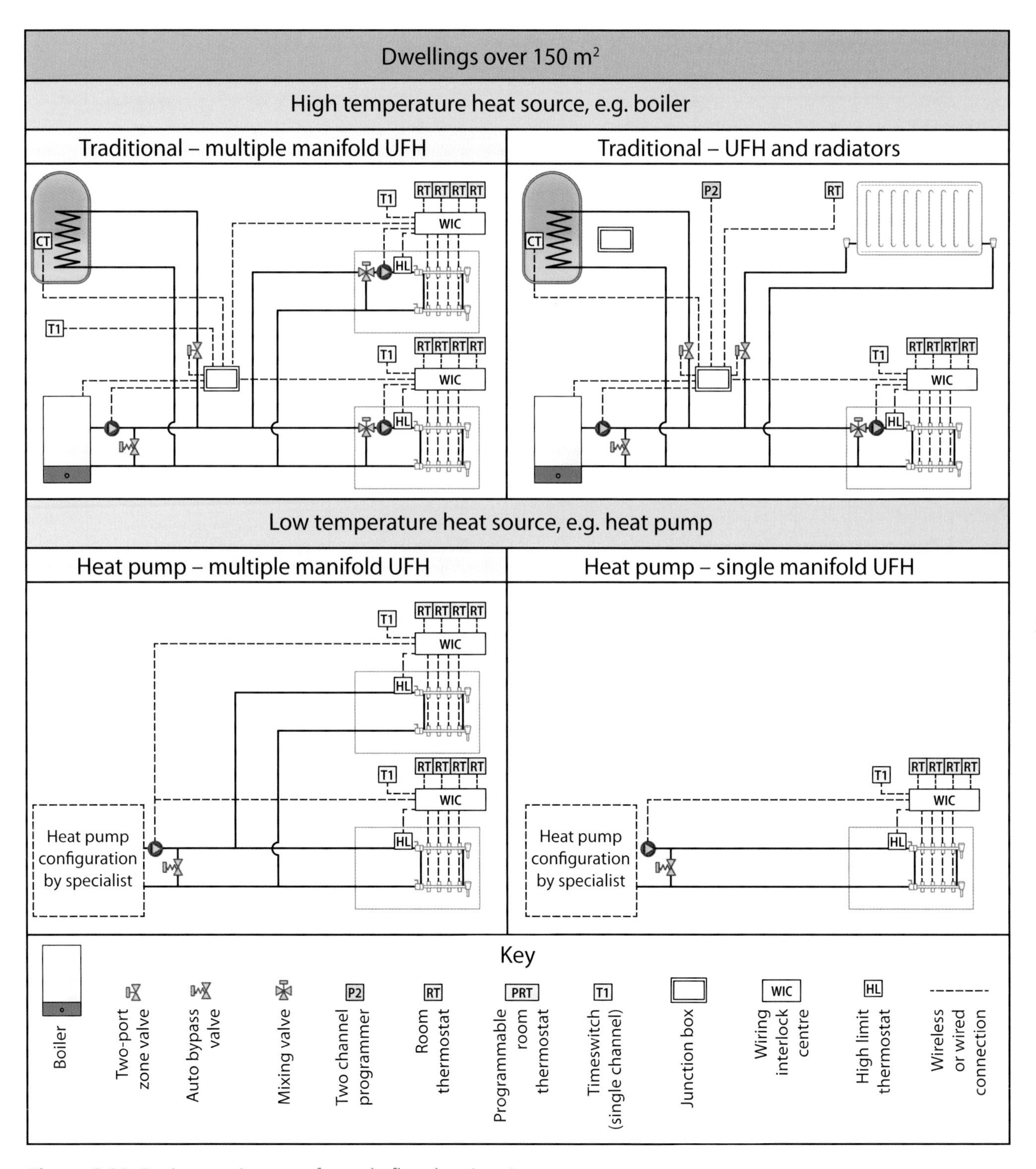

Figure 6.29: Zoning requirements for underfloor heating: 1

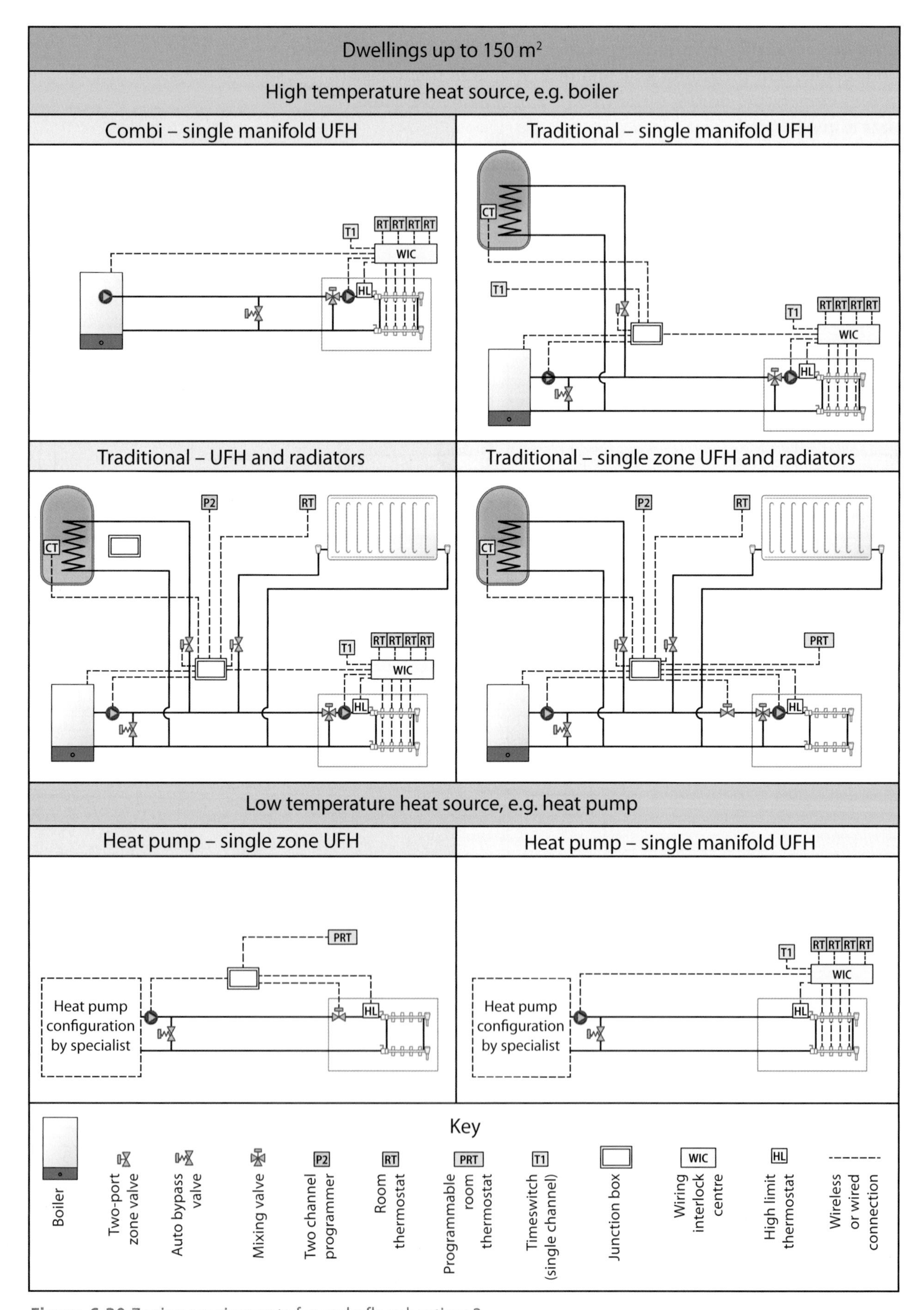

Figure 6.30 Zoning requirements for underfloor heating: 2

Underfloor pipework arrangements from manifold to room

Underfloor heating distribution boards (manifolds) should be located in a central position between rooms being heated. This is in order to minimise the lengths of interconnecting pipe services.

In the event that long lengths cannot be avoided, pipes should be insulated or routed by conduits to reduce distribution losses and the risk of overheating in the rooms through which the pipes run.

Designing an underfloor central heating systems

Underfloor heating designs are based on BS EN 1264, which employs results from years of testing the construction materials used in underfloor heating and the involvement of the Underfloor Heating Manufacturers Association (UHMA).

BS EN 1264

Remember

Manufacturers' associations are good sources of technical information to help installers get the best out of their systems.

When starting to design an underfloor heating system, you need to consider:

- room and building heat loss through floors, walls, ceilings, glazing and air changes
- maximum floor surface temperature of 29°C
- distances to manifolds around the perimeter
- pipe spacing and pipe runs to and from manifolds
- number of circuits per zone or area
- flow rates
- floor constructions and thermal resistance of floor finishes
- heat source
- manifold position.

Most underfloor heating systems are designed by the manufacturers themselves, using expensive CAD systems. This section aims to give you a good understanding of the design principles so that, if you wish to specialise in underfloor heating, you are ready for a specialist course operated by groups such as BPEC and LOGIC.

The spacing of the pipes within the floor changes the heat output of the floor: the closer together the pipes are, the more heat is delivered into that area of flooring and radiated into the room.

The starting point to calculate the design of the underfloor heating is to work out the heat loss from each room. To do this you can calculate the heat loss through walls and windows and ceilings, as for a radiator system. Do not calculate heat loss through the floor, as this will be warmer than your room temperature and will be well insulated. Even so, there will be some heat loss through the floor, so you will need to add 10 per cent to your boiler size to cover this loss at the end of your calculating.

Worked example 1

Look at Figure 6.6 Bungalow floor plan on page 288.

Imagine you need to make a calculation from the lounge minus the floor area.

Fabric loss = 97.8 + 810.5 + 75.6 + 75.6 + 112.1= 1171.6
Ventilation loss = 392.0
Total loss = 1563.6 W

Room size 5 m x 3 m = 15 m^2
Room temperature = 20°C
Flow temp = 55°C

Total heat requirement is 1.563.6 ÷ 15 = 104.24 W/m^2 (check against Table 6.9, below modular wood systems).

At this rate a modular floor system at 55°C with a 6 mm carpet would be ideal.

Worked example 2

Imagine you have to work out a calculation from this information.

- A living room at a designed temperature of 21°C
- Heat loss from room = 1200 W
- Room size = 13 m^2
- Flow temp = 55°C
- 18 mm timber floor with foiled polystyrene underfloor heating

Heat requirement will be 1200 ÷ 13 = 92.30 W/m^2

Using Table 6.9, you would look up the figure for foiled polystyrene floor.

Timber floor would require pipe spacing of 200 mm at a flow temperature of 55°C.

As most manufacturers give a maximum of 100 W/m^2, the working temperature of the system is down to water temperature and flow rate through the floor type. This is maintained by the balancing procedures carried out by the commissioning engineer on site.

Control systems

Control systems

Air temperature control

A thermostat is positioned in the room which is being heated, and set to the required temperature. Once that room gets up to temperature, the thermostat will send a signal to the wiring centre, which will close the actuator fitted on the manifold. This stops the flow of water around the underfloor heating circuit until there is a further heat demand from the thermostat.

Foiled Polystyrene Systems, W/m²

Floor finish	Resistance of Floor Finish m²K/W	Flow/Return Temperatures & UFH Pipe Centres					
		65–55°C		55–45°C		45–35°C	
		200 mm	300 mm	200 mm	300 mm	200 mm	300 mm
7 mm Laminate	0.044	83	69	62	51	41	34
18 mm Timber	0.113	73	61	55	45	37	30
6 mm Carpet	0.075	78	65	59	49	39	32
12 mm Carpet	0.150	69	57	51	43	34	28
4 mm Vinyl	0.016	87	72	65	54	44	36
10 mm Tiles	0.007	89	73	66	55	44	37
25 mm Stone	0.015	87	72	65	54	44	36
18 mm Timber*	0.113	96	79	72	59	48	40

All coverings allow for an 18 mm chipboard deck to be installed beneath except for 18 mm Timber*

Modular Wood Systems, W/m²

Floor finish	Resistance of Floor Finish m²K/W	Flow/Return Temperatures & Installation Type					
		65–55°C		55–45°C		45–35°C	
		Below	Above	Below	Above	Below	Above
7 mm Laminate	0.044	116	107	87	81	58	54
18 mm Timber	0.113	95	88	71	66	47	44
6 mm Carpet	0.075	106	98	80	73	53	49
12 mm Carpet	0.150	85	78	64	59	42	39
4 mm Vinyl	0.016	126	117	95	87	63	58
10 mm Tiles	0.007	130	120	97	90	65	60
25 mm Stone	0.015	127	117	95	88	63	59

When 'Modular Wood' systems are installed from above, approximately 7.5% of the 'Heated Area' comprises unheated access panels. All heat output values are based on a 20°C room air temperature

Table 6.9: (Top) Foiled Polystyrene Systems, W/m² (Bottom) Modular Wood Systems, W/m²

To have extra control, use a separate circuit and thermostat for each room or occupied area. The heat demands and losses will differ throughout the property, depending on usage and external conditions.

Water temperature control

A range of factors will determine the water temperature required for the underfloor heating system, including:

- floor construction
- heat requirement of the space
- floor coverings
- pipe diameter and pipe centres used.

Generally, underfloor heating systems will run at temperatures ranging from 35° to 50°. If the boiler or heat source cannot supply the water at the required temperature, either thermostatic or actuated blending controls can be used. Typically, a secondary circulating pump would also be required where there is blending control on the system.

As previously mentioned, the underfloor heating requires a low flow temperature; at design conditions, there will be approximately a 7° temperature drop through the underfloor heating circuits. Most manufacturers supply a three-port thermostatic blending valve to blend the primary flow from the boiler to mix with the underfloor heating return water, in order to maintain the required temperature for the underfloor heating system. Most manufacturers will supply a preassembled thermostatic mixing valve and pumping unit that fits directly onto the underfloor heating manifold.

An alternative method of blending the water is to use an actuated blending valve and weather compensation controller. This is generally a slightly more expensive control method; however, it does offer a more efficient way of controlling the water temperature. As the outside temperature decreases, the heating requirement of the building will increase, and you have to put more energy back into the building if comfort conditions are to be maintained.

Boiler interlock

The controls should be interlocked with the boiler and hot water system to prevent the boiler firing when there is no demand, thus preventing energy wastage.

Filling and flushing

- Ensure each circuit is complete.
- If filling through the manifold, this is done one circuit at a time.
- Connect one hose to the filling point on the manifold and a short hose to the drain point on the return arm (this can go into a bucket).
- Fill the system circuit until the water runs clear and free from bubbles.
- Close this off at both ends and repeat for all circuits.
- Shut the valves and remove the hosepipes.

Pressure testing

- Connect the hydraulic tester to one fill/drain point.
- Open the first circuit valves, raise the pressure to 1 bar and hold for 45 minutes.
- Inspect for leaks. If there are any fittings present, it is recommended that they are flexed during this inspection.
- Increase the pressure to 6 bar and hold this for 15 minutes.
- Reduce the pressure to 2 bar and hold for a further 45 minutes again inspecting for leaks.

Recording

You must keep a record of the test, with details of who carried out the test and the time and date written on the record sheet.

Commissioning

Initial heating should commence with the flow water temperature between 20°C and 25°C. This temperature should be maintained for at least three days. The flow water temperature should then be increased to the system's design temperature. This temperature should be maintained until the moisture content of the floor and air are both stable (minimum four days).

Screed floors

For screed floors, the maximum recommended flow temperature is 55°C. A standard sand/cement screed should be allowed to cure for 21 days after being laid, before starting the initial heating procedure. Under no circumstances should the underfloor heating be used to speed up the curing process. For other coverings, please refer to the manufacturer's instructions.

Timber floors

For timber floors, the maximum recommended flow temperature is 60°C. Before laying timber flooring, it should be acclimatised to the room with the underfloor system at design temperature, until such time as its moisture content is stable (typically around 10 per cent). Heated floor surfaces should not exceed 9°C above the design room temperature (15°C for peripheral areas).

Products with restricted operating temperatures should not be installed without first ensuring they are suitable for use and will not adversely restrict the system performance.

- Commission the circulator in accordance with the manufacturer's instructions and set it to speed 3.
- Adjust the flow water temperature by turning the thermostatic actuator. Refer to the manufacturer's instructions/or the temperature gauge on the flow arm of the manifold for the correct setting.
- Set the flow thermostat 10°C higher than the setting of the thermostatic actuator.

Did you know?

Some heat sources will have an operational requirement for the difference in flow and return temperatures to be greater than a specific value. As a result, it may be necessary to reduce the pump speed to increase the temperature drop.

Post commissioning

Once the system is fully commissioned, thermostats should be rechecked to ensure they are operating the correct thermoelectric actuator. If the heated property is newly constructed or if it has had substantial work carried out on it, the moisture in the air and fabric of the property will significantly increase the heat losses. As a result it may not be possible to achieve the desired temperatures until the moisture content has normalised.

You can use an infrared thermometer to check the floor temperature. The heating should typically exhibit a surface temperature of 27°–29°C, when the ambient air temperature is about 18°–20°C.

Most manufacturer's programmable thermostats incorporate optimised start, and will calculate the correct time to turn the underfloor heating on in each zone, in order to achieve the set temperature at the set time. For example, if a temperature of 20°C is required at 7:00 a.m., this time and temperature is what would be set on the thermostat. If the thermostat calculates the room will take 45 minutes to reach 20°C from the current temperature, the heating would be turned on at 6:15 a.m.

Where standard thermostats are used, it is recommended that from a cold start the system should be programmed to start its heating cycle 1–1½ hours before the set room temperature is required.

Warm-up times for underfloor heating systems vary according to the following factors:

- external temperature
- target internal temperature
- level of insulation
- ventilation rate
- mean water temperature
- floor construction
- floor covering
- maintenance.

During all construction activities, you should cover the manifold with a polyethylene sheet or an enclosure to prevent damage.

- Clean the manifold with a soft cloth.
- Periodically inspect the system for leaks and erosion of brass and plastic components.
- Follow the manufacturer's recommendations with regard to flushing and additives.

Fault finding from manifold

System trouble shooting		
Manifold checks		
Symptom	**Problem**	**Solution**
No heat in any zone	UFH system not turning on	Ensure the UFH controls are programmed correctly, and the heat source is available to provide hot water for the programmed period.
	Heat source/UFH pump not running	Ensure at least one thermostat is calling for heat and that the switched lives to the boiler and the circulations become live according to demand.
	Primary flow and return pipes crossed	Check the flow and return pipes from the heat source are correctly connected to the manifold.
	Valves closed	Check the isolation valves are open, the balancing valves are in their balanced positions and that the thermoelectric actuators are opening on demand (a white band will be visible on the raised cap).
UFH keeps switching off	Flow Water Protection thermostat is activating	Check the flow temperature from the manifold is correct and that limit thermostat is set 10° higher. If flow temperature is not responding correctly check thermostatic actuator for fault.
Some zones do not become warm	Air trapped within pipework	Set the UFH pump to speed setting H1, open the balancing valve fully for the problem zone ensuring all other zones are isolated. Air should automatically vent from the system.
	Manifold incorrectly balanced	Perform quick and if necessary Advanced Circuit Balancing.
Zone takes a long time to warm up	Manifold incorrectly balanced	Perform quick and if necessary Advanced Circuit Balancing.
	Flow temperature set too low	Check the blending valve is set correctly and that the primary flow temperature into the mixing valve is equal to or warmer than the required secondary flow water temperature.
	High heat losses	Some rooms will have higher heat losses than others, such as a conservatory. The effects can be compensated for by setting the heating to come on for longer in these zones.
	Thermally resistive floor finish	Some floor constructions work more efficiently with underfloor heating. For example stone or tiled floors will have a greater heat output than carpeted ones (check floor manufacturer's details).

Figure 6.31 Fault finding from manifold

Control faults

Control trouble shooting		
Below is an example of a table of symptoms and solutions for problems regarding a control system. This is not a definitive list and should therefore be read in conjunction with any control system installation guides which contain a similar table regarding its operation from the manufacturer's guide/instructions.		
Symptom	**Problem**	**Solution**
One or more channel indicators are flashing green and the heating comes on for 10 minutes every hour	Connection to an enrolled thermostat has been lost	Check and replace faulty BUS cable Replace battery in appropriate wireless thermostat Remove any non-CE approved radio frequency devices
One or more channel indicators are flashing red (rapidly)	Channel outputs have been overloaded	Ensure only 24 V thermoelectric actuators are connected to outputs Ensure only one actuator is connected to each channel Check for faulty actuator by measuring its electrical resistance
Heating does not appear to be working and the Mode indicator is green	Control system is on holiday mode	Switch off the holiday mode switch which has been connected using ◀ or ▶ on the control centre select MODE and press RES
Heating does not appear to be working and there are no indicators illuminated	Thermostats have not been enrolled / there is no power	If indicators flash red after pressing ◀ or ▶, enroll thermostats Ensure power supply is connected and turned on Wiring/hardware fault has blown a fuse, check fuses and locate fault
Heating turns off moments after becoming active	Flow Watch thermostat on manifold is active	See manifold instructions
A floor area is not operating in time with the thermostat in that zone	Channel is enrolled to another thermostat	Determine which circuit is supplying the floor area and re enroll correctly

Figure 6.32: Control faults

Underfloor heating general faults

- Air in the loops is a common fault on underfloor systems, often caused by poor flushing of the system. Further flushing of the loop is needed to remove the air.
- Check the correct thermostat is connected to the right actuators: this is a common problem when rooms are not getting the correct temperature.
- Underfloor heating is designed for continuous use. If a system is used intermittently, it will not heat up the building adequately and the customer will report the heating taking too long to warm up.

PROGRESS CHECK

1. What temperature does the water in underfloor heating systems work at?
2. What temperature should the floor on an underfloor heating system not exceed?
3. What is meant by the set-back temperature?
4. With underfloor systems, what does the floor act as?
5. How does underfloor heating mainly transfer heat?
6. On a stapled system, there are two pipe configurations. Name them.
7. Manifolds draw off water from the primary circuit which is often at a _________?
8. What are the four specific developed controls for underfloor heating systems?
9. What is the usual maximum floor heat output?
10. When commissioning an underfloor heating system, what can you use to check the floor temperature?

Professional Practice

Ben has been contacted by a customer who has had an underfloor heating system fitted to his conservatory but it is not getting warm. What is the possible problem and how should Ben correct it?

Check your knowledge

1. SAP procedures also produce a TER. What does TER stand for?
   - **a** Target CO_2 Emission Rate
   - **b** Temporary Energy Review
   - **c** Targeted Energy Reduction
   - **d** Temporary Energy Results
2. All new heating systems have to be:
   - **a** reverse return
   - **b** one pipe
   - **c** gas-powered
   - **d** fully pumped.
3. What does ASHP stand for?
   - **a** Anodised System Heating Pump
   - **b** Air-Source Heat Pump
   - **c** Available Seasonal Heating Phase
   - **d** Assured Sealed Heating Pipes
4. What is a low loss header?
   - **a** A pipe that sits at the lowest point of a heating system
   - **b** A pipe that creates a primary circuit that allows the water velocity to be maintained at the required flow rate
   - **c** A manifold for a microbore system
   - **d** The pressure loss that occurs in a pipe header due to frictional resistance
5. What is a competent person's scheme?
   - **a** A scheme that allows anyone to install a heating system
   - **b** A scheme that allows a qualified person to register a compliance certificate to Building Control
   - **c** A scheme to register incompetent plumbers to allow them to keep on working
   - **d** A method for heating engineers to register heating systems with the YMCA
6. What is the principle of heat loss?
   - **a** Heat migrates to colder areas
   - **b** Cold migrates to hotter areas
   - **c** Heat travels by conduction only
   - **d** Heat is attracted to solid objects
7. What must a boiler used with a sealed system have fitted to it?
   - **a** Service valve
   - **b** Cold water cistern
   - **c** High-limit thermostat
   - **d** Low-pressure control valve
8. On completion of commissioning, what should the customer be given?
   - **a** A DIY handbook
   - **b** Manufacturer user's manual
   - **c** Manufacturer's installation instructions
   - **d** A thank you present
9. What can cause boiler noise?
   - **a** Damaged heat exchanger
   - **b** A blown fuse
   - **c** A newly fitted pump
   - **d** Soft water in the system
10. What are floating floors?
    - **a** Floors designed to float on water in case of flooding
    - **b** A new type of lift
    - **c** Chipboard flooring sitting on underfloor heating pipework
    - **d** Polystyrene underfloor heating with screeded flooring

Preparation for assessment

The information in this unit of the book, as well as the continued practical assignments that you will carry out in your college or training centre, will help you with preparing for both your end-of-unit test and the diploma multiple-choice test. It will also support you in preparing for the practical assignments you will need to complete to demonstrate your understanding in this which will enable you to carry out and apply **domestic heating systems** installation, commissioning, service and maintenance techniques.

There are opportunities throughout the unit for you to test your progress in and understanding of the required underpinning knowledge; this will enhance your preparation for the forthcoming assessments, so make good use of them.

The unit will be assessed by the following assessment method:

- externally set knowledge assessment
- externally set assignments.

With regards to the mechanical services industry, you will need to know:

- the types of central heating systems and their layout requirements
- the installation requirements of central heating systems and components
- the commissioning requirements of central heating systems and components
- the fault diagnosis and rectification procedures for central heating systems and components
- the design techniques for central heating systems.

Using your knowledge from this unit practise by testing yourself on the following key points:

Space heating zoning requirements, Boiler Efficiency database, SEDBUK, SAP, TER, DER, Energy change, Fully pumped systems, Water heating zones, Three-port or two-port motorised valves, Programmable room thermostat, Single multi-channel programmer, Domestic Heating Compliance Guide, Optimiser, Ambient air temperature, Weather compensation, Home automation system, Heat Pumps, Ground-source heat pumps, Air-source heat pumps, Micro combined heat and power, Stirling engine, Internal combustion engine, Fuel-cell, low loss headers, water velocity, Thermal shock, Completion Certificate, (W/m/°C), U-value, F1 Correction factor, Pipe sizing calculations, Frictional resistance, European Ecodesign Directive, Glandless circulators, Heating load, Filling loop, Double check valve assembly, Expansion vessel, Soundness test, Power flushing, Hot flushing, Mains pressure cleansing, Anti-gravity, Gravity cleansing, Boiler commissioning, Radiator thermometer, Inhibitor, Benchmark, Underfloor heating, Actuator, Screeded floors, Floating floors, Joisted and Battened floors, Staple systems, Series or Serpentine pattern, Bifilar, Snail or Spiral pattern, Manifold and Pocketed polystyrene.

7 Understand and carry out electrical work on domestic plumbing and heating systems and components

In this unit you will cover the essential electrical concepts, principles and techniques that you need to know in order to be able to do your job.

This unit covers the following learning outcomes:

- the principles of electricity supply to buildings
- know the layout features of electrical circuits in buildings
- types of cables and conductors used for the installation of electrical equipment
- the function of electrically operated components
- the purpose and requirements of earthing systems
- the operating principles of electrical circuit protection devices
- electricity and safe isolation procedures
- installation and connection requirements of electrically operated mechanical services components
- selecting cables for components and circuits
- requirements for protecting cables installed in the building fabric and terminating in enclosures
- electrical inspection and testing requirements
- safely diagnosing and rectifying faults in electrically operated mechanical services components.

Note on standards

As with plumbing, the electrical industry is governed by a range of standards – particularly British Standards – that are there to ensure that systems, components and installations meet certain levels of quality and safety. Rather than dealing with these in a separate section, references to relevant standards appear where appropriate throughout this unit. The references to each standard will appear in the margin applying the principles to your work. There are questions relating to them as you work through the unit.

1. The principles of electricity supply to buildings

What is electricity?

Key term

Electricity – the flow of electrons through a conductor.

To understand what **electricity** is, you need to look at how atoms are made up. Each atom is made up of three types of particle: electrons, protons and neutrons.

At the centre of each atom is the nucleus, which is made up of protons and neutrons. Protons have a positive charge (+); neutrons are electrically neutral and act as a type of 'glue' that holds the nucleus together. The remaining particles in the atom orbit at varying distances around the nucleus. These are electrons, which have a negative charge (–).

Normally, atoms have an equal number of protons and electrons, so the positive and negative charges balance out – they are electrically neutral. However, while the electrons orbiting nearest the nucleus are generally held tightly in place, those furthest away are only loosely attached. Sometimes it is possible to remove or add an electron to a neutral atom, leaving it with a net positive or negative charge. Such 'unbalanced' atoms are called ions.

Key terms

Conductor – a material through which electrons flow easily.

Insulator – a material through which electrons cannot move easily.

Current – the rate of flow of electron charges through a conductor.

Since all atoms want to be balanced, the atom that is unbalanced will look for a 'free' electron to fill the place of the missing one. It is these wandering or 'free' electrons moving about the molecular structure of a material that give rise to electricity. A material that allows the easy movement of these free electrons is called a **conductor**, while one that does not is called an **insulator**. The rate of flow of electron charges through a conductor is called **current**.

Electrical circuits were covered at Level 2. Remind yourself about the fundamentals by reading *Level 2 NVQ Diploma Plumbing* pages 190–195 before reading on.

The a.c. generator (alternator)

The operating principle of the alternating current (a.c.) generator is that by rotating a loop of wire inside a magnetic field, an electromotive force (emf or voltage) is induced in the loop (see Figure 7.1). If we were to plot this full 360° revolution (cycle) of the loop as a graph, we would see the emf induced in the loop as a sine wave (see Figure 7.2).

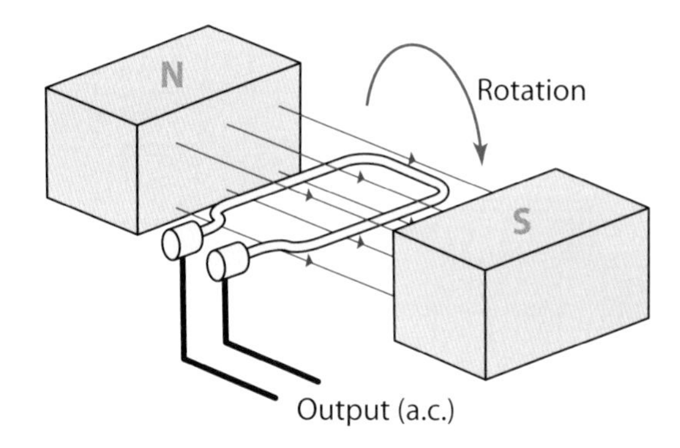

Figure 7.1 An a.c. generator

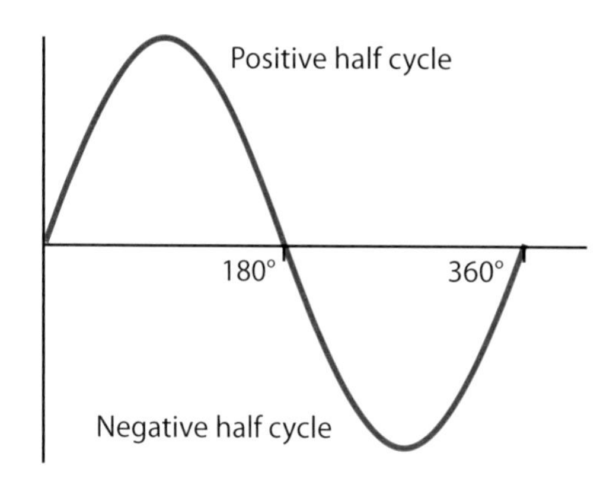

Figure 7.2 A sine wave graph

The sine wave represents one complete cycle of alternating emf. During the top half of the cycle, the current will flow in one direction, and during the bottom half of the cycle, the current will flow the other way – hence, alternating.

As you can see, the most magnetic lines will be cut through when the loop has moved through 90°. The maximum induced emf or 'peak value' in this direction will occur at this point. After another half a cycle, emf will be induced in the opposite direction, with the maximum at 180°, until the loop returns to its original starting position.

The opposite directions of the induced emf will still drive a current through a conductor, but that current will alternate as the loop rotates through the magnetic field. The current will be flowing in the same direction as the induced emf and consequently the current will rise and fall in the same way as the induced emf. When this happens, we say that they are in phase with each other.

To access this a.c. output, the ends of the loop are connected via slip rings as shown in Figure 7.1. Because the two brushes contact two continuous rings, the two external terminals are always connected to the same ends of the coil, hence the output in the form of a sine wave.

Did you know?

In the UK, the electricity has 50 of these cycles produced every second, which is known as the frequency of the supply, measured in hertz (Hz).

How electricity is generated

There are a few methods of generating electricity. A Van de Graaff generator, for example, can be used to generate static electricity and in a small number of power stations, nuclear reaction can be converted directly into electricity. However, in the main, electricity is generated using electromagnetic induction, in which mechanical energy is used to rotate the shaft of an a.c. generator.

Figure 7.3 on page 344 shows water heated by a fuel until it becomes high-pressure steam. At this point the steam is forced onto the vanes of a steam turbine, which in turn rotate the alternator. A variety of energy sources can be used to heat the water in the first place and the more common ones are coal, gas, oil and nuclear power.

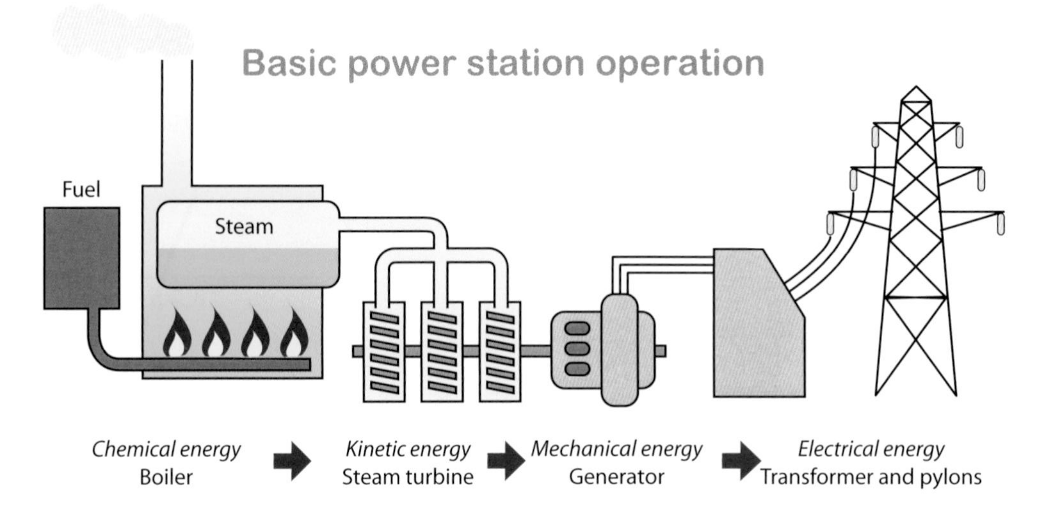

Figure 7.3 How electricity is generated

Transmission and distribution

A power station generator output is transformed upwards before transmission. Electricity is then transmitted at very high voltage (400 kV, 275 kV for the super grid) to compensate for the power losses that occur in the power lines; transmission at low voltage would need very large cables and switchgear.

The National Grid

At this point, the electricity is fed into the National Grid system (132 kV), a network of nearly 5000 miles of overhead and underground power lines that link power stations together and is interconnected throughout the country. The idea behind the National Grid is that, should a fault develop in any one of the contributing power stations or transmission lines, electricity can be requested from another station on the system to maintain supplies.

Electricity is transmitted around the grid, mainly via steel-cored aluminium conductors, which are suspended from steel pylons. This is done for three main reasons:

- The cost of installing cables underground is excessive.
- Air is a very cheap and readily available insulator.
- Air also acts as a coolant for the heat being generated.

Electricity is then 'taken' from the National Grid via a series of sub-stations that sequentially transform the grid supply down as:

- 66 kV and 33 kV for secondary transmission to heavy industry
- 11 kV for high-voltage distribution to lighter industry
- 415/400 V to commercial consumer supplies
- 240/230 V to domestic consumer supplies.

At the 11 kV stage, electricity is transmitted to a series of local sub-stations, which transform it down to 400 V and distribute this, usually via a network of underground radial circuits, to customers. At this point the neutral conductor is introduced.

Did you know?

Sub-stations are dotted throughout our cities. These small brick buildings are normally connected together on a ring circuit basis.

Final distribution to the customer

Electricity finally comes into the customer's premises at the 'main intake position', which usually has the following features:

- **fuse** – a sealed overcurrent device that protects the supply company's cable
- **meter** – an energy measuring system to determine the customer's electricity usage
- **consumer unit** – a type of distribution board which supplies the subsidiary circuits in the building
- **main earth terminal** – a terminal block where earthing and bonding conductors are connected together (see page 383).

Generally, domestic installations in the UK will be provided with a single-phase supply: a 2 core (1 × line plus 1 × neutral) cable at 230 V. This is usually more than adequate to meet the lighting and small power needs of a domestic consumer or small commercial user. Where commercial or industrial installations require a larger capacity, it is normal to provide a three-phase supply: 3 × line plus neutral. See page 382 for more details.

PROGRESS CHECK

1. At what frequency do UK power stations generate electricity of alternating current (in Hz)?
2. Electricity leaves a sub-station and arrives at a customer's premises. Within the customer's premises, what is this position called?
3. What is a consumer unit?
4. What does the electricity meter at the supply intake position measure?

2. Know the layout features of electrical circuits in buildings

Circuits

All electrical circuits run on the same basic principle: if a conductor such as copper wire is bent into a loop and connected to a battery across the two ends, it creates a complete circuit and a current will flow. In this section you will look at the main types of electrical circuit you will come across in your work as a plumber.

Power (final) circuits

A final circuit is one connected directly to current-using equipment, a socket or other outlet point. The equipment converts electrical energy into another form, such as light, heat or motive power.

BS 7671

This section covers the use of socket outlets and fused connection units, and the installation of rings, radials and spurs.

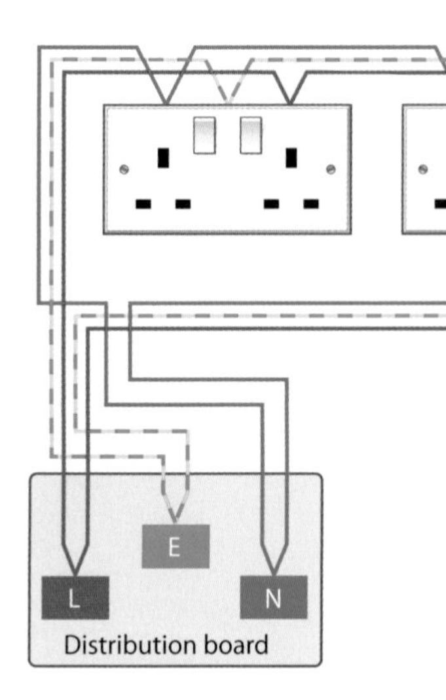

Figure 7.4 Ring final circuit

BS 7671, Appendix 15

Ring final circuit

The ring main circuit design allows a ring to be created by simply linking two existing radial circuits together. Figure 7.4 shows how, in the ring final circuit, the live, neutral and **circuit protective conductor (cpc)**:

- start from their respective terminals at the distribution board
- pass through the corresponding terminals of each socket outlet
- return to the same terminals at the distribution board.

They are protected by a 30 A or 32 A protective device.

Although an unlimited number of socket outlets can be installed, the load current in any part of the circuit should be unlikely to exceed the current-carrying capacity of the cable for long periods.

This can generally be achieved by:

- not supplying immersion heaters, electric space heaters or similar from the ring
- positioning the sockets so that the load is reasonably shared around the ring
- connecting cookers, ovens and hobs with a rated power greater than 2 kW on their own dedicated radial circuit
- taking account of the floor area being served by the ring (<100 m^2 is a good rule of thumb).

Key term

cpc – connects exposed-conductive parts to the protective earthing terminal.

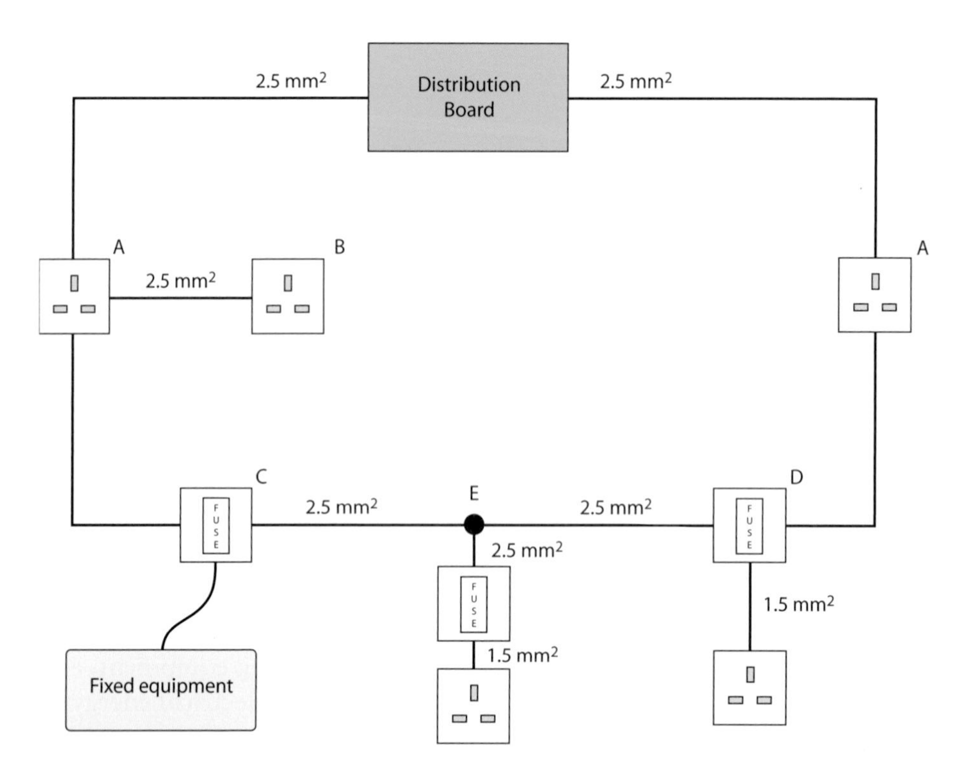

Figure 7.5 Requirements of a ring final circuit

In Figure 7.5, please note the points made in Table 7.1.

A	A BS 1363 single or twin 13 A socket outlet. Each outlet of twin or multiple sockets is regarded as one socket outlet.
B	A BS 1363 socket fed via an unfused spur. An unfused spur should only feed one single or twin socket outlet. The unfused spur may also be connected to the origin of the circuit in the distribution board.
C	A fused connection unit (FCU) supplying fixed equipment. The fuse in the FCU must not exceed 13 A.
D	A BS 1363 socket outlet fed via a fused connection unit. The number of sockets supplied from the FCU depends on the load characteristics, taking diversity into account.
E	A spur made using a junction box and FCU that is not directly connected to the ring.

Table 7.1: Notes on Figure 7.5

The conductor sizes shown in the diagram relate to the live conductors only within flat 'twin and earth' cable; however a reduced circuit protective conductor size is permitted (such as 2.5 mm^2 live conductors with 1.5 mm^2 cpc).

Wherever possible, make sure cables are not covered by thermal insulation. If a cable is partially or completely covered, refer to Regulation 523.9.

Where more than one ring main is installed in the same premises, it is good practice to share the load over the ring main circuits, so that the assessed load is balanced. Take care to meet the regulation requirements in terms of additional protection by a **residual current device (RCD)** (see page 392).

Key term

RCDs – a group of devices that provide extra protection to people and livestock by reducing the risk of electric shock.

BS 7671 Reg 523.9

BS 7671 Reg 411.3.3
Regs 522.6.101 to 522.6.103

Radial circuit

In a radial circuit, the live, neutral and cpc:

- start at the distribution board connected via an appropriate overcurrent protective device
- pass through the respective terminals of each socket outlet in the circuit
- finish at the last outlet.

There are two types of radial circuit: one protected by a 20 A overcurrent protective device and the other by a 30 A/32 A overcurrent protective device, as shown in Figure 7.6.

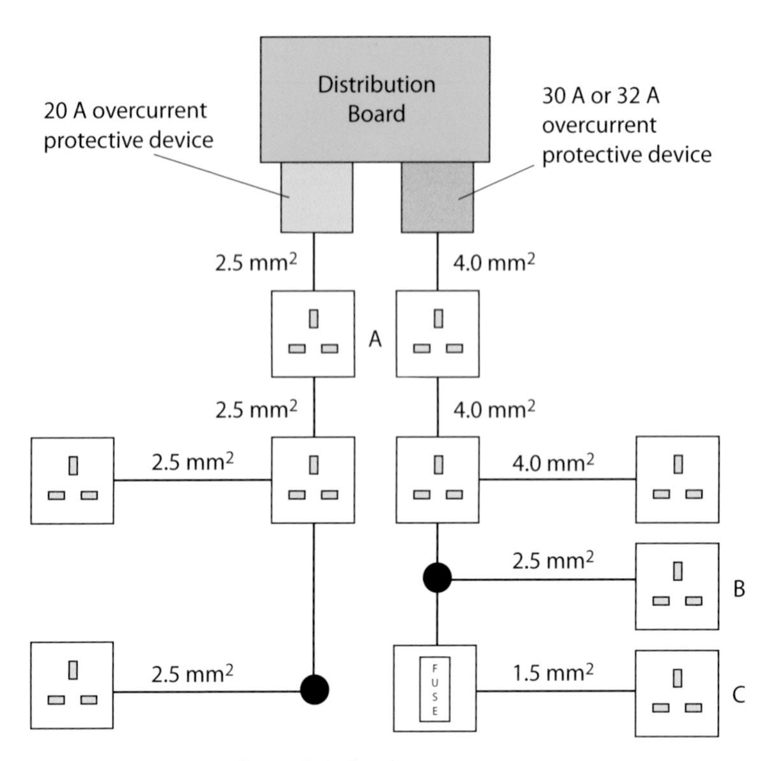

Figure 7.6 Requirements of a radial circuit

In Figure 7.6, please note the points made in Table 7.2.

A	A BS 1363 single or twin 13 A socket outlet. Each outlet of twin or multiple sockets is regarded as one socket outlet.
B	A BS 1363 socket fed via an unfused spur. An unfused spur should only feed one single or twin socket outlet and only 2.5 mm^2 cable should be used. The unfused spur may also be connected to the origin of the circuit in the distribution board.
C	A BS 1363 socket outlet fed via a fused connection unit. The number of sockets supplied from the FCU depends on the load, taking diversity into account.
●	A junction box. Junction boxes with screw terminals must be accessible for inspection, testing and maintenance. Alternatively, use maintenance-free terminals/connections (Regulation 526.3).

Table 7.2 Notes on Figure 7.6

The conductor sizes shown relate to the live conductors only within flat 'twin and earth' cable; however, reduced circuit protective conductor size is permitted (for example, 4.0 mm^2 live with 1.5 mm^2 cpc).

Where the radial circuit is protected by a 20 A overcurrent protective device, the floor area served by the ring is generally limited to 50 m^2; with a 30 A/32 A overcurrent protective device, the limit is generally 75 m^2.

PROGRESS CHECK

1. An unlimited number of socket outlets can be installed on a ring, but what is the generally accepted maximum floor area covered by the ring?
2. What are the two types of radial circuit?
3. In a radial circuit, where do the line, neutral and cpc start?

3. Types of cables and conductors used for the installation of electrical equipment

The following cable types are most commonly used in domestic situations. Note that all of these will have copper conductors.

Circuit wiring

PVC insulated PVC sheathed flat profile cable with integral CPC (6241Y, 6242Y, 6243Y)

PVC insulated PVC sheathed flat profile cable with integral CPC

- Designed for domestic and general wiring where a circuit protective conductor (cpc) is required for the circuit (two-core version known as 'twin and earth').
- Normally concealed in walls and ceilings; can be clipped to a surface where there is no risk of mechanical damage.
- Solid or stranded copper conductors with PVC insulation laid parallel with a plain and uninsulated copper cpc and then overall PVC sheath (-15° to +70°C).
- Available as single-core (6241Y), two-core (6242Y) and three-core (6243Y) in sizes from 1.0 mm^2 through to 16 mm^2.
- Insulation only brown, blue, black and grey; sheath normally only grey.

Single-core thermoplastic insulated unsheathed cable (6491X)

If you need to extend an existing circuit that has been wired in conduit or trunking, the cable will almost certainly be 6491X, as explained next.

Earthing and bonding

Single-core thermoplastic insulated unsheathed cable (6491X)

- Designed for installing into conduit, trunking and, when protected, within lighting fittings and control panels.
- Solid or stranded copper conductor with insulation of PVC (-15° to +70°C), so cheap and easy to install.
- Available in sizes from 1 mm^2 to 630 mm^2.
- Insulation colours include grey, black, blue, yellow, brown, white, orange, violet and green/yellow stripes for use as circuit protective conductor or bonding conductor.

Final connections

Circular flexible cords (3182Y, 3183Y, 3184Y, 3185Y)

- Designed for general purpose indoors or outdoors in dry/damp situations. Suitable for portable tools, immersion heaters, washing machines, vacuum cleaners, lawn mowers and refrigerators.
- Not for use with heating appliances or where sheath can come into contact with hot surfaces.
- Stranded copper conductors, PVC insulated, two, three, four or five cores laid up and PVC-sheathed (0°C–70°C).
- Available with conductor sizes of 0.75 mm^2 to 4.0 mm^2, with core colours:
  - 3182Y two-core: brown and blue
  - 3183Y three-core: brown, blue and green/yellow
  - 3184Y four-core: black, grey, brown and green/yellow
  - 3185Y five-core: black, grey, brown, blue and green/yellow.

Remember

Immersion heater flex should be 1.5 mm^2, 85°C rubber high temperature cord.

Circular flexible cords (3092Y, 3093Y, 3094Y)

- Designed for general purpose indoors or outdoors in dry/damp situations. Suitable for portable tools, washing machines, vacuum cleaners, lawn mowers and refrigerators especially in higher temperature zones, and heating system connections.
- Stranded copper conductors, heat-resistant PVC insulated, two, three or four cores laid up and heat-resistant PVC sheathed (0°C–90°C). Available with conductor sizes of 0.5 mm^2 to 2.5mm^2, the core colours are:
  - 3092Y two-core: brown and blue
  - 3093Y three-core: brown, blue and green/yellow
  - 3094Y four-core: black, grey, brown and green/yellow.

4. The function of electrically operated components

Classification

There are many types of electrical component available and here you will look at the most common ones. Be aware that all components generally fall under one of two IEC classes for earthing requirements.

Class I	Class I components must have their case/chassis connected to earth	⏚
Class II	A Class II or double-insulated electrical appliance is designed not to need a safety connection to earth. No single failure can result in dangerous voltage becoming exposed, partly by having two layers of insulating material around live parts or by using reinforced insulation.	⧈

Table 7.3 Earthing requirements under IEC protection classification codes

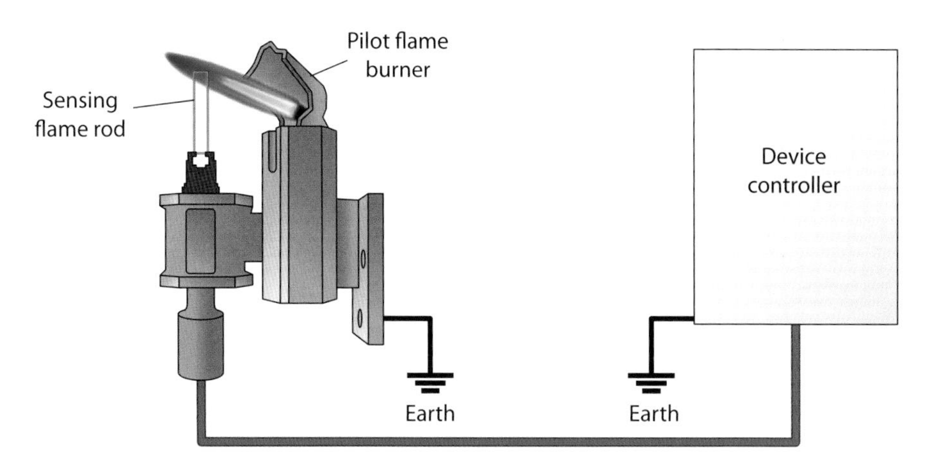

Figure 7.7: Flame rectification device

General components

Flame rectification device

Heat in a flame makes molecules in and around the flame collide, freeing some of the outer electrons of the atoms and allowing a small current to be conducted through the flame – a process called flame ionisation.

The device controller applies an a.c. supply between a sensing flame rod and the base of the flame via the burner; the ions in the flame provide a high-resistance current path between the two. Because the burner surface at the flame base is larger than the sensing flame rod, more electrons flow in one direction than the other, resulting in a small d.c. current (2–4 micro amps (μA)) – hence 'rectification' (a.c. to d.c.).

If there is a flame present, the d.c. is detected by the controller, which tells the gas valve to stay open. If there is no current flowing, the controller will close the gas valve and the system will purge itself of any remaining gas before trying to reignite or lock out.

Flame supervision device (FSD)

A Flame Supervision Device (also known as a Flame Failure Device) is a generic term for any device designed to stop gas going to a burner if the flame is extinguished, thus preventing any dangerous build-up of gas within the appliance; they are to be found on all modern gas appliances. Causes of flame failure include draughts, interruption of the gas supply or even spillage of liquids onto the cooker surface.

There are many types of FSD, each using different means of operation, such as the previously mentioned flame rectification. Another popular method is using a thermocouple. Utilising what is known as the 'Seebeck effect', a thermocouple is a junction of two dissimilar metals that produce a voltage that responds to temperature and, consequently, changes in temperature affect the voltage produced. In the FSD, if the flame is extinguished then the thermocouple cools down and stops producing a voltage which in turn closes a spring-operated gas valve.

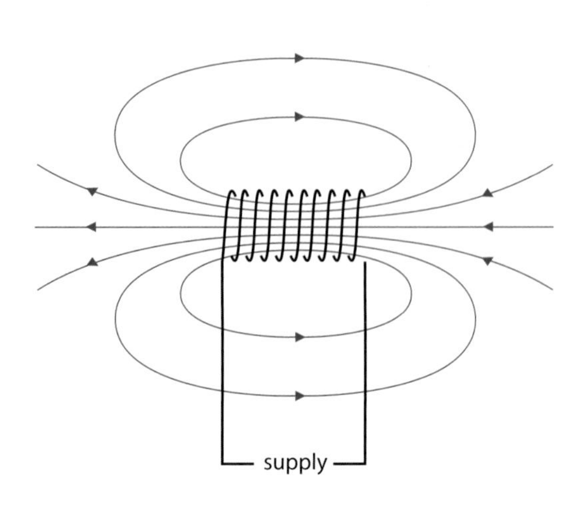

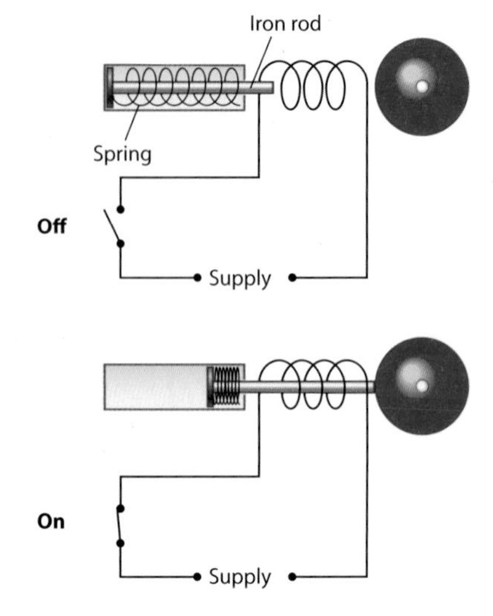

Figure 7.8 How a solenoid works

Solenoid valves

The workings of solenoid valves were covered in Unit 3, on page 71. Look back now before reading on.

In a solenoid valve, when the switch energises the magnetic field, the rod is moved against an elastic diaphragm. When operated the pressure on the diaphragm will change and fluid within the valve's cavity will flow out to an output port.

Solenoid valves can have two or more ports. In a two-way valve the flow is switched off and on; in a three-way valve the flow is switched between the two different output ports. It is possible to combine multiple valves together to build a more complex system.

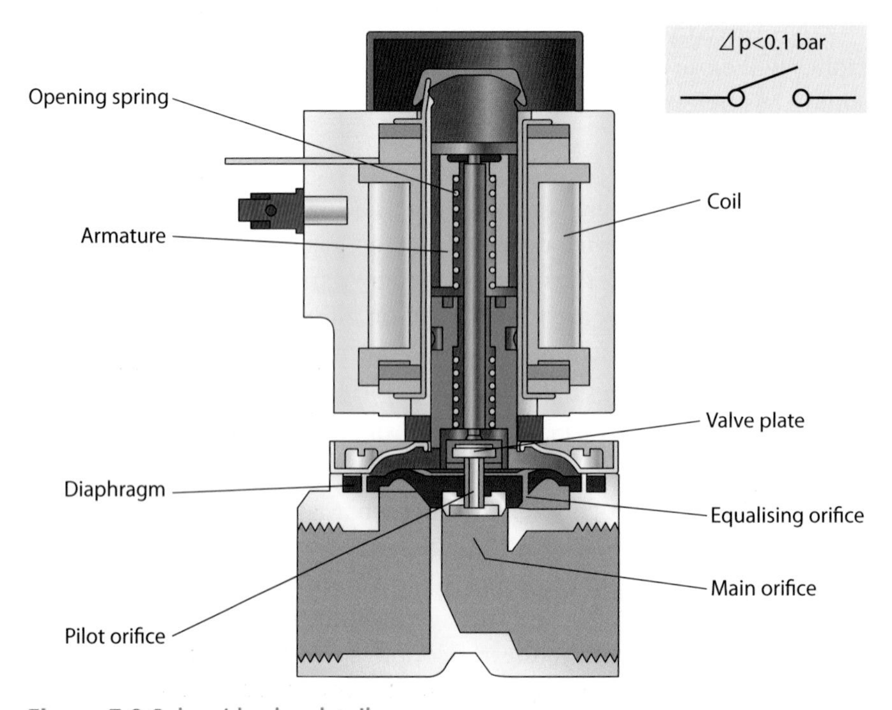

Figure 7.9 Solenoid valve details

As you can see in Figure 7.9, when voltage is applied to the coil, the valve plate is pressed down against the pilot orifice. The pressure across the diaphragm is built up via the equalising orifice. The diaphragm closes the main orifice as soon as the pressure across the diaphragm is equivalent to the inlet pressure. The valve will be closed for as long as there is voltage to the coil.

Thermistors

Thermistors are heat-sensitive resistors that undergo a precise change in electrical resistance when subjected to a corresponding change in temperature. Negative temperature coefficient (NTC) thermistors exhibit a decrease in electrical resistance when subjected to an increase in temperature; positive temperature coefficient (PTC) thermistors exhibit an increase in electrical resistance when subjected to an increase in temperature.

Thermistors are capable of operating over the temperature range of -100°F to +600°F. Because of their predictability and stability, they are seen as the best type of sensor for many applications, including temperature measurement and control, so are widely used in devices such as thermostats.

Thermocouples

Using the Seebeck effect, a thermocouple is a junction of two different conductors (usually metal **alloys**) that produce a voltage; the voltage responds to temperature and changes in temperature affect the voltage.

In contrast to most other methods of temperature measurement, thermocouples are self-powered and require no external form of excitation. The main limitation with thermocouples is accuracy: system errors of less than one degree Celsius (C) can be difficult to achieve.

Central heating boilers use thermocouples as a safety switch to tell whether the pilot light is lit. The central heating boiler detects that there is no voltage so either tries to relight the pilot light or turns the gas supply to the pilot light off.

Key term

Alloy – a mixture of two (or more) metallic elements.

Activity

Do some research to find out what the Seebeck effect is.

Micro-switches

A micro-switch is a miniature switch that is actuated by very little physical force. Micro-switches are common due to their low cost and durability, often with greater than 1 million cycles and up to 10 million cycles for heavy-duty models. They may be found inside a motorised valve to control between outputs.

The main design feature of micro-switches is that a small movement at the actuator button produces a relatively large movement at the electrical contacts, which occurs at high speed. Most successful designs also exhibit **hysteresis**; there must be a significant movement in the opposite direction. Both of these characteristics help to achieve a clean and reliable interruption to the switched circuit.

Key term

Hysteresis – when a small reversal of an actuator is insufficient to reverse the contacts.

Relays

A relay is an electrically operated switch. Many relays use an electromagnet to operate a switching mechanism mechanically, but other operating principles are also used. Relays are used where it is necessary

Remember

Relays are useful as they can be used to switch current between circuits or turn a circuit on and off.

Remember

Since relays are switches, the terminology applied to switches is also applied to relays.

to control a circuit by a low-power signal (with complete electrical isolation between control and controlled circuits), or where several circuits must be controlled by one signal.

A relay will switch one or more poles, each of whose contacts can be thrown by energising the coil in one of three ways. This is similar to what happens with a one-way switch; you can think of a relay as being an assembly that contains a one-way switch and a coil. This is exactly the same: electricity will pass from one terminal to the other when the contact is closed.

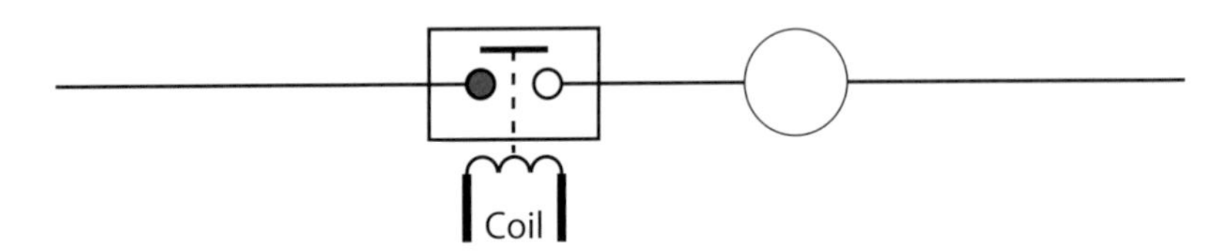

Figure 7.10: Relay switch 1

Did you know?

In a relay we can operate a high-voltage circuit with a very low, safe, switching circuit, because the coil-energising circuit is completely separate to the contact circuit(s).

Relays are a bit like switches. Consider what would happen if you wanted to control a light by using a relay, rather than a switch. If you were to energise the coil, the resulting magnetic field would pull the contact across the two terminals, closing the circuit, and the light would come on. Instead of the switch contact being held in place mechanically, it is being held by the magnetic field produced by the coil in the relay. It will only remain this way while the coil is energised.

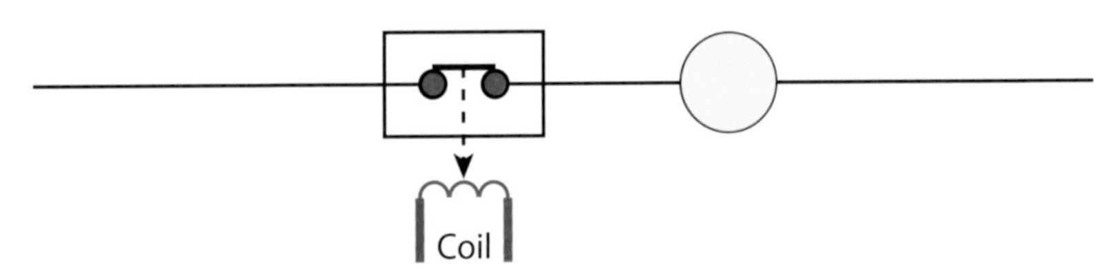

Figure 7.11: Relay switch 2

Did you know?

The production and soldering of PCBs can be done by totally automated equipment.

This type of relay has 'normally open' contacts: when the coil is de-energised, the contact opens and no electricity can pass through the relay.

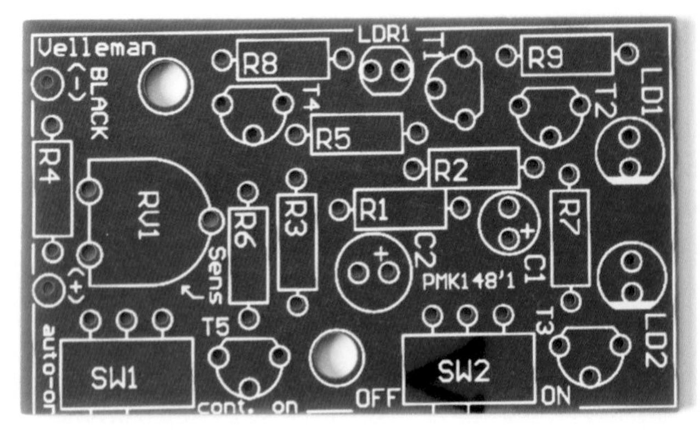

A printed circuit board

Printed circuit boards (PCBs)

A printed circuit board is used to mechanically support and electrically connect electronic components using conductive pathways, tracks or signal traces etched from copper sheets laminated onto a non-conductive substrate. Printed circuit boards are used in virtually all but the simplest commercially produced electronic devices.

PCBs are inexpensive, and can be highly reliable. They require much more layout effort and higher initial cost than either wire-wrap or point-to-point construction, but are much cheaper and faster for high-volume production.

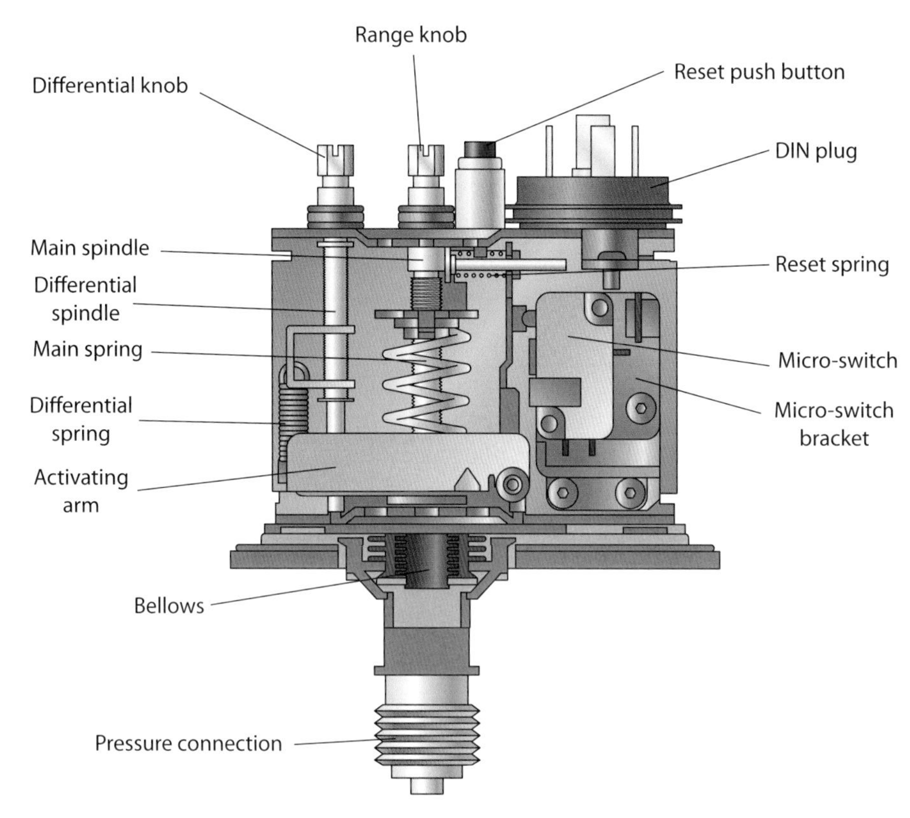

Figure 7.12: Typical pressure switch

Pressure switches

A pressure switch is a form of switch that makes or breaks electrical contact when it senses a change in pressure. It is used to provide on/off switching from a pneumatic or hydraulic source. The switch may be designed to make contact either on pressure rise or on pressure fall via a diaphragm moving against a micro-switch operating arm.

A pressure switch is configured to sense a change in a certain pressure on something and then react in a certain manner. Pressure switches are used on devices such as electric and gas heaters, pumps, small air compressors and electric strip heaters. It is the safety device that checks the operation of the fan: when the fan is not running, they should not be firing or in use.

Pumps

In many modern condensing combi boilers an automatic variable speed circulating pump is fitted. This automatically adjusts its speed to meet the pressure fluctuations caused by differing heating demands. As thermostatic radiator valves close in response to set room temperatures the pump speed automatically adjusts, reducing the differential pressure as the resistance of the system rises. This saves energy (electricity) and reduces the possibility of system noise such as radiator valve noise and pipe flow noise.

Circulating pumps installed by the manufacturer within the combi boiler casing require a 230 V mains supply and have a low power consumption ranging between 35–100 watts (similar to a standard light bulb) and are electrically protected by the combi boiler fuse normally rated at 3 amps. The pumps are always earthed by the combi boiler manufacturer.

As the motor rotates the shaft a circular veined wheel, known as an impellor, draws water in through its centre and throws it out at its edges by centrifugal force. As the water is thrown outward it creates a pressure drop causing more water to be drawn in at the centre. The faster the impellor spins the greater the centrifugal force created, which increases the water pressure in the system.

Did you know?

Fan-assisted boilers are room-sealed, because combustion is independent of any air supply within the room.

Pumps are controlled by the time and temperature controls of the central heating system or by flow switches (micro-switches) in the secondary hot water circuit. Circulating pumps will pump in one direction only. The flow direction is usually indicated with an arrow moulded to the base. If a pump is installed the wrong way around the combi boiler will overheat and often go to lockout, so ALWAYS ensure the pump is fitted correctly.

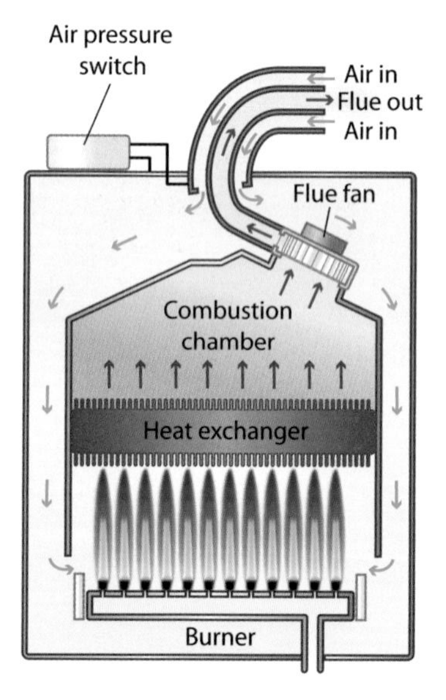

Figure 7.13: Fan-assisted boiler

Fans

A fan consists of a rotating arrangement of blades, usually powered by an electric motor, which are normally housed in a case that directs the air flow and prevents objects contacting the blades.

Most boilers are now fan-assisted, with small, cross-sectional flue ducts. The flue fan provides the fresh air to the burner and removes the products of combustion from the combustion chamber.

The most popular style of kitchen or bathroom extractor fan uses ducts to draw out the polluted air and dispel it outside. The other, less efficient option available does not use ducts, but uses a filter to clean the stale air and then recirculate it.

Fans in bathrooms

At the most basic level, two distinct types of extractor fan are available.

- **Axial extractor fans** have blades that force air to move along the axis of the fan, parallel to the shaft about which the blades rotate. Standard axial flow fans have diameters of 300–400 mm or 1800–2000 mm, and work under pressures up to 800 Pa.
- **Centrifugal fans** have a moving impellor – a central shaft about which a set of blades, or ribs, are positioned. The impellor rotates, causing air to enter the fan near the shaft and move at right angles from the shaft, spinning it towards the outlet in the scroll-shaped fan casing. Centrifugal extractor fans are often mounted into a ceiling, as they can overcome the air pressure resistance generated by long lengths of ducting.

Did you know?

Centrifugal fans are often called 'squirrel cages' because of their similarity to a hamster or squirrel exercise wheel.

Leak detection devices

Nearly all modern, low-cost, combustible gas detection sensors are of the electro-catalytic type. They have a small sensing 'bead', and are made of an electrically heated platinum wire coil, covered with a ceramic base and an outer coating of palladium or rhodium catalyst. When a combustible gas/air mixture passes over the hot catalyst surface, combustion occurs, increasing the bead's temperature. This alters the resistance of the platinum coil. The resistance change is directly related to the gas concentration in the surrounding atmosphere, and can be displayed on a meter or similar indicating device.

Gas-specific electrochemical sensors can be used to detect the majority of common toxic gases, including carbon monoxide (CO), hydrogen sulfide (H_2S), chlorine (Cl_2) and sulphur dioxide (SO_2), in a wide variety of safety applications. They are compact, require little power and have a typical lifespan of one to three years.

Reliable, hard-wired carbon monoxide alarms with battery backup are also available, designed for use in domestic and light commercial environments.

Control components

Thermostats

In its simplest form, a thermostat is nothing more than a switch, but one that automatically reacts to changes in temperature. However, there are varied operating systems and you will need to establish which one is being used.

Did you know?

Bimetallic thermostats are low cost but work slowly. Although they are still manufactured, electronic versions are more likely to be found on newer systems.

In the case of thermostats with a bimetallic strip, temperature change causes mechanical movement. This is because the strip is made up of two joined strips of different metals rolled into a coil. The two metals have different coefficients and expand at different rates as they are heated. This makes the strip bend one way if heated, and the other way if cooled. As the room heats up or cools down, the metal reacts. Once the thermostat reaches a specific set level, it sends a signal to the central heating boiler to switch it on or off. Although low cost, these thermostats can be slow to react to changes in room temperature.

Electronic thermostats use a thermistor to sense room temperature and a micro-controller (a micro-computer on a single chip) that measures the resistance change and converts this to a temperature

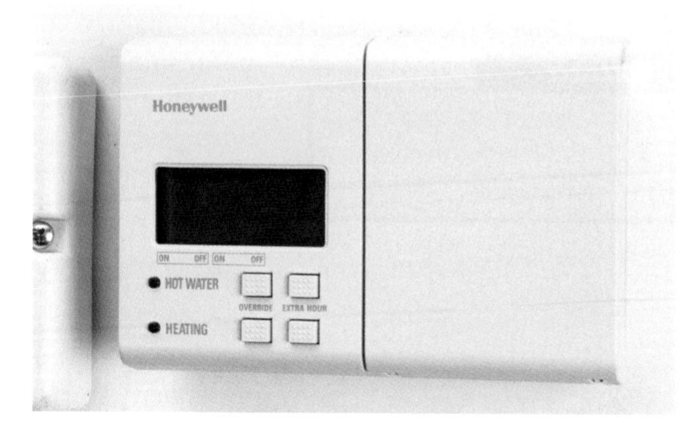

Bimetallic and electronic thermostats

A cylinder thermostat

Remember

Even if the timer is ON or the heating switch is set to CONSTANT, the heating system will not work if the room or radiator thermostats are turned down.

Key terms

Valve – a device that regulates, directs or controls the flow of a gas or liquid by opening, closing or partially obstructing various passageways.

Torque – rotating force.

Remember

All actuators should have overload protection to prevent stuck valve failures.

reading. Once the temperature in the room moves above or below the set temperature, the thermostat sends an electric signal to the boiler to switch it on or off.

Hot water cylinder thermostats are attached between one-quarter and one-third of the way up the hot water cylinder and control the temperature of the water in it. When the water temperature reaches the set level – usually about 60°C – the motorised valve (either two- or three-port assuming the water and space heating can be controlled separately) stops the flow of hot water to the cylinder.

The room or space thermostat and the hot water thermostat should be wired up to the boiler in what is known as an interlock; if both the house and hot water are at the required temperature, the boiler will be switched off.

Programmers/timers

The electronic timer or programmer is basically a timed switch. You can set periods when the heating can be ON (subject to the other controls) and times when the heating will be OFF.

Some switches have separate controls for hot water and heating. Timers usually incorporate an 'override' – flick a switch and the heating and/or hot water will be on constantly. However, many systems use electronic programmers that incorporate a seven-day timer that makes it possible to have different settings for any day of the week if required.

Electrically operated control valves (actuators)

An actuator is a mechanism or device to automatically or remotely control a **valve** from outside. Valves can also be controlled by actuators attached to the system. They can be electromechanical actuators, such as an electric motor or solenoid.

Designs vary greatly: they come in a variety of sizes to fit different valve applications. Power can be a.c. or d.c., depending on the system, but the key requirement is to ensure that there is sufficient **torque** from the electric motor to open and close the valve.

Valves that are not used often may be hard to turn at first. Plant designers will specify valve actuators with extra torque to prevent this, or add manual control wheels or handles. Sticking valves can cause the motor to overheat and burn out. There are different ways to protect the motor (including over-torque, motor overheat or electric current limiting sensors), which monitor for excessive valve opening force, motor winding temperature or electric flow, and shut off the motor to protect it.

Sensors

A sensor (or detector) is a device that measures a physical quantity and converts it into a signal that can be read by an observer or by an instrument. For example, a mercury-in-glass thermometer converts the measured temperature into expansion and contraction of a liquid, which can be read on a calibrated glass tube. A thermocouple converts temperature to an output voltage that can be read by a voltmeter. For accuracy, most sensors are calibrated against known standards.

Wiring centres

A wiring centre is the central point to which all the wiring from your central heating controls goes. It is an enclosure containing connector strips that have been configured with links to the system you are installing. A wiring centre allows connection to all wiring to and from the system components and can ease the testing process considerably.

Switches

In its most basic format, a switch is a device that allows the operator to make, break or divert a signal in an electric circuit. Although you are most unlikely to install these switches, their concepts are applied within many heating system components and therefore the most common switches in domestic properties are as follows.

The one-way switch

Single-pole devices are so called because they only interrupt the phase conductor (L).

In the system in Figure 7.14, one terminal (A) receives the supply, while the switch wire leaves from the other terminal (B) and goes directly to the equipment, an unlit ceiling light. Note that actual switch terminals are not marked in this way; these letters are for ease of explanation.

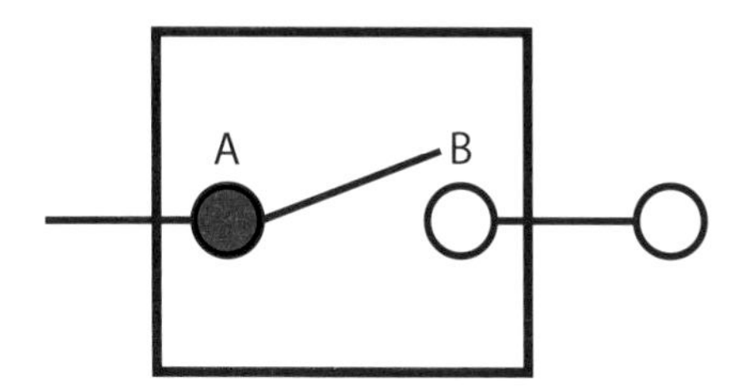

Figure 7.14: One-way switch in off position

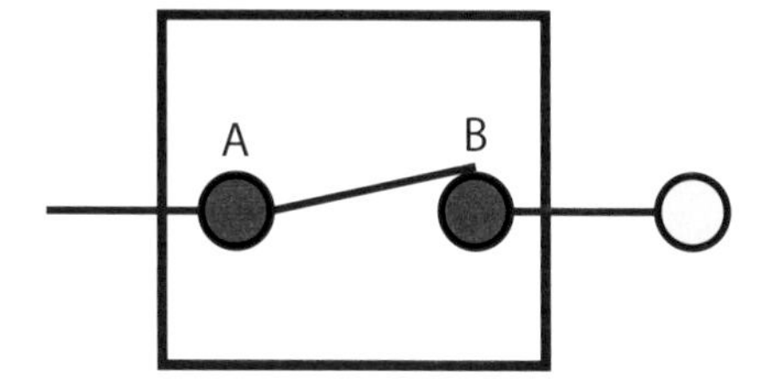

Figure 7.15: One-way switch in on position

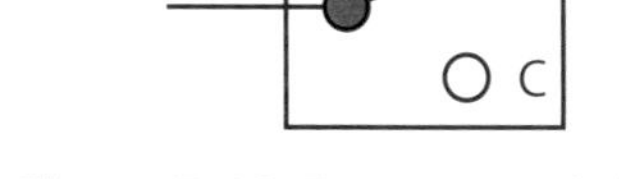

Figure 7.16: Two-way switch

The light in Figure 7.14 is 'off'. To switch it on, you operate the switch with your finger, which moves the contact and then mechanically holds it in the 'on' position (Figure 7.15).

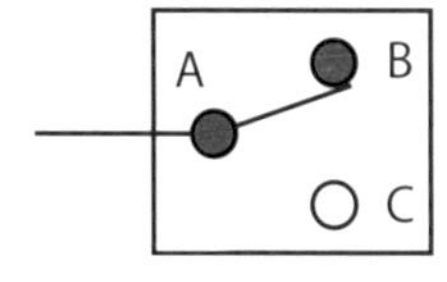

Figure 7.17: Two-way switch with terminal B energised

The two-way switch

Although primarily used by electricians to control lighting circuits in corridors, this switch has a relevance to equipment control. In Figure 7.16, you supply the switch feed terminal, point (A).

Depending on the switch contact position, the electricity will come out on either terminal B or terminal C. In Figure 7.17, it is energising terminal B.

If you now operate the switch, the contact moves across to energise terminal C, as in Figure 7.18.

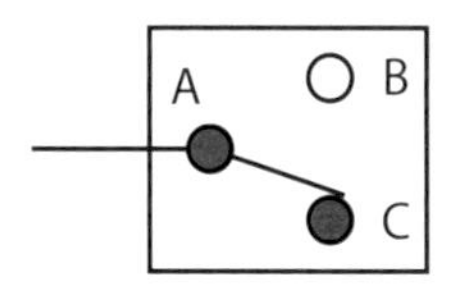

Figure 7.18: Two-way switch with terminal C energised

Remember

Isolators are mechanical switching devices that separate an installation from every source of electrical energy.

Double-pole switch (DP switch)

For certain situations (for example, because the equipment fed by the switch can be accessed and removed for repair or replacement) you use a double-pole switch. This acts like a one-way switch in its operation except that, because the contacts are linked, it is breaking both live conductors (phase and neutral) at the same time when you operate it.

Where the switch is placed local to the equipment and within the control of the person carrying out the work, it can also be regarded as an isolating switch.

Any of the previously mentioned switch types can take the style of a rocker plate switch or ceiling mounted pull-cord switch.

Safety tip

With any lockable option, the key should stay with the person who locked the device and who is carrying out the work.

BS 7671 537.2.1.5
BS 7671 Regulation 537.2.2.4

Lockable isolator switch

Where an isolating device for a circuit is a distance away from the equipment it controls, you must be able to lock it in the open position, and it must be protected from interference. This can be done by placing the isolating device in a lockable space or enclosure or, more commonly, having a locking mechanism or device on it.

Electrics and wiring for controlling central heating

Central heating is one of the areas of plumbing where you will need a solid understanding of the electrics involved, even if you do not have to wire them yourself. If you fully understand how central heating system controls work, mechanically and electrically, you will be in a far better position to maintain them properly.

This section gives you the basic learning that is needed to wire central heating systems and components.

Controls systems

Programmer

Figure 7.19 shows a terminal strip to which the external wires are connected. The internal components of the programmer – a clock or timing device and two simple switches – are connected to this strip. The purpose of the clock is to automatically control the operation of the switches, independently of each other, when we require them.

You should already be aware that a heating system must have an isolation point – usually a double-pole switch or a socket outlet and plug. This is also the point at which the overcurrent protection device (a fuse) is located. The fuse is rated at 3 amps for domestic work.

The isolation point is usually located near to the programmer. A feed wire three-core (minimum size 1.0 mm^2) is taken from the isolation

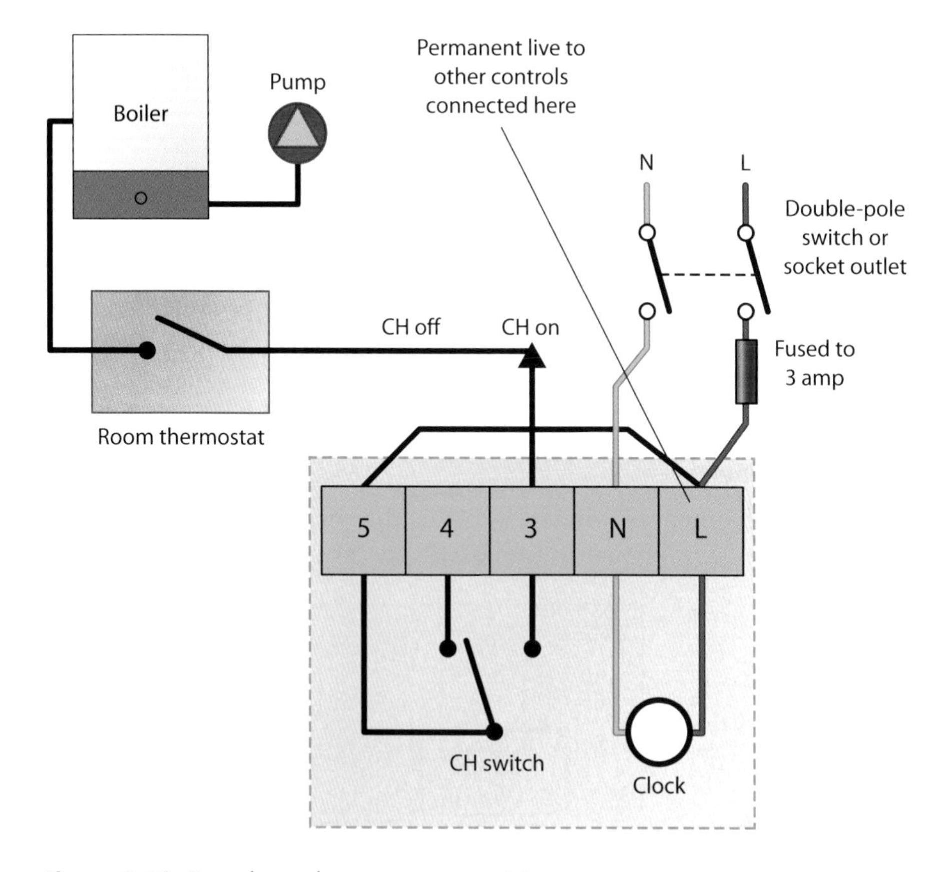

Figure 7.19: Two-channel programmer – wiring

point directly into the programmer (L-N-E). Some programmers do not have an earth connection. However, other system controls will require an earth, so a connection is made in the programmer using a terminal strip with a further earth wire that connects to the main earth point on the system wiring centre. So the power supply comes into the programmer, connecting to the L, N and E points. With this programmer, you are required to provide live link wires across terminals L-5. These provide the power supply to make the switches function correctly. The switches are classed as 'two-way' switches.

To operate a circuit, you may need power for other components, even when the circuit is off (for example, room thermostat or motorised valve). Wires are therefore taken from the required circuits to the main wiring centre to provide the key on/off functions for the components to operate.

You also take a connection from the live or L terminal, known as a permanent live. This stays permanently live without being affected by any programmer switching functions. You need this to get the boiler and pump to work properly (for a 2 × two-port valve system only).

Isolation point of a heating system: 3 amp fuse

A room thermostat

Pumped heating only systems

The wiring of a pumped heating only system operates through a programmer. The room thermostat will turn off the boiler and pump simultaneously.

Pumped heating system with combination boilers

The combi boiler has been very popular with plumbers and customers alike for many years, partly due to the ease of wiring the system. From a fused spur, three wires (live, neutral and earth) at 3 amps are taken into the combi control unit, which may have a programmer built in, or have a remote one wired to it as well as a room thermostat (see Figure 7.20).

The wiring becomes even easier with a programmable room thermostat, as shown in Figure 7.21.

With the development of wireless systems, the wiring is just a case of connecting the combi boiler to the fused spur, as shown in Figure 7.22.

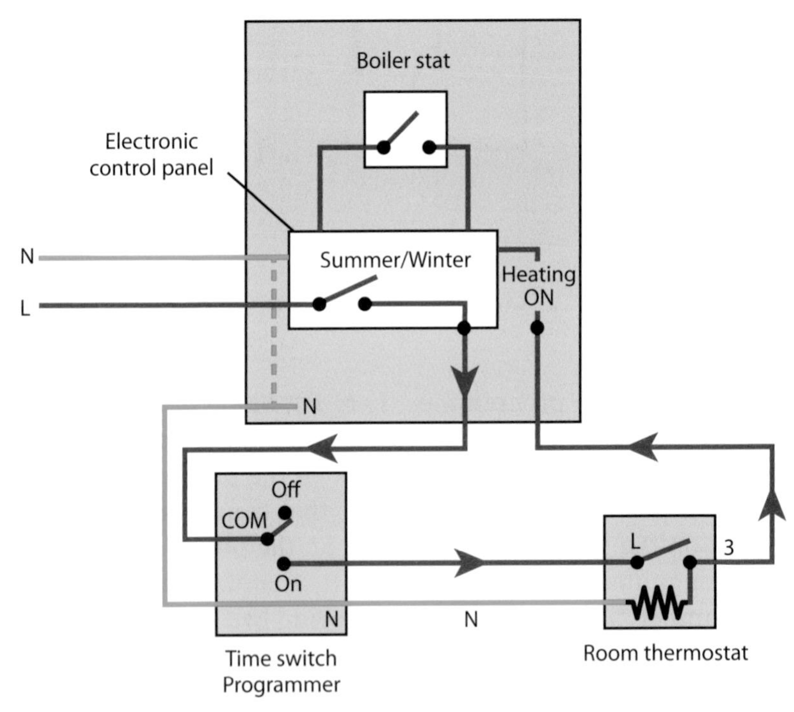

Figure 7.20: Combi boiler with room stat fitted

Pumped heating with gravity hot water systems

A pumped heating system with a gravity hot water system is rare these days. If you do come across one, it is important that you know how it functions. The boiler is operated by a time clock that will fire the boiler controlled by the boiler thermostat. The water in the boiler heats up and rises into the primary flow to the cylinder. To have any heating, the time

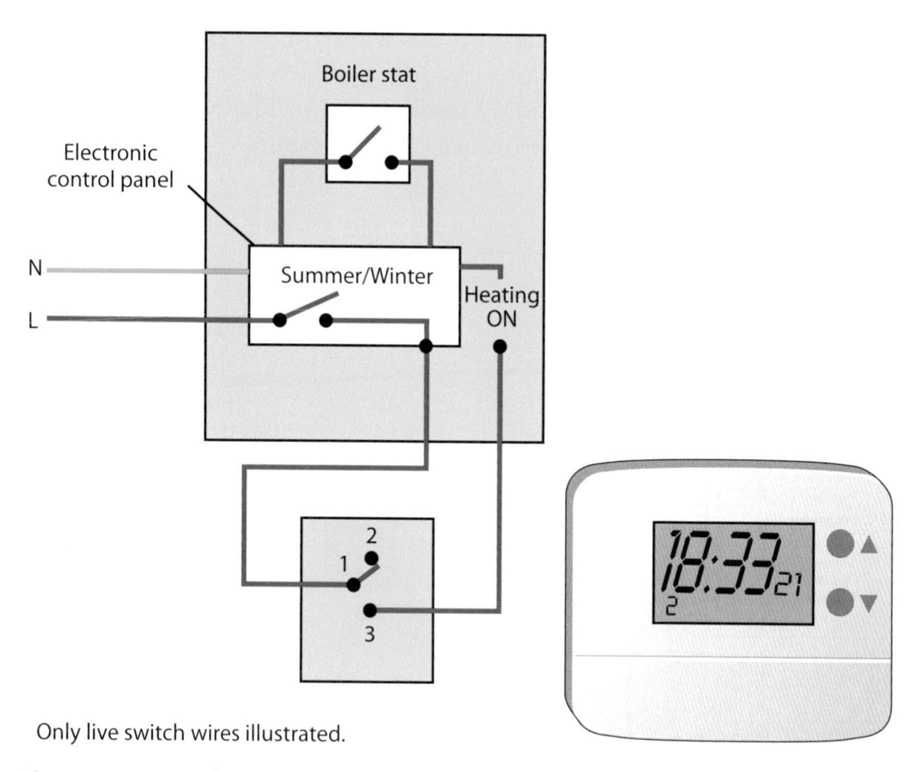

Figure 7.21: Combi wiring with programmable room thermostat

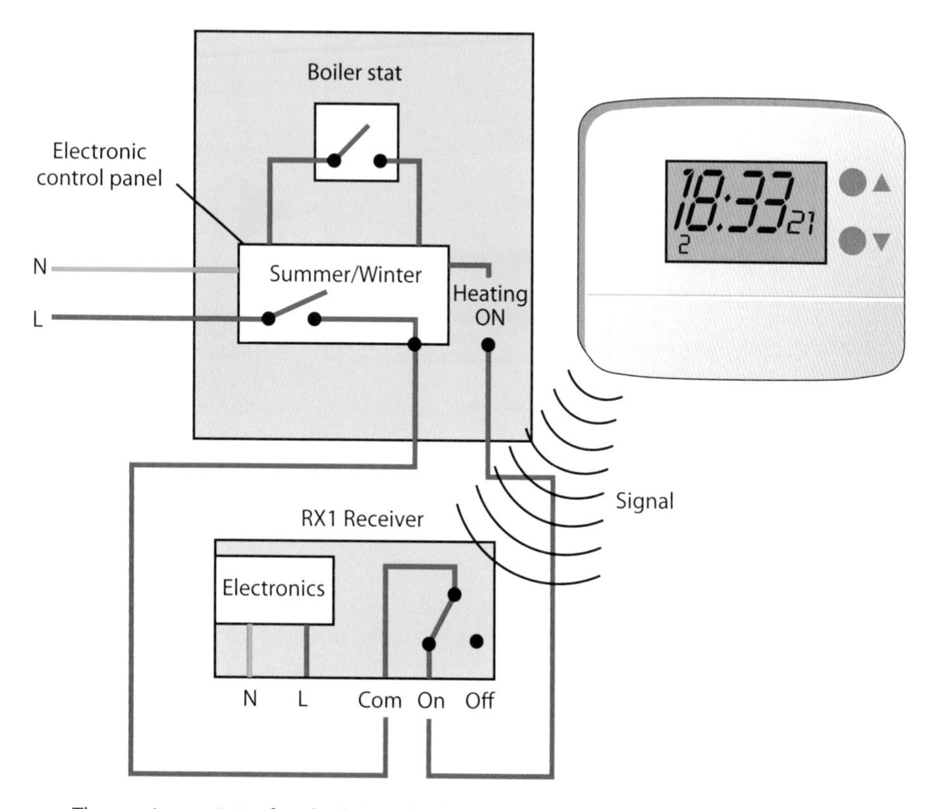

Figure 7.22: Combi wiring with wireless room thermostat and programmer

switch can be turned to 'heating' and this will send power to a room thermostat, which in turn operates the pump, which will then pump heated water around the secondary heating circuit. The wiring diagram in Figure 7.23 illustrates this primitive control system.

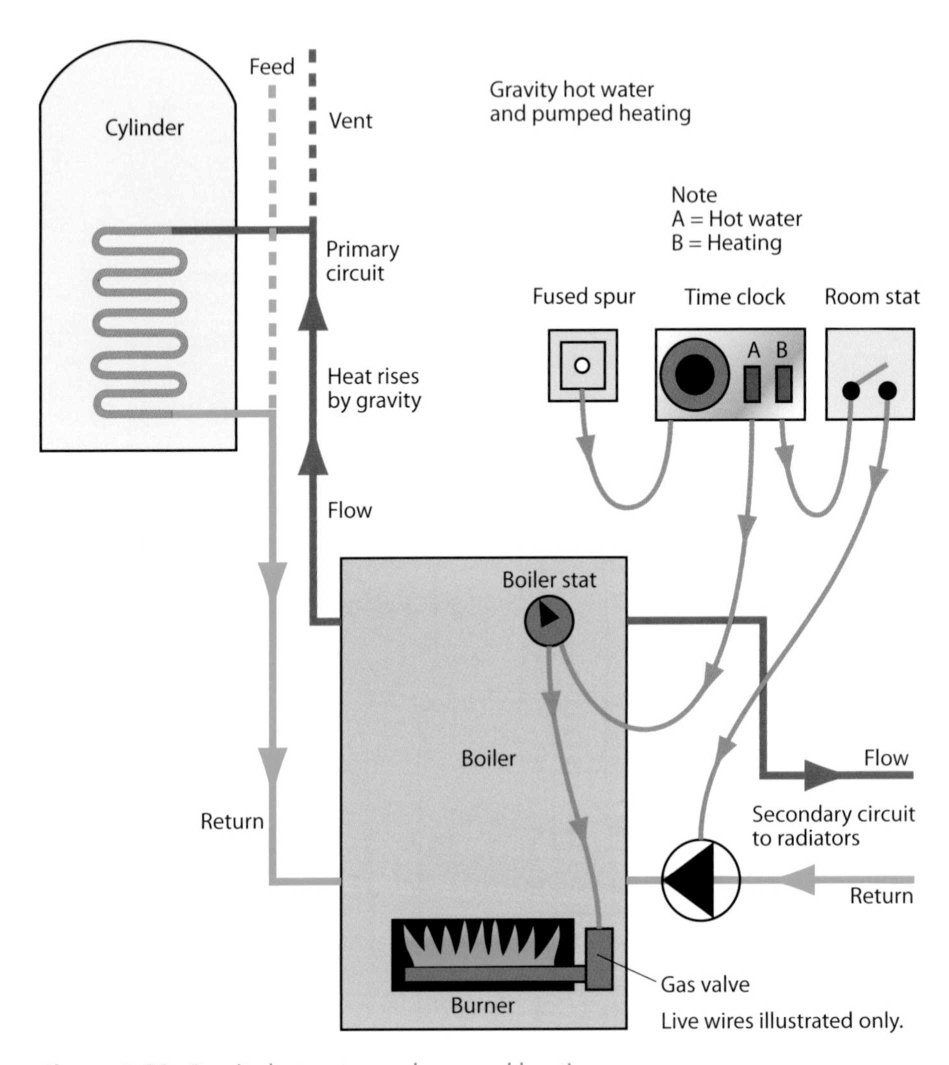

Figure 7.23: Gravity hot water and pumped heating

Fully pumped

With 2 x two-port valves

Here a L-N-E goes into the programmer from the isolation point. See Figure 7.25 on page 367. Leaving the programmer to go to the main wiring centre are:

- a permanent live (L connection)
- a neutral (N connection)
- a heating circuit on (switched live connection terminal 3)
- a hot water circuit (switched live connection terminal 6)
- an earth (E connection not shown but from isolation point to system main earthing).

The 2 × two-port valve has three-core cable into the programmer and five-core cable to the wiring centre (L-N-E and two switched lives). The wires must be properly colour-coded; neutral colours must not be used for live, and so on. Uninsulated earth cable cores must never be used for neutral or live connection.

With mid-position valve

Here the same L-N-E goes into the programmer from the isolation point, but leaving the programmer to go to the wiring centre are:

- a neutral (N connection)
- a heating circuit on (switched live connection terminal 3)
- a hot water circuit on (switched live connection terminal 6)
- a hot water circuit off (switched live connection terminal 7)
- an earth (E connection not shown but from isolation point to system main earthing).

The mid-position valve has three-core cables into the programmer and five-core cable to the wiring centre.

A room thermostat is a switch, so it has a live connection. This comes via the wiring centre from the 'on' switch connection for central heating (terminal 3 at the programmer) to the live in terminal 1 on the room thermostat. A live out connection back from terminal 3 on the room thermostat connects via the wiring centre to the heating motorised valve; the neutral and earth connect back to the main neutral and earth connection points at the wiring centre.

This means that there is a simple switch in the line across terminals 1 and 3. Four-core cable is required to feed the room thermostat from the wiring centre (N-E and two live switch wires).

Cylinder thermostat

With the Honeywell thermostat, no earth or neutral connection is required; other manufacturers may have different requirements. The situation varies for the different system types.

With 2 x two-port valves

There is a live in connection via the wiring centre from the 'on' switch connection for hot water (terminal 6 on the programmer) to the 1 terminal in the cylinder thermostat. There is a live out connection from the C terminal, which connects via the wiring centre to the hot water motorised valve.

This means that there is a simple switch in the line across terminals 1 and C. Two-core cable is required to feed the cylinder thermostat from the wiring centre.

With mid-position valve

The situation with a mid-position valve is quite different. It has a live in connection via the wiring centre from the 'off' switch connection (terminal 7 on the programmer) to the 2 terminal on the cylinder thermostat. It has a live in connection via the wiring centre from the 'on' switch connection for hot water (terminal 6 on the programmer) to the C terminal in the cylinder thermostat. A live out connection from the 1 terminal connects via the wiring centre to the mid-position valve. This means there are two live feeds in and one out (as at the programmer). Three-core cable is required to feed the cylinder thermostat from the wiring centre.

Boiler and pump

These normally just take a L-N-E from the wiring centre. The N and E come straight from the main wiring centre terminals; the live feeds to each come from one of the wires on the motorised valve (for both 2 × two-port and mid-position valve systems).

Two-port motorised valve

Two of these are included on a standard, fully pumped system, one controlling heating and one controlling hot water. These valves bring the key circuits together to get the system to function fully.

Spring-return two-port valves are usually used for this function. The valve always returns to the closed position by means of a spring when there is no electrical supply to the valve motor.

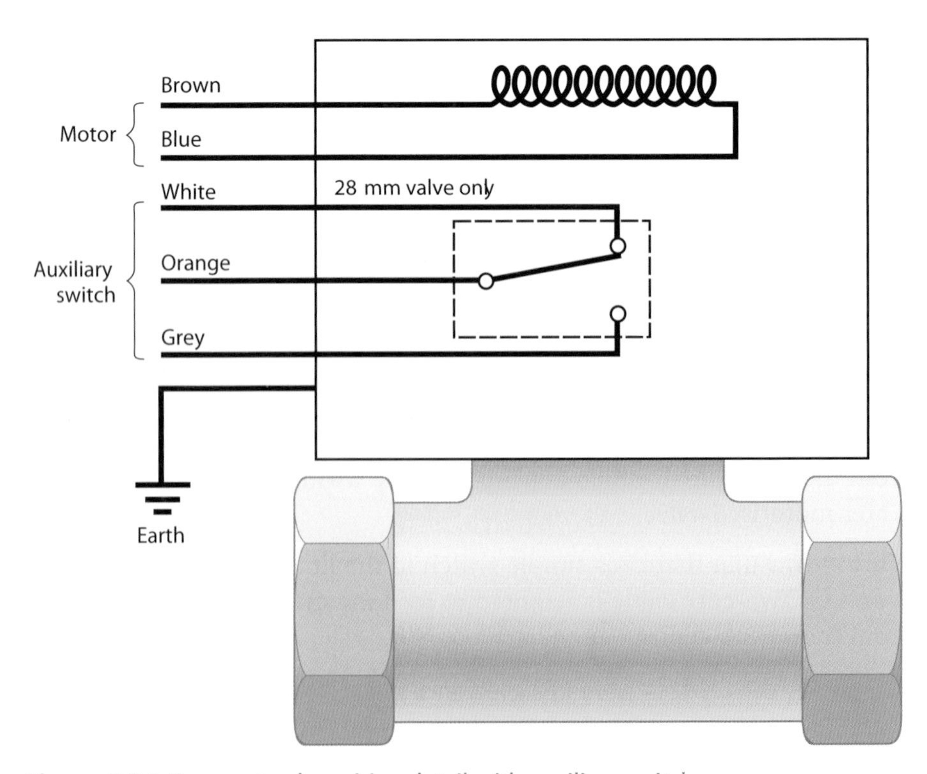

Figure 7.24: Two-port valve wiring detail with auxiliary switch

With regard to the electrics, the valve has two basic parts: an electric motor used to drive the valve open so that water can flow through it, and an auxiliary switch (often called a micro-switch) that powers the supply to the boiler and pump. The auxiliary switch is there to ensure the effective separation of the electrical supply from each circuit to feed the pump and boiler, and ensures that hot water and central heating can work independently of each other.

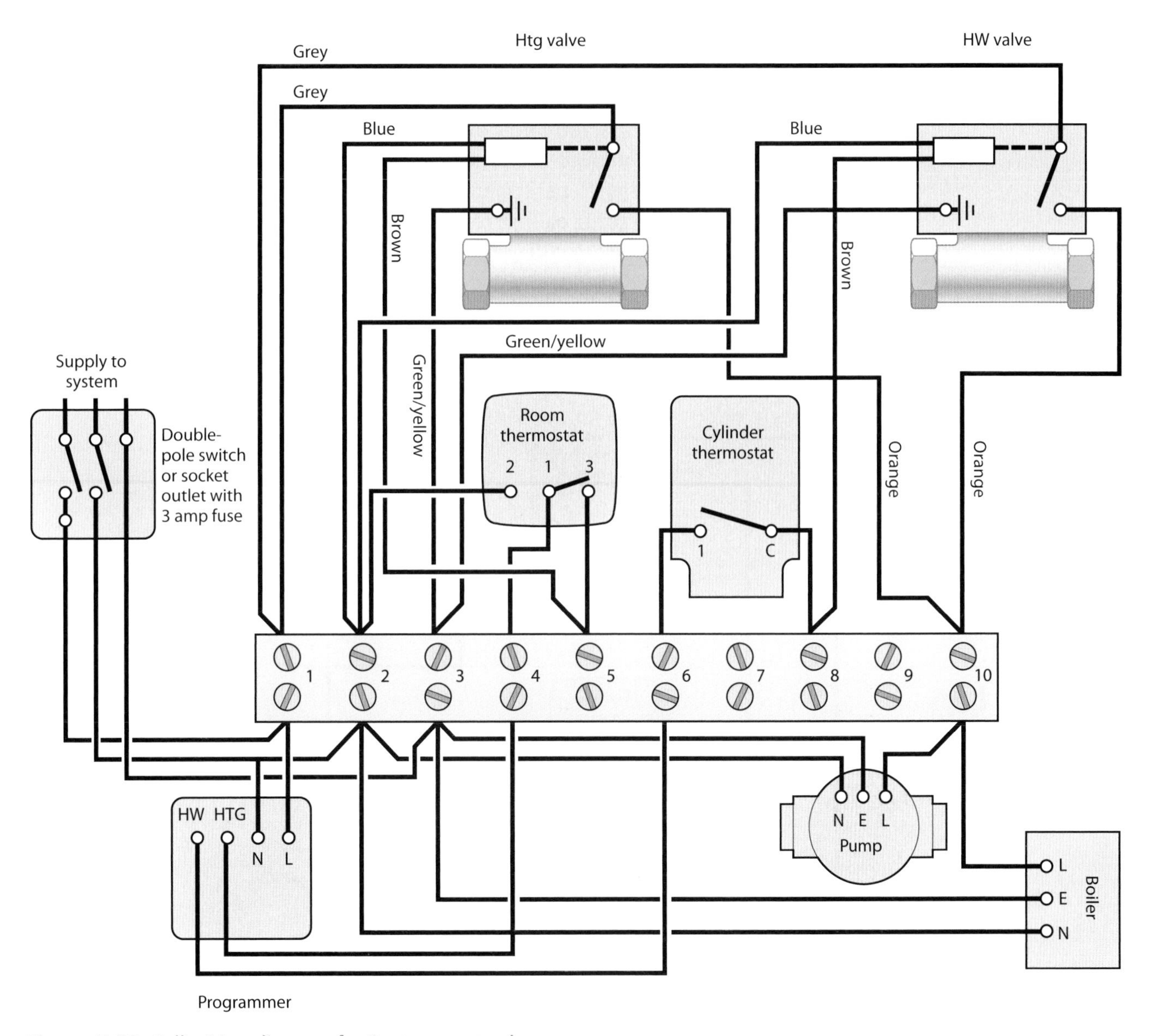

Figure 7.25: Full wiring diagram for 2 x two-port valves

Essentially, the live out connection from either the cylinder or the room thermostat connects to the brown wire on the motorised valve. This drives the motor open. As the motor moves to its fully open position, a mechanical connection is made with the auxiliary switch (like turning on a light switch). The auxiliary switch receives a permanent live supply (grey wire) from the L connection – permanent live on the programmer via the wiring centre. As the switch is turned on by the valve motor, the permanent live supply is allowed to flow through the orange wire from the motorised valve that goes to the live supply (L), to feed both the pump and boiler. When the programmer turns off or the thermostat reaches temperature, the live supply to the valve motor (brown) is cut and the valve shuts due to the action of the spring return. The auxiliary switch is disconnected and power stops flowing from the permanent live (grey) wire through to the pump and boiler feed (orange wire). If that is the only circuit currently operating, it will shut the pump and boiler off. If the other circuit is calling for heat, the pump and boiler will stay on – so both circuits work independently of each other.

Figure 7.25 on page 367 draws all this information together in the form of a full 2 × two-port valve wiring diagram.

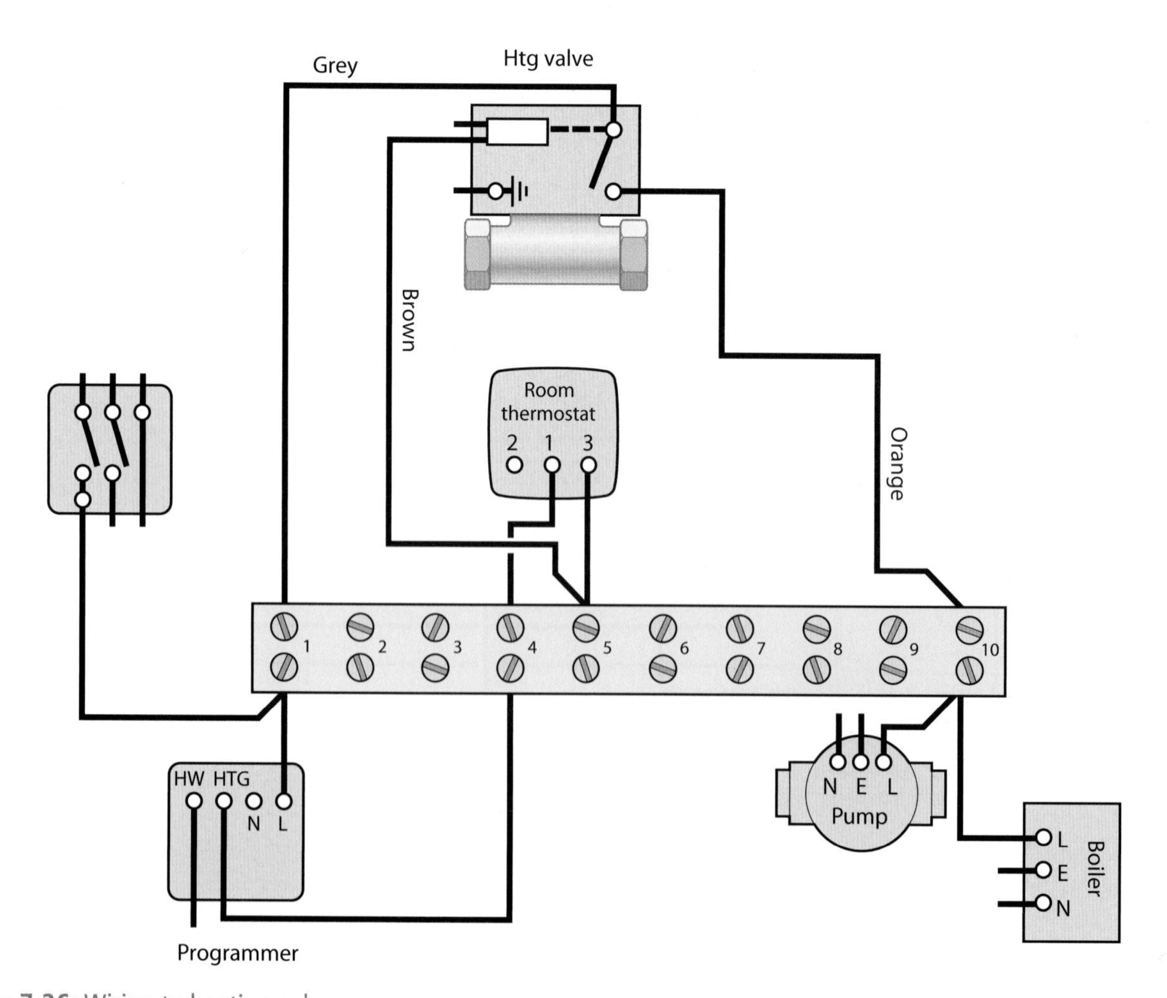

Figure 7.26: Wiring to heating valve

By now you will know that the components that take a neutral and earth, connect together to feed back via the main isolation point to the electrical circuit, so they are not involved in any of the switching arrangements. In Figure 7.26, the neutral and earth connections have been removed to make things clearer.

How each circuit works

Central heating circuit

Figure 7.26 shows how the main feed to the isolation point is via the wiring centre, and not the programmer. Either is acceptable, but you will need to connect a permanent live neutral and earth back to the programmer. Here is how the circuit works.

- The programmer calls for heating and energises the heating on terminal.
- Power is supplied to wiring centre terminal 4, which in turn feeds terminal 1 on the room thermostat.
- If the thermostat is calling for heat, the circuit will be made and terminal 3 will be live, feeding back to terminal 5 in the wiring centre.
- The brown (live motor) wire to the motorised valve is connected to terminal 5, so will be live if the thermostat is calling for heat.
- The brown wire drives the motor open. On fully opening, the motor mechanically makes the contacts on the auxiliary switch.
- Power is then allowed to flow via the grey permanent live wire, which connects to terminal 1 in the wiring centre (permanent live direct from isolation point or programmer terminal) through the orange wire to terminal 10 in the wiring centre, which feeds both live connections to the pump and boiler.
- The boiler and pump begin to operate.

When either the programmer reaches the end of its timing period or a thermostat reaches the desired temperature:

- the live feed to the brown wire in the motorised valve is cut
- the valve motor returns to the closed position, breaking the connection to the auxiliary switch and cutting the live feed to the pump and boiler by the orange wire
- the boiler and pump turn off if it is the last of the circuits to close – if another circuit is calling for heat, they remain operational.

Remember

If the heating is up to temperature, the circuit will not be made and the rest of the circuit remains dead.

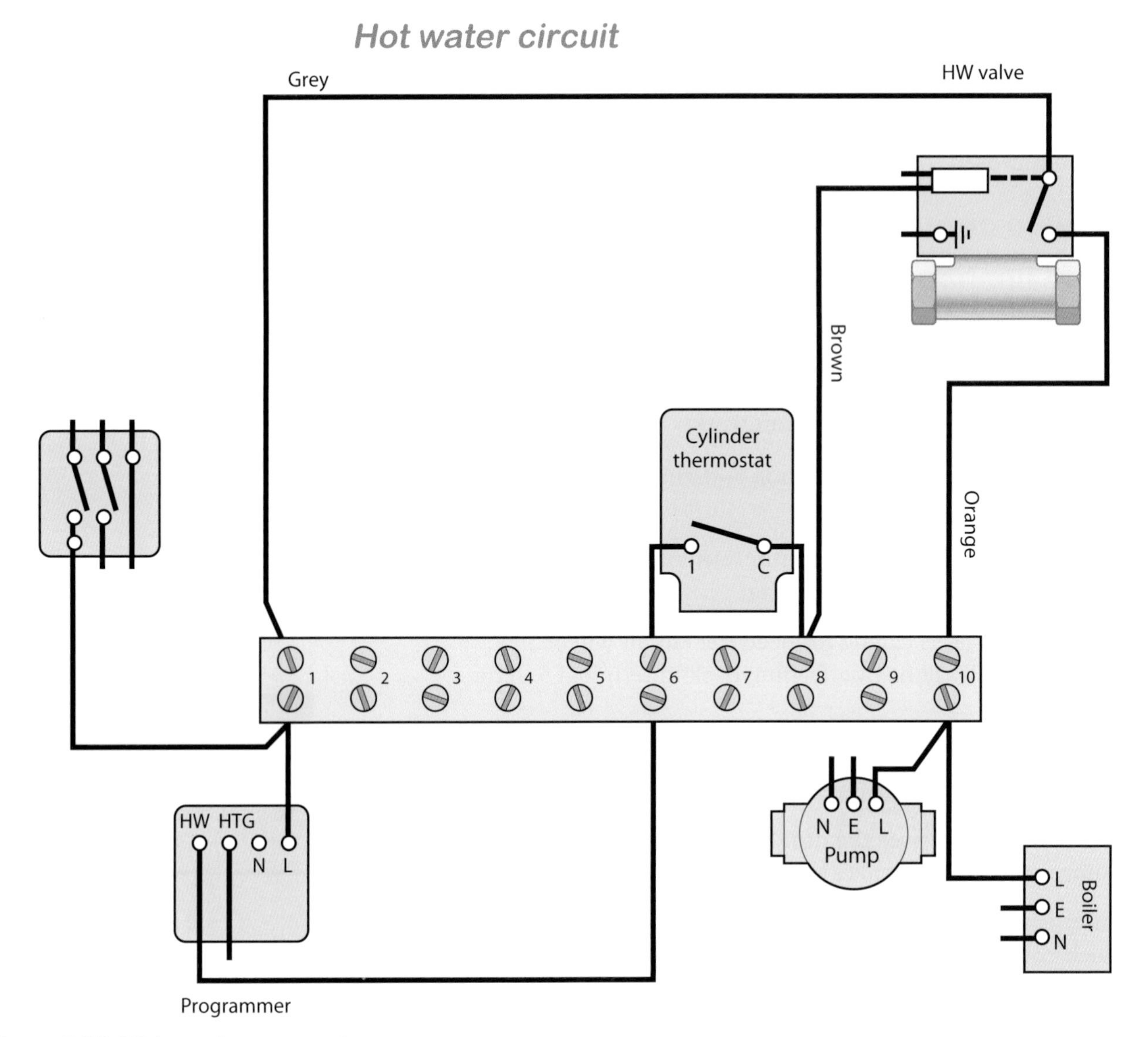

Figure 7.27: Wiring to hot water valve

- The programmer calls for heating and energises the heating on terminal.
- Power is supplied to wiring centre terminal 6, which in turn feeds terminal 1 on the cylinder thermostat.
- If the thermostat is calling for heat, the circuit will be made and terminal C will be live, feeding back to terminal 8 in the wiring centre. The brown (live motor) wire to the motorised valve is connected to terminal 8 so will be live if the thermostat is calling for heat.
- The brown wire drives the motor open. On fully opening, the motor mechanically makes the contacts on the auxiliary switch.
- Power is then allowed to flow via the grey permanent live wire, which connects to terminal 1 in the wiring centre (permanent live direct from isolation point or programmer terminal) through the orange wire to terminal 10 in the wiring centre, which feeds both live connections to the pump and boiler.
- The boiler and pump begin to operate.

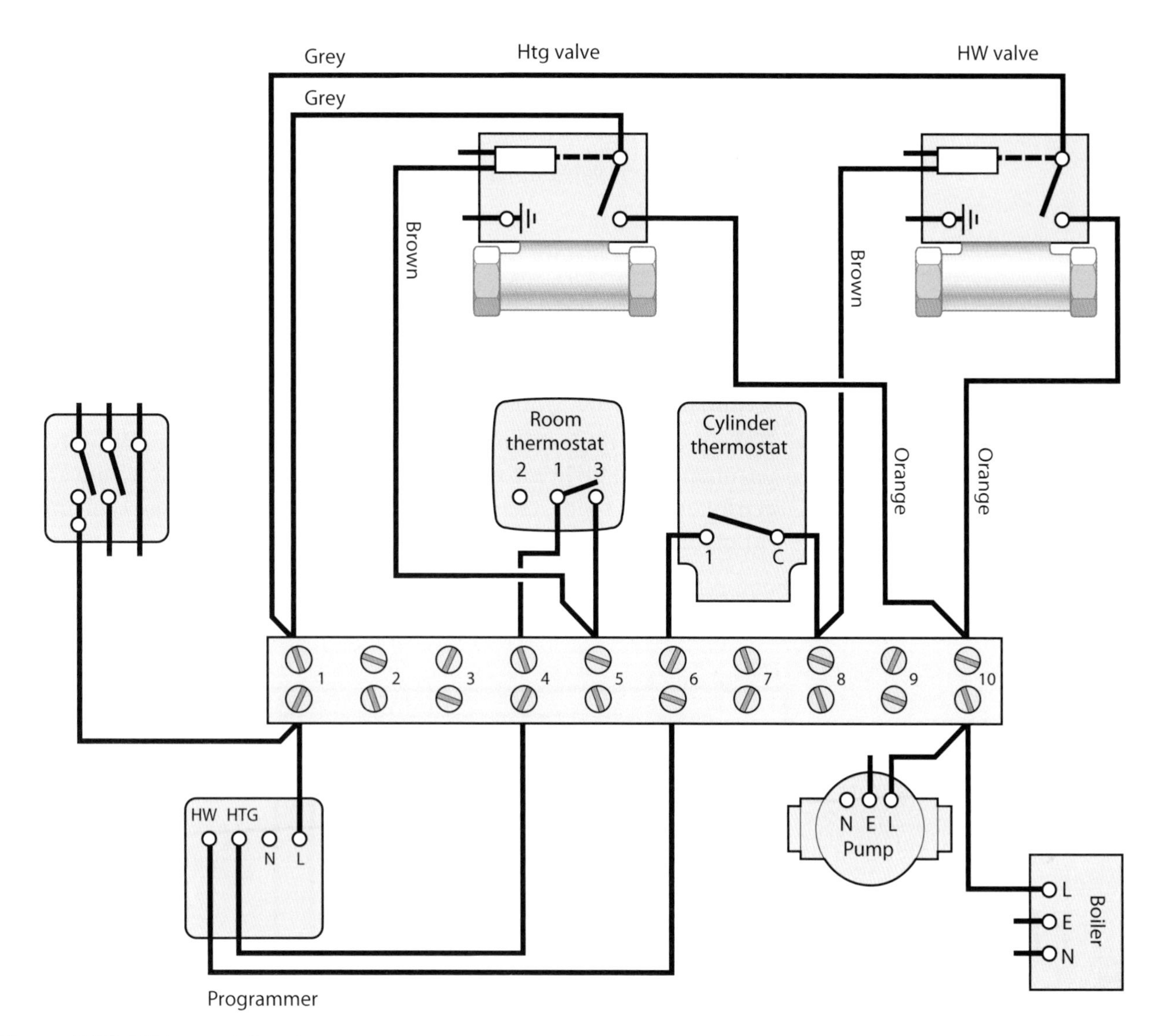

Figure 7.28: Wiring to full system

When either the programmer reaches the end of its timing period or a thermostat reaches the desired temperature:

- the live feed to the brown wire in the motorised valve is cut
- the valve motor returns to the closed position, breaking the connection to the auxiliary switch and cutting the live feed to the pump and boiler by the orange wire
- the boiler and pump turn off if that is the last of the circuits to close – if another circuit is calling for heat, they remain operational.

Figure 7.28 shows the wiring diagram with both circuits together, but with the live and neutral connections removed to make things clearer, and show how both circuits can work together and independently of each other.

Mid-position valve

Figure 7.29 also has the live and neutral connections removed, but shows both heating and hot water together.

The mid-position valve in its de-energised state is open to the hot water-only port.

Here is what happens for hot water only.

- The programmer calls for hot water and energises the hot water on terminal.
- Power is supplied to wiring centre terminal 6, which in turn feeds terminal C on the cylinder thermostat.
- If the thermostat is calling for heat, the circuit will be made and terminal 1 will be live, feeding back to terminal 8 in the wiring centre. At this point, terminal 2 on the thermostat is not receiving supply from the hot water off terminal.

The pump and boiler are directly connected to terminal 8, so they become operational; the connection to the motorised valve orange wire

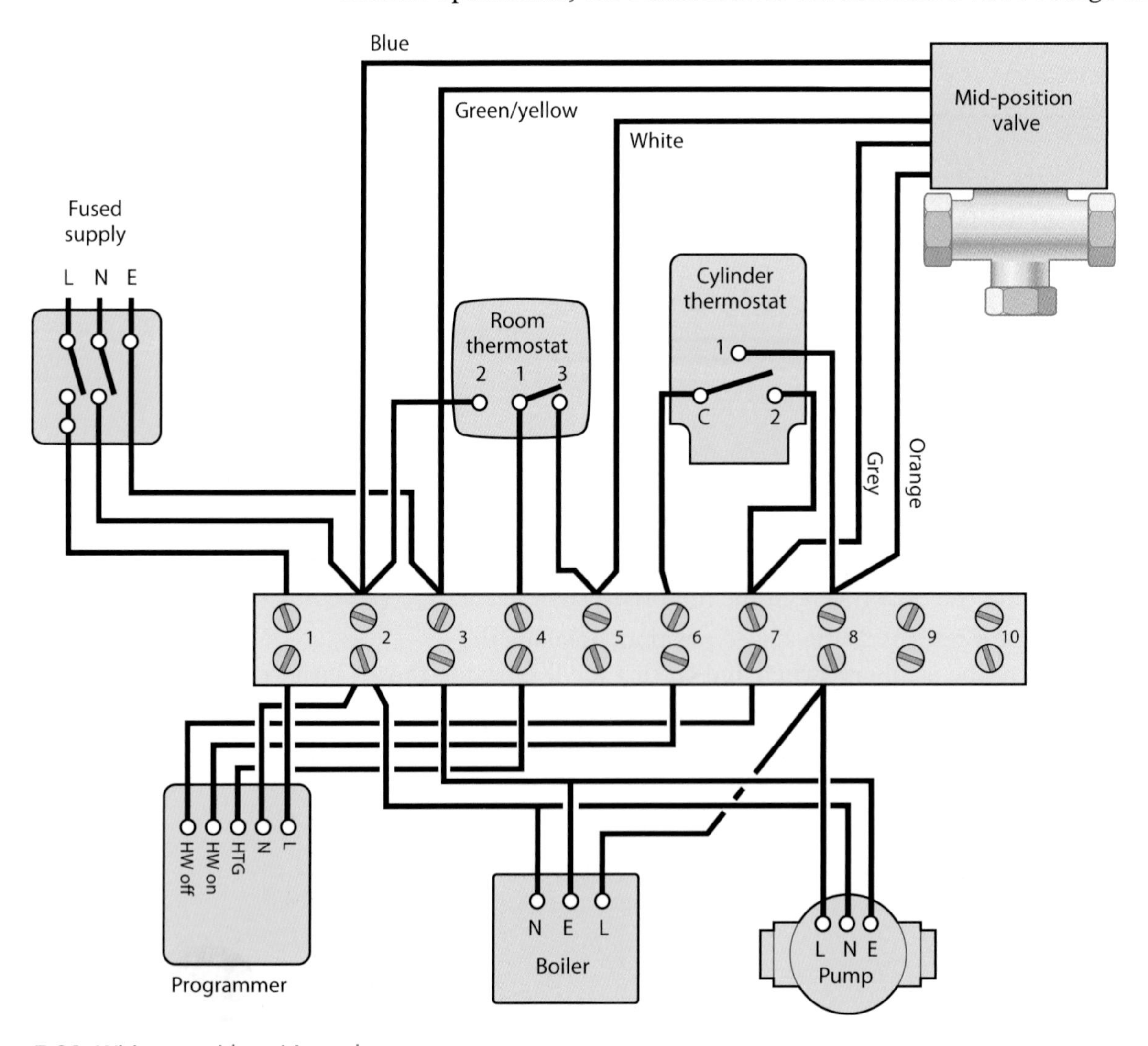

Figure 7.29: Wiring to mid-position valve

is live but performs no function at this stage. The valve position does not alter, as it is normally open to hot water.

Here is what happens for heating and hot water.

- The programmer calls for heat and energises the hot water on and heating on terminals.
- Power is supplied to wiring centre terminal 6, which in turn feeds terminal C on the cylinder thermostat. Power is supplied to wiring centre terminal 4, which in turn feeds terminal 1 on the room thermostat.
- Assuming both circuits are calling for heat, the pump and boiler are activated by the live connection from the cylinder thermostat (terminal C).

The motorised valve is driven to its mid-position (supplying heating and hot water) by the white wire connected to the outlet of the room thermostat terminal 3 (via wiring centre connection 5).

Here is what happens when pump and boiler supply both heating and hot water (see Figure 7.30), for heating only.

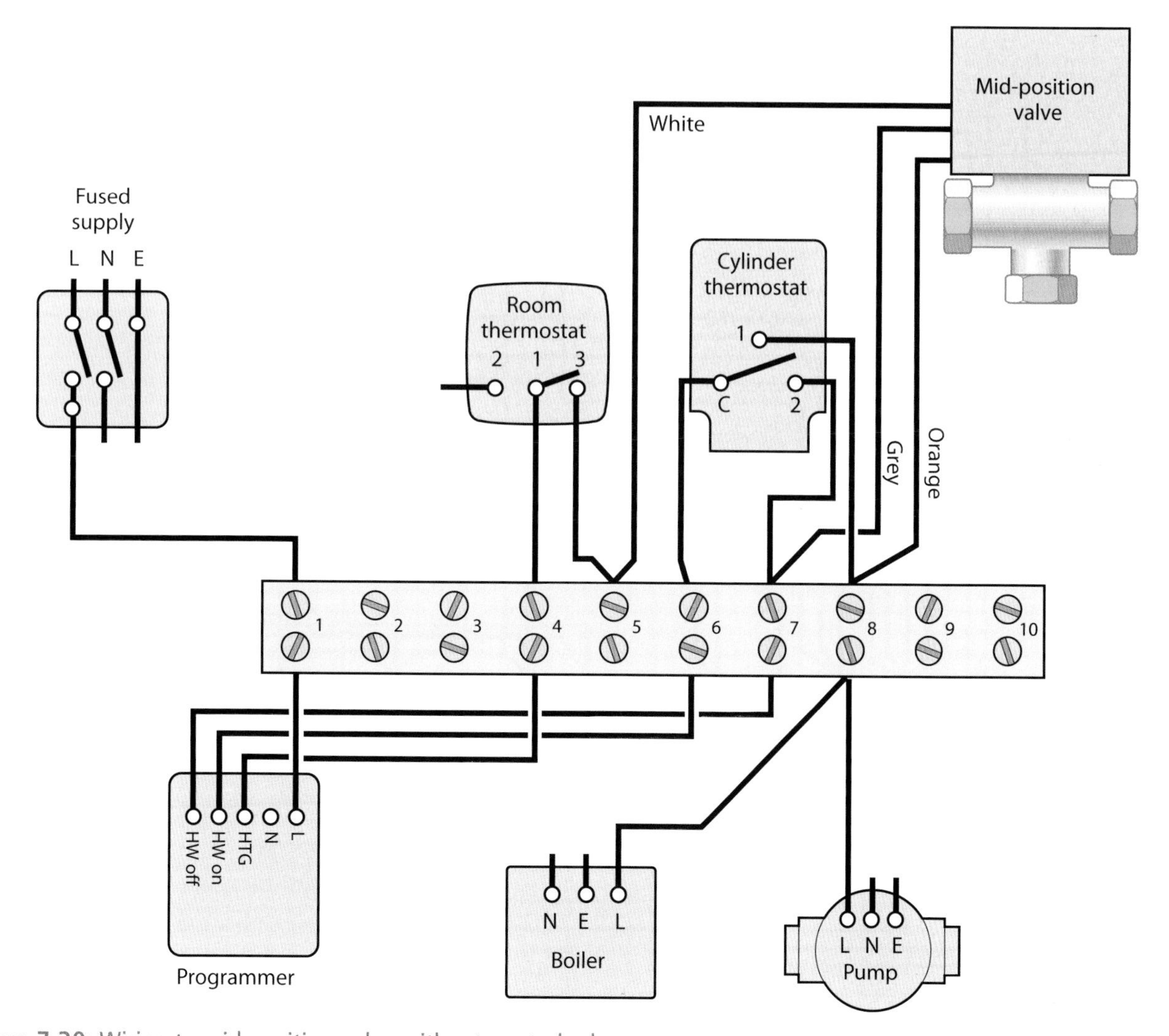

Figure 7.30: Wiring to mid-position valve without neutrals shown

- With the programmer calling for heating only or the cylinder thermostat up to temperature, the grey wire in the valve is activated (either by the hot water off terminal at the programmer or the 2 terminal at the cylinder thermostat).
- The valve travels fully across to the heating only port where an auxiliary switch connection is made by a mechanical connection. This in turn energises the orange wire that feeds the boiler and pump.

The boiler and pump are operational. On this principle, the valve can move backwards and forwards through its operating sequence to respond to the demands of whichever circuits are requiring heat. This set-up is more complicated than 2 × two-port valves and can be a bit more difficult to diagnose faults on, but it is important for you to know the principles involved.

Fully pumped incorporating hot water and multiple space-heating zones

To be compliant with the Building Regulations, most heating systems will need to have control over two heating zones as well as hot water. This can be achieved using two-port zone valves to control each circuit via a thermostat (see Figure 7.31).

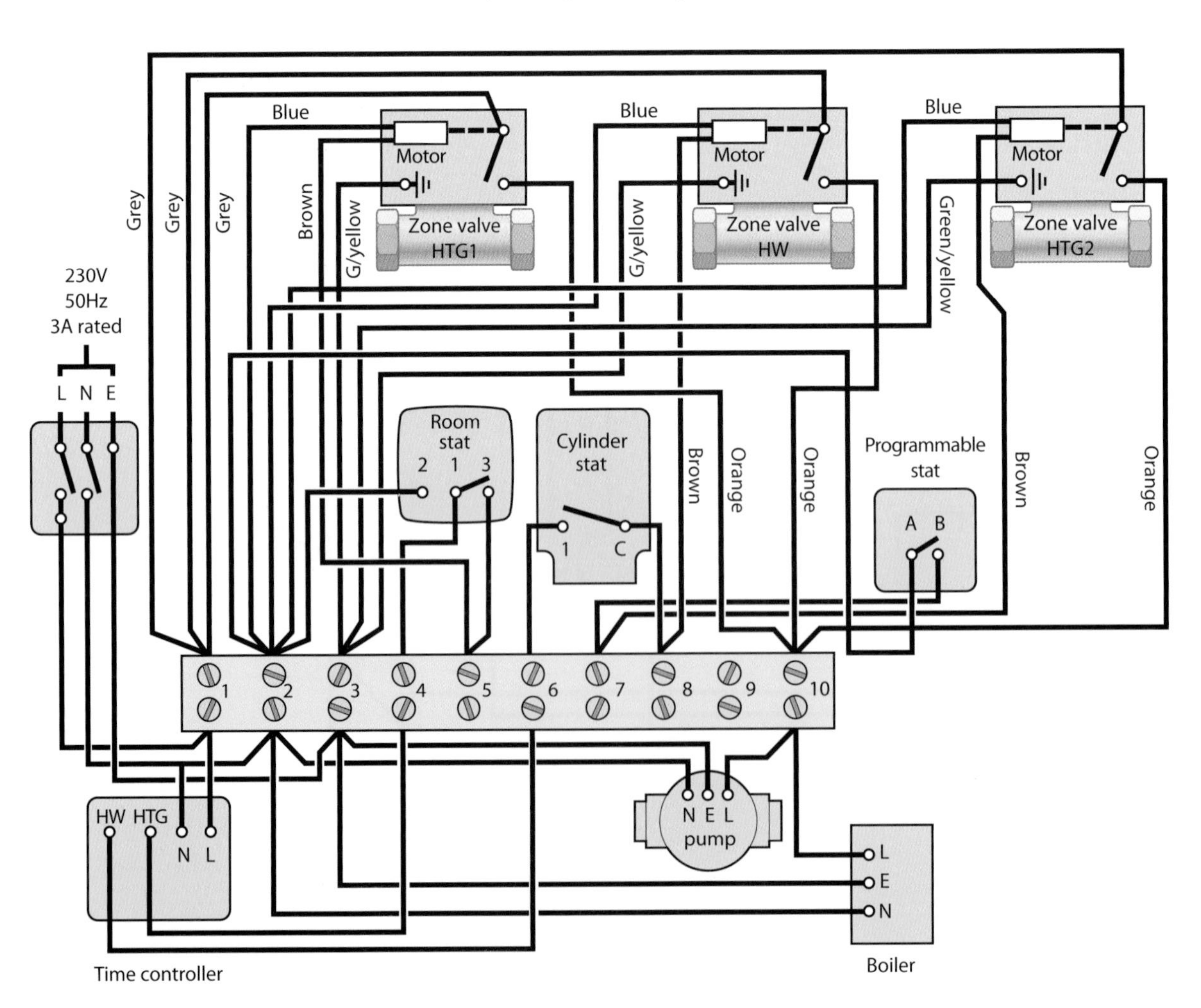

Figure 7.31: Multi-zone control valve system

Fully pumped multiple-boiler controls

The modern multi-boiler installation often includes a low loss header, so either shunt pumps are used or there is a normal circuit with a pump for each boiler. Figure 7.33 shows a low loss header. Each manufacturer will have a different wiring diagram, so just one example is shown in Figures 7.32 and 7.33.

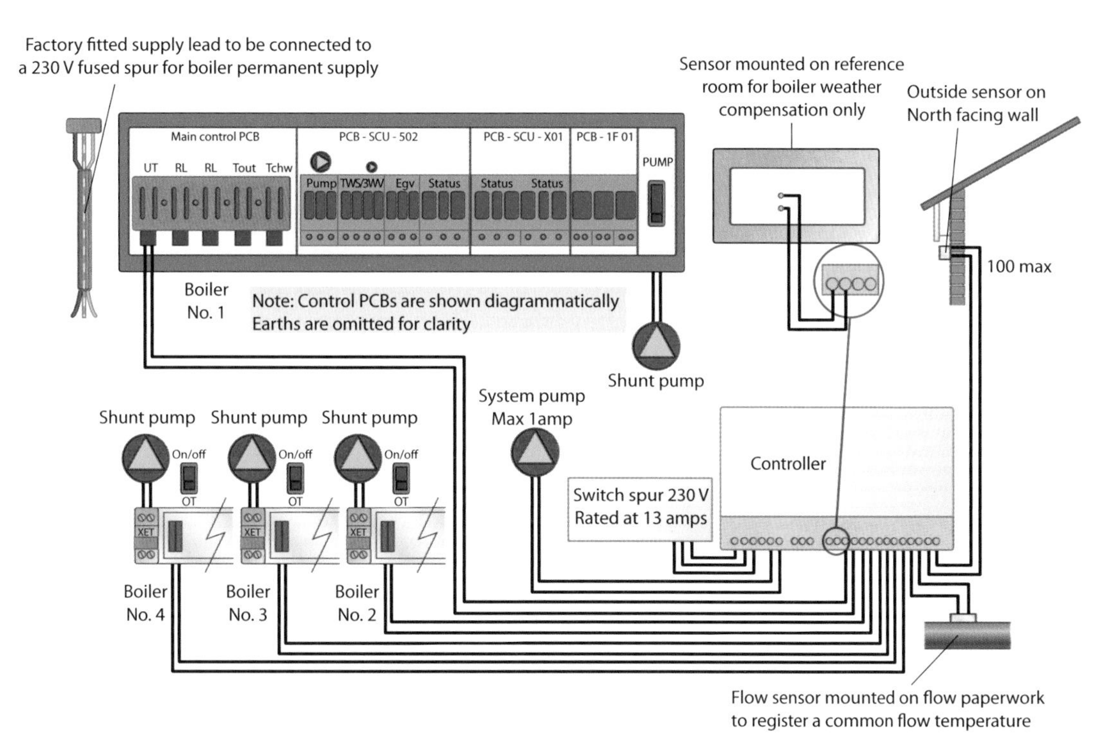

Figure 7.32: Wiring installation for multi-boiler installation

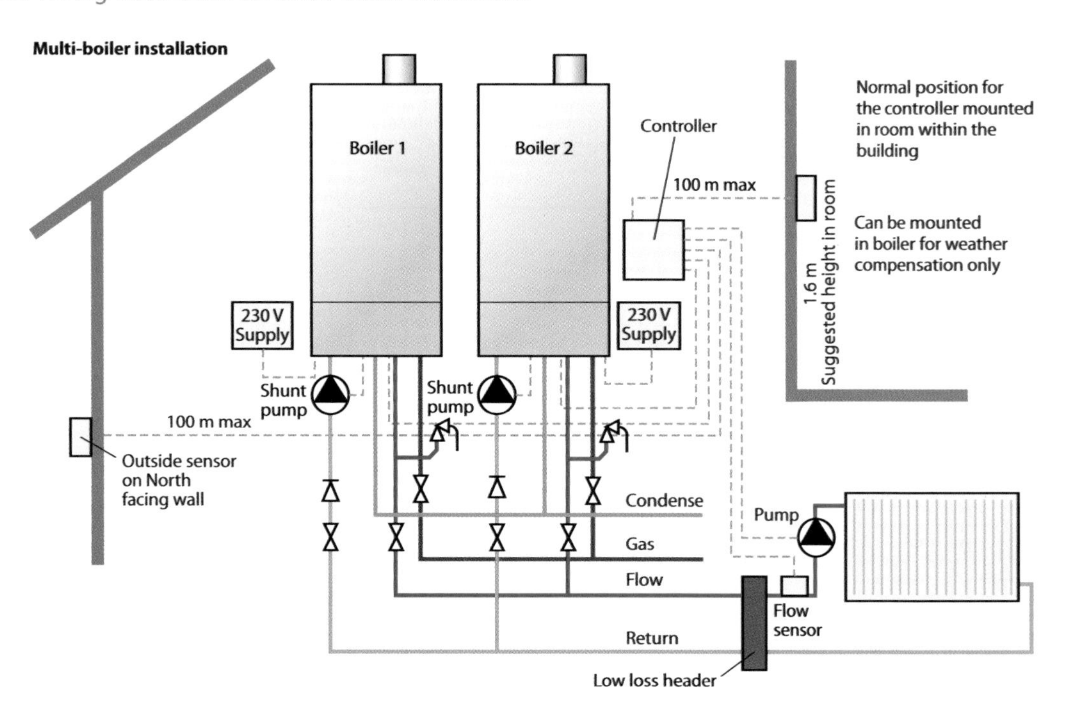

Figure 7.33: Multi-boiler installation

Frost thermostats

The frost thermostat is similar to the room thermostat except that it reads much lower temperature levels and has a tamper-proof cover. It is sited near to the components that need guarding against freezing, such as in an external boilerhouse, and is usually set to about 4°C, the point at which freezing can begin. The thermostat is designed to override the time and temperature controls to a particular circuit (usually one of the heating zones, if there are more than one). A permanent live connection is taken from the wiring centre to the frost thermostat, which feeds the motorised valve (see page 377 for differences with mid-position and 2 × two-port valve systems).

If the frost thermostat activates, the pump and boiler are activated via the feed to the motorised valve until the temperature set on the pipe thermostat is reached. At that point the frost thermostat is still activated, but the pipe thermostat is not. The water in the pipework is well above freezing. The temperature in the pipework will begin to fall, the boiler and pump are reactivated by the pipe thermostat, and so on. Figure 7.35 shows the wiring layout for these components for a 2 × two-port valve system. Figure 7.36 shows a mid-position valve system.

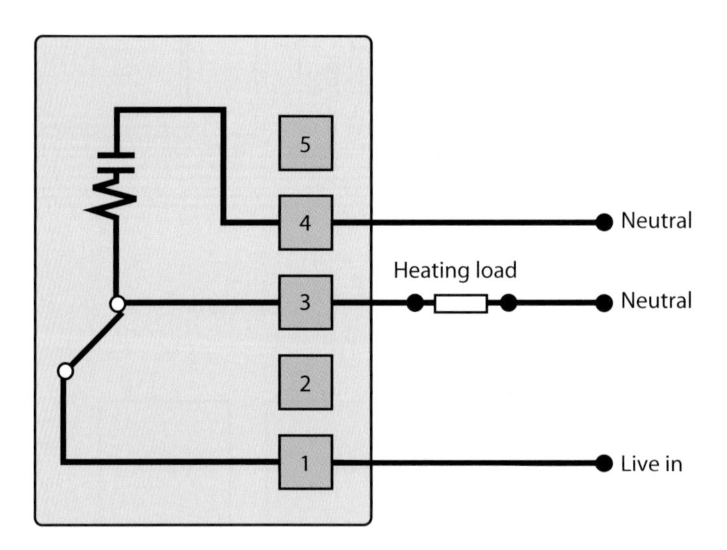

Figure 7.34: Frost thermostat

Pump overrun thermostats

Pump-overrun thermostats are rare on new boilers, but you might come across them on existing installations. They have a slightly different wiring arrangement, and the boiler is more involved in the controls function. The differences for a 2 × two-port valve system are:

- the boiler takes a neutral and earth connection as normal
- a permanent live supply is taken direct from the wiring centre to the boiler
- the orange wires from the motorised valves are connected via the wiring centre, directly to a terminal on the boiler
- the pump live connection is supplied via the wiring centre by the boiler.

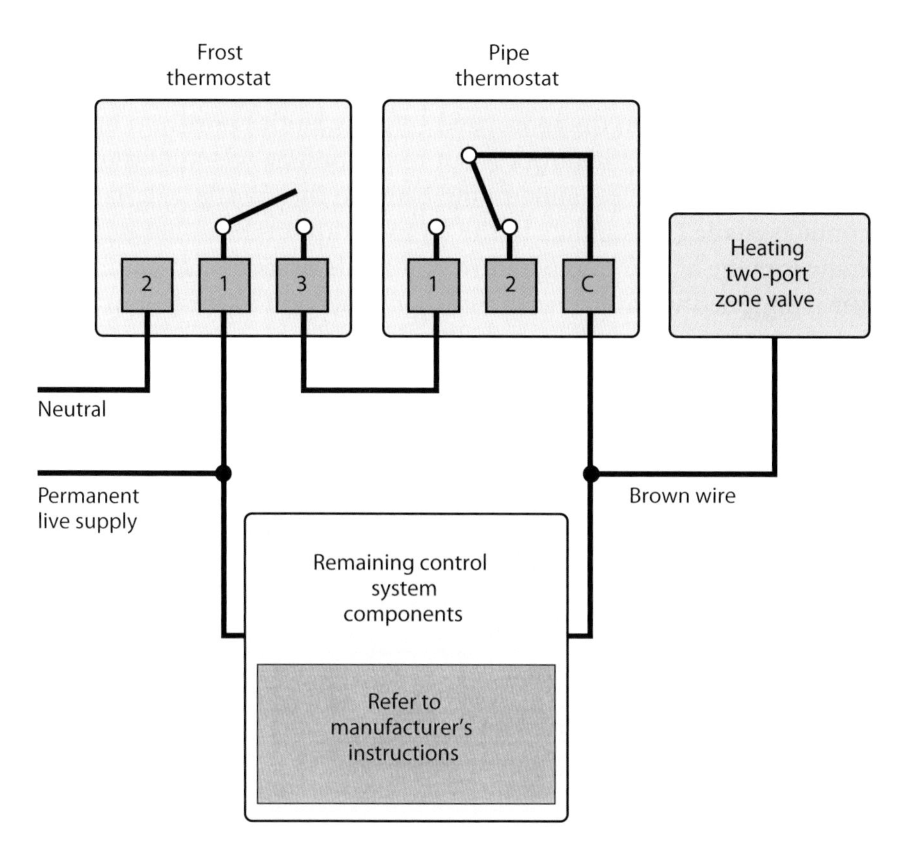

Figure 7.35: Frost thermostat and pipe thermostat wiring detail with two-port zone valve

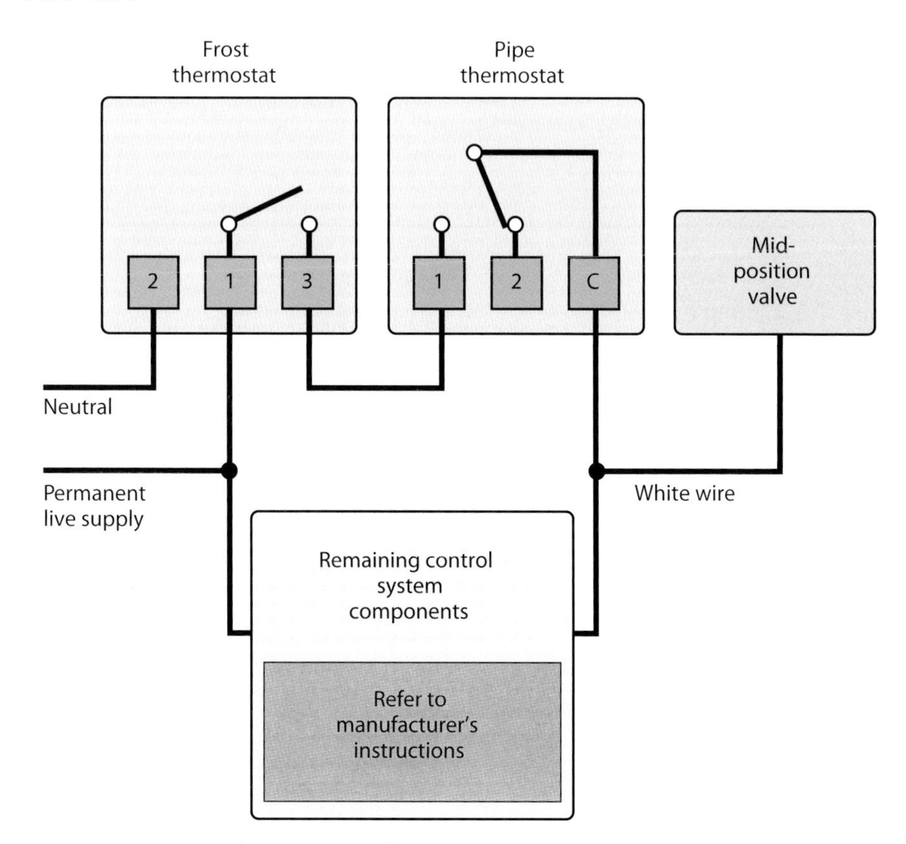

Figure 7.36: Frost thermostat and pipe thermostat wiring detail with mid-position valve

Five-core cable to the boiler is needed: N-E and live from orange wires, permanent live and live to pump. When the boiler receives power via the orange wire(s) from the motorised valve(s), both pump and boiler are operational. When the power is cut to the orange wire and the boiler is up to temperature, a separate pump overrun thermostat will have its contacts made (it responds to the temperature in the boiler heat exchanger made at high temperature, not made at low temperature). Power is supplied via the permanent live connection to the pump only, and the pump continues to operate. The boiler is not firing; it continues to operate until the water cools to the temperature setting of the overrun thermostat, then it turns off (see Figure 7.37).

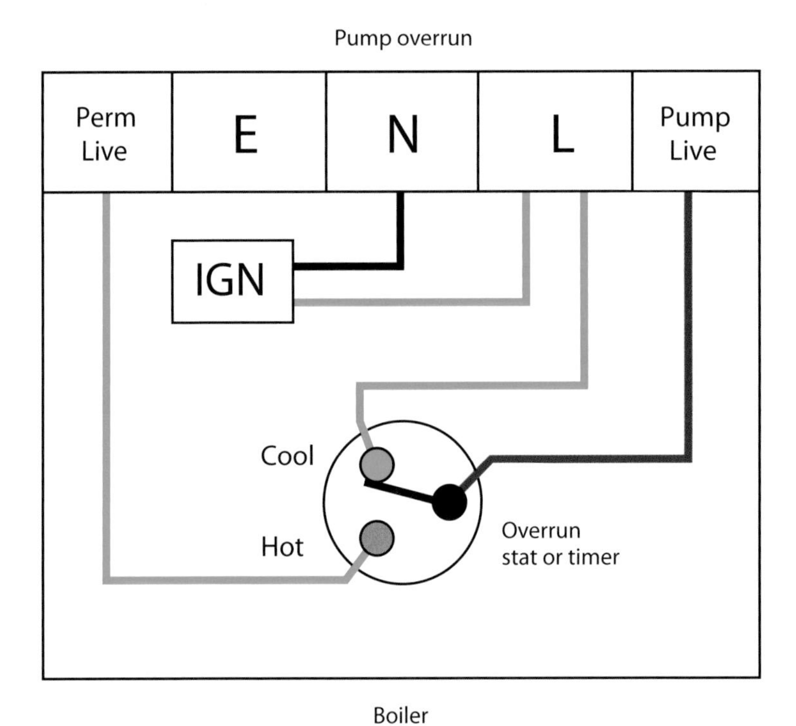

Figure 7.37: Pump overrun

PROGRESS CHECK

1. What is a programmer?
2. What is the fuse rate for domestic heating wiring?
3. What are the switches on a programmer classed as?
4. What should uninsulated earth cable cores never be used for?
5. On a master room thermostat that requires an earth, what must the wire feeding it have?
6. The cylinder thermostat normally has only three wires. Which wires are these?
7. On a two-zone valve system, what is the grey wire?
8. Which port is open in the mid-position valve when in its de-energised state?
9. What is a frost thermostat similar to?
10. A boiler with a pump overrun thermostat must be fitted with which permanent feature?

5. The purpose and requirements of earthing systems

British Standards classifications

Chapter 31 of BS 7671, Regulation 312 deals with conductor arrangement and system earthing and goes on to classify the types of system earthing using a series of letters as shown in Table 7.4.

BS 7671 Regulation 312

First letter (relationship between supply and earth)	
T	The supply is connected directly to earth ('T' coming from the Latin word **T**erra, meaning earth) at one or more points
I	The supply is **i**solated from earth, or one point is connected to earth through a high impedance
Second letter (relationship between earth and the exposed-conductive-parts of the installation)	
N	Direct electrical connection of the exposed-conductive-parts to the earthed point of the supply system (remember that for an a.c. system, the earthed point of the supply system is normally the **n**eutral point)
T	Direct electrical connection of exposed-conductive-parts to earth, independent of the earthing of any point of the supply system
Subsequent letters (arrangement of neutral and protective conductors)	
C	The neutral and protective functions are **c**ombined in a single conductor (the PEN conductor)
S	The protective function provided by a conductor **s**eparate from the neutral conductor or from the earthed line (or, in a.c. systems, earthed phase) conductor

Table 7.4: Letters used in BS classification of systems earthing

Main earthing systems

Here are the most common variations of these letter codes.

TT system

The first letter T means that the supply is connected directly to earth at the source (for example, the generator or transformer at one or

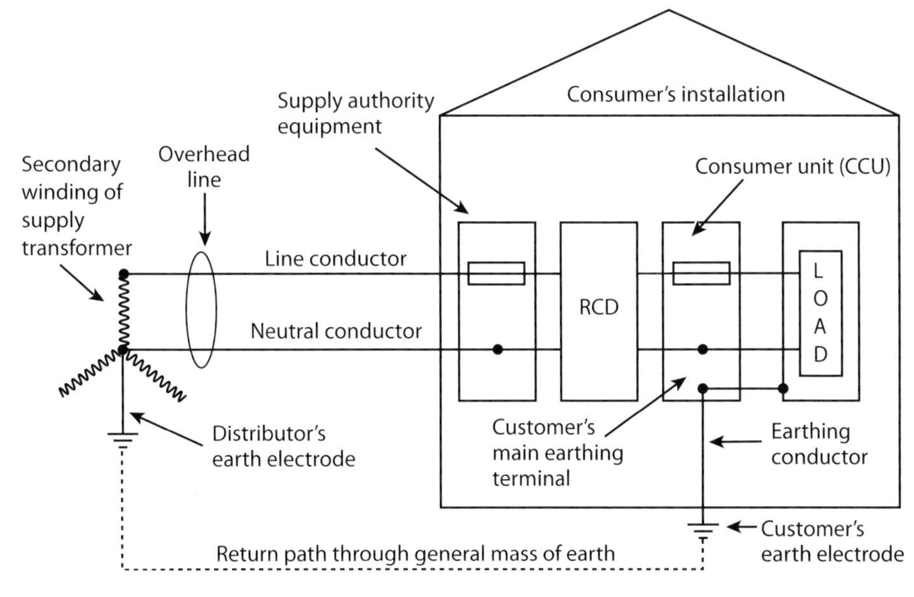

Figure 7.38: TT system

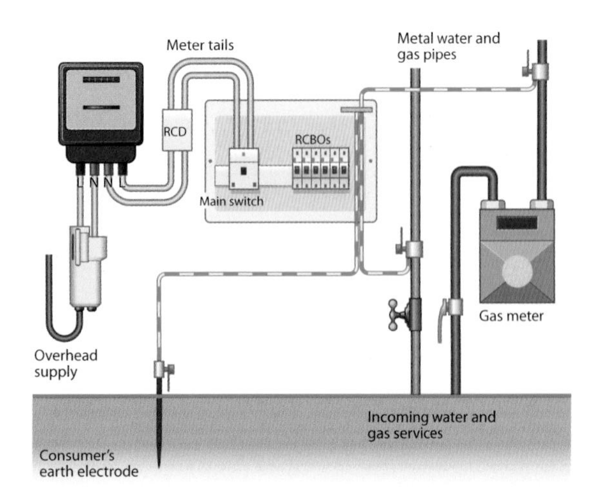

Figure 7.39 Customer's intake position for a TT system

BS 7671 411.5.2

more points). The second letter T means that the exposed metalwork of the installation is connected to the earth by a separate earth electrode. The only connection between these two points is the general mass of earth (soil and so on) as shown in Figure 7.38.

This system is used where the customer installation has not been provided with an earth terminal by the electricity supply company, so usually in rural areas where it is easier to provide an overhead supply.

Figure 7.39 shows a typical consumer's intake position for a TT system where the BS 7671 RCD requirement for TT systems has been met by the provision of RCBOs in the consumer unit.

TN-S system

Here T means that the supply is connected directly to earth at one or more points; N means that the exposed metalwork of the installation is connected directly to the earthing point of the supply; and S that separate neutral and protective conductors are being used throughout the system from the supply transformer all the way to the final circuit, to provide the earth connection, as in Figure 7.40.

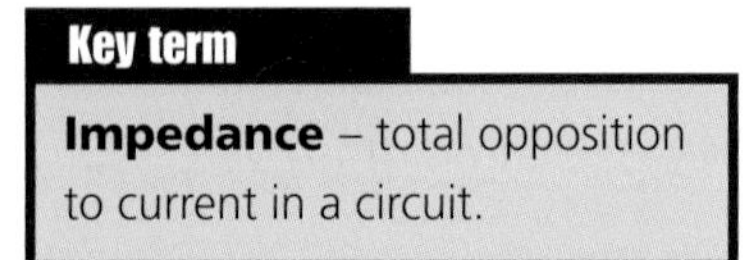
Key term

Impedance – total opposition to current in a circuit.

This is probably the most common system in the UK. The earth connection is usually through the sheath or armouring of the supply cable and then by a separate conductor within the installation. A conductor is used throughout the system to provide a return path for the earth fault current, so the return path should have a low **impedance**. Figure 7.41 shows a typical consumer's intake position for a TN-S system.

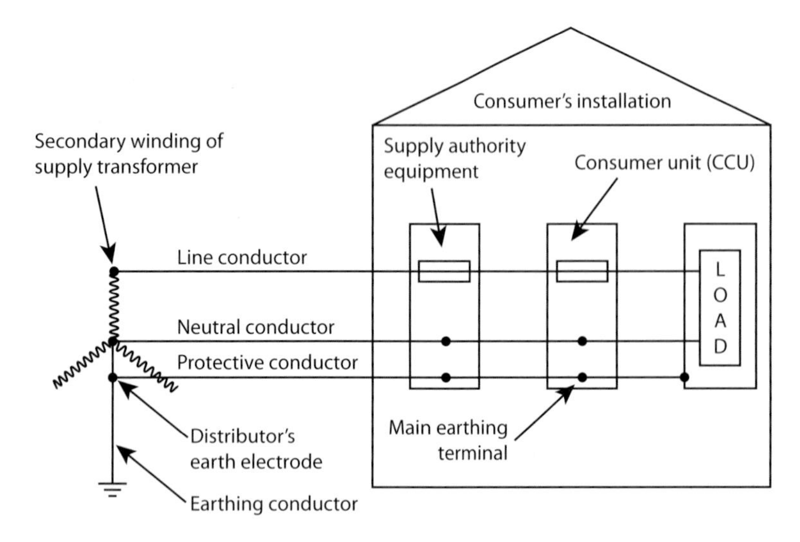

Figure 7.40 Earth connection for a TN-S system

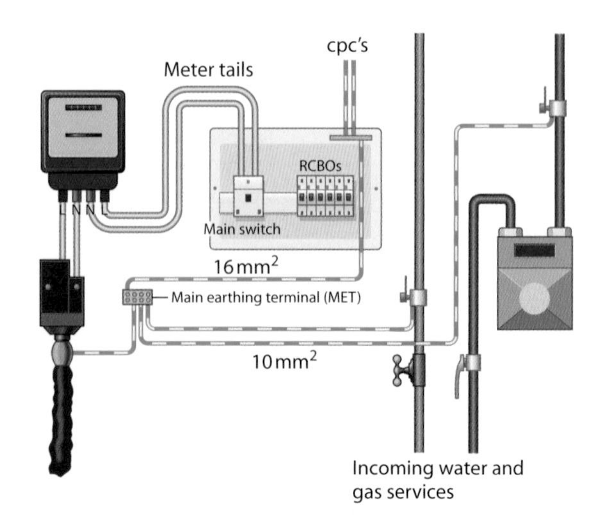

Figure 7.41 Customer's intake position for a TN-S system

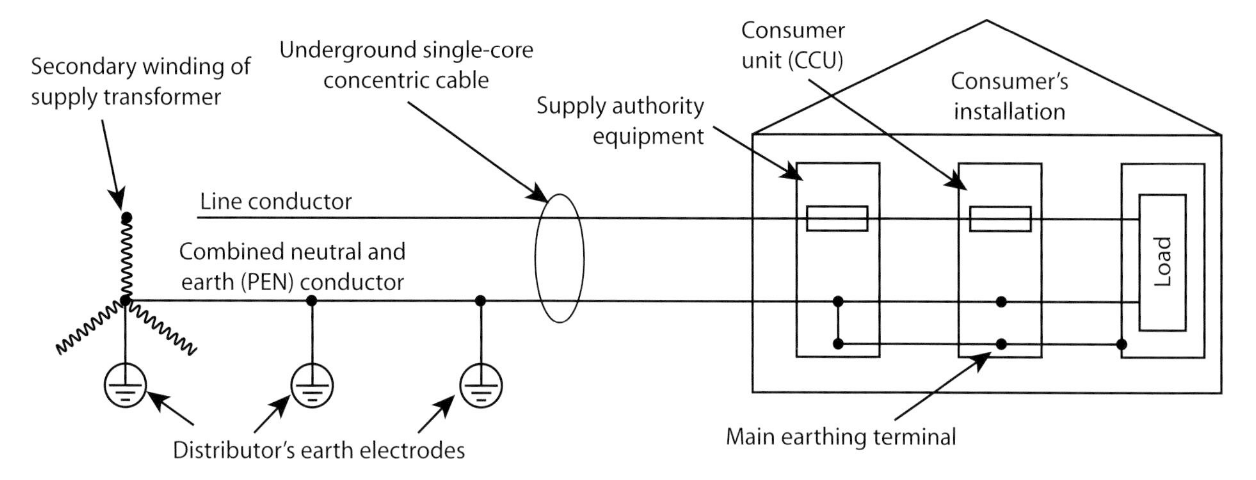

Figure 7.42: Earth connection for a TN-C-S system

TN-C-S system (PME)

For the system shown in Figure 7.42, T means the supply is connected directly to earth at one or more points; N means the exposed metalwork of the installation is connected directly to the earthing point of the supply; C means that for some part of the system (generally in the supply section) the functions of neutral conductor and earth conductor are combined in a single common conductor; and S means that for some part of the system (generally in the installation), the functions of neutral and earth are performed by separate conductors.

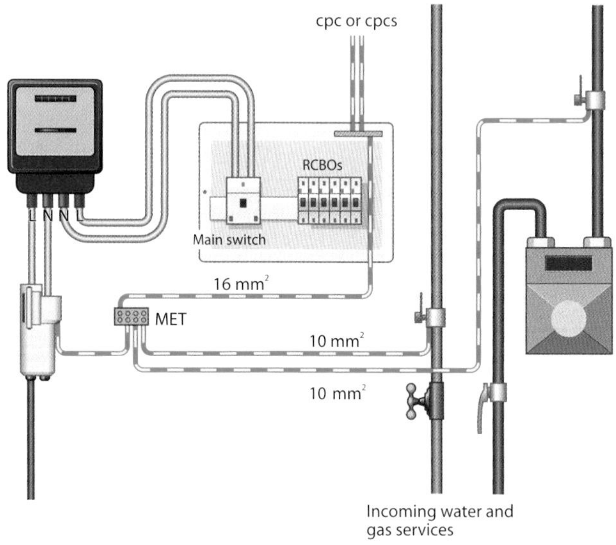

In a TN-C-S system, the supply uses a common conductor for both the neutral and the earth.

Figure 7.43: Customer's intake position for a TN-C-S system

The supply PEN needs to be earthed at several points, so this type of system is also known as protective multiple earthing (PME). However, this effectively means that the distribution system is TN-C and the consumer's installation is TN-S – a combination giving a TN-C-S system. Figure 7.43 shows a typical consumer's intake position for a TN-C-S system.

Did you know?

The combined conductor in a TN-C-S system is known as the protective earthed neutral (PEN) or combined neutral and earth (CNE) conductor.

Protective multiple earthing (PME)

PME is an extremely reliable system of earthing, so is becoming increasingly more common as a distribution system in the UK.

With the PME system, the neutral of the incoming supply is used as the earth point and all CPCs connect all metalwork, in the installation to be protected, to the consumer's earth terminal. All line-to-earth faults are converted into line-to-neutral faults, which ensures that under fault conditions a heavier current will flow, operating protective devices rapidly.

However, this increase in fault current can produce two hazards:

- the increased fault current results in an enhanced fire risk during the time the protective device takes to operate
- if the neutral conductor ever rose to a dangerous potential relative to earth, then the resultant shock risk would extend to all the protected metalwork on every installation that is connected to this particular supply distribution network.

Because of this, certain conditions are laid down before a PME system is used.

- PME can only be installed by the supply company if the supply system and the installations it will feed meet certain requirements.
- The neutral conductor must be earthed at a number of points along its length.
- The neutral conductor must have no fuse or link that can break the neutral path.
- Where PME conditions apply, the main equipotential bonding conductor must be selected in accordance with the neutral conductor of the supply and Table 54.8 of BS 7671.

BS 7671, Table 54.8

Supply systems

In 1988, the EU decided to harmonise voltage levels throughout Europe to 230 V between phase and neutral conductors. European nominal voltage was 220 V and the UK voltage was 240 V – but the agreement gave a voltage tolerance of 230 V +10% – 6%, so in many parts of the UK you will find your nominal supply voltage is still 240 V.

Most equipment is designed to operate satisfactorily between 230 V and 240 V. However, if an item of equipment was designed to operate at less than 240 V, this could cause it to operate at a reduced efficiency, use more power or operate at a higher temperature and subsequently reduce operational life.

Key terms

Single-phase – supplied by a 2 core (1 × line plus 1 × neutral) cable at 230 V.

Three-phase – 3 × line plus neutral, normally at 400 V.

Generally, domestic installations in the UK will be provided with a **single-phase** supply, the most modern ones rated at 80 A where there is no electric heating, or 100 A where there is. This will normally be more than adequate to meet the small power needs of a domestic consumer or small commercial user. Where commercial or industrial installations require a larger capacity, a **three-phase** supply is usually provided.

BS 7671, Chapter 31

With the arrangements of current-carrying conductors under normal operating conditions taken into account in Chapter 31 of BS 7671, the most common systems are shown in Table 7.5.

Three-phase & neutral	400 V	4 wire	Commercial/industrial installations
Single-phase & neutral	230 V	2 wire	Domestic/light commercial installations
Three-phase	400 V	3 wire	Motors

Table 7.5: Most common systems

The purpose of earthing and bonding

In electrical supply terms, we reference potential against the potential of the earth. In terms of our standard UK domestic supply, we talk about it having a potential of 230 V. By this we mean the difference in potential between the supply (230 V) and the general mass of our planet earth, which is taken as being at zero potential (0 V) – we need there to be a difference in potential for there to be current.

The earth is an important part of the UK supply system, as we make a connection between the two at the supply transformer, where the neutral conductor is connected to the star point of the transformer, which is in turn connected to earth using an earth electrode or the metal sheath and armouring of a buried cable. This effectively removes the chance of a **'floating' neutral** and locks it to zero potential – in other words, 0 V.

> **Key term**
>
> **'Floating' neutral** – where the star point of the transformer is not connected to the general mass of earth.

Using a 230 V domestic situation as a reference point, there is a three wire supply system where: one conductor (line) is at a 'high' potential and 'supplies' the property; one conductor (neutral) is maintained at zero potential because of its connection to the star point of the transformer, where it acts as a 'return' conductor and gives the potential difference (230 V – 0 V); and the third conductor is directly connected to earth and not normally intended to carry current.

If someone touches a live conductor, they complete the electrical circuit and a current flows through them, the floor and the actual earth, back to the supply transformer via one or more earth connections of the transformer neutral. If a fault occurs, the current will always head for earth, taking the easiest route it can find to get there. It is this difference in potential (230 down to 0) that can be deadly.

The purpose of earthing is to minimise the risk of electric shock if someone touches a metal part that is not intended to be live, but which has become live due to a fault. For example, if there is a fault with a washing machine and the case becomes live, the circuit protective conductor (cpc) provides a path for that fault current to flow safely to earth.

Effectively, this provides a path for the fault current, which can be detected and interrupted by protective devices such as fuses or circuit breakers – the earth fault loop.

Figure 7.44 shows the earth fault loop, which, starting from the point of the fault comprises:

- circuit protective conductor (CPC)
- consumer's main earth terminal (MET) and earth conductor
- return path (may be via electrodes or the cable armouring/sheath)
- earthed neutral of the supply transformer and then the transformer winding (phase)
- line conductor.

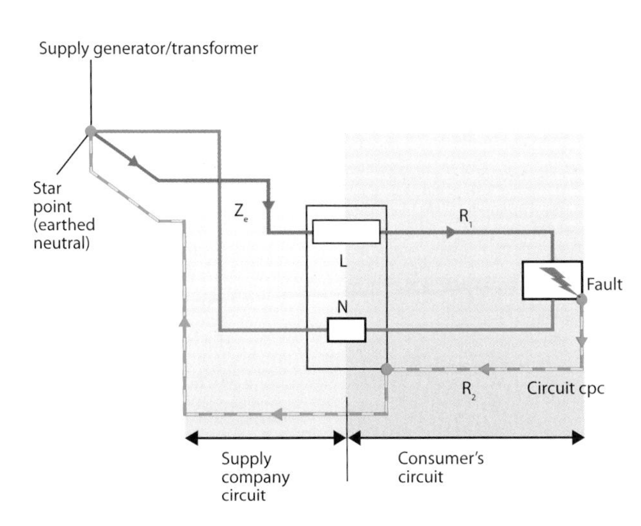

Figure 7.44: Earth fault loop

In the event of a fault, you want the protective device to operate as quickly as possible. Its ability to do this will be affected by the size of the fault current, which is affected by the value of the combined resistances within the earth fault loop – so this must be as low as possible.

If you were to connect all such metalwork together and then connect it to earth using the third conductor, it would all be at 0 V, and this is the principle behind bonding.

Bonding (correctly known as equipotential bonding) aims to limit the size of the fault by generally equalising the potential between separate conductive parts. As an example, let's say the previously mentioned fault on the washing machine came into play while the user was resting one hand on the case and the other on a nearby metal tap fitted to a metal sink.

Although the fault will go to earth via the cpc and trip the protective device, if all metalwork has been connected together it will all rise to a potential of 230 V. However, as there is no potential difference (because we have an equal potential with all things at 230 V) then no dangerous 'shock' current can exist.

Temporary continuity bonding

In the event of a fault, you need a low earth fault loop impedance value to operate the protective device. To stop dangerous potentials in the event of an electrical fault, all the conductive extraneous metalwork is connected together as shown in Figure 7.45.

However, this means that should any maintenance or repair work require the removal of any metal pipework, even for the shortest period of time, then it is essential that the electrical continuity be maintained.

You achieve this by 'bridging' any gap in the pipework with a temporary bonding conductor held in place with the correct earthing clamps. These are designed to indicate the importance of the connection and show that it ensures a safe electrical connection. Please note that this 'bridging' connection must be made **before** removal of any pipework.

High-risk rooms in dwellings

BS 7671 Part 7

Supplementary equipotential bonding may be found in locations containing a bath or shower (see Part 7 of BS 7671).

A specific example would be within a bathroom, when a fault develops on an electrically heated towel rail. If correctly earthed, this in conjunction with its fuse or circuit breaker will limit the duration of that shock risk by causing the fuse/circuit breaker to trip. However, without bonding, the fault may result in the towel rail being at 230 V for a time, while taps on the sink next to it may be offering a path to earth (0 V) via the water pipework. This would be very dangerous if someone was touching both the rail and tap – there would be a potential difference of 230 V between the rail and earth via the tap and the person – and could cause severe injury or even death.

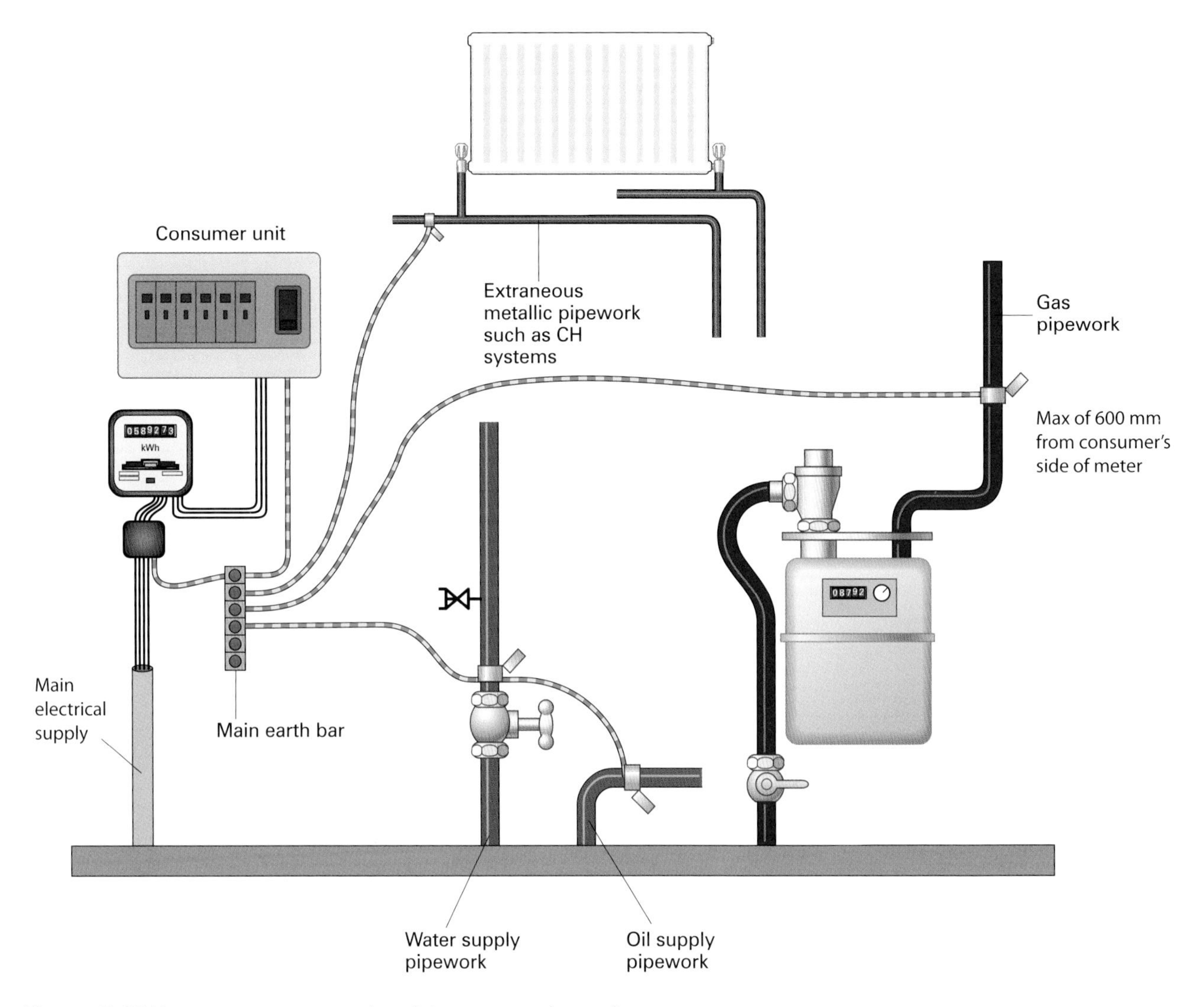

Figure 7.45 How extraneous metalwork is connected together

However, if the pipework feeding both hot and cold taps is bonded together with that of the earth of any electrical circuits supplying the room, the fault on the towel rail will try to bring the towel rail, the taps and any other touchable metalwork to 230 V. Anyone touching the rail and the tap at the same time is exposed to a potential difference of zero volts.

BS 7671 Section 701 says the following about rooms with a bath or shower.

- All circuits supplying the bathroom (irrespective of the points they are serving) have to be protected by 30 mA RCDs.
- Supplementary equipotential bonding is not required when all of the following are met:
  - All circuits for the bathroom meet the requirements for automatic disconnection of supply (ADS).
  - All final circuits have 30 mA RCD protection.
  - All extraneous-conductive-parts of the bathroom are connected to the protective equipotential bonding of the installation (411.3.1.2).

Key term

ADS – a protective measure that provides both basic protection and fault protection.

BC 7671 Section 701

- **SELV** socket outlets and shaver sockets are permitted outside Zone 1. Socket outlets of 230 V are permitted provided they are more than 3 m horizontally from Zone 1.

Key term

SELV – Safety extra-low voltage.

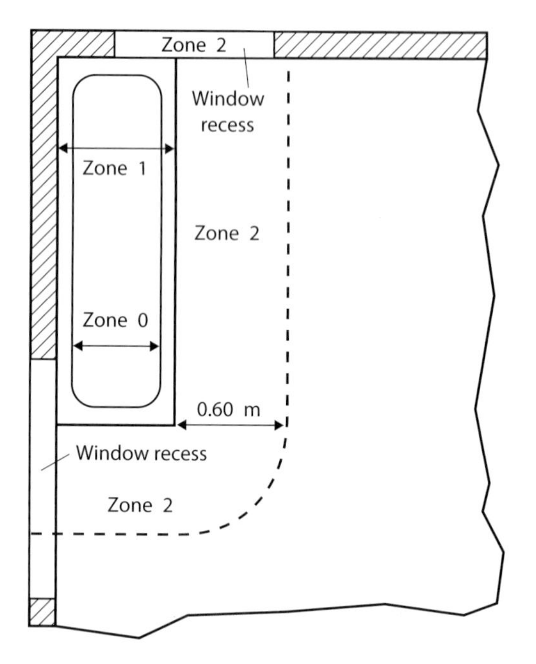

(a) Bath tub

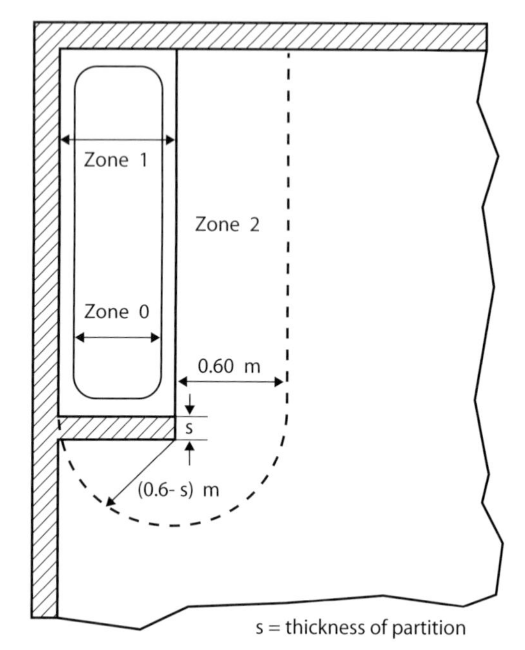

(b) Bath tub with permanent fixed partition

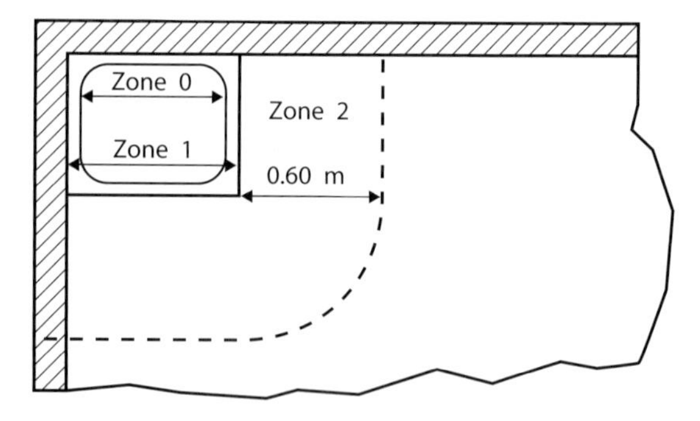

(c) Shower basin

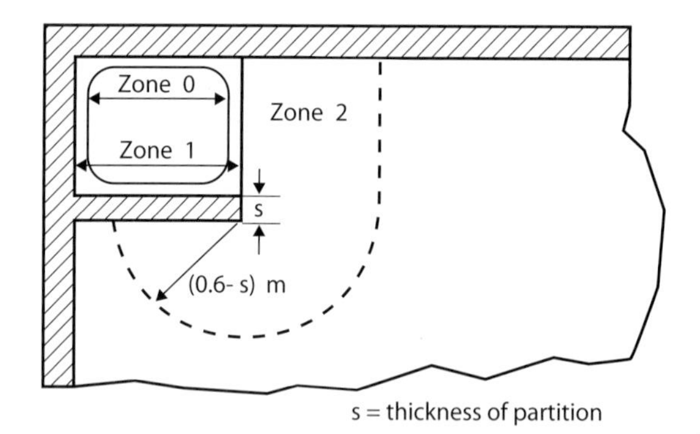

(d) Shower basin with permanent fixed partition

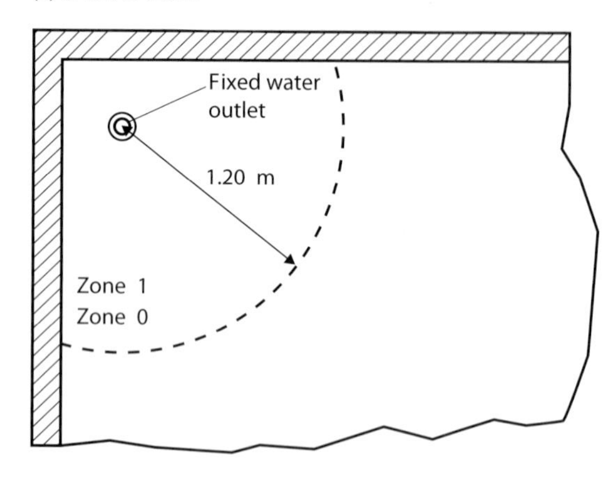

(e) Shower without basin

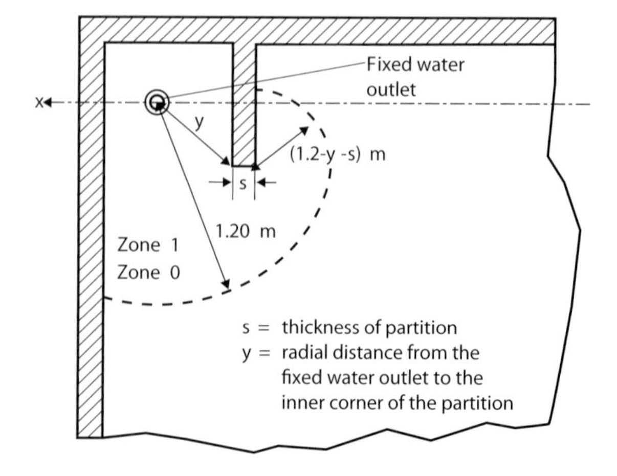

Figure 7.46: Zone arrangements in plan view

Figure 7.46 shows the zone arrangements and dimensions, in plan view.

Figure 7.47 gives the same information, but in elevation view. Note that * denotes a Zone 1 if the space is accessible without the use of a tool. Spaces under the bath that are only accessible with the use of a tool are outside the zones.

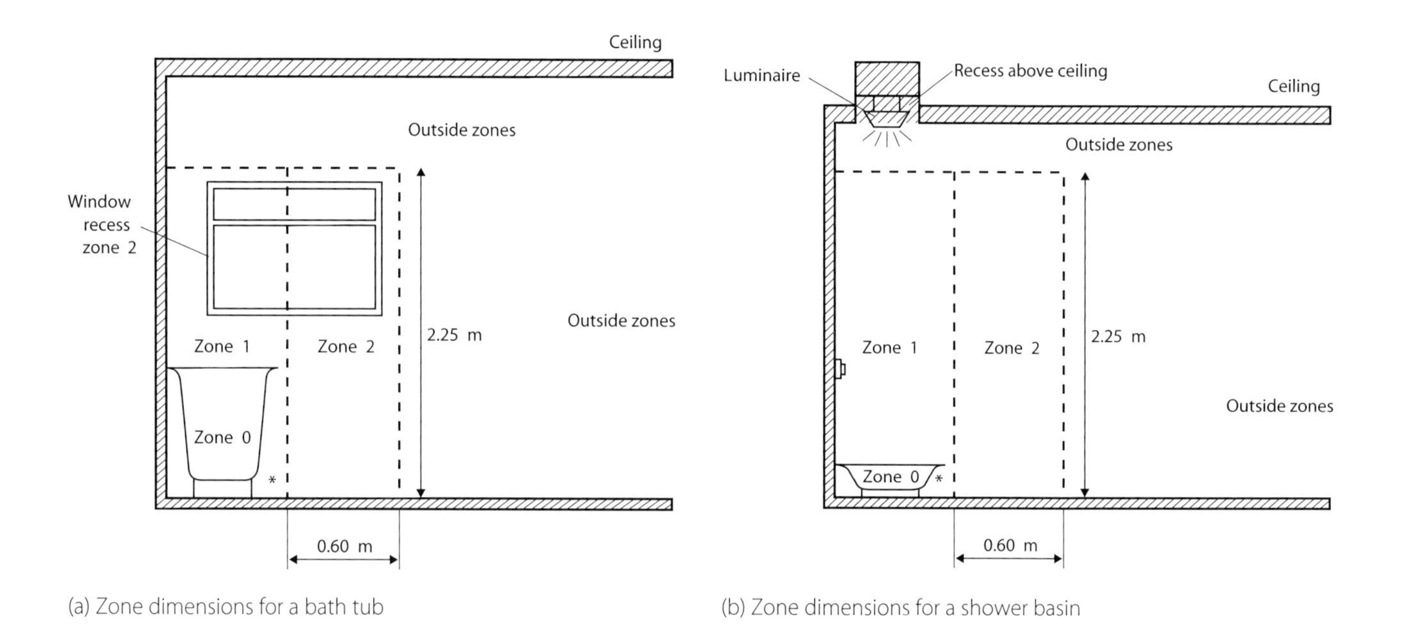

(a) Zone dimensions for a bath tub

(b) Zone dimensions for a shower basin

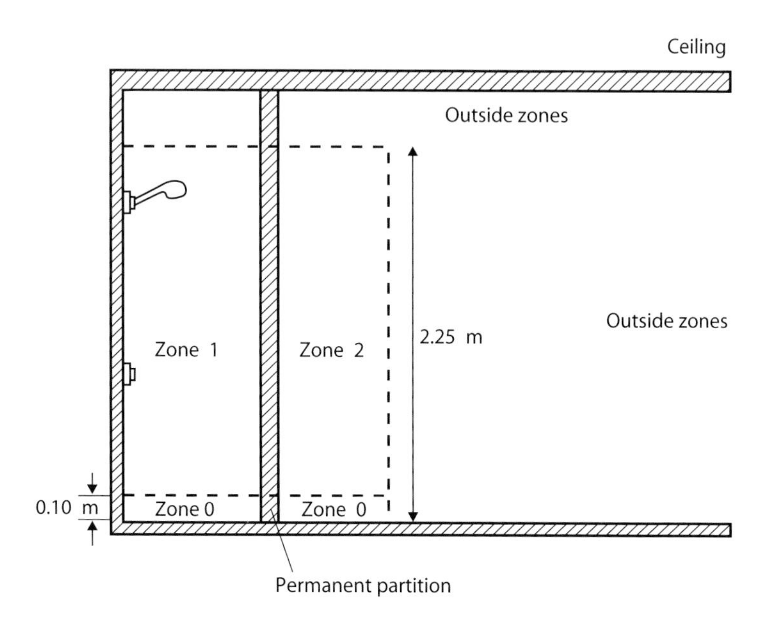

(c) Zone dimensions for a shower basin with no permanent fixed partition

Figure 7.47: Zone arrangements in elevation view

Warning notices

Labels should be applied next to every fuse or circuit breaker indicating the size and type of fuse or the nominal current rating of the circuit breaker and details of the circuit they protect. Other notices and labels required by the Regulations are as follows, where the Regulations also state minimum physical sizes for same.

a) At the origin of every installation as shown below:

IMPORTANT
This installation should be periodically inspected and tested and a report on its condition obtained as prescribed in BS 7671 (formally the IET Wiring Regulations for Electrical Installations) published by the Institute of Electrical Engineers.

Date of last inspection : ....

Recommended date of next inspection : ....

b) Where different voltages are present:

- in equipment or enclosures within which a voltage exceeding 250 volts exists but would not be expected
- terminals between which a voltage exceeding 250 volts exists which, although contained in separate enclosures, are within arm's reach of the same person
- the means of access to all live parts of switchgear or other live parts where different nominal voltages exist.

c) At earthing and bonding connections, a label as shown below to BS.951 must be permanently fixed in a visible position at or near:

- the point of connection of every earthing conductor to an earth electrode
- the point of connection of every bonding conductor to an extraneous-conductive-part
- the main earth terminal of the installation where separate from the main switchgear.

Safety Electrical Connection – Do Not Remove

d) Where RCDs are fitted within an installation, a suitable permanent durable notice as shown below must be permanently fixed in a prominent position at or near the main distribution board (DB).

Important

This installation, or part of it, is protected by a device that automatically switches off the supply if an earth fault develops. Test quarterly by pressing the button marked 'T' or 'Test'. The device should switch off the supply and should then be switched on to restore the supply. If the device does not switch off the supply when the button is pressed, seek expert advice.

e) If alterations/additions are made such that an installation contains both the current harmonised wiring colours and the old colours of previous regulations, then the following sign should be fixed to the distribution board.

CAUTION

This installation has wiring colours to two versions of BS 7671.
Great care should be taken before undertaking extension, alteration or repair that all conductors are correctly identified.

f) All caravans and motor caravans should have the following durable, permanently fixed label near the main switch giving instructions on the connection and disconnection of the caravan installation to the electricity supply.

INSTRUCTIONS FOR ELECTRICITY SUPPLY

TO CONNECT

1. Before connecting the caravan installation to the mains supply, check that:
   a) the supply available at the caravan pitch supply point is suitable for the caravan electrical installation and appliances, and
   b) the voltage and frequency and current ratings are suitable, and
   c) the caravan main switch is in the OFF position.

   Also, prior to use, examine the supply flexible cable to ensure there is no visible damage or deterioration.
2. Open the cover to the appliance inlet provided at the caravan supply point, if any, and insert the connector of the supply flexible cable.
3. Raise the cover of the electricity outlet provided on the pitch supply point and insert the plug of the supply cable.

   THE CARAVAN SUPPLY FLEXIBLE CABLE MUST BE FULLY UNCOILED TO AVOID DAMAGE BY OVERHEATING
4. Switch on at the caravan main isolating switch.

   IN CASE OF DOUBT, OR IF CARRYING OUT THE ABOVE PROCEDURE THE SUPPLY DOES NOT BECOME AVAILABLE, OR IF THE SUPPLY FAILS, CONSULT THE CARAVAN PARK OPERATOR OR THE OPERATOR AGENT OR A QUALIFIED ELECTRICIAN
5. Check the operation of residual current devices (RCDs) fitted in the caravan by depressing the test button(s) and reset.

TO DISCONNECT

6. Switch off at the caravan main isolating switch, unplug the cable first from the caravan pitch supply point and then from the caravan inlet connector.

PERIODIC INSPECTION

Preferably, not less than once every three years and annually if the caravan is used frequently, the caravan electrical installation and supply cable should be inspected and tested and a report on their condition obtained as prescribed in BS 7671 Requirements for Electrical Installation published by the Institution of Engineering and Technology and BSI.

PROGRESS CHECK

1. In a TN-S system, what does the letter 'S' denote?
2. What is the purpose of equipotential bonding?
3. Where maintenance or repair work requires the removal of any metal pipework, temporary continuity bonding is required. Should this be applied before or after the removal of any pipework?
4. When an RCD is fitted within an installation, a suitable permanent durable notice should be fitted how and where?
5. Which part of BS 7671 relates to special installations or locations?

6. Operating principles of electrical circuit protection devices

Types of protection

BS 7671 Part 4

BS 7671 Part 4 gives guidance on what we are offering protection against:

- Chapter 41 – Protection against electric shock
- Chapter 42 – Protection against thermal effects
- Chapter 43 – Protection against overcurrent
- Chapter 44 – Protection against voltage disturbance and electromagnetic disturbance.

Protection against electric shock

BS EN.61140

BS EN.61140, a basic safety standard that applies to the protection of people and livestock, says that:

- hazardous live parts must not be accessible
- accessible conductive parts must not be hazardous-live, without a fault or in single-fault conditions.

There also needs to be some additional means of protection in the event of a fault. So you must provide the following two levels of protection for persons and livestock:

- basic protection – protection when in normal use, without a fault, such as insulating live parts; and
- fault protection – protection under fault conditions (BS 7671 Chapter 41).

BS 7671 Chapter 41

To provide this, British Standards say a protective measure must consist of:

- an appropriate combination of provision for basic protection and an independent provision for fault protection **or**
- an enhanced protective provision (such as reinforced insulation) that provides both basic and fault protection.

BS 7671 Regulation 410.3.2

It is important that you understand some of the key terms used here.

- **Exposed-conductive-parts** are the conductive parts of an installation that can be touched and, although not normally live, could become live under fault conditions. Examples are metal casings of appliances such as boilers or wiring enclosures such as steel conduit.
- **Extraneous-conductive-parts** are the conductive parts within a building that do not form part of the electrical installation, but could become live under fault conditions. Examples are copper water pipes and metal ductwork.
- **Live parts** are conductors or conductive parts that are meant to be live in normal use, such as a neutral conductor.

Regulation 410.3.3 says that in each part of an installation, bearing in mind the external influences, one or more protective measures must be applied, which are generally:

BS 7671 Regulation 410.3.3

- automatic disconnection of supply (ADS)
- double or reinforced insulation
- electrical separation from one item of current using equipment
- extra low voltage (SELV and **PELV**).

Key term

PELV – Protected extra-low voltage.

In electrical installations ADS will probably be the most commonly used protective measure as it provides both basic and fault protection.

The regulations give the requirements for basic and fault protection for each of the protective measures. However, the following information is given as a guide, where the particular section or regulation within BS 7671 is shown in brackets.

Automatic disconnection of supply (ADS) (411)

Automatic disconnection of supply (ADS) is a protective measure that provides both basic and fault protection. Basic protection is given by basic insulation of live parts (416.1), barriers and enclosures (416.2) or, where appropriate, by obstacles (417.2)/ placing out of reach (417.3). Fault protection is given by protective earthing (411.3.1.1), protective equipotential bonding (411.3.1.2) and by automatically disconnecting the supply in the event of a fault.

Be aware that protection by obstacles and placing out of reach is only providing basic protection. Consequently they are also only to be used in installations that are under the control or supervision of a skilled person (417.1 and 410.3.5).

In terms of fault protection, 'protective earthing' is where exposed-conductive-parts are connected to the protective earthing terminal by means of the circuit protective conductors (cpcs). Protective equipotential bonding is then the interconnection of extraneous-conductive-parts that are then connected to the protective earth terminal by means of protective bonding conductors. These conductors and their connection points are shown in Figure 7.48.

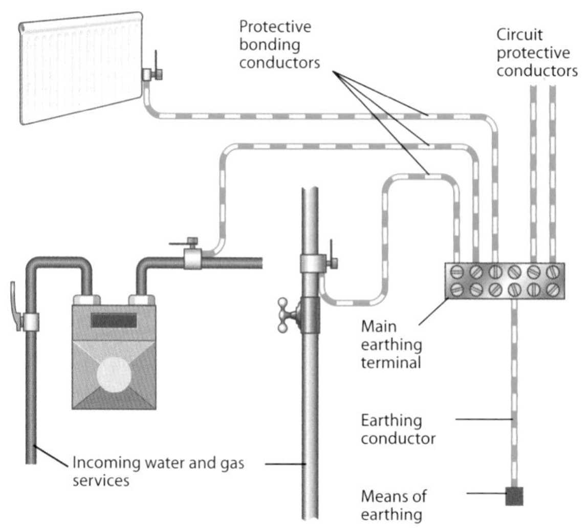

Figure 7.48: Protective bonding conductors and their connection points

So that ADS can automatically disconnect the supply, there needs to be a large enough earth fault current to operate the protective device (such as a fuse, circuit breaker or RCD) within the times specified in Table 41.1 of BS 7671.

However, you must provide 'additional protection' (411.3.3) on a.c. systems for socket outlets with a rated current not exceeding 20 A for use by ordinary persons and for general use. This additional protection is afforded by the use of a 30 mA RCD that must operate within 40 ms (415.1.1). The same applies to mobile equipment with a current rating not exceeding 32 A for use outdoors.

However, there are exceptions to the socket requirements:

- where socket outlets are to be used under the supervision of a skilled or instructed person
- where socket outlets are labelled or identified as being only for use with a particular piece of equipment, such as a fridge freezer
- where minor works associated with existing socket-outlet circuits are not provided with additional protection by means of an RCD, where the designer is satisfied that there would be no increased risk from the installation of the addition or alteration. The decision here must be recorded under Part 2 of the Minor Works certificate or the comments section of the Electrical Installation Certificate.

Additional protection (415)

In addition to the four protective measures mentioned, additional protection may be specified, particularly under certain external influences and in some special locations. Two methods are given:

- residual current devices (RCD) operating at 30 mA within 40 ms (415.1.1)
- supplementary equipotential bonding (415.2).

Note: an RCD is not recognised as the sole means of protection so does not remove the need to apply one of the four protective measures (411 to 414). However, supplementary equipotential bonding is considered as an addition to fault protection.

The various protective measures are summarised in Table 7.6.

Residual current devices (RCD)

Residual current devices (RCDs) are a group of devices that provide extra protection by reducing the risk of electric shock. Although RCDs operate on small currents, there are circumstances where the combination of operating current and high earth fault loop impedance could result in the earthed metalwork rising to a dangerously high potential.

The Regulations draw attention to the fact that, if the product of operating current (A) and earth fault loop impedance (W) exceeds 50 V, the potential of the earthed metalwork will be more than 50 V above earth potential, and hence dangerous. An RCD monitors the

Protective measure							
Automatic disconnection of supply (411)		Double or reinforced insulation (412)		Electrical separation (413)		Extra low voltage (414)	
Basic	**Fault**	**Basic**	**Fault**	**Basic**	**Fault**	**Basic**	**Fault**
Gives basic protection by basic insulation of live parts or by barriers/ enclosures	Gives fault protection by protective earthing, protective equipotential bonding and ADS in case of fault	Gives basic protection by basic insulation	Gives fault protection by supplementary insulation	Provides basic protection by one or two methods in 416, i.e. basic insulation of live parts of barriers/ enclosures	Provides fault protection by simple separation of the separated circuit from other circuits and earth		
There is an additional protection requirement by use of an RCD with socket outlets not exceeding 20 A		Gives basic and fault protection by reinforced insulation between live and accessible parts				Gives basic and fault protection by limiting voltage to the upper limit of Band 1 using a supply such as an isolating transformer with circuits in accordance with Regulation 411	

Table 7.6: Protective measures

current flowing in a circuit using a small transformer, specially designed to detect earth fault currents via a winding taken to a trip mechanism.

Figure 7.49 shows how all live conductors pass through the transformer. The currents flowing in the live conductors of a healthy circuit will balance, so no current will be induced in the **toroid**.

Key term

Toroid – a form of ring-shaped solenoid.

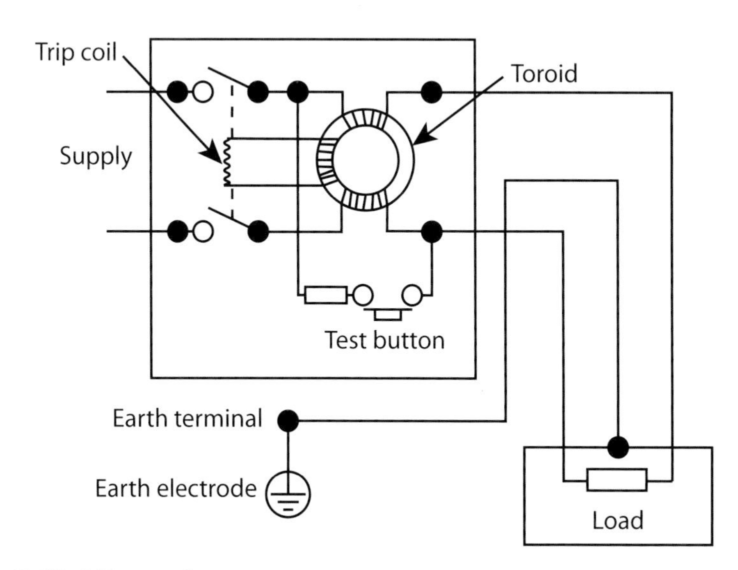

Figure 7.49: RCD transformer

Live conductors of a circuit include all line and neutral conductors. Under normal safe working conditions, the current flowing in the line conductor into the load will be the same as that returning via the neutral conductor from that same load. When this happens, the line and neutral conductors produce equal and opposing magnetic fluxes in the transformer, resulting in no voltage being induced into the trip coil.

However, if more current flows in the phase side than the neutral side (or less returns via the neutral because the current has 'leaked' to earth within the installation), an out-of-balance flux will be produced in the transformer, which will induce a voltage into the fault detector coil. This earth fault path could be through a person in contact with live parts or where basic insulation has failed at an exposed conductive part. The fault detector coil then energises the trip coil, which in turn opens the double-pole (DP) switch, due to the residual current produced by the induced voltage in the trip coil.

You should carry out functional testing of an RCD by operating the test button at regular intervals; this simulates a fault by creating an imbalance in the detection coil. Initial and periodic inspection and testing procedures also require the RCD to be tested using the appropriate instrument. RCDs are available for single- and three-phase applications.

Practical application of RCDs (Regulation 531.2)

- An RCD must disconnect all line conductors of the circuit at substantially the same time.
- The residual operating current of the RCD must comply with Regulation 411 relative to the type of earthing system.
- The RCD must be selected and circuits subdivided so that any normally occurring earth leakage (such as with some computers) does not cause nuisance tripping of the RCD.
- The use of a 30 mA RCD with a circuit having a protective conductor is not sufficient to afford fault protection.
- An RCD must be positioned so that its operation is not affected by the magnetic fields of other equipment.
- Where an RCD is used for fault protection with, but separately from, an overcurrent device, then it must be capable of withstanding (without damage) any stresses likely to be encountered in the event of a fault on the load side of the point at which it is installed.
- RCDs must be selected and installed so that, if two or more RCDs are connected in series, and discrimination is required to avoid danger, that discrimination is achieved.
- Where an RCD can be operated by someone other than a skilled or instructed person, it must be selected and installed so that its settings cannot be altered without using a tool and the result of any alteration will be clearly visible.

Sensitivity rated residual operating current	Level of protection	Applications
10 mA	Personnel	High risk areas: schools, colleges, workshops, laboratories. Areas where liquid spillage may occur
30 mA	Personnel	Domestic, commercial and industrial
100 mA	Personnel, fire	Only limited personnel protection Excellent fire protection
300 mA	Fire	Commercial and industrial

Table 7.7: Sensitivity rated residual operating current; level of protection; applications

Disadvantages of MCBs	Advantages of MCBs
• They have mechanical moving parts • They are expensive • They must be regularly tested • Ambient temperature can change performance	• They have factory-set operating characteristics which cannot be altered • They will maintain transient overloads and trip on sustained overloads • Easily identified when they have tripped • The supply can be quickly restored

Table 7.8: Disadvantages and advantages of MCBs

The RCBO

The RCBO is a new type of MCB that incorporates earth fault protection. Generally a combination of a thermal-magnetic Type B MCB and integrated RCD, it enables both overcurrent protection and earth fault current protection to be provided by a single unit.

This allows earth fault protection to be restricted to a single circuit, so only the circuit with the fault is interrupted. Often a distribution board will be protected by an RCD, in which case all circuits would be disconnected in the event of a fault.

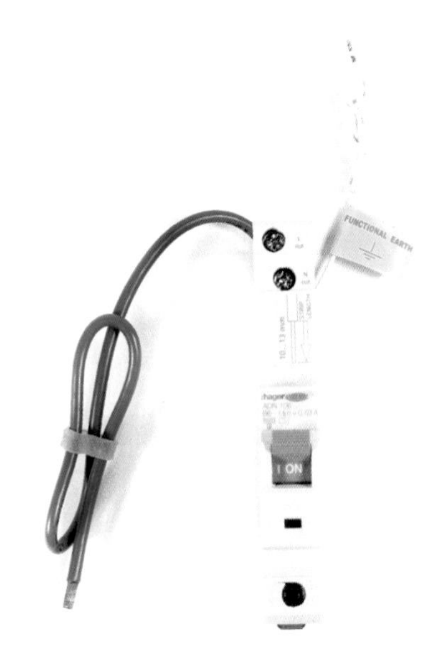

An RCBO is a type of MCB that incorporates earth fault protection

Did you know?

RCBO stands for residual current operated circuit breaker with integral overload protection.

Note on basic protection

'Basic protection' means providing protection against electric shock under fault-free conditions. You would do this by preventing people or livestock from touching any live parts by insulating them or preventing access to them. The protective measures previously described and summarised in Table 7.6 on page 393 are applicable for offering basic protection.

Protection against thermal effects

Electrical fires are common. The cause can be:

- heat radiation, hot components or equipment
- failure of equipment such as protective devices, thermostats, seals and wiring systems
- overcurrent
- insulation faults
- arcs and sparks
- loose connections
- external influences such as lightning surge.

BS 7671 Chapter 42

BS 7671 Chapter 42 requires us to provide protection for people and livestock against:

- the effects of heat or thermal radiation from electrical equipment
- the ignition, combustion or degradation of materials
- fire hazard spread from an electrical installation
- safety services being cut off by failure of electrical equipment.

Fire caused by electrical equipment

- Comply with manufacturer's instructions.
- Select and erect equipment so that its normal temperature cannot cause a fire.
- If the equipment's surface temperature could cause a fire, mount it inside something that can withstand the temperature.
- Place any equipment that causes a focusing of heat (such as an electric fire) far enough away from any object (such as curtains) to avoid them reaching a dangerous temperature.
- Take precautions wherever there are flammable liquids.
- Make sure any termination, connection or joint of a live conductor is contained within an enclosure in accordance with Regulation 526.5.

BS 7671 Regulation 526.5

Protection against overcurrent

BS 7671 Regulation 430.3

Regulation 430.3 requires a protective device to be provided to break any **overcurrent** in a circuit conductor before it could cause danger due to thermal or mechanical effects.

An overcurrent happens when an otherwise healthy circuit is carrying more current than it was designed to. This may be as a result of faulty equipment such as a stalled motor or it may occur when many pieces of equipment are added to the circuit, such as when several 13 A plugs are connected to an adaptor and then plugged into a 13 A socket outlet.

Fault current can arise from either a short circuit or an earth fault.

- A short circuit happens when there is a fault of negligible impedance between live conductors.
- An earth fault current is a fault of negligible impedance between a line conductor and an exposed-conductive-part or protective conductor.

Regulation 431 requires detection of overcurrent to be provided for all line conductors, causing disconnection of the conductor in which the overcurrent is detected, although not necessarily in other line conductors (unless disconnecting one line conductor could cause danger). One example is with a magnetic crane in a foundry, where you would not wish it to stop being a magnet while carrying heavy loads at height!

This regulation also says that, as the neutral conductor in standard single- and three-phase circuits will normally have the same cross-sectional area (csa) as the line conductors, it is not necessary to provide overcurrent detection and an associated disconnecting device. Where there is a reduced neutral, overcurrent detection will be required in line with the cable csa.

Protective devices for overcurrent should then be installed at the point where a reduction occurs in the current-carrying capacity of the circuit conductors caused by a change in csa, or perhaps method of installation.

Here you will look at protective devices designed to disconnect the supply automatically.

Key terms

Overcurrent – a current in excess of the rated value or the current-carrying capacity of a conductor, resulting from an **overload** or a fault current.

Overload – an overcurrent occurring within a circuit that is electrically sound.

Fuses

BS 3036 rewireable fuses

Early rewireable fuses had a low short-circuit capacity so were dangerous under fault conditions: the melting fuse element would 'splash' melted copper around and could cause fires. Later rewireable fuses incorporated an asbestos pad to protect the fuse holder, reducing the risk of fire.

A rewireable fuse consists of a fuse and fuse element, plus a holder and fuse carrier both made of porcelain or Bakelite. These fuses have a colour-code on the fuse holder, indicating the rating of the circuits they are designed for.

Rewireable fuse

5 A	White
15 A	Blue
20 A	Yellow
30 A	Red
45 A	Green

Table 7.9: Rewireable fuse colouring and coding

This once-popular type of fuse is still available, but is not normally used because of the disadvantages shown in Table 7.10.

Disadvantages of rewireable fuses	Advantages of rewireable fuses
• Easily abused when the wrong size of fuse wire is fitted • Fusing factor of around 1.8–2.0 means they cannot be guaranteed to operate up to twice the rated current is flowing. As a result cables protected by them must have a larger current-carrying capacity • Precise conditions for operation cannot be easily predicted • Do not cope well with high short circuit currents • Fuse wire can deteriorate over time • Danger from hot metal scattering if the fuse carrier is inserted into the base where the circuit is faulty	• Low initial cost • Can easily see when the fuse has blown • Low element replacement cost • No mechanical moving parts • Easy storage of spare fuse wire

Table 7.10: Disadvantages and advantages of rewireable fuses

Cartridge fuses have 3A and 13A fuse ratings

BS 1361 (to be discontinued)/1362 cartridge fuses

Here the cartridge fuse consists of a porcelain tube with metal end caps to which the element is attached and the tube is then filled with granulated silica. The BS 1362 fuse is generally found in domestic plug tops used with 13 A BS 1363 domestic socket outlets.

There are two common fuse ratings available:

- 3 A – for use with appliances up to 720 watts (such as radios, table lamps, TVs, domestic central heating controls, boilers)
- 13 A – for use with appliances rated over 720 watts (such as irons, kettles and fan heaters).

Other sizes are available –1, 5, 7 and 10 A – though these are hard to find. The physically larger BS 1361 fuse can be found in distribution boards and at main intake positions.

Disadvantages of cartridge fuses	Advantages of cartridge fuses
• They are more expensive to replace than rewireable fuses. • They can be replaced with an incorrect size fuse (plug top type only). • The cartridge can be shorted out with wire. • It is not possible to see if the fuse has blown. • They require a stock of spare fuses to be kept.	• They have no mechanical moving parts. • The declared rating is accurate. • The element doesn't weaken with age. • They have a small physical size and no external arcing, which permits their use in plug tops and small fuse carriers. • They have a low fusing factor – around 1.6–1.8. • They are easy to replace.

Table 7.11: Disadvantages and advantages of cartridge fuses

BS 88 high breaking-capacity (HBC) fuses

An HBC fuse protects motor circuits and industrial installations

The HBC fuse (sometimes called an HRC or high rupturing-capacity fuse) is a more sophisticated variation of the cartridge fuse and is normally found protecting motor circuits and industrial installations. It consists of a porcelain body filled with silica, a silver element and lug-type end caps. Another feature is the indicating bead, which shows when the fuse element has blown.

It is quite fast-acting and can discriminate between a starting surge and an overload. These types of fuse would be used when an abnormally high prospective short circuit current exists.

Disadvantages of BS 88 fuses	Advantages of BS 88 fuses
• These are very expensive to replace. • Stocks of these spares are costly and take up space. • Care must be taken when replacing them to ensure that the replacement fuse has the same rating and also the same characteristics as the fuse being replaced.	• They have no mechanical moving parts. • The element doesn't weaken with age. • Operation is very rapid under fault conditions. • It is difficult to interchange the cartridge, since different ratings are made to different physical sizes.

Table 7.12: Disadvantages and advantages of BS 88 fuses

Miniature circuit breakers (MCBs)

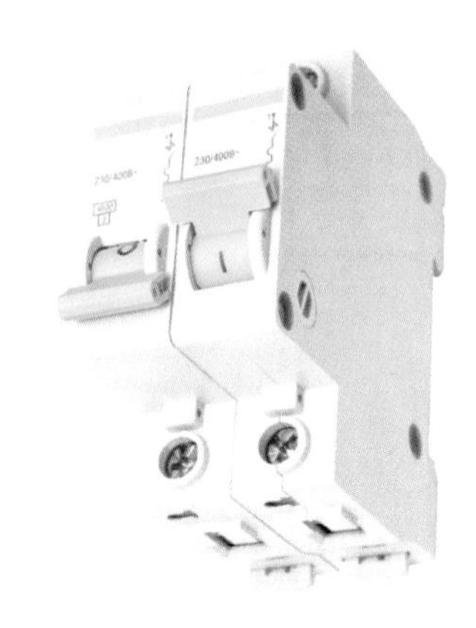

An MCB forms an essential part of most installations

In essence, a circuit breaker is a switch with contacts that automatically open to break an electrical circuit when a fault occurs, thus protecting the attached equipment and wiring.

Often they are a more flexible alternative to fuses, as they can be easily reset and do not need replacement if an electrical overload or fault occurs (usually overloads and short circuits).

Circuit breaker contacts must carry the load current without excessive heating. However, once a fault is detected, the contacts inside the circuit breaker must open to break the circuit. When a current is interrupted, an arc is generated; this arc must be contained, cooled and extinguished. Different circuit breakers use a vacuum, air, insulating gas or oil as the medium in which the arc forms, with MCBs traditionally using arc division and air to extinguish the arc. Once the fault condition has been cleared, the contacts can again be closed to restore power to the interrupted circuit.

The modern MCB now forms an essential part of most installations at the final distribution level. There are several different types, including thermal tripping and thermal-magnetic tripping.

Thermal tripping

A simple thermal trip-operating breaker incorporates a bimetallic strip, through which passes the current of the circuit to be protected. The current heats the strip, which bends by an amount commensurate with the current. However, under overload conditions the bimetallic strip bends further than usual and activates the breaker's trip mechanism.

Thermal breakers are simple and inexpensive, but they provide only limited protection against short circuits. They should only be used where overload protection is the most important requirement and where short circuit currents will not exceed 1000 A.

Thermal-magnetic (combined) tripping

More versatile and more frequently used, thermal-magnetic MCBs have a similar thermal trip mechanism to that just described, which provides effective protection against small and moderate overloads. However, they also have an electromagnetic mechanism that provides virtually instantaneous tripping for large overloads and short circuits.

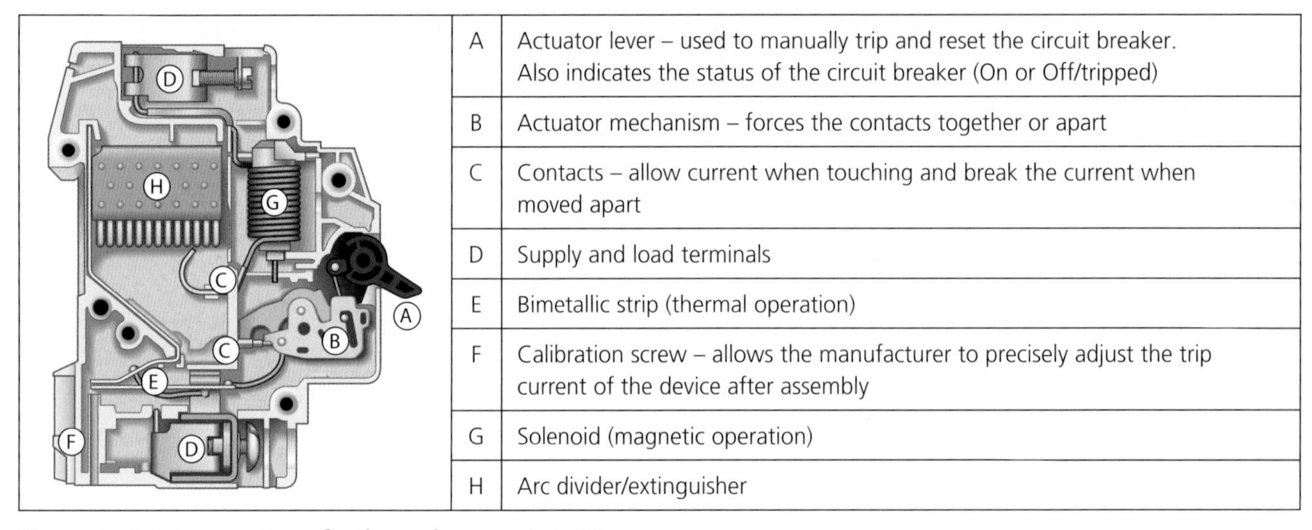

A	Actuator lever – used to manually trip and reset the circuit breaker. Also indicates the status of the circuit breaker (On or Off/tripped)
B	Actuator mechanism – forces the contacts together or apart
C	Contacts – allow current when touching and break the current when moved apart
D	Supply and load terminals
E	Bimetallic strip (thermal operation)
F	Calibration screw – allows the manufacturer to precisely adjust the trip current of the device after assembly
G	Solenoid (magnetic operation)
H	Arc divider/extinguisher

Figure 7.50: Construction of a thermal-magnetic MCB

This type of MCB can be designed to provide high breaking capacities, and to have various types of trip characteristic. They are ideal for applications where comprehensive protection is required at a moderate price, and where high short circuit levels may be encountered.

The current rating of an MCB is the maximum current that it will carry continuously without tripping and MCBs should always be chosen so that their current rating matches, as closely as possible, the maximum load current of the circuit they are protecting.

Trip characteristics

An MCB that trips as quickly as possible under fault conditions sounds ideal and in truth this is what we want for short-circuit faults. However, for overload protection, many items of equipment such as fluorescent lights, transformers and motors, draw a high peak current for a short period when they are switched on.

An MCB that reacts instantaneously would trip every time such a peak occurred, which would make it unusable. Fortunately, the thermal element in MCBs doesn't react instantaneously as the bimetallic strip takes time to heat up, so is hardly affected by short-term current peaks.

By changing the design of the bimetallic elements, MCB manufacturers can determine what size of peak current a particular MCB will ignore, and for what length of time.

This relationship between current and tripping time is usually shown as a curve, known as the MCB trip characteristic. BS EN 60898 defines several types of standard characteristics, the most important of which are Types B, C and D. There is no Type A to avoid confusion with A for Amperes, but you may comes across older MCBs classed as Types 1, 2, 3 and 4.

BS EN 60898

MCB type	Instantaneous trip current	Application
1 B	2.7 to 4 × I_n 3 to 5 × I_n	Domestic and commercial installations having little or no switching surge
2 C 3	4.0 to 7.0 × I_n 5 to 10 × I_n 7 × 10 I_n	General use in commercial/industrial installations, where the use of fluorescent lighting, small motors, etc. can produce switching surges that would operate a Type 1 or Type B circuit breaker. Type C or Type 3 may be necessary in highly inductive circuits such as banks of fluorescent lighting.
4 D	10 to 50 × I_n 10 to 20 × I_n	Suitable for transformers, X-ray machines, industrial welding equipment, etc. where high in-rush currents may occur

Table 7.13: MCB types

Fuse characteristics

Fuse utilisation categories are used to define a fuse's particular application or characteristic and the fuse is marked accordingly. These replace the previous fuse classes of Class P, Q1 (similar to gG), Q2 and R (similar to aM).

gG	Full range breaking capability, general applications
gM	Full range breaking capability, motor circuit protection (dual rating)
aM	Partial range breaking capability, motor applications
aR/gR	Semiconductor protection, fast acting
gS	Semiconductor protection including cable overload protection

Table 7.14: UK utilisation categories for fuses

Category gG fuses can protect motor circuits and, when selected correctly, can withstand motor starting surges and full load currents without deterioration, as well as offering short circuit protection to associated motor starter components.

Category gM type fuse links have a dual rating with two current values. The first is the maximum continuous current of the fuse and associated fuse holder; the second indicates the equivalent electrical characteristic to which the fuse conforms.

These two ratings are normally separated by the letter M, which defines the application. For example, a 20M32 fuse link is intended for use in the protection of motor circuits and has a maximum continuous rating of 20 A but the electrical characteristics of a 32 A rating. This means that the associated equipment need only be rated at 20 A, providing significant economies against 32 A equipment.

Breaking capacity

On fuses you may have noticed a number followed by the letters kA stamped onto the end cap of an HBC fuse or printed onto the body of a BS 1361 fuse. This is known as the breaking capacity of fuses and circuit breakers.

When a short circuit occurs, the current may, for a fraction of a second, reach hundreds or even thousands of amperes. The effects of a short circuit current are thermal, which can cause melting of conductors and insulation, fire or the alteration of the properties of materials and mechanical whereby large magnetic fields can build up, resulting in distortion of conductors and breaking of supports and insulators.

Key term

Arcing – a plasma discharge as the result of current flowing between two terminals through a normally non-conductive medium (such as air), producing light and heat.

Each protective device must therefore be able to safely break or make such a current without damage to its surroundings by **arcing**, overheating or the scattering of hot particles.

BS 7671 Reg 434.5

Regulation 434.5 of BS 7671 requires that the breaking capacity of every fault current protective device shall be no less than the maximum prospective fault current (this includes both short circuit and earth fault conditions) at the point at which it is installed; this value of prospective fault current having been established by measurement, calculation or enquiry.

Did you know?

When making an enquiry to the electricity supplier, a design value of 16 kA could be expected for single-phase 230 V supplies up to 100 A.

Most manufacturers offer ranges of MCBs with breaking capacities of 10 kA–15 kA, extended to 25 kA for certain products. This higher breaking capacity can, in some applications, reduce costs substantially by allowing MCBs to be used where previously more costly moulded case breakers would have been required.

The breaking capacities of MCBs are indicated by an 'M' number indicating thousands of amperes: for example, M6 means that the breaking capacity is 6 kA or 6000 A. The breaking capacity will be related to the prospective short circuit current.

Time-current characteristics of overcurrent protective devices

The time-current characteristic is essentially a curve showing the operating time of a fuse in relation to the prospective fault current under set conditions.

Tables in BS 7671 Appendix 3 give time-current characteristics for overcurrent protective devices and RCDs. These take the form of graphs with a horizontal axis representing the prospective current in amps and a vertical axis representing the time in seconds. Both scales are logarithmic: the scale increases by multiples of 10.

BS 7671 Appendix 3

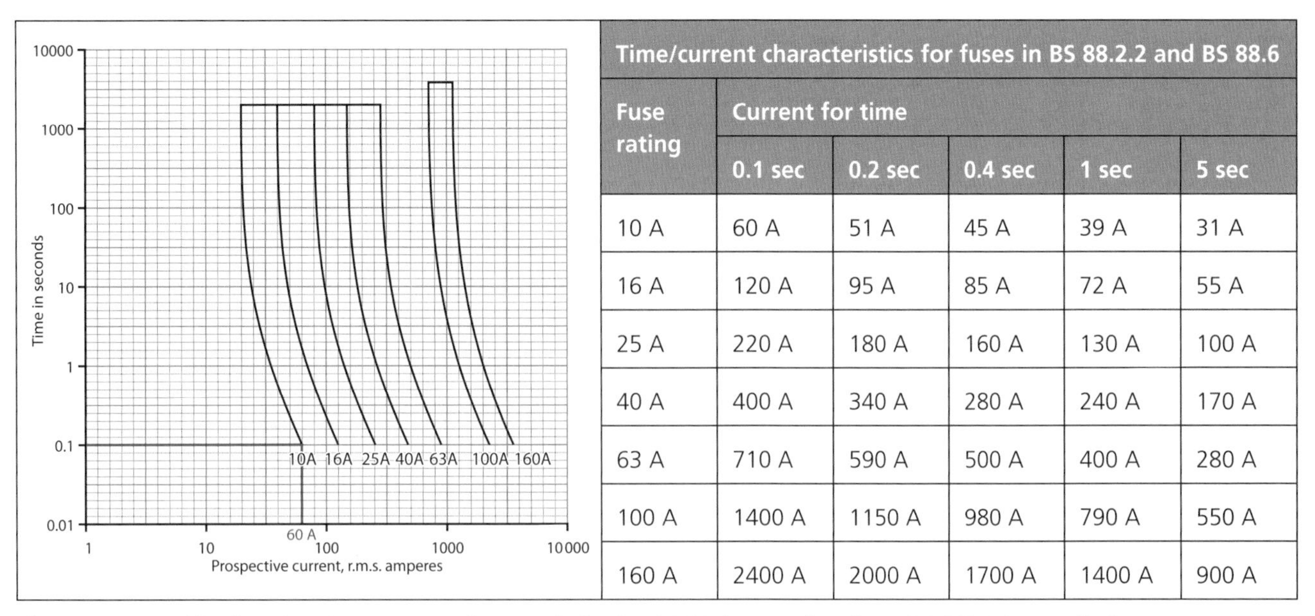

Time/current characteristics for fuses in BS 88.2.2 and BS 88.6					
Fuse rating	Current for time				
	0.1 sec	0.2 sec	0.4 sec	1 sec	5 sec
10 A	60 A	51 A	45 A	39 A	31 A
16 A	120 A	95 A	85 A	72 A	55 A
25 A	220 A	180 A	160 A	130 A	100 A
40 A	400 A	340 A	280 A	240 A	170 A
63 A	710 A	590 A	500 A	400 A	280 A
100 A	1400 A	1150 A	980 A	790 A	550 A
160 A	2400 A	2000 A	1700 A	1400 A	900 A

Figure 7.51: Table showing time-current characteristics for BS 88 fuses, taken from BS 7671 Appendix 3

In Figure 7.51, you can see that, for a 10 A rated BS 88 fuse, a fault current of 60 A will cause the device to operate in 0.1 seconds.

Protection against overload current

Every circuit must be designed so that a small overload of a long duration is unlikely to happen (see BS 7671 Regulation 433.1). If an overload occurs, the protective device operates and disconnects the circuit, protecting the cable insulation, termination or equipment from any great rise in conductor temperature.

BS 7671 Reg 433.1

To do this, the circuit has to be designed to co-ordinate the current-carrying capacity, the load current and the characteristics of the protective device. So the device must satisfy the following conditions.

(a) The rated current of the protective device (I_n) is not less than the design current (I_b) of the circuit, and

(b) the rated current of the protective device (l_n) does not exceed the lowest of the current-carrying capacities (I_Z) of any of the conductors of the circuit, and

(c) the current (I_2) causing effective operation of the protective device does not exceed 1.45 times the lowest current-carrying capacity (I_Z) of any of the conductors in the circuit.

Where:

I_b = design current for that circuit
I_Z = current-carrying capacity of the conductor
I_n = rated current or current setting of the protective device
I_2 = current ensuring effective disconnection by the protective device in the conventional time of 0.4 or 5 seconds.

Generally the formula for this is:

$$I_b \leq I_n \leq I_Z$$

However, for point (b), co-ordination will be met with the following formula:

$$I_2 \leq 1.45 \times I_Z$$

Where the protective device is either a general-purpose type (gG) fuse to BS 88, a fuse to BS 1361, a circuit breaker to BS EN 60898, a circuit breaker to BS EN 60947-2 or an RCBO to BS EN 61009-1, compliance with conditions (a) and (b) will also result in compliance with condition (c).

Where the protective device is a semi-enclosed fuse to BS 3036, compliance with condition (c) is met if the rated current (I_n) of the fuse does not exceed 0.725 times the current-carrying capacity (I_Z) of the lowest rated conductor in the circuit being protected.

Protection against fault current

For fault current protective devices, the breaking capacity must be no less than the maximum prospective fault current (pfc), including both short circuit and earth fault conditions at the point at which it is installed; this value of prospective fault current must be established by measurement, calculation or enquiry (BS 7671 Regulation 434.1 and 434.5.1).

BS 7671 Regs 434.1 and 434.5.1

Remember

The further away from the intake position, the greater the resistance and the lower the value of pfc. Some designers simply check that the breaking capacity of the lowest rated fuse in the installation exceeds the pfc at the intake position.

Bear in mind that the cable must be able to carry the full amount of current that can pass through the protective device before it operates, not just the design current. A fault occurring at any point in a circuit must be interrupted before the fault current causes the limiting temperature of any conductor to be exceeded.

Effectively giving the maximum disconnection time, Regulation 434.5.2 states that the maximum permitted time in which a fault current can raise live conductors from the highest operating temperature to the limiting temperature is calculated using the formula:

$$t = \frac{k^2 S^2}{I^2}$$

Where:

t = duration in seconds
S = conductor csa in mm^2
I = fault current in amperes
k = a value used to denote conductor insulation, resistivity and heat capacity, given in Table 43.1 of BS 7671.

Activity

What is the value of k as for a copper conductor with 70°C thermoplastic (PVC) insulation?

BS 7671 Reg 434.5.2
Table 43.1

Discrimination

In most installations there is a series of fuses and/or circuit breakers between the incoming supply and the electrical outlets; the relative rating of the protective devices used will decrease the nearer they are to the current-using equipment.

Discrimination is the ability of protective devices to operate selectively to ensure that, in the event of a fault, only the faulty circuit is isolated from the system, allowing healthy circuits to remain in operation. For example, in a house, a fault on an appliance should cause the fuse in the plug top connected to the appliance to operate before the protective device in the consumer unit. Similarly, a fault on a lighting circuit should only result in the MCB in the consumer unit for that circuit to trip; all other MCBs in the consumer unit and the main DNO fuse should remain intact and energised.

However, having a 3 A fuse in the plug top and a higher-rated device in the consumer unit does not cover all that you need to consider.

Discrimination between fuses

It is standard practice to find HRC fuse links in series with one another to provide protection at different levels in an electrical installation. Discrimination between fuse links can be checked by ensuring that the time/current characteristics do not overlap at any point.

Discrimination between fuses and other protective devices

Protective devices other than fuses, such as circuit breakers, are generally electromechanical devices that have operating times longer than those of similarly rated fuses, except at low values of overcurrent. To achieve discrimination, make sure that the tripping time characteristic curves of the electromechanical device do not intersect with the time current characteristics of the fuse.

Remember

When using BS 7671 Appendix 3, you will see a difference in appearance between the characteristics for a fuse and a circuit breaker, in that the circuit breaker curves (see Figure 7.52 on page 406) have a vertical section to them.

Did you know?

You may have noticed that fuses have different values of current to achieve the different disconnection times from 0.1 to 5 seconds, whereas circuit breakers only have one value of current as represented by the vertical section. This is because circuit breakers have both magnetic and thermal components. The vertical part of the graph shows the operation of the magnetic element dealing with short circuits (faults), whereas the actual curved part shows the thermal component that deals with overloads.

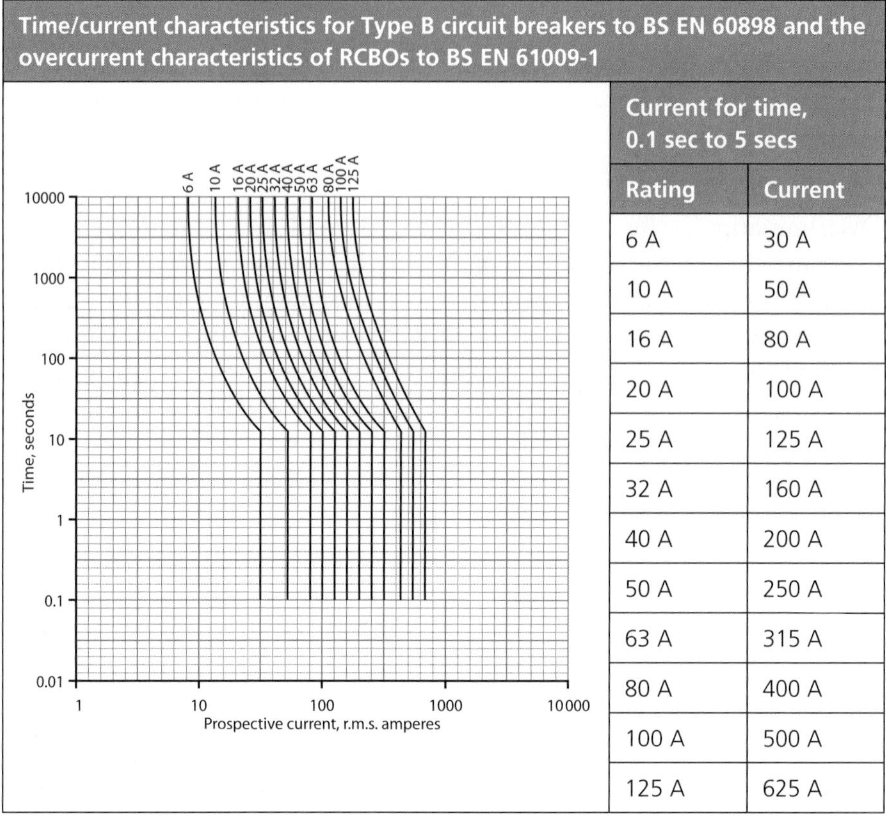

Current for time, 0.1 sec to 5 secs	
Rating	Current
6 A	30 A
10 A	50 A
16 A	80 A
20 A	100 A
25 A	125 A
32 A	160 A
40 A	200 A
50 A	250 A
63 A	315 A
80 A	400 A
100 A	500 A
125 A	625 A

Figure 7.52: Disconnection times for circuit breakers

BS 7671 Reg 411.3.2.1
Table 41.1
Reg 411.3.2.4

Disconnection times

Regulation 411.3.2.1 requires that, when a fault of negligible impedance occurs between the line conductor and an exposed-conductive-part or a protective conductor in the circuit or equipment, the protective device should operate in the required time. Table 41.1 applies to final circuits not exceeding 32 A and requires a disconnection time of 0.4 seconds on a final circuit supplied by a TN system at a nominal voltage of 230 V and 0.2 seconds for a TT system.

Should the final circuit be in excess of 32 A, a disconnection time of 5 seconds is permitted where supplied by a TN system at a nominal voltage of 230 V, and of 1 second for a TT system.

PROGRESS CHECK

1. One of the following does not provide overload protection. Which is it?
   - **a** RCD, 63 amp BS 88 Part 2 fuse
   - **b** 32 A Type C
   - **c** RCBO to BS EN 61009
   - **d** 16 A Type B
   - **e** BS EN 60898 circuit breaker?
2. What is meant by the term 'discrimination'?
3. State two positive features of a BS 88 fuse.
4. Define what is meant by an overload.
5. What is meant by the term 'short circuit'?
6. In the formula $I_b \le I_n \le I_z$ what does I_b denote?

7. Electricity and safe isolation procedures

Why safe isolation is important

Electricity can kill. Even non-fatal shocks can cause severe and permanent injury. Shocks from faulty equipment may lead to falls from ladders or other work platforms. Those using electricity may not be the only ones at risk, as poor electrical installations and faulty electrical appliances can lead to fires. Yet most of these accidents can be avoided with careful planning and straightforward precautions.

The main hazards are:

- contact with live parts causing shock and burns (mains voltage at 230 volts a.c. can kill)
- faults, which could cause fires
- fire or explosion where electricity could be the source of ignition in a potentially flammable or explosive atmosphere, such as in a spray-paint booth.

Electrical equipment used on building sites (particularly power tools and other portable equipment and their leads) faces severe conditions and rough use. It is likely to be damaged and therefore become dangerous. Modern double-insulated tools are well protected, but their leads are still vulnerable to damage and should be regularly checked. Where cables are needed for temporary lighting or mains-powered tools, run these at high level, particularly along corridors. Alternatively, you can use special abrasion-resistant or armoured flexible leads.

Cordless tools or tools which operate from a 110 V supply system (which is centre-tapped to earth so that the maximum voltage to earth should not exceed 55 V) will effectively eliminate the risk of death and greatly reduce injury in the event of an electrical accident. For other purposes such as lighting, particularly in confined and wet locations, still lower voltages can be used and are even safer.

If mains voltage has to be used, the risk of injury is high if equipment, tools, leads and so on are damaged, or there is a fault. Residual current devices with a rated tripping current no greater than 30 mA with no time delay will be needed to ensure that the current is promptly cut off if contact is made with any live part.

RCDs must be installed and treated with great care if they are to save life in an accident. They have to be kept free of moisture and dirt and protected against vibration and mechanical damage. They need to be properly installed and enclosed, including sealing of all cable entries. They should be checked daily by operating the test button.

Mains equipment is more appropriate to dry indoor sites where damage from heavy or sharp materials is unlikely, and if mains voltage is to be used, make sure that tools can only be connected to sockets protected

Did you know?

Each year about 1000 accidents at work involving electric shock or burns are reported to the Health and Safety Executive. Around 30 of these are fatal. Most of these fatalities arise from contact with overhead or underground power cables.

Safety tip

Many hazards can be reduced by precautions: safety guards and fences can be put on or around machines, safe systems of work introduced, safety goggles, helmets and shoes issued. Your employer should also provide any additional PPE required, such as ear defenders, respirators, eye protection and overalls.

Safety tip

With all aspects of health and safety, wherever possible, eliminate the risks.

by RCDs. By installing an RCD at the start of the work, immediate protection can be provided.

Even so, RCDs cannot give the assurance of safety that cordless equipment or a reduced low-voltage (such as 110 V) system provides. Electrical systems should be regularly checked and maintained. Everyone using electrical equipment should know what to look out for. A visual inspection can detect about 95 per cent of faults or damage.

Before you use any 230 V hand tool, lead or RCD follow this checklist.

Checklist

- No bare wires are visible.
- The cable covering is not damaged and is free from cuts and abrasions (apart from light scuffing).
- The plug is in good condition: for example, the casing is not cracked, the pins are not bent and the key way is not blocked with loose material.
- There are no taped or other non-standard joints (such as connector strips) in the cable.
- The outer covering (sheath) of the cable is gripped where it enters the plug or the equipment. The coloured insulation of the internal wires should not be visible.
- The equipment outer casing is not damaged and all screws are in place.
- The cables and equipment are appropriate to the environment they are being used in.
- There are no overheating or burn marks on the plug, cable or the equipment.
- Check any equipment to be used in a flammable atmosphere – it must not cause ignition!
- RCDs are working effectively – press the 'test' button every day.

Safety tip

Do not carry out makeshift repairs. It is essential that site workers are not exposed to danger when working on or near live electrical systems or equipment.

EaWR Regs 12, 13, 14

Safety tip

When fault finding, certain activities require the circuit to be live. However, danger is always present so you must take precautions to ensure safety and prevent injury.

Did you know?

There can be commercial pressure to carry out work on or near live conductors, especially in areas such as banking and high-cost manufacturing premises or retail premises that operate 24 hours per day.

Workers should be instructed to report any faults immediately and stop using the tool or cable as soon as any damage is seen. Managers should also arrange for a formal visual inspection of 230 V portable equipment on a weekly basis, and damaged equipment should be taken out of service as soon as the damage is noticed.

The following guidance is based on information published by The Electricity Safety Council on low voltage installations, and takes into account the requirements of the Electricity at Work Regulations (EaWR) 1989. Here is what the most relevant EaWR Regulations say.

- Regulation 12 of the EaWR says that, where necessary to prevent danger, there should be suitable ways (including, where appropriate, methods of identifying circuits) to a) cut off the supply of electrical energy to any electrical equipment; and b) isolate any electrical equipment. 'Isolating' means disconnecting and separating the electrical equipment from every source of electrical energy in a secure way.
- Regulation 13 reinforces the safety aspect, saying that precautions need to be taken with equipment that has been made dead, including securing the means of disconnection in the OFF position, putting a warning notice or label at the point of disconnection, and proving 'dead' at the point of work with an approved voltage indicator.
- Under Regulation 14, 'dead' working should be the normal method of carrying out work on electrical equipment or circuits, with live work only where it is unreasonable to work 'dead'.

- Regulation 16 requires that no one should work with electricity unless they are trained and competent to do so.

Safe isolation

To ensure compliance with Regulations 12 and 13 of the EaWR, you must follow the working principles.

- The correct point of isolation has been identified.
- An appropriate means of isolation is used.
- The point of isolation is ideally under the control of the person who is carrying out the work on the isolated conductors.
- Warning notices should also be applied at the point(s) of isolation.
- Conductors must be proved dead at the point of work before they can be touched.
- The supply cannot be knowingly/unknowingly re-energised while the work is in progress.

Before starting work for components fed with an electrical supply you will have to isolate the supply, at either a consumer unit or a fused spur outlet.

- How you isolate at the consumer unit depends which type of fuse/circuit breaker is used. If it is a cartridge type, remove the fuse and keep that fuse with you. If the circuit breaker is an MCB or RCD, switch it off and lock it off.
- When the supply is from a fused spur, switch off and remove the cartridge fuse, then lock off the fuse holder.

After isolating the supply you will then need to perform the safe isolation procedure using a voltage-indicating device.

Any circuit you work on *must* be tested to ensure it is dead. You will need to follow the correct testing procedure, and use the correct test equipment for this. Details are on pages 413–415.

When using a voltage indicator to test if a circuit is dead, remember these points.

- Test the voltage indicator on a proven supply before you start; this will confirm that the kit is working. The best piece of equipment for doing this is a proving unit.
- Only now can you use the voltage-indicating device to establish that the circuit you are about to work on is dead. You should check phase (live)-to-neutral conductors, phase-to-earth conductors and neutral-to-earth conductors to make sure all connections are dead.
- *You are not quite ready to begin work* – you should again check the test equipment on a known supply to make sure it is working correctly and has not become damaged during the testing procedure.

Safety tip

Live working should only be carried out when justified using the risk assessment criteria explained in HSE document HSG85.

EaWR Regs 12 and 13

Activity

What does HSG85 cover? How could you use it in your work?

An electrical lock off

Safety tip

When isolating, always leave a notice saying 'ELECTRICITY ISOLATED: DO NOT SWITCH ON'.

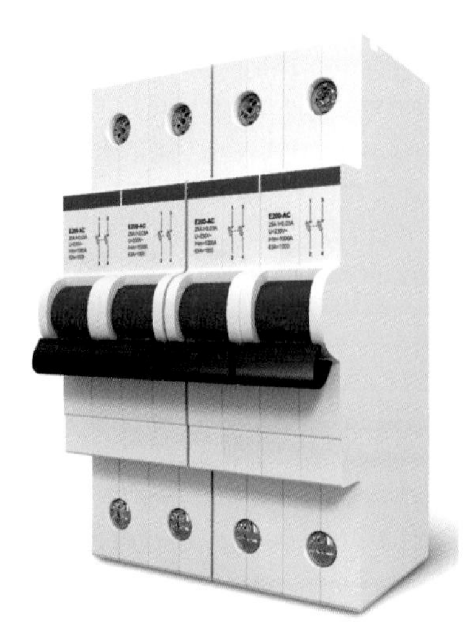

A switch disconnector

Did you know?

All live conductors must be isolated before work can be carried out, including the neutral conductor as this is a live conductor. This may mean removing the conductor from the neutral block in the distribution board. Not all distribution boards are fitted with double-pole isolators, so the connecting sequence for neutral conductors needs to be verified and maintained.

Did you know?

Table 53.2 of BS 7671 gives comprehensive guidance on the selection of protective, isolation and switching devices.

Isolating a complete installation

Sometimes circumstances mean that you have to isolate either a whole installation or large parts of an installation. Here the normal method is to use the main switch or the DB (distribution board) switch disconnector mounted within the DB.

In either case, you should lock the locking device with a unique key or combination that stays in your possession. If the switch has no locking-off facilities, you can use a locked DB that prevents access to that switch, as long as it uses a unique lock/combination.

Isolating individual circuits or items of equipment

Sometimes it is impractical to isolate a whole section of a building just to work on one item of equipment: for example, if you needed to repair a wall light in a hospital ward, you would not isolate the whole ward, just the relevant circuit.

Types of equipment used to provide switching and isolation of circuits, and indeed complete installations, can be categorised as having one or more of the following functions:

- control
- isolation
- protection.

Figure 7.53 gives a simple example of these three functions, showing a one-way lighting circuit supplied from a distribution board with a mains switch and circuit breakers.

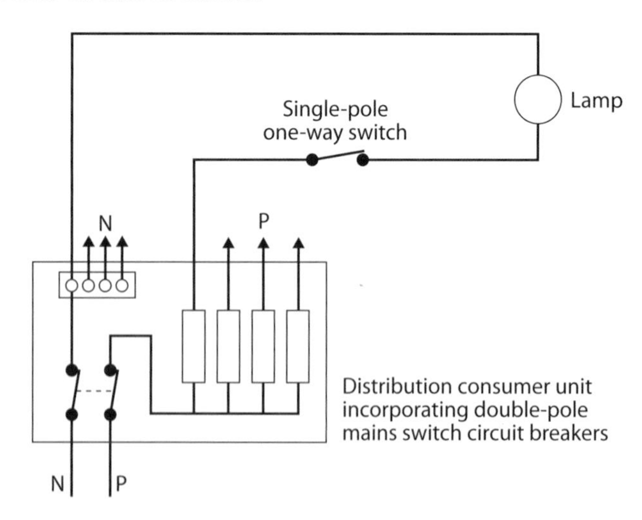

Figure 7.53: One-way lighting circuit

From Figure 7.53 you can see that:

- the distribution consumer unit combines all three functions of control, protection and isolation
- the main double-pole switch can provide the means for switching off the supply and, when locked off, can provide complete isolation of the installation

- the circuit breakers in the unit provide protection against faults and overcurrents in the final circuits and, when switched off and locked off, the circuit breakers can provide isolation of each individual circuit
- the one-way switch has only one function, which is to control the circuit enabling the luminaire to be switched 'on' or 'off'.

On- and off-load devices

Did you know?

When a current is flowing in a circuit, the operation of a switch (or disconnector) to break the circuit will result in a discharge of energy across the switch terminals – you may have seen this when switching on a light, where there is a blue flash from behind the switch plate. This is the arcing of the current as it **dissipates** and makes contact across the switch terminals. A similar arcing takes place when circuits are switched off or when protective devices operate, breaking fault current levels.

Key term

Dissipate – get rid of, disperse or scatter.

Not all devices are designed to switch circuits 'on' or 'off'. An isolator is designed as an off-load device, and is usually only operated after the supply has been made dead and there is no load current to break. An on-load device can be operated when current is normally flowing, and is designed to make or break load current.

An example of an on-load device could be a circuit breaker, which is designed not only to make and break load current but also to withstand high levels of fault current.

It is important to install a device that meets the needs at a particular part of a circuit or installation. Some devices can meet the needs of all three functions – control, isolation and protection – but others can only meet those of a single function.

All portable appliances should be fitted with the simplest form of isolator, a fused plug: unplugging from the socket outlet completely isolates the appliance from the supply. For equipment isolation, this should be mounted near the equipment and fitted with a means of being locked off.

The means of isolation could also be an adjacent local isolation device, such as a plug and socket, fused connection unit or circuit breaker, or fuse. For this to be allowed, it must be under the direct control of the competent person carrying out the work and must visible to them at all times, to stop anyone interfering with it.

Remember

A plug is not designed to make or break load current. The appliance should be switched off before removing the plug.

Professional Practice

A plumber tells you he has received an electric shock while connecting a boiler on site. On investigation, it is clear that the supply is on.

- What requirements do the Electricity at Work Regulations 1989 place on operatives working on circuits?
- What procedure should have been followed?
- What essential steps would be taken to ensure the circuit is made safe for work to be carried out?

Did you know?

When there is no local means of isolation, the method of isolating circuits or equipment is to use the main switch or DB switch disconnector, as if for isolating a whole installation. It helps if each circuit can have a suitable locking device and padlock fitted.

Sometimes more than one person is working on different circuits supplied from the same DB. Here you should ideally isolate and lock off each individual circuit using the appropriate devices. However, sometimes this may not be possible, in which case you would normally use a multi-lock hasp on the main switch or DB switch disconnector. This device holds one padlock for each person working, each with a unique key/combination, and cannot be removed until all the locks have been removed by those persons.

Safety tip

The practice of placing insulating tape over a circuit breaker to prevent inadvertent switching on is not a safe means of isolation.

If there is no facility to isolate the circuit like this, it is permissible to disconnect the circuit from the DB, provided that the disconnected conductors are made safe against inadvertent re-energising of the circuit and suitable labels are put up. However, work carried out inside a live DB is classed as live working when there is access to exposed live conductors, so you will need to take the appropriate precautions (HSG85 with respect to Regulation 14 of the EaWR).

EaWR Reg 14

Making sure circuits cannot be reactivated

Individual circuits protected by circuit breakers

Where circuit breakers are used as the means of isolation, they should be locked using an appropriate locking-off clip and padlock that can only be operated with a unique key or combination, both of which must be retained throughout by the individual working on the circuit. A warning notice should also be fitted at the point of isolation.

Did you know?

In the case of a new installation, make sure that all protective devices are correctly identified at the DB before the circuit is energised. Equally, on older installations, make sure that all records of the installation are available and that circuits have been correctly identified.

Individual circuits protected by fuses

Where a fuse is the means of isolation, it must be removed and retained by the person carrying out the work. Additionally, a lockable fuse insert should be fitted in the gap remaining and should be locked using a padlock that can only be operated with a unique key or combination. A warning notice must also be fitted. If lockable fuse inserts are not available then consider:

- fitting a 'dummy' fuse (a holder with no fuse in)
- padlocking the DB door (retain unique key and fit notice)
- disconnection of the circuit (fit warning notice).

Safety tip

Despite the fact that BS 7671 forbids it, the practice of using the neutral of one circuit to supply another still remains with lighting and control circuits being the favourite culprits. This is not safe practice.

Did you know?

In BS 7671 a neutral is referred to as a 'live conductor' because in these circumstances, a neutral can be live if disconnected and the circuit borrowed from has its load energised.

Proving isolated equipment or circuits are 'dead'.

Just because you think you have isolated something doesn't mean that you have. Never assume that just because you locked what you thought was the right circuit in the off position, that it is actually 'dead'. Always assume it is live, and treat something as being live, until you have proved otherwise.

Remember

Test instruments are expensive and so you need take care not to damage them. If you have any doubt about an instrument or its accuracy, ask for assistance.

Did you know?

Guidance Note GS 38 published by the Health and Safety Executive outlines who would be competent to do electrical testing, diagnosis and repair: electrically competent people may include electricians, electrical contractors, test supervisors, technicians, managers or appliance repairers.

Voltage indicating devices

Instruments used solely for detecting a voltage come in two categories:

- detectors that rely on an illuminated lamp (test lamp) or a meter scale (test meter). Test lamps are fitted with a 15 watt lamp and should not give rise to danger if the lamp is broken. A guard should also protect it
- detectors that use two or more independent indicating systems (one of which may be audible) and limit energy input to the detector by the circuitry used. An example is a two-pole voltage detector: a detector unit with an integral test probe, an interconnecting lead and a second test probe.

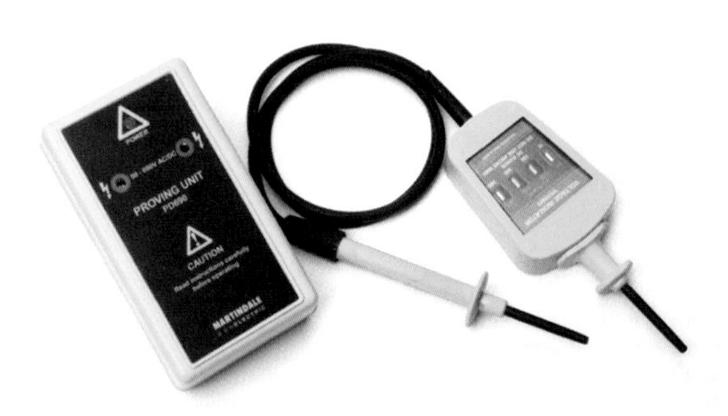

A voltage indicating device

Both these detectors are designed to limit the current and energy that can flow into the detector, through circuit design using the concept of protective impedance, and current-limiting resistors built into the test probes.

The detectors also have in-built test features to check the functioning of the detector before and after use. The interconnecting lead and second test probes are not detachable components. These types of detector do not need additional current-limiting resistors or fuses as long as they are made to an acceptable standard and the contact electrodes are shrouded. Test lamps and voltage indicators should be clearly marked with the maximum voltage they can test and any **short-time rating**, if applicable.

Key term

Short-time rating – the recommended maximum current that should pass through a detector device for a few seconds.

Restoration of the supply

Rectifying a fault can result in parts being replaced or simple reconnection of conductors, so the circuit must be tested for functionality. These tests may be simple manual rotation of a machine or the sequence of tests prescribed in BS 7671. For example, a simple continuity test will check resistance values, open and closed switches and their operation.

BS 7671

Colour-coding of conductors

BS 7671 Table 51

Since April 2004, all new installations in the UK have had to use cables whose conductors complied with Table 51 of BS 7671, as shown in Table 7.15.

Function	Colour
Protective conductors	Green and yellow
Functional earthing conductor	Cream
a.c. power circuit[1]	
Phase of single-phase circuit	Brown
Neutral of single- or three-phase circuit	Blue
Phase 1 of three-phase a.c. circuit	Brown
Phase 2 of three-phase a.c. circuit	Black
Phase 3 of three-phase a.c. circuit	Grey
Two-wire unearthed d.c. power circuit	
Positive of two-wire circuit	Brown
Negative of two-wire circuit	Grey
Two-wire earthed d.c. power circuit	
Positive (of negative earthed) circuit	Brown
Negative (of negative earthed) circuit	Blue
Positive (of positive earthed) circuit	Blue
Negative (of positive earthed) circuit	Grey
Three-wire d.c. power circuit	
Outer positive of two-wire circuit derived from three-wire system	Brown
Outer negative of two-wire circuit derived from three-wire system	Grey
Positive of three-wire circuit	Brown
Mid-wire of three-wire circuit[2]	Blue
Negative of three-wire circuit	Grey
Control circuits, ELV and other applications	
Phase conductor	Brown, black, red, orange, yellow, violet, grey, white, pink or turquoise
Neutral or mid-wire[3]	Blue

Table 7.15: Post-2004 insulation colours

NOTES

(1) Power circuits include lighting circuits.

(2) Only the middle wire of three-wire circuits may be earthed.

(3) An earthed PELV conductor is blue.

Before 2004, all UK installations used the insulation colours shown in Table 7.16. You will come across conductors with these colours of insulation, but only in existing installations.

Conductor	Old colour
Phase	Red
Neutral	Black
Protective conductor	Green and yellow
Phase 1	Red
Phase 2	Yellow
Phase 3	Blue
Neutral	Black
Protective conductor	Green and yellow

Table 7.16: Pre-2004 insulation colours

Remember

If you are extending an existing installation, you may well have cables of both colour systems. Take great care, as black and blue insulation is used in both systems.

Isolation flow charts

Here is a safe isolation flow chart for isolating a complete installation or for isolating an individual circuit or piece of equipment.

Complete installation and individual circuits

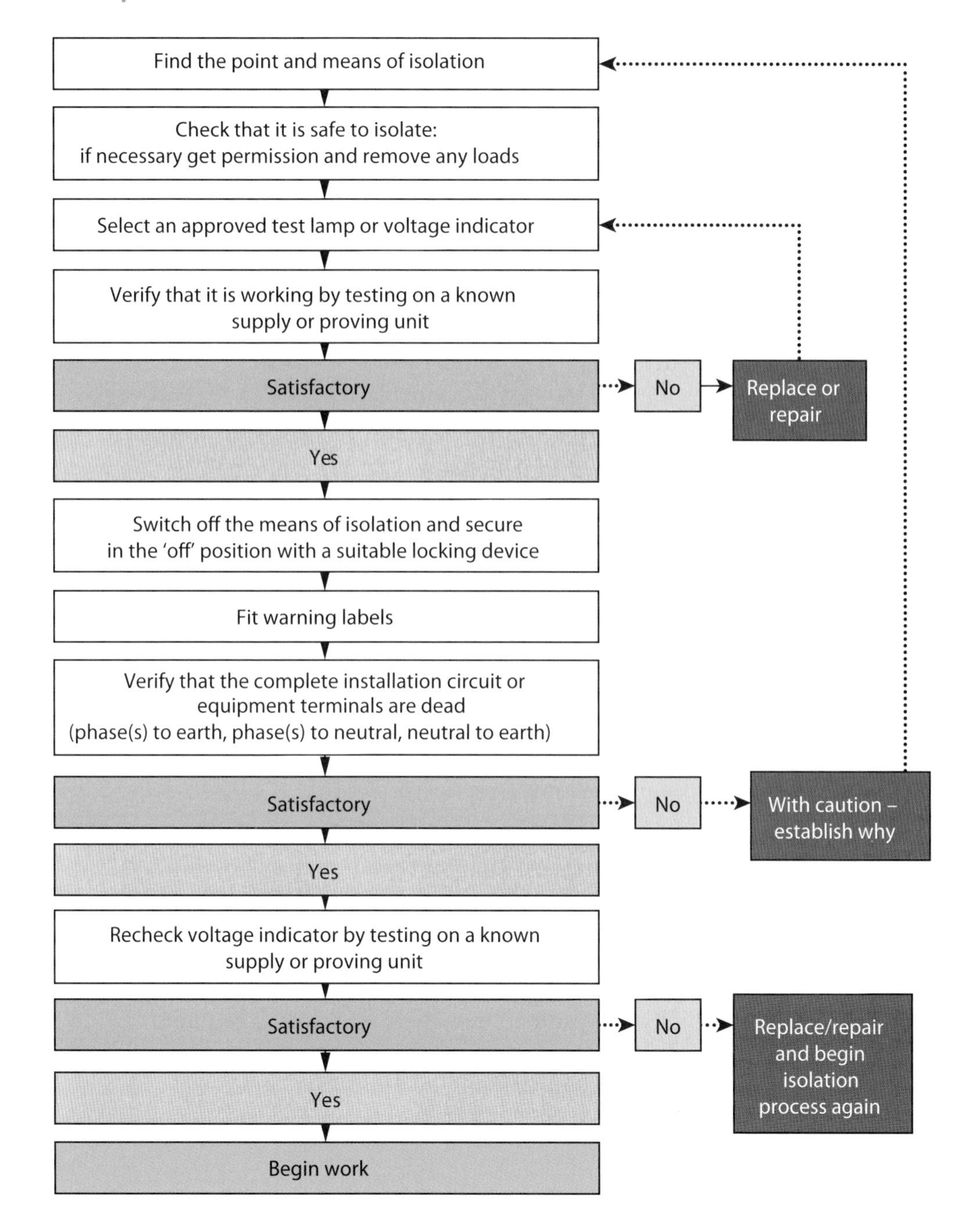

PROGRESS CHECK

1. In the UK before April 2004, what colour was the insulation on the live conductors in an installation?
2. Published by the Health and Safety Executive, what document gives guidance on electrical test equipment used by persons working on electrical equipment/ installations?

8. Installation and connection requirements of electrically operated mechanical services components

Sources of information

In Unit 1 of this book, it was explained how both employer and employee have legal duties to establish and maintain a safe working environment. Once you understand what the particular installation requirements are, you should undertake risk assessments (using the HSE '5 steps to risk assessment' model) to ensure that:

- you have safe access and egress to, from and around the work area
- you eliminate/control all work area hazards to property, personnel and livestock.

Remember

Make sure your findings are implemented before any work begins.

Did you know?

Although BS 7671 is not a statutory document itself, it can be used in a court of law as evidence of compliance with a statutory document, such as the EaWR or Building Regulations.

To do this you will use statutory documents (HASAWA, EaWR), site drawings, wiring diagrams, the specification and relevant technical data such as manufacturer's instructions. These were covered in Unit 3 on pages 118-121. Look back now to refresh your memory.

You will also need to use relevant British Standards, particularly BS 7671 – more commonly referred to as the IET Wiring Regulations – which applies to the design, erection and verification of electrical installations.

Preparatory work

Before starting work, as well as being aware of legislative requirements, you also need to consider how to plan the work. Good preparation to create a work programme involves:

- interpreting drawings and specifications to produce accurate material and equipment requisites
- identifying and selecting material and equipment that meet the specification (where not specified)
- identifying suitable installation, fitting and fixing methods
- identifying where co-ordination with other trades is required
- confirming that the site is ready for the installation including suitable access and storage facilities
- confirming that all tools, equipment and instruments are ready and fit for purpose
- checking that the right cables, materials, components and fittings are available
- identifying suitable access equipment lifting equipment, if required.

You also need to consider any potential damage to the client's property, relative to the type of building (new or existing), the type of wiring systems and enclosures being used and the contractual arrangements.

Here are two examples where different preparation is needed.

Worked example 1

You are asked to provide a heating installation within a newly constructed sheltered housing building, specified as being concealed within plastered walls and ducting within ceiling voids.

Most of your installation will take place as the building develops; ultimately walls will be plastered – the responsibility of the main contractor. However, if you do not pay due care and attention, your work could damage walls, floors and ceilings as the project develops.

Worked example 2

You are asked to use concealed cabling to install a heating control system in a 200-year-old stately home. The contract states that you are responsible for all repairs to the building structure to maintain its appearance. This will mean carefully selecting cable routes and being aware of the type of decorative plaster, paints and materials used throughout the building.

In such a contract, it pays to have established what **pre-damage** exists in case you find yourself carrying out restoration work! You would need to check for existing damage to walls, floors, ceilings and artefacts.

It would also usually be a contractual requirement that the prospective contractor visits the site, inspects the property and allows for any restorative work costs in their estimate. In the example, an initial visit should mean:

- walking around the property
- looking at the intended location of the equipment
- planning your intended cable routes to accommodate these locations and establishing alternatives if necessary
- marking your routes on the location drawings
- establishing what protective measures may be needed for the building or items within it
- noting any items that may need to be removed and stored and how you might do that
- taking photographs or video footage of areas affected by your installation
- establishing any special decorative techniques or finishes that may be required
- checking the style of existing installation components and establishing whether any of your modifications need to match them
- establishing through risk assessment whether the building or the required installation present any hazards.

Remember

Make sure you work closely with other trades for the satisfactory completion of each stage and to reduce damage to the building's structure and fabric.

Key term

Pre-damage – the condition a building or area is in before you arrive on site.

Establish any special decorative techniques or finishes that may be required

The estimator will then research any costs and include them in the quotation.

The estimator will consider whether walls and floors must be protected with a covering material, such as wood or sheeting; whether areas need to have restricted access and barriers; and whether items need to be carefully removed and stored.

You may also need to consider costs for the decommissioning and removal of any parts of the existing installation, if materials or equipment prove to be redundant or in need of replacement due to age or any other identified reason.

Remember

Continual monitoring of damage is crucial. During your work you may uncover unexpected problems, and will need to take careful account of these as they arise.

The estimator should include a covering letter that outlines any damage noted during the initial inspection, including supporting evidence and, where possible, any alternatives to avoid further damaging the property.

If you win the contract, it would be sensible to carry out a second visit immediately before starting work, to see whether conditions have changed within the property.

9. Selecting cables for components and circuits

The type of cable, its installation, termination and connection will depend on the item of equipment, the manufacturer involved and the installation circumstances. There are too many different scenarios to cover them individually here, but each will be a variation of the types covered in this section.

Suitability of existing supplies and circuits

An electrical installation must be designed to provide for:

- the protection of people, livestock and property
- the proper functioning of the installation for its intended use.

When looking at any installation and connection work, BS 7671 Chapter 13 requires you to:

- gather information on the characteristics of the supply, such as nature of current (a.c.) and number of conductors (line, neutral, protective, PEN) (Regulation 132.2)
- know the nature of the demand: that is, the location of power-consuming equipment and the demand it will place on circuits (Regulation 132.3)
- establish correct conductor sizes (Regulation 132.6)
- determine the size and nature of protective devices (Regulation 132.8).

BS 7671 Regs 132.2, 132.3, 132.6, 132.8

You will need to ask:

- does the existing supply have sufficient spare capacity to handle any additional electrical load?

- are the earthing and protection arrangements adequate to allow safe operation?
- are there any special installation requirements?

Different supply systems (230 V/400 V) and earthing arrangements (TN-S/PME) that can exist in properties as well as the different types of protective devices and their operating concepts (fuses/MCBs) have already been covered. Look back at pages 379–406 before reading on.

Sizing cables and circuit protection devices

Terminology and symbols

A range of special codes, symbols and terms are used when sizing cables and circuit protection devices. It is important that you have a good grasp of these.

Symbol		Definition
I_z		The maximum current-carrying capacity of the cable in the situation where it is installed
I_t		The value of current given in BS 7671 Appendix 4 for a single circuit of the relevant cable at an ambient temperature of 30°C
I_b		The design current of the circuit, i.e. the current intended to be carried by the circuit in normal use
I_n		The rated current or current setting of the protective device
I_2		The operating current (i.e. the fusing current or tripping current for the conventional operating time) of the device protecting the circuit against overload
$C_?$		A rating factor to be applied where the installation conditions differ from those shown in the tables of Appendix 4. Shown as the letter C followed by a subscript letter, the various rating factors are identified as follows:
	C_a	Ambient temperature
	C_g	Grouping
	C_i	Thermal insulation
	C_t	Operating temperature of conductor
	C_f	Semi-enclosed fuse to BS 3036
	C_s	Thermal resistivity of soil
	C_c	Buried circuits
	C_d	Depth of buried circuit

Table 7.17: Important symbols and definitions

External influences

Appendix 5 of BS 7671 lists the 'Classification of External Influences', where each condition of external influence is designated with a code that comprises a group of two capital letters and a number.

- The first letter relates to the general category of the external influence (for example, the environment).
- The second letter relates to the circumstances or nature of that external influence (for example, the presence of water).
- The number relates to the specific degree or level of that condition within each external influence (for example, splashes).

Appendix 5 of BS 7671

Using BS 7671 Table 5, the code for the example above would be AD3:

General category of influence	Environment	A
Nature of the influence	Water	D
Level of that influence	Splashes	3

Here are three more codes and their explanations.

AA4
A Environment
AA Environment – Ambient temperature
AA4 Environment – Ambient temperature – in the range –5°C to +40°C

BC1
B Utilisation
BC Utilisation – Contact with earth
BC1 Utilisation – Contact with earth – None

CA2
C Construction of buildings
CA Construction of buildings – Materials
CA1 Construction of buildings – Materials – Combustible

Sizing the conductor

Now you must establish the size of conductor to be used, starting by looking at the design current.

Design current (Ib)

The design current is the normal resistive load current designed to be carried by the circuit. You will have to calculate this using formulae that apply to single- and three-phase supplies.

Single-phase supplies (U_o = 230 V)

$$I_b = \frac{\text{Power}}{U_o}$$

where U_o is the line voltage to earth (supply voltage)

Three-phase supplies (U_o = 400 V)

$$I_b = \frac{\text{Power}}{\sqrt{3} \times U_o}$$

In a.c. circuits, the effects of highly inductive or highly capacitive loads can produce a poor power factor (PF), and you will have to allow for this. To find the design current, you may need to use the following equations where PF is the power factor of the circuit concerned:

Single-phase circuits:

$$I_b = \frac{\text{Power}}{U_o \times \text{PF}}$$

Three-phase circuits:

$$I_b = \frac{\text{Power}}{\sqrt{3} \times U_o \times \text{PF}}$$

Rating of the protective device (In)

The next stage is to work out the required current rating or setting (I_n) of the protective device. Regulation 433.1.1 of BS 7671 says that current rating (I_n) must be no less than the design current (I_b) of the circuit. The device must be able to pass enough current for the circuit to operate at full load, but without the protective device operating and disconnecting the circuit.

Protective devices are supplied with standard values, such as 13 A or 20 A, which you can get from manufacturers or find in BS 7671, Tables 41.2, 41.3 and 41.4, or in Appendix 3.

Activity

What would you choose as the nominal (unadjusted) rated device for a load producing a design current of 39A to be protected by a Type B circuit breaker? Use BS 7671 Table 41.3 to find out.

BS 7671 Regulation 433.1.1
Tables 41.2, 41.3, 41.4
Appendix 3

Remember

Which table you use to select the value of your protective device will depend on the type of equipment or circuit to be supplied and the requirements for disconnection times.

Installation and reference methods

The method of installation can affect the chosen cable's ability to get rid of any heat generated in normal operation; this in turn can affect its current-carrying capacity. It is not practical to calculate the current ratings for every single installation method, and many have the same rating. BS 7671 Table 4A2 offers a range of calculated current ratings covering all of the installation methods, known as reference methods.

To use these, you first need to look at Table 4A1 (Table 7.18 on page 422) to find out whether the proposed installation method will be allowable for the type of cable that needs to be installed. For example, if you want to install a PVC/PVC twin and earth cable, of a yet to be determined size, by clipping it directly to a wall, Table 4A1 shows that this installation method is acceptable (circled in red).

Table 4A1 Schedule of installation methods in relation to conductors and cables									
Conductors and cables		Installation methods							
		Without fixings	Clipped direct	Conduit systems	Cable trunking systems*	Cable ducting systems	Cable ladder, cable tray, cable brackets	On insulators	Support wire
Bare conductors		np	np	np	np	np	np	P	np
Non-sheathed cable		np	np	P[1]	P[1 2]	P[1]	np[1]	P	np
Sheathed cables (including armoured and mineral insulated)	Multicore	P	(P)	P	P	P	P	n/a	P
	Single-core	n/a	P	P	P	P	P	n/a	P

P Permitted
np Not permitted
n/a Not applicable or not normally used in practice
* = including skirting, trunking and flush floor trunking
[1] = non-sheathed cables which are used as protective conductors or protective bonding conductors need not be laid in conduits or ducts
[2] = non-sheathed cables are acceptable if the trunking system provides at least the degree of protection IP4X or IPXXD and if the cover can only be removed by means of a tool of deliberate action

Table 7.18: Table 4A1 from Appendix 4 of BS 7671

Now use Table 4A2 (Table 7.19 below) to establish the specific installation method and resultant reference method that will govern the current-carrying capacity of your cable. For our example, the specific installation method is 20, classed as falling within reference method C (circled in red on the extract shown in Table 7.19).

Table 4A2			
Installation method			Reference method to be used to determine the current-carrying capacity
Number	Examples	Description	
15		Non-sheathed cables in conduit or single-core or multicore cable in architrave	A
16		Non-sheathed cables in conduit or single-core or multicore cable in window frames	A
(20)		Single-core or multicore cables fixed on (clipped direct), or spaced less than 0.3 × cable diameter from a wooden or masonry wall	(C)

21		Single-core or multicore cables fixed directly under a wooden or masonry ceiling	C Higher than standard ambient temperatures may occur with this installation method
22		Single-core or multicore cables spaced from a ceiling	E, F or G Higher than standard ambient temperatures may occur with this installation method

Table 7.19: Extract from Table 4A2 from Appendix 4 of BS 7671

BS 7671 Tables 4A1 and 4A2

Rating factors (C)

Next you need to understand the rating factors, when they are required and how to apply them to the nominal rating of your selected protective device (I_n). Table 7.20 shows the available factors from BS 7671.

C_a	Ambient temperature	Tables 4B1 and 4B2
C_g	Grouping	Tables 4C1 to 4C5
C_i	Thermal insulation	Regulation 523.9
C_t	Operating temperature of conductor	
C_f	Semi-enclosed fuse to BS 3036	0.725 – Regulation 433.1.101
C_s	Thermal resistivity of soil	Table 4B3
C_c	Buried circuits	0.9 – where the cable installation method is 'in a duct in the ground' or 'buried direct'. For cables installed above ground C_c = 1
C_d	Depth of buried circuit	
n/a	Mineral insulated cable	0.9 – Table 4G1A

Table 7.20: Available rating factors from BS 7671

The rated current or current setting of the protective device (I_n) must not be less than the design current (I_b) of the circuit, and the rated current or current setting of the protective device (I_n) must not exceed the lowest of the current-carrying capacities (I_Z) of any of the conductors of the circuit.

This is summarised as: $I_b \leq I_n \leq I_Z$

Where the overcurrent device is intended to protect against overload, I_2 must not exceed $1.45 \times I_Z$ and I_n must not exceed I_Z. However, where the overcurrent device is intended to afford fault current protection only, I_n can be greater than I_Z and I_2 can be greater than $1.45 \times I_Z$.

Remember

The protective device must be selected for compliance with Regulation 434.5.2.

Ambient temperature

Ambient temperature is the temperature of the environment surrounding the cable, such as the room or ground in which the cable is installed. When a cable is in operation, it gives off heat. The hotter the surroundings, the harder it is to get rid of that heat. If the temperature of the surroundings is low, the cable can get rid of its heat more easily and could carry more current.

BS 7671 Appendix 4

In Appendix 4, the current-carrying capacities given for cables direct in the ground or in ducts in the ground are based on an ambient temperature of 20°C. However, the factor of 1.45 applied in Regulation 433.1.1 when considering overload protection assumes that the tabulated current-carrying capacities are based on an ambient temperature of 30°C. So to achieve the same degree of overload protection for cables directly in ground or in ducts as with other installation methods, the tabled current-carrying capacity is multiplied by a rating factor of 0.9.

A lower ambient temperature will allow the cable to get rid of heat more easily, so you may be able to reduce the size of your cable. If a cable passes through areas of different ambient temperature, apply a correction factor based on the highest temperature.

Grouping

When cables are installed so that they are touching, it is harder for them to dissipate any heat generated in normal use. The rating factors in Appendix 4 are based on groups of similar and equally loaded cables. However, with groups of different sizes of sheathed/non-sheathed cable installed in conduit, trunking or ducting systems, you use the following formula to calculate the group rating factor:

$$F = \frac{1}{\sqrt{n}}$$

Where:

F = the group rating factor
n = the number of circuits in that group.

This formula can also be used when these cables are installed on a tray although this is not ideal.

Thermal insulation

Thermal insulation is used throughout the construction industry, but for a cable, it makes it harder to dissipate any heat being generated. Regulation 523.9 of BS 7671 states that:

- for a cable installed in a thermally insulated wall or above a thermally insulated ceiling, where the cable has one side in contact with the thermally conductive surface, current-carrying capacities are given in Appendix 4 of BS 7671

- for a single cable that is totally covered by thermal insulation for more than 0.5 m and where no more precise information is available, the current-carrying capacity is taken as half the value of that cable when clipped to a surface and open (reference method C)
- for a single cable that is totally covered by thermal insulation for less than 0.5 m, the current-carrying capacity shall be reduced as appropriate depending on the size of the cable, its length in the insulation and the thermal properties of that insulation. De-rating factors are given in BS 7671 Table 52.2 (see Table 7.21) for conductor sizes up to 10 mm^2.

Length in insulation (mm)	De-rating factor
50	0.88
100	0.78
200	0.63
400	0.51

Table 7.21: BS 7671 Table 52.2: Length in insulation and de-rating factor

BS 7671 Reg 523.9
Appendix 4
Table 52.2

Protective device

According to Regulation 433.1.100, where overload protection is given by either a fuse to BS 88 or 1361, an MCB to BS EN 60898 or an RCBO to BS EN 61009, the requirements for co-ordination of conductor and protective device in Regulation 433.1.1 have been complied with – so the rating factor (C_f) is given as 1.0. However, if the protective device is a rewireable fuse to BS 3036, due to their potentially poor performance, you use a rating factor for C_f of 0.725.

BS 7671 Regs 433.1.100 and 433.1.1

If you have a cable installed directly in the ground or in ducts, you apply a rating factor for C_f of 0.9. If such a cable is protected by a BS 3036 fuse, both rating factors would apply and the new rating would be:

$$C_f = 0.725 \times 0.9 \quad \text{and therefore} \quad C_f = 0.653$$

Application of rating factors

Applying these factors ensures that a cable is large enough to carry the current without too much heat being generated. If external conditions reduce the cable's ability to give off heat, you have to increase the size of the cable.

Appendix 4 of BS 7671 requires that the current-carrying capacity of the cable (I_z) is not less than the circuit's design current (I_b). Equally, the rating of the fuse or circuit breaker (I_n) must be at least as big as the circuit design current (I_b). The reason for doing this is to make sure that you don't overload the cable or operate the protective device as soon as you turn the load on.

BS 7671 Appendix 4

To find I_z, start by dividing the rated current of the protective device (I_n) by the applicable factors that can de-rate the cable. This gives the true rated current of the cable required, which you use to select your cable from the appropriate table in Appendix 4. The value of current obtained from the tables in Appendix 4 of BS 7671 is given the symbol I_t and it must be greater than the value resulting from the above calculation (I_z).

You use the following formula to establish I_z:

$$I_z = \frac{I_n}{C_a \times C_i \times C_s \times C_d \times C_f \times C_c}$$

The best way to see all this in operation is through some worked examples.

Note: you will still need to check for voltage drop, shock protection and thermal constraints. All these subjects will be covered as you go through the examples.

Worked example 1

A single-phase circuit has a design current of 30A. It is to be wired in flat two-core 70°C PVC insulated and sheathed cable with cpc to BS 6004. It will have copper conductors and the cable is enclosed in trunking with four other similar cables. If the ambient temperature is 30°C and the circuit is to be protected by a BS 3036 fuse, what should be the nominal current rating of the fuse and the minimum csa of the cable conductor?

In answering these types of questions, it is a good idea to construct a table of information. This will help you understand and retain the process of cable selection. It also makes cable selection easier to understand.

Installation Method number will be 6/7 (installed in trunking), therefore Reference Method = B

As ambient temperature equals 30°C, then from Table 4B1 for 70°C thermoplastic, $C_a = 1.0$.

As grouped, then from Table 4C1, $C_g = 0.6$ (there are five circuits and Reference Methods A to F apply).

Design current (I_b) was given as 30A.

As the protective device rating (I_n) must be ≥ the rating of the design current (I_b) you can say that:

I_n = 30 A.

But as the protective device is a BS 3036 fuse, a factor C_c equal to 0.725 must be applied

$$I_z = \frac{I_n}{C_a \times C_g \times C_c} = \frac{30}{1.0 \times 0.6 \times 0.725} = 69\text{A}$$

As said above I_t must be ≥ I_z, therefore using Table 4D2A (column 4), you see that a 16 mm² conductor size has a tabulated current rating of 69A.

Therefore the minimum conductor csa that can be used is 16 mm².

Checking voltage drop

The voltage drop between the origin of the installation (usually the supply terminals) and a socket outlet (or the terminals of fixed current using equipment) should not exceed that stated in Appendix 12 (Regulation 525.101 of BS 7671).

Also, for low-voltage installations supplied from a public distribution system, voltage drop should be no greater than 3 per cent of the supply voltage for lighting circuits and no greater than 5 per cent for other circuits (Appendix 12).

So where the supply voltage is 230 V, a maximum voltage drop of 6.9 V for lighting or 11.5 V for other circuits is allowed. In unusual circumstances where the final circuit has a length in excess of 100 m, BS 7671 now allows an increase in voltage drop of 0.005 per cent per metre of that circuit above 100 m, up to a maximum value of 0.5 per cent.

The voltage drops either because the resistance of the conductor becomes greater as the length of the cable increases or the cross-sectional area (csa) of the cable is reduced. On long cable runs, the cable csa may have to be increased, reducing the resistance allowing current to 'flow' more easily and reducing the voltage drop across the circuit.

You can calculate the maximum allowed voltage drop like this.

1. For single-phase 230 V lighting circuits:

$$3\% = 230 \times \frac{3}{100} = 6.9 \text{ volts}$$

2. For three-phase 400 V systems:

$$5\% = 400 \times \frac{5}{100} = 20 \text{ volts}$$

Remember

When carrying out circuit calculations, you are allowed to have a minimum system voltage of as low as 223.1 V on lighting circuits and 218.5 V on other circuits.

It is better to keep the voltage drops as low as possible, because low voltages can reduce the efficiency of the equipment being supplied: lamps will not be as bright and heaters may not give off full heat. The values for cable voltage drop are given in the accompanying tables of current-carrying capacity in Appendix 4 of the IET Wiring Regulations. The values are given in millivolts per ampere per metre (mV/A/m).

You should use the formula below to calculate the actual voltage drop.

$$\text{Voltage drop (VD)} = \text{mV/A/m} \times \frac{I_b \times L}{1000}$$

Where:

mV/A/m = the value given in the Regulation Tables
I_b = the circuit's design current
L = the length of cable in the circuit measured in metres.

As mV/A/m uses the prefix 'milli' (thousandths), you are going to have to convert I_b and L into a common unit, by dividing them by 1000.

Did you know?

In some areas of the country people prefer the following equation:

Voltage drop (VD) = mV/A/m × $I_b \times L \times 10^{-3}$

Shock protection

On page 391, you looked at automatic disconnection of supply (ADS). With this measure, all metalwork including extraneous-conductive-parts and exposed-conductive parts is connected to earth. If an earth fault occurs, you need the protective device to work quickly. The current must be large enough to operate the device within given times, and the earth fault loop impedance must be low enough to let this happen.

Under BS 7671 Regulation 411.3.2.1, a protective device must automatically interrupt the supply to the line conductor (of a circuit or of equipment) in the event of a fault of negligible impedance between the line conductor and an exposed-conductive-part or protective conductor in the circuit or equipment, within the times given in Table 41.1 of BS 7671 (see Table 7.22 overleaf).

BS 7671 Reg 411.3.2.1
Tables 41.1–41.4

From this table you can see that, for a 230 V final circuit (such as lighting or socket outlets) on a TN system, the disconnection time is 0.4 seconds, but on a TT system it is 0.2 seconds. A time on TN of 5.0 seconds is only acceptable for distribution circuits (circuits supplying a DB or switchgear). To achieve these times, the value of earth fault loop impedance (Z_s) should not be larger than the values given in Tables 41.2, 41.3 and 41.4 of BS 7671.

Worked example 2

To reinforce understanding, this is the same question as in Example 1, but this time you are also required to work out the voltage drop for the cable and consider its effect.

A single-phase circuit has a design current of 30A. It is to be wired in flat two-core 70°C PVC insulated and sheathed cables with cpc to BS 6004. It will have copper conductors and the cable is enclosed in trunking with four other similar cables. If the ambient temperature is 30°C, the circuit is to be protected by a BS 3036 fuse and the circuit length is 27m, what should be the nominal current rating of the fuse (I_n) and the minimum csa of the cable conductor?

Installation Method number will be 6/7 (installed in trunking) therefore Reference Method = B

As ambient temperature = 30°C then, from Table 4B1, for 70°C thermoplastic: $C_a = 1.0$

As grouped, then from Table 4C1, $C_g = 0.6$ (there are five circuits and Reference Methods A to F apply)

Design current (I_b) was given as 30A.

As the protective device rating (I_n) must be ≥ the rating of the design current (I_b) you can say that $I_n = 30A$.

But as the protective device is a BS 3036 fuse, the factor $C_c = 0.725$ must be applied

$$I_z = \frac{I_n}{C_a \times C_g \times C_c} = \frac{30}{1.0 \times 0.6 \times 0.725} = 69\text{ A}$$

As stated, I_t must be ≥ I_z, therefore using Table 4D2A (column 4), you will see that a 16mm² conductor size has a tabulated current rating of 69A and is therefore acceptable in this respect. You must now check the voltage drop for this cable.

From Table 4D2B (column 3) you will see that the voltage drop is given as 2.8 mV/A/m. You need to work in common units, so you must divide I_b and L by 1000.

$$\text{Voltage drop} = \text{mV/A/m x} = \frac{I_z \times L}{1000} = \frac{(30 \times 27)}{1000} = 2.27\,\text{V}$$

BS 7671 states that for other than a lighting circuit, voltage drop must not exceed 5 per cent of the supply voltage, which in this case would be 5 per cent of 230 V or 11.5 V. As the actual voltage drop of the proposed cable was calculated to be 2.27 V, this means that the minimum csa of the cable conductor will be 16 mm^2.

System	50 V < U_o < 120 V seconds		120 V < U_o < 230 V seconds		230 V < U_o < 400 V seconds		U_o < 400 V seconds	
	a.c.	d.c.	a.c.	d.c.	a.c.	d.c.	a.c.	d.c.
TN	0.8	NOTE 1	0.4	5	0.2	0.4	0.1	0.1
TT	0.3	NOTE 1	0.2	0.4	0.07	0.2	0.04	0.1

Table 7.22: Maximum disconnection times (Table 41.1 BS 7671)

You need to do a calculation to check that the circuit protective device will operate within the required time. You do this by checking that the actual Z_s is lower than the maximum Z_s given in the relevant table, for the protective device you have chosen.

Maximum Z_s can also be found from IET Wiring Regulations or manufacturers' data. Use the following formula to calculate the actual Z_s:

$$\text{Actual } Z_s = Z_e + (R_1 + R_2) \times (\text{m.f.}) \times \frac{\text{length}}{1000}$$

Where:

Z_e = the external impedance on the supply authority's side of the earth fault loop.

You can get this value from the supply authority. Typical maximum values are: TN-C-S (PME) system 0.35 ohms; TN-S (cable sheath) 0.8 ohms; TT system 21 ohms.

$R_1 + R_2$ = the resistance of the phase conductor plus the cpc resistance.

You can find values of resistance/metre for $R_1 + R_2$, for various combinations of phase and cpc conductors up to and including 50 mm^2 in Table G1 of the *Unite Guide to Good Electrical Practice*.

Table 7.23 on page 430 gives you multipliers (mf) to apply to the values given in Table G1 to calculate the resistance under fault conditions. If the conductor temperature rises, resistance in the conductors will increase. This table is based on the type of insulation used and whether the cpc is not incorporated in a cable or bunched with cables (54.2), or is incorporated as a core in a cable or bunched with cables (54.3).

	Insulation material		
Conductor installed as	70° thermoplastic (pvc)	85° thermosetting (rubber)	90° thermosetting
54.2 (not incorporated)	1.04	1.04	1.04
54.3 (incorporated)	1.20	1.26	1.28

Table 7.23: Multipliers to be applied in Table G1

Worked example 3

Again this is the same question that was used in Examples 1 and 2, but this time you are also required to work out the requirement for shock protection and consider its effect.

A single-phase circuit supplying a domestic cooker has a design current of 30 A and is to be wired in flat two-core 70°C PVC insulated and sheathed cables with cpc to BS 6004. It will have copper conductors and the cable is enclosed in trunking with four other similar cables. If the ambient temperature is 30°C, the circuit is to be protected by a BS 3036 fuse and the circuit length is 27 m, what should be the nominal current rating of the fuse (I_n) and the minimum csa of the cable conductor? You are informed that Z_e is 0.8 Ω.

Installation Method number will be 6/7 (installed in trunking) therefore Reference Method = B.

As ambient temperature = 30°C then, from Table 4B1, for 70°C thermoplastic: $C_a = 1.0$

As grouped, then from Table 4C1, $C_g = 0.6$ (there are five circuits and Reference Methods A to F apply)

Design current (I_b) was given as 30 A.

As the protective device rating (I_n) must be ≥ the rating of the design current (I_b) you can say that $I_n = 30$ A.

But as the protective device is a BS 3036 fuse, the factor $C_c = 0.725$ must be applied.

$$I_Z = \frac{I_n}{C_a \times C_g \times C_c} = \frac{30}{1.0 \times 0.6 \times 0.725} = 69 \text{ A}$$

I_t must be ≥ I_z, therefore using Table 4D2A (column 4), you will see that a 16 mm² conductor size has a tabulated current rating of 69 A and is therefore acceptable in this respect. You must now check the voltage drop for this cable.

From Table 4D2B (column 3) you will see that the voltage drop is given as 2.8 mV/A/m. You need to work in common units, so you must divide I_b and L by 1000.

$$\text{Voltage drop} = \text{mV/A/m x} = \frac{I_z \times L}{1000} = \frac{(30 \times 27)}{1000} = 2.27 \text{ V}$$

BS 7671 states that for other than a lighting circuit, voltage drop must not exceed 5 per cent of the supply voltage, which in this case would be 5 per cent of 230 V or 11.5 V. As the actual voltage drop of the proposed cable was calculated to be 2.27 V, this is acceptable. You must now check for shock protection.

From Table 41.2, the maximum permitted Z_s for our 30 A BS 3036 fuse is 1.09 Ω and from Table G1 of the Unite guide, for a 16 mm² with 6 mm² cpc, the resistance/m ($R_1 + R_2$/m) is given as 4.23 mΩ/m.

Additionally you need the multiplier for a grouped cable, which from Table 7.23 on page 430 is 1.2. You also need to convert length from m to mm by dividing by 1000. Therefore:

$$Z_s = Z_e + ([R_1 + R_2] \times \text{mf} \times \frac{\text{length})}{1000} = 0.8 + ([4.23] \times 1.2 \times \frac{27}{1000}) = 0.937\ \Omega$$

As actual Z_s (0.937 Ω) is less than the permitted value (1.09 Ω), this means that 16 mm² cable with 6 mm² cpc is the minimum csa of the cable conductor.

This information, together with the length of the circuit, can now be applied to the formula. Then you can check to see that the actual Z_s is less than the maximum Z_s given in the appropriate tables.

Now you can prove that the protective device will disconnect the circuit in the time that is specified in the appropriate table: if the actual Z_s is less than the maximum Z_s then you have compliance for shock protection.

Thermal constraints

Now that you have chosen the type and size of cable to suit the conditions of the installation, you must check to make sure that the size of the cpc – the 'earth conductor' – complies with the IET Wiring Regulations.

If there is a fault on the circuit, which could be a short circuit or earth fault, a fault current of hundreds or thousands of amperes could flow. Imagine that this is a 1 mm² or 2.5 mm² cable; if this large amount of current was allowed to flow for a few seconds, the cable would melt and a fire could start.

You need to check that the cpc will be large enough to carry this fault current without causing any damage; for this you use the adiabatic equation (see page 432). The cpc will only need to carry the fault current for a short period of time, until the protective device operates.

Regulation 543.1.1 states that 'the cross-sectional area of every protective conductor shall be calculated in accordance with Regulation 543.1.3 (adiabatic equation) or selected in accordance with Regulation 543.1.4 (Table 54.7)'. You will see that Regulation 543.1.4 asks that reference be made to Table 54.7. This table shows that, for cables 16 mm² and below, with the cpc made from the same material as the line conductor, the cpc should be the same size as the line conductor. A line conductor between 16 mm² and 35 mm² requires a cpc to be 16 mm².

A line conductor above 35 mm^2 requires a cpc to be at least half the cross-sectional area.

Multicore cables have cpcs smaller than their respective line conductors, except for 1 mm^2, which has the same-sized cpc. Regulation 543.1.2 of BS 7671 gives two options: calculation or selection. Applying option (ii) of this Regulation would make these cables contravene the Regulations.

BS 7671 Reg 543.1.1–4

It is not intended that composite cables should have cpcs increased in accordance with the table, so you should apply the adiabatic equation required in option (i) of the same Regulation.

The adiabatic equation enables a designer to check the suitability of the cpc in a composite cable. If the cable does not incorporate a cpc, a cpc installed as a separate conductor may also be checked. The equation is as follows:

$$S = \sqrt{\frac{I_f^2 \times t}{k}}$$

Where:

S = the cross-sectional area of the cpc in mm^2
I_f = the value of the fault current in amperes
t = the operating time of the disconnecting device in seconds
k = a factor that takes into account resistivity, temperature coefficient and heat capacity

To apply the adiabatic equation, first calculate the value of I_f (fault current) from the following equation:

$$I_f = \frac{U_o}{Z_s}$$

Where:

U_o is the nominal supply line voltage to earth
Z_s is the earth fault loop impedance.

If you are using method (i) from Regulation 543.1.2 and applying the adiabatic equation, you must find out the time/current characteristics of the protective device. A selection of time/current characteristics for standard overcurrent protective devices is given in Appendix 3 of the IET Wiring Regulations. You can get the time (t) for disconnection to the corresponding earth fault current from these graphs.

If you look at the time/current curve, you will find that the scales on both the time (seconds) scale and the prospective current (amperes) scale are logarithmic and the value of each subdivision depends on the major division boundaries into which it falls.

For example, on the current scale, all the subdivisions between 10 and 100 are in quantities of 10, while the subdivisions between 100 and 1000 are in quantities of 100, and so on. This also occurs with the time scale, subdivisions between 0.01 and 0.1 being in hundredths and the subdivisions between 0.1 and 1 being in tenths.

Look at the graph in Figure 7.54. You will see that, for a BS 88 fuse with a rating of 32 A, a fault current of 200 A will cause the fuse to clear the fault in 0.6 seconds. The IET has also produced a small table showing some of the more common sizes of protective devices and the fault currents for a given disconnection time.

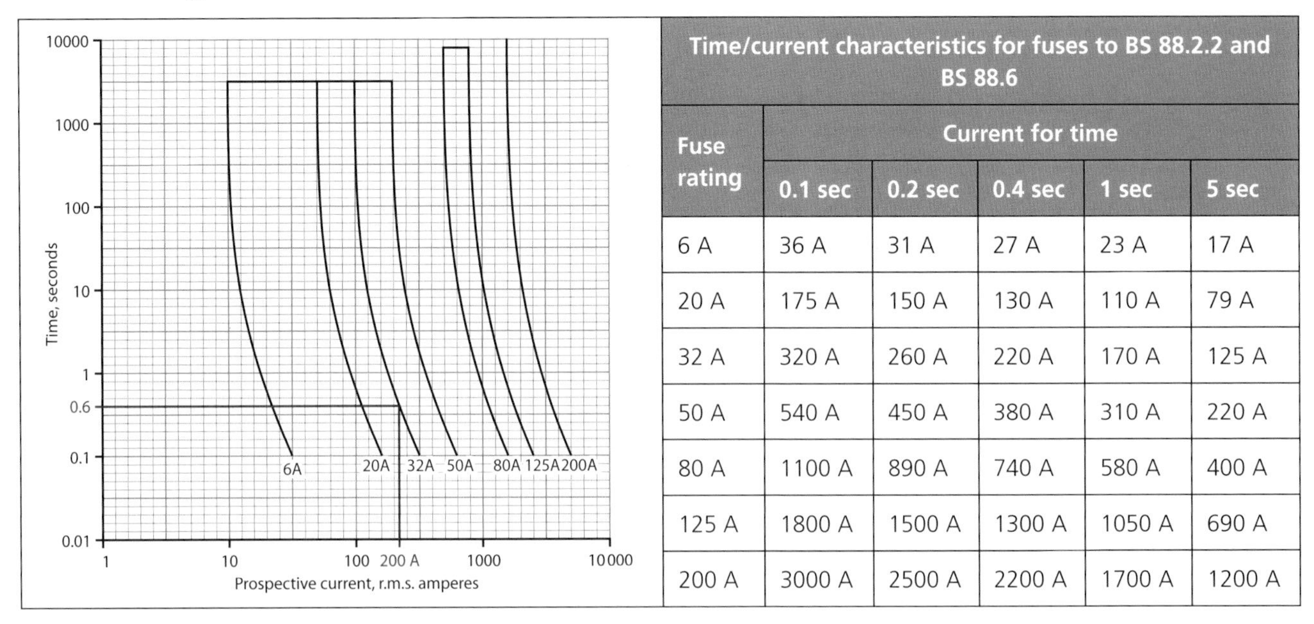

Time/current characteristics for fuses to BS 88.2.2 and BS 88.6					
Fuse rating	Current for time				
	0.1 sec	0.2 sec	0.4 sec	1 sec	5 sec
6 A	36 A	31 A	27 A	23 A	17 A
20 A	175 A	150 A	130 A	110 A	79 A
32 A	320 A	260 A	220 A	170 A	125 A
50 A	540 A	450 A	380 A	310 A	220 A
80 A	1100 A	890 A	740 A	580 A	400 A
125 A	1800 A	1500 A	1300 A	1050 A	690 A
200 A	3000 A	2500 A	2200 A	1700 A	1200 A

Figure 7.54 Time/current characteristics graph

BS 7671 Tables 54.2 and 54.6

Next you need to select the k factor using Tables 54.2 to 54.6. The values are based on the initial and final temperatures in each table. You may need to refer to the cable's operating temperature shown in the cable tables.

Now substitute the values for I, t and k into the adiabatic equation. This will give you the minimum cross-sectional area for the cpc. If your calculation produces a non-standard size, you must use the next largest standard size. Using the calculation method may lead to savings in the size of cpc.

Now try the following example to give you a complete understanding of cable selection. This builds on Examples 1, 2 and 3, but here you also need to complete calculations for thermal constraints.

Worked example 4

To reinforce understanding, this is the same question that was used in Examples 1, 2 and 3, but this time we will only concern ourselves with the calculations relevant to thermal constraints. You are also required to work out the requirement for shock protection and consider its effect.

A single-phase circuit supplying a domestic cooker has a design current of 30 A and is to be wired in flat two-core 70°C PVC insulated and sheathed cables with cpc to BS 6004. It will have copper conductors and the cable is enclosed in trunking with four other similar cables. If the ambient temperature is 30°C, the circuit is to be protected by a BS 3036 fuse and the circuit length is 27 m, what should be the nominal current rating of the fuse (I_n) and the minimum csa of the cable conductor?

You are informed that Z_e is 0.8 Ω.

From the previous calculations in Example 3, we know that I_n = 30 A and that 16 mm² cable with 6 mm² cpc is the minimum csa of the cable conductor.

For thermal constraints, we need to know the fault current before we can apply the adiabatic equation. We find this by:

$$I_f = \frac{U_o}{Z_s} = \frac{230}{0.937} = 245.5 \text{ A}$$

We also need to know the value of t and the value of k. Therefore from Table 3.2A within Appendix 3 of BS 7671 we can see that our fault current of 245.5 A will operate the device in a time of 0.3 seconds. We therefore state t as being 0.3 seconds.

From Table 54.3 (BS 7671) we can see that the value of k will be 115, as the cable is under 300 mm².

Now using the adiabatic equation:

$$S = \frac{\sqrt{I_f^2 \times t}}{K} = \frac{\sqrt{245.5^2 \times 0.3}}{115} = \frac{\sqrt{18081.08}}{115} = \frac{134.5}{115} = 1.17 \text{ mm}^2$$

As 1.17 mm² is less than our 6 mm² cpc, our choice of cable remains as a cable with 16 mm² conductors and a 6 mm² cpc to ensure compliance with thermal constraints.

Diversity

BS 7671 Chapter 31

In this section you will look at maximum demand and diversity as considered in BS 7671 Chapter 31. The current demand for a final circuit is determined by adding up the current demands of all points of utilisation and equipment in the circuit and, where appropriate, making an allowance for diversity. Diversity makes allowances on the basis that not all of the load or connected items will be in use at the same time.

In most cases, main or sub-main cables will supply a variety of final circuits. You need to consider use of the various loads; otherwise, if all the loads are totalled, you will select a larger cable than necessary, at extra cost.

Section 3 of the *Unite Guide to Good Electrical Practice* offers a method for assessing the load, which allows diversity to be applied depending on the type of load and installation premises. You add the individual circuit/load figures together to determine the total 'assumed current demand' for the installation. This value can then be used as the starting point to determine the rating of a suitable protective device and the size of cable, considering any influencing factors in a similar manner to that applied to final circuits.

Remember

The calculation of maximum demand is not an exact science, and a suitably qualified electrical engineer may use other methods of calculating maximum demand.

Worked example

A 230 volt domestic installation consists of the following loads:

15 × filament lighting points

6 × fluorescent lighting points, each rated at 40 watts

4 × fluorescent lighting points each rated at 85 watts

3 × ring final circuits supplying 13 A socket outlets

1 × radial circuit protected by a 20 A device supplying 13 A sockets for the adjoining garage

1 × 3 kW immersion heater with thermostatic control

1 × 13.6 kW cooker with a 13 A socket outlet incorporated in the control unit.

Determine the maximum current demand for determining the size of the sub-main cable required to feed this domestic installation. The circuit protection is by the use of BS 1361 fuses.

Answer

Lighting

Tungsten light points (See page 40 of the *Unite Guide to Good Electrical Practice*)

15 × 100 W minimum	1500

Fluorescent light points (See page 40 of the *Unite Guide to Good Electrical Practice*)

6 × 40 W with multiplier of 1.8 (40 × 1.8 = 72)	432
4 × 85 W with multiplier of 1.8 (85 × 1.8 = 153)	612
Total	2544 W

Using Item 1 of the table on page 41 of the *Unite Guide to Good Electrical Practice*, we can apply diversity as being 66 per cent of the total current demand. Therefore:

66% of 2544 = 1679 W and since $I = \frac{P}{V}$ this $\frac{1679}{230}$ gives = 7.3 A

Power (Item 9 on page 41 of the *Unite Guide to Good Electrical Practice*)

3 ring final circuits

1 × ring at 100% rating (30 A)	30 A
2 × ring at 40% rating (40% of 30 A)	24 A

20 A radial circuit (see Item 9 on page 41 of the *Unite Guide to Good Electrical Practice*)

1 × radial at 40% rating (40% of 20 A)	8 A

3 kW immersion heater (See Item 6 on page 41 of the *Unite Guide to Good Electrical Practice*)

3 kW heater with no diversity $I = \frac{P}{V} = \frac{3000}{230}$ which gives us 13 A

13.6 kW cooker with socket outlet (See Item 3 on page 41 of the *Unite Guide to Good Electrical Practice*)
The first 10 A, plus 30% of the remainder of the overall rated current, plus 5 A for the socket.

$I = \frac{P}{V} = \frac{136000}{230}$ giving a total rated current of 59 A

59 A – 10 A = 49 A and therefore 30% of 49 A = 14.7 A

The allowable total cooker rating is therefore:

10 A + 14.7 A + 5 A = 29.7 A

Our total assumed current demand is therefore:

7.3 + 30 + 24 + 8 + 13 + 29.7 giving a total of **112 A**

Other factors affecting cable selection

The correct selection of cable for an electrical installation is very important. You need to consider:

- conductor material
- conductor size
- insulation
- wiring system
- environmental conditions.

Conductor material

For conductors, the choice generally is between copper and aluminium. Copper has better conductivity for a given cross-sectional area, but its cost has risen over recent years. Aluminium conductors are now sometimes preferred for medium and larger cables, but all cables smaller than 16 mm^2 cross-sectional area must have copper conductors.

Whatever the material, conductors will usually be either stranded or solid.

- Solid conductors are easier and cheaper to manufacture, but are harder to install because they are not very pliable.
- Stranded conductors are made up of individual strands brought together in set numbers (such as 3, 7, 19 or 37). With the exception of the three-strand conductor, all have a central strand surrounded by the other strands.

Conductor size

As you have seen, many factors affect the choice of size of conductor. Here are some of the factors you need to consider.

- **Load and future development:** the current the cable is expected to carry can be found from the load, taking into account its possible future development, such as change in use of premises, extensions or additions.
- **Ambient temperature:** the hotter the surrounding area, the less current the cable is permitted to carry.
- **Grouping:** if a cable is run with other cables, its current-carrying capacity must be reduced.
- **Type of protection:** special factors must be used when BS 3036 (semi-enclosed) fuses are employed.
- **Thermal insulation:** if cables are placed in thermal insulation, de-rating factors must be applied.
- **Voltage drop:** the length of circuit, the current it carries and the csa of the conductor will combine to affect the voltage drop. Appendix 12 of BS 7671 states that the maximum voltage drop between the installation origin and load for an LV installation fed from the public supply must not exceed 3 per cent for lighting and 5 per cent for other uses.

Conductor insulation

To insulate the conductors of a cable from each other and to insulate the conductors from surrounding metalwork, you must use materials with excellent insulating properties. The three main types you will find in domestic situations are PVC, XLPE and synthetic rubber.

- PVC or polyvinyl chloride is a thermoplastic polymer which is a good insulator, and is flexible and cheap, easy to work with and easy to install. However, it does not stand up to extremes of heat and cold. BS 7671 recommends that ordinary PVC cables should not constantly be used in temperatures above 60°C or below 0°C.
- XLPE or cross-linked polyethylene is a thermosetting compound. The cable has a high softening temperature, small heat distortion and high mechanical strength under high temperature. XLPE is also lighter than its PVC counterpart. At high voltage there is a potential of failure because of 'treeing', when moisture penetrates the cable and causes a reaction that breaks down the insulation. Heat-shrink tubing, which also provides stress control, will aid this situation.
- Synthetic rubbers, such as butyl rubber, will withstand high temperatures much better than PVC. They are normally used for the flexible final connection to items such as immersion heaters, storage heaters and boiler house equipment.

Environmental factors

Many factors affect cable selection, some of which have already been mentioned:

- risk of excessive ambient temperature
- effect of any surrounding moisture
- risk of electrolytic action

- proximity to corrosive substances
- risk of damage by animals
- exposure to the elements
- risk of mechanical stress or of mechanical damage
- aesthetic considerations.

Ambient temperature

On page 424, you read how the rate at which current-carrying cables get rid of heat is affected by the temperature surrounding the cable. Typical problem areas are boiler houses and plant rooms, thermally insulated walls and roof spaces. Low temperatures can also damage PVC cables. PVC cables stored in areas where the temperature has dropped to 0°C should be warmed slowly before being installed. However, if cables have been left out in temperatures below 0°C (perhaps during a heavy frost), you must report this to the person in charge of the installation.

Safety tip

Having two different metals together in the presence of moisture can cause an electrolytic action, resulting in the deterioration of the metal: for example, where brass glands are used with galvanised steel boxes. Take care to prevent this.

Moisture

Water and electricity do not mix, and you should take care at all times to avoid the movement of moisture into any part of an electrical installation, using watertight enclosures where appropriate. Any cable with an outer PVC sheath will resist the penetration of moisture and will not be affected by rot. However, you should use suitable glands for termination of these cables. Mineral-insulated cable can be affected by moisture even when indoors; if you cannot terminate the cable, seal the end of the cable to stop any moisture getting in.

Corrosive substances

Metal cable sheaths, cable armouring, glands and fixings of cables can also suffer from corrosion when exposed to certain substances, including:

- magnesium chloride, used in the construction of floors
- plaster undercoats containing corrosive salts/lime
- unpainted walls of lime or cement
- oak and other types of acidic wood.

Metalwork should be plated or given a protective covering. You may also need to use special materials such as a PVC-coated tray or PVC-sheathed cables and accessories.

Damage by animals

Rodents gnaw through cables and can leave them in a dangerous condition, so where necessary you should give cables additional protection or install them in conduit or trunking. Consider the same precautions where there are domestic animals, if possible keeping any installation out of their reach to prevent the effects of rubbing, gnawing and urine.

Exposure to the elements

Cables sheathed in PVC should not be installed in positions where they are exposed to direct sunlight as this causes them to harden and crack: the ultraviolet rays leach out the plasticiser in the PVC, making it brittle. Similarly, PVC cables should not be installed where they will be operating for long periods at temperatures below 0°C, such as when exposed to snow.

Mechanical stress and mechanical damage

Cables can be subject to mechanical stress (for example, when providing an overhead supply between buildings) and a **catenary wire** is used to support them. Cables installed in this way should have **drip loops** at either end, to allow for a degree of movement should the system be hit.

Flexible cables are often used to suspend luminaires from ceilings. To avoid stress, the maximum weight that can be accommodated is given in Appendix 4 of BS 7671 in Table 4F3A, as shown in Table 7.24. Bear in mind that cables can also suffer from stress when subjected to excessive vibration, which can cause breakdown of the insulation.

Key terms

Catenary wire – a wire hanging from two points on the same level horizontally.

Drip loop – a loop of cable, installed at a slight angle, that lets water run off the cable and away from walls.

Conductor (cross-sectional area mm²)	Current-carrying capacity		Maximum mass
	1-phase a.c.	3-phase a.c.	
0.5	3 A	3 A	2 kg
0.75	6 A	6 A	3 kg
1	10 A	10 A	5 kg
1.25	13 A	10 A	5 kg
1.5	16 A	16 A	5 kg
2.5	25 A	20 A	5 kg
4	32 A	25 A	5 kg

Table 7.24: Table 4F3A of Appendix 4 of BS 7671

Selecting the correct cable for protection from mechanical damage depends on the type of installation and the anticipated level of damage.

- In a domestic installation where the cables are to be concealed in walls and ceilings at a depth no less than 50 mm from any surface, the main function of the cable sheath is to protect the cable from light mechanical damage. Here you would probably use PVC/PVC and cpc (circuit protective conductor) cable.
- Where cables are to be installed on the surface of a building fabric or in an underground trench between buildings, they should have a metal sheath (such as MICC) or armouring (such as SWA) that will resist any likely mechanical damage.

Aesthetic considerations

Although not necessarily electrically relevant, where possible it makes sense to choose electrical systems that are pleasing to the eye and sympathetic to the building. For example, you would not choose to install surface-mounted, grey-enamelled metal-clad switches on the wall of a living room when all of the wiring is concealed in the walls. White, flush-mounted plate switches would be much neater.

10. Requirements for protecting cables installed in the building fabric and terminating in enclosures

Note on older wiring

Nearly all modern domestic wiring is recessed into the walls, but on older buildings you may find:

- wiring installed in vulcanised rubber insulation (VRI) with a tough rubber sheath (TRS) or a lead sheath – this is greatly affected by temperature and eventually the rubber becomes brittle, usually breaking off as it is handled
- wiring installed in slip-gauge conduit (light duty conduit with an open seam), which is not to be used as an earth return under any circumstances.

Protecting cables

This section covers some of the techniques required for the installation and rewiring of an existing building, and the regulations that apply for the protection of installed cables.

Cables run under floors or above ceilings

BS 7671 Regulations 522.6.100–10

Remember

At the end of the job when you replace the floorboards, screwing them back in place carefully will stop them squeaking.

In line with Regulation 522.6.100, a cable installed under a floor or above a ceiling should be run in such a way it is not liable to be damaged. A cable passing through a joist within a floor or ceiling construction or through a ceiling support must do one of the following:

- be at least 50 mm measured vertically from the top/bottom as appropriate, of the joist or batten
- incorporate an earthed metallic armour or screen

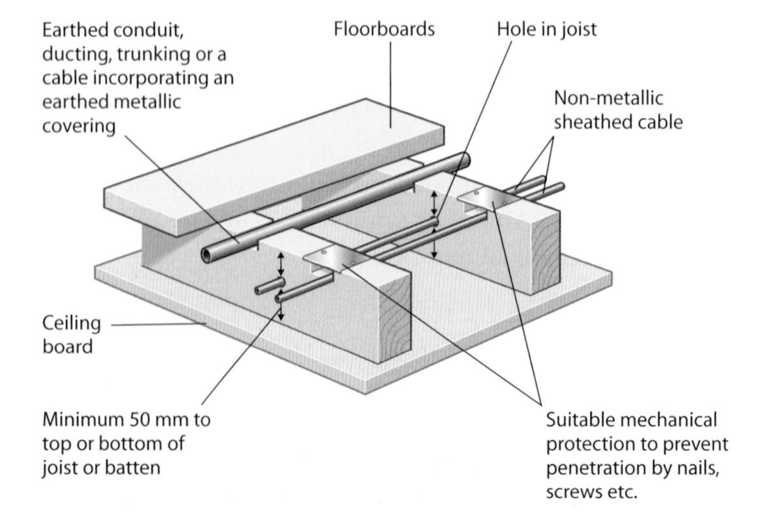

Fig 7.55: Dealing with cables under floorboards

- be enclosed in earthed conduit/trunking
- be mechanically protected against damage sufficient to prevent penetration of the cable by nails, screws and the like.

Take care to ensure that any holes or notches do not weaken the integrity of any load-bearing part of the structure (see Figure 7.55).

Cables run into walls

When cables have been installed in a wall at a depth less than 50 mm from any surface (for example, to feed a lighting switch or socket outlet), in accordance with Regulation 522.6.101 they must do one of the following:

- incorporate an acceptable earthed metallic covering
- be enclosed in acceptable earthed conduit, trunking or ducting
- be mechanically protected to prevent penetration by nails or screws
- be installed in a zone. Note that a zone formed on one side of a wall of less than 100 mm thickness extends to the reverse side, but only if the location of that accessory can be determined from the reverse side.

Where the finished installation will not be under skilled or instructed supervision:

- cables that are not mechanically protected and are installed in a designated zone (Regulations 522.6.102 and 522.6.101) must be provided with additional protection via a suitable RCD
- whatever depth the cable is from the surface of the wall or partition, when that wall is constructed using metallic parts other than nails and screws, cables must do one of the following:
    - incorporate an acceptable earthed metallic covering
    - be enclosed in acceptable earthed conduit, trunking or ducting
    - be mechanically protected to prevent penetration by nails or screws
    - be provided with additional protection by means of an RCD (Regulation 522.6.103).
- if the cable is installed at a depth of 50 mm or less from the wall or partition surface, Regulation 522.6.101 also applies.

Regs 522.6.101–103

Chasing

Chasing is cutting slots into a wall for conduit or cable. A chasing tool or attachment (available on most electric drills) makes this job easy; all that is required is a line on the wall as a guide to work to. Using a bolster chisel and hammer takes longer, but does the same job.

The back boxes for wall-mounted flush fittings must be mounted in a hole cut into the brickwork. The hole should be made slightly bigger than the box to allow plaster or filler to be applied around the edge and back of the hole, to make sure the box remains firm in the wall. The box can then be fitted in and secured. The front edge of the box must not protrude from the surface of the wall.

In new buildings, electrical wiring and back boxes are usually fitted before the plasterers render the walls – a process known as first fixing. Once all plastering and other decoration is complete, the accessories are then normally fitted – known as second fixing.

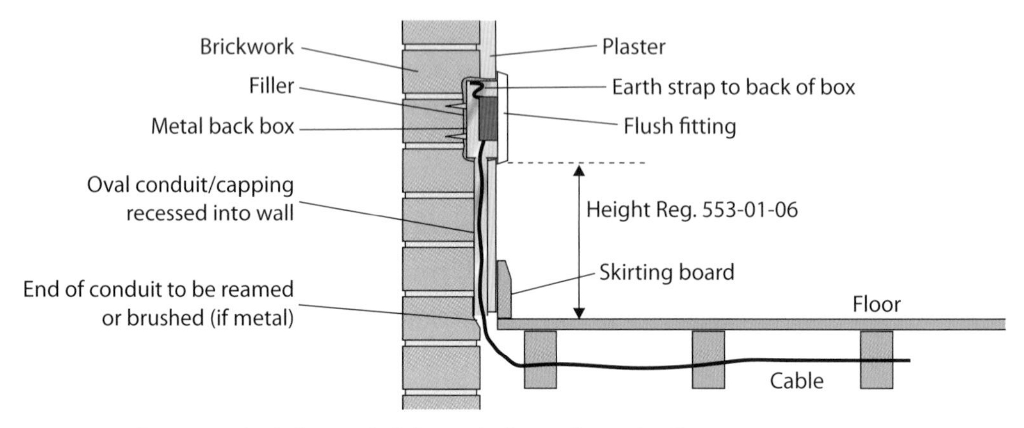

Fig 7.56: Flush fitting fed through the wall via the floor

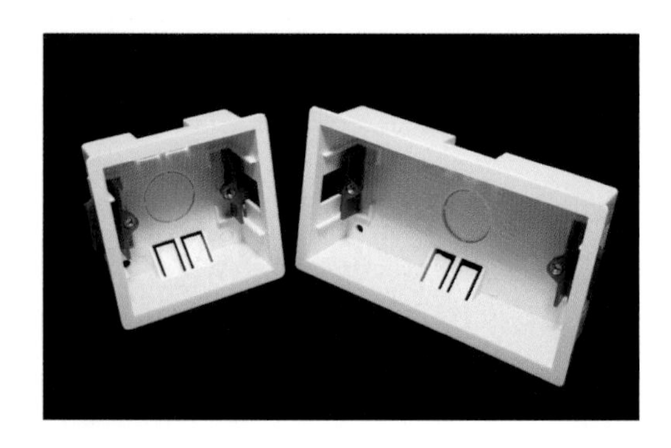

Dry lining box for wiring in partitions

Wiring in partitions

Wiring must be done before any lining (surface) is fixed in position. BS 7671 has the same requirements for partitions as for cables in walls.

The back boxes for fitting must be securely fixed, preferably to timber noggins in the structure of the partition. However, dry lining boxes can be used if the partition lining is of sufficient strength and thickness.

Ceiling fittings

Boxes for ceiling outlets must be securely fixed to either a joist or to a noggin between the joists as shown in Figure 7.57.

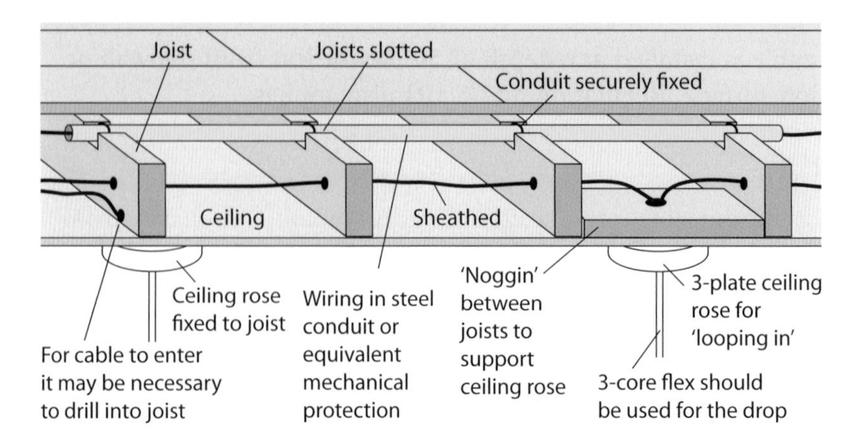

Fig 7.57: Cables through joists

Considerations when installing

You have seen how cables should be run through floor joists or through walls. Regulation 522.6.1 states that precautions should be taken to prevent abrasion of any cable. When cables pass through metalwork, this can be achieved using an insulated grommet. Where significant solar or ultraviolet radiation is experienced or expected, select or install a wiring system suitable for the conditions or provide adequate shielding.

Protection against heat damage

To avoid the effects of heat from external sources, such as hot water systems, equipment, manufacturing processes or solar gain, one or more of the following should be used (Regulation 522.2.1 applies):

- shielding
- placing sufficiently far from the heat source
- selecting a wiring system with regard for the additional temperatures
- local reinforcement or substitution of insulating materials.

Parts of a cable in any accessories, appliances or lights need to be able to cope with the temperatures they will face, or have extra insulation so that they can (Regulation 522.2.100 and Regulation 522.1.1).

BS 7671 Regs 522.2.100 and 522.1.1

Protection against the spread of fire

The first way you must minimise the risk of fire spreading is by choosing appropriate materials and equipment. Where a wiring system passes through building elements such as floors, walls, roofs, ceilings, partitions or cavity barriers, you need to seal the remaining openings to the same degree of fire resistance as before the wiring was installed (Regulation 527.2.10).

During installation, temporary sealing arrangements might also be necessary (Regulation 527.2.1.1), and any disturbances made to them during alterations must be repaired as soon as possible (Regulation 527.2.1.2).

Regulation 527.2.2 also states that, when a wiring system such as a conduit, cable ducting or cable trunking, busbar or busbar trunking penetrates elements of building construction that have specific fire-resistance, it should be internally sealed to maintain the degree of fire resistance, as well as being externally sealed (to comply with Regulation 527.2.1).

Mini-trunking

You can cause damage in your efforts to conceal wiring. For some types of domestic property (for example, with concrete floors), the costs may be disproportionate to the end look. In such cases it is often possible to conceal cabling within PVC mini-trunking.

Trunking is normally a square/rectangular casing system with a removable lid, through which cables are pulled between various pieces of electrical equipment. Electrical trunking provides good mechanical protection and

Did you know?

Some smaller types of mini-trunking are self-adhesive.

Remember

Patching up afterwards will probably be the job of the main contractor (builder). However, on smaller jobs you may have this responsibility.

is traditionally installed to carry many cables down a common route, branching off to supply individual circuits or items of equipment.

PVC trunking is manufactured in many styles and sizes. Unlike conduit, which is traditionally used as a complete system with unsheathed cables, PVC mini-trunking is often used as surface-mounted protection for short drops to accessories from ceilings for other cables, such as data cabling or PVC/PVC cables. This is especially useful when circuits are added to an existing installation and cannot be concealed without damaging the building fabric.

The repair needed will depend on the work done. In domestic installations, you may only need to use substances such as all-purpose filler to repair damage around ceiling outlets or switch boxes. However, in a domestic rewire, you should also think about flooring. Replace floorboards with care, screwing them back in place after you work to stop them squeaking.

PROGRESS CHECK

1. When cables have been installed in a wall at a depth less than 50 mm from any surface, what must happen?
2. If cables are to be used in situations such as to provide an overhead supply from one building to another, they will be subjected to mechanical stress. How can this be avoided?
3. What problem will occur when two different metals are joined together in the presence of moisture?

Terminating and connecting

Here you will use statutory documents (HASAWA, EaWR), site drawings, wiring diagrams, the specification and relevant technical data such as manufacturer's instructions. These were covered in Unit 3 on pages 118–121. Look back now to refresh your memory.

Did you know?

Means of connection include the terminal of a wiring accessory, such as a socket, a junction box, a wiring centre, a distribution board or compression lugs. It is the entry of the cable end into an accessory that is known as a termination.

Again, the most important British Standard here is BS 7671 – more commonly referred to as the IET Wiring Regulations. The requirements for electrical connections are given in Part 5, Chapter 52, Section 526. Here are the main points.

BS 7671 Chapter 52 Section 526

- Regulation 526.1 requires that any connection between conductors or equipment must provide durable electrical continuity as well as adequate mechanical strength and protection.
- Regulation 526.2 requires the means of connection to take into account the conductor material and csa, its insulation, the number of conductors to be connected and the terminal temperature in normal service.
- Regulation 526.3 generally requires every connection to be accessible for inspection, testing and maintenance.

A cable termination of any kind should securely anchor all the wires of the conductor that may put mechanical stress on the terminal or socket, to prevent the risk of disconnection.

In the case of a stranded conductor, one or more strands or wires left out of the terminal or socket will reduce the effective cross-sectional area of the conductor at that point. This may result in increased resistance and probably overheating.

You should twist strands together with pliers before terminating, taking care not to damage the wires. When terminating flexible cords into a conduit box, the flex should be gripped with a flex clamp. Any cord grips used to secure flex or cable should be clamped onto the protective outer sheathing. Table 7.25 shows the most common terminal types.

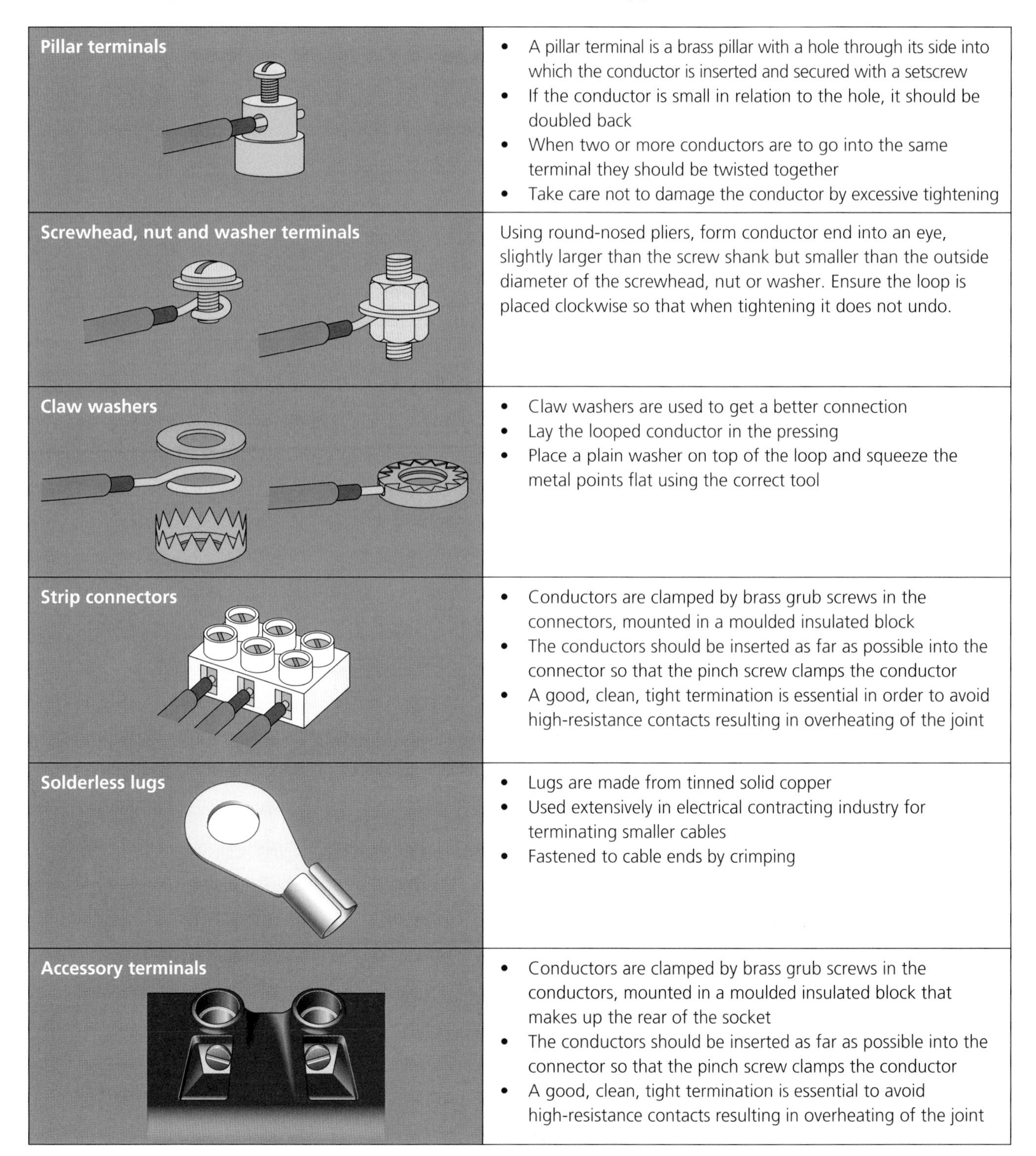

Terminal type	Description
Pillar terminals	• A pillar terminal is a brass pillar with a hole through its side into which the conductor is inserted and secured with a setscrew • If the conductor is small in relation to the hole, it should be doubled back • When two or more conductors are to go into the same terminal they should be twisted together • Take care not to damage the conductor by excessive tightening
Screwhead, nut and washer terminals	Using round-nosed pliers, form conductor end into an eye, slightly larger than the screw shank but smaller than the outside diameter of the screwhead, nut or washer. Ensure the loop is placed clockwise so that when tightening it does not undo.
Claw washers	• Claw washers are used to get a better connection • Lay the looped conductor in the pressing • Place a plain washer on top of the loop and squeeze the metal points flat using the correct tool
Strip connectors	• Conductors are clamped by brass grub screws in the connectors, mounted in a moulded insulated block • The conductors should be inserted as far as possible into the connector so that the pinch screw clamps the conductor • A good, clean, tight termination is essential in order to avoid high-resistance contacts resulting in overheating of the joint
Solderless lugs	• Lugs are made from tinned solid copper • Used extensively in electrical contracting industry for terminating smaller cables • Fastened to cable ends by crimping
Accessory terminals	• Conductors are clamped by brass grub screws in the conductors, mounted in a moulded insulated block that makes up the rear of the socket • The conductors should be inserted as far as possible into the connector so that the pinch screw clamps the conductor • A good, clean, tight termination is essential to avoid high-resistance contacts resulting in overheating of the joint

Table 7.25: The most common terminal types

Safety tip

Remember that, when current flows through a resistance, heat is developed. At a terminal, the consequent expansion and contraction due to heating and cooling can loosen the conductor in the terminal, particularly if it is under tension.

Connections

The joints in non-flexible cables must be made by soldering, brazing or welding mechanical clamps, or be of a compression type and be insulated. The devices used should relate to the size of cable and be insulated to the voltage of the system being used. Choose connectors of the appropriate British Standard, consider the temperature of the environment, and install them in a way that will prevent corrosion.

Plastic connector

Plastic connectors are common, and often come in a block of 10 or 12, nicknamed a 'chocolate block'. The size is important and should relate to the current rating of the circuit. They are available in 5, 15, 20, 30 and 50 A ratings.

Porcelain connectors

Porcelain connectors are used where a high temperature is expected. They are often found inside appliances such as water heaters, space heaters and luminaires. They should also be used where fixed wiring has to be connected into a totally enclosed fitting.

Compression joints

Compression joints include many connectors that are fastened onto the conductors, usually using a crimping tool. The connectors may be used for straight-through joints or special end configurations. If the conductors are not clean when making the joint, this may result in a high-resistance joint that could cause a build-up of heat, leading to a fire risk.

Uninsulated connectors

Uninsulated connectors are used to connect earth cables and protective conductors. They are often required inside wiring panels, fuse boards, and so on.

Junction boxes and wiring centres

The two important factors when choosing junction boxes are the current rating and the number of terminals. Junction boxes are usually either for lighting or socket outlet circuits.

Six-terminal junction box

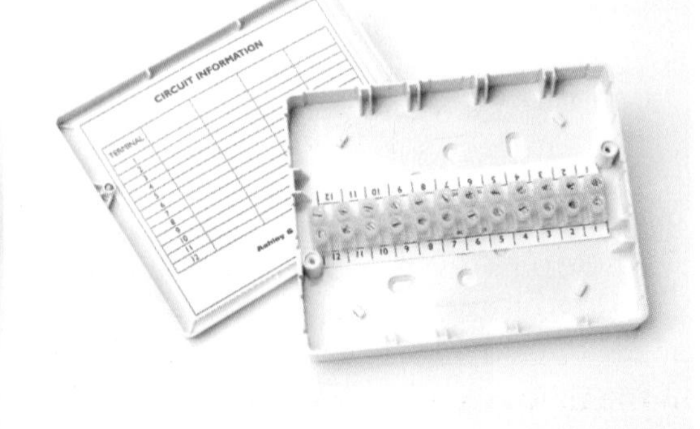

Modern RB4 junction box

Wiring centres are similar to junction boxes, except their terminals are pre-arranged and partially pre-connected to align with popular heating system arrangements. When a conductor end is formed into an 'eye', you should place the conductor over the post or terminal so that the screw or nut draws the eye with it when tightened. If not, tightening the nut or screw will unwind the eye of the conductor, resulting in an imperfect connection.

Consumer units and distribution boards

A consumer unit and a distribution board are basically the same thing: an assembly that allows the control and distribution of electricity to circuits and systems.

The distribution board tends to be used in more commercial and industrial situations and consequently is metal-clad and heavy-duty. The domestic consumer unit will typically be made from plastic and will comprise a double-pole isolation switch on the incoming supply from the electricity company, and an assembly of one or more fuses, circuit breakers, residual current operated devices or RCBOs. In the modern consumer unit these items will be mounted on DIN-rail.

Remember

Make sure that any additional or replacement circuit breakers are of the correct type and style for the consumer unit concerned.

With the changes introduced by the 17th Edition IET Wiring Regulations (BS 7671), consumer units in the UK will generally have to provide RCD protection to all cables embedded in walls. This can be done by having all circuits protected by one RCD, individual circuits protected by an RCBO, or a dual split-load consumer unit as shown below could be arranged with each RCD protecting circuits as follows.

RCD1
Upstairs lights, downstairs sockets, garage sockets, shower

RCD2
Downstairs lights, upstairs sockets, cooker, central heating

Double-pole main switch

Dual split-load consumer unit

By arranging circuits like this, there will always be lighting and power available on one floor if either RCD trips out.

The image to the right shows the inside of a consumer unit where certain circuits of an existing installation (old colour wiring used) are being protected by an RCD. Notice that all cables have extra length on them, which is to allow for any possible future alterations that may mean connecting the cable to a terminal which is further away within the consumer unit.

Consumer unit

Proving terminations and connections are electrically and mechanically sound

You can use a low-resistance ohmmeter as part of the process of testing to see whether a joint or termination is of acceptable quality. Other, more basic methods include visual inspection and tugging the cable to see if it is secure.

Look at Figure 7.58, which shows two pieces of metal joined together, fitting over an insulated post terminal and held in place by a nut.

If you take an insulation resistance tester, acting as a low-resistance ohmmeter, and apply a lead to each piece of metal, the meter shows the effect of a 'dead short': you have zero resistance. There is nothing to impede the meter signal from input at one side, through the metals to the lead on the other. This is the perfect joint.

If the two metals are not tightly connected, a resistance will be present and the signal will struggle to get from one side to the other. The opposite situation of no resistance is a resistance so high that its value is infinite – infinity. This means that the gap between the two metals is so great that the signal from one lead can never reach the other lead, as shown in Figure 7.59 below.

Resistance, whether caused by loose connections or the effects of corrosion, will be a value between zero and infinity. Even so, this is called a high-resistance joint.

Safety tip

If a large current is passing through a high-resistance joint, a lot of heat can be generated. This can lead to arcing and damage to the equipment or cable, and can pose a serious fire risk.

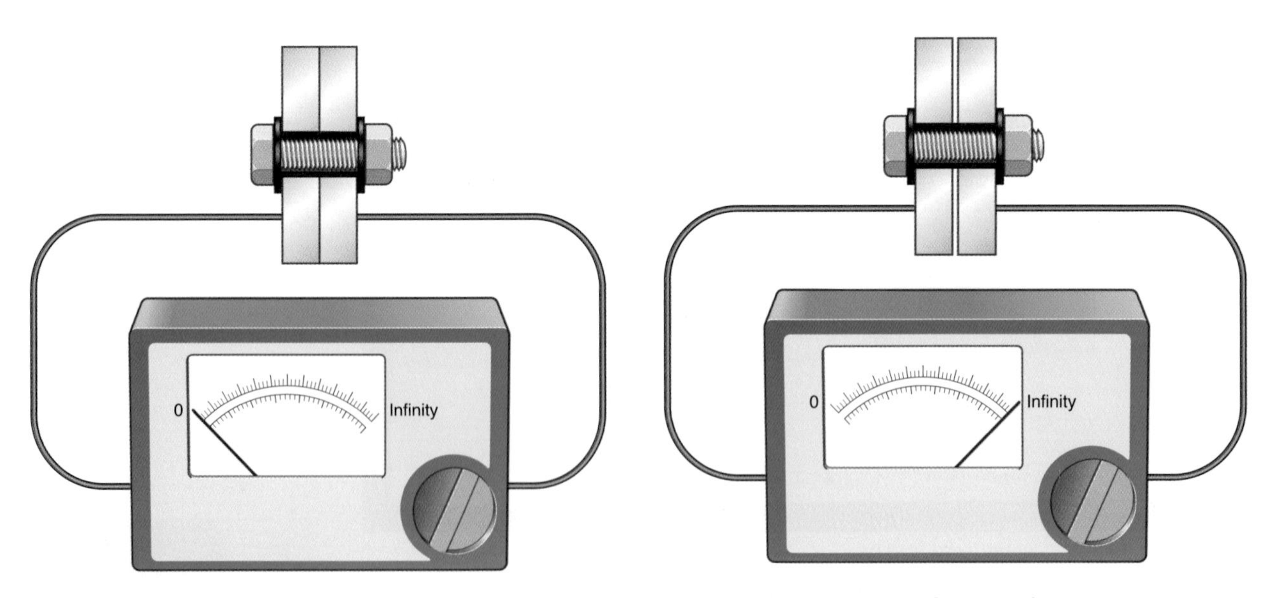

Figure 7.58: The perfect joint

Figure 7.59: Joint not tight enough

Remember

Sleeve the CPC with green and yellow sleeving before connecting to the equipment and back box.

Terminating PVC/PVC flat profile cable

Once the cable ends are prepared, the conductor should be doubled over before connecting to the equipment. This gives the equipment grub screw a large surface area to grip.

Terminating PVC/PVC cables

Step 1 Nick the cable at the end with your knife and pull apart.

Step 2 When the required length has been stripped, cut off the surplus sheathing with the knife.

Step 3 The insulation can be stripped from the conductors with the knife or with a pair of purpose-made strippers. Examine the conductor insulation for damage.

11. Electrical inspection and testing requirements

Safety

It is the responsibility of the person carrying out the test to ensure the safety of themselves and others. Where testing does not require any part of the installation to be made live, it should be made dead and isolated safely (see pages 407–415).

When using any test instrument, you must:

- have a thorough understanding of the equipment being used and its rating
- follow all safety procedures, such as putting up warning notices and barriers
- make sure the instrument conforms to the appropriate British Standard safety specifications (BS EN 61010 or, for older instruments, BS 5458 if it is in good condition and has been recently calibrated)
- check that test leads, probes and clips are in good condition, are clean and have no cracked insulation (observing where appropriate the requirements of GS38).

BS EN 61010
BS 5458

Pay particular attention when using instruments that can generate a test voltage in excess of 50 volts, such as insulation resistance testers. If you touch the live terminals, you will get a shock and even though this may not be harmful itself, it may make you lose concentration, which can be dangerous, especially if working at height.

Take care with instruments that use the supply voltage for the test, such as when earth loop impedance testing or when testing a residual current device (RCD). These can impose a voltage on associated earthed metalwork, so take care to avoid an electric shock.

Competence and responsibility

Anybody inspecting and testing an installation must be skilled and experienced and have enough knowledge of the installation type to make sure there is no risk of injury or damage (see page 452).

When carrying out inspection and testing you must:

- ensure no danger occurs to persons, property or livestock
- compare the installation design against the results of the inspection and testing
- take a view on the installation's condition and advise remedial work if necessary
- recommend immediate isolation of any defective part and inform the client.

What do we test and when?

Apart from when fault finding, there are various times when inspection and testing are needed. Before carrying out any inspection and test procedure, make sure that all the required information is available, the person carrying out the procedure is competent to do so, and all safety requirements have been met.

Periodic inspection and testing

EAWR
BS 7671 Reg 621.1
The Landlords and Tenants Act 1985

Under the Electricity at Work Regulations (1989), all systems must be maintained to prevent danger, as far as is practical. BS 7671 then requires every electrical installation to undergo regular inspection and testing to make sure that it remains in a satisfactory and safe condition (Regulation 621.1). On top of this, The Landlords and Tenants Act 1985 requires landlords to keep installations in proper working order; this affects rented domestic and residential accommodation, such as student housing.

Initial verification

BS 7671 Reg 610.1

BS 7671 requires every electrical installation to be inspected and tested during its construction and on completion, to make sure it meets all requirements (Regulation 610.1). Initial verification is meant to confirm that the installation complies with the designer's requirements and has been constructed, inspected and tested in accordance with BS 7671.

Inspection is a key part of this procedure and should be carried out before electrical testing, usually with that part of the installation disconnected from the supply. The inspector must record the results of all such inspections and tests and compare them with the relevant design criteria (mainly BS 7671, but also any specific design requirements).

Information

When you carry out an inspection and test you need to know:

- maximum demand of the installation expressed in amperes per phase together with details of the number and type of live conductors, both for the source of energy and for each circuit to be used within the installation (for example, single-phase two-wire a.c. or three-phase four-wire a.c.)
- general characteristics of the supply, such as:
  - nominal voltage (U_o)
  - nature of the current (I) and its frequency (Hz)
  - prospective short circuit current at the origin of the installation (kA)
  - earth fault loop impedance (Z_e) of that part of the system external to the installation
  - type and rating of the overcurrent device acting at the origin of the installation (if not known it must be established either by calculation, measurement, inquiry or inspection)
- type of earthing arrangement used for the installation, such as TN-S, TN-C-S, TT
- type and composition of each circuit (i.e. details of each sub-circuit, what it is feeding, the number and size of conductors and the type of wiring used)
- location and description of all devices installed for the purposes of protection, isolation and switching (such as fuses/circuit breakers)
- details of the method selected to prevent danger from shock in the event of an earth fault (invariably protection by earthed equipotential bonding and automatic disconnection of the supply)
- presence of any sensitive electronic device susceptible to damage by the application of 500 volts d.c. when carrying out insulation resistance tests.

You can get this information from sources such as the project specification, contract drawings, as-fitted drawings or distribution board schedules. Otherwise approach the person ordering the work.

Remember

BS 7671 requires that you record this information on an on-site record of the design. This would usually be in the Health and Safety File required by the Construction (Design and Management) Regulations.

BS 7671
CDMR

Initial inspection

Your initial inspection should verify:

BS 7671 611.0, 611.1, 611.2

- all equipment and material is of the correct type and complies with applicable British Standards or acceptable equivalents
- all parts of the fixed installation are correctly selected and erected
- no part of the fixed installation is visibly damaged or otherwise defective
- the equipment and material used are suitable for the installation relative to the environmental conditions.

Remember

You must have suitable inspection checklists prepared and the certification document available for completion.

Activity

You are one of the tools used as part of an inspection, so you need to learn to use your senses. What might you see, smell or hear that could indicate a problem?

Remember

Only protective conductors can be identified by a combination of green and yellow.

Many new installations will be hidden once the building fabric has been completed, so it is best to do some visual inspection throughout the installation process. For example, conduit, cable tray or trunking is often installed above the ceiling or below the floor; once tiles have been fitted, it can be difficult and expensive to access it for inspection purposes. The same applies to some tests, such as earth testing.

Regulation 611.3 Inspection requirements

BS 7671 Regulation 611.3 states that, as a minimum, inspection must include all particular requirements for special installations or locations (BS 7671 Part 7) of the items listed below.

All references to Tables and Chapters in this section relate to BS 7671.

1) Connection of conductors

Check that every connection between conductors or between conductors and equipment is electrically continuous and mechanically sound. All connections must be adequately enclosed but accessible where required by the Regulations.

2) Identification of conductors

Check that each conductor is identified as per Table 51. Numbered sleeves can be used, but coloured insulation or sleeving (not green) is more common.

3) Cable routes

Check cables are safely routed and protected against mechanical damage where necessary. Permitted cable routes must be clearly defined (note the RCD situation) or cables must be installed in earthed metal conduit or trunking.

4) Current-carrying capacity

Where practicable, check the cable size for current-carrying capacity and voltage drop, based on information from the designer.

5) Verification of polarity

Check that all single-pole devices are connected in the phase conductor only.

6) Accessories and equipment

Check to ensure these have been connected correctly, including correct polarity.

7) Selection and erection to minimise the spread of fire

Check, preferably during construction, that fire barriers, suitable seals and/or other means of protection against thermal effects have been provided as per the Regulations.

8) Protection against electric shock

Check that the requirements of Chapter 41 have been met.

9) Prevention of mutual detrimental influence

Take account of the proximity of other electrical services in a different voltage band and of non-electrical services and influences. For example, fire alarm and emergency lighting circuits must be separated from other cables and from each other, and Band 1 and Band 2 circuits must not be present in the same enclosure or wiring system unless they are either segregated or wired with cables suitable for the highest voltage present.

Band 1 circuits are circuits that are nominally extra-low voltage, i.e. not exceeding 50 volts a.c. or 120 volts d.c., for example telecommunications or data and signalling. Band 2 circuits are circuits that are nominally low voltage, that is exceeding extra-low voltage but not exceeding 1000 volts a.c. between conductors or 600 volts a.c. between conductors and earth.

10) Isolating and switching devices

There must be an effective, suitably positioned way to cut off all voltage from every installation, every circuit within the installation and all equipment, if necessary. Switches and/or isolating devices of the correct rating must be installed per these requirements. Where practicable, you should carry out an isolation exercise to check this, including switching off, locking-off and testing to verify that the circuit is dead and there is no other source of supply.

11) Under (or no-volt) voltage protection

Take suitable precautions where a loss or lowering of voltage or a subsequent restoration of voltage could cause danger. The most common situation would be where a motor-driven machine stops due to a loss of voltage and unexpectedly restarts when the voltage is restored, unless precautions are employed such as the installation of a motor starter containing a contactor. Where unexpected restarting of a motor may cause danger, a motor starter designed to prevent automatic restarting must be provided.

12) Labelling

Check that labels and warning notices have been fitted, such as labelling of circuits, MCBs, RCDs, fuses and isolating devices, periodic inspection notices advising of the recommended date of the next inspection, and warning notices about earthing and bonding connections.

13) Selection of equipment appropriate to external influences

Select equipment that is suitable for the environment in which it is likely to operate. You should consider: ambient temperature, external heat sources, water, likelihood of corrosion, ingress of foreign bodies, impact, vibration, flora, fauna, radiation, building use and structure.

14) Access to switchgear and equipment

Every piece of equipment that needs operation or attention must be installed with adequate and safe means of access and working space.

15) Presence of danger notices and other signs

Appropriate notices must be suitably located and give warnings about voltage, isolation, periodic inspection and testing, RCDs and earthing and bonding connections.

16) Presence of diagrams, charts and other similar information

All distribution boards should have a schedule attached in or near them with information on types of circuits, number and size of conductors, type of wiring, and so on.

17) Erection methods

The correct correct methods of installation should be checked. Fixings of switchgear, cables, conduit and so on, must be adequate and suitable for the environment.

PROGRESS CHECK

1 What is the most important British Standard regarding terminating and connecting?

2 Which are the only conductors that can be identified by a combination of green and yellow?

3 What four things should an initial inspection verify?

Inspection schedule and checklist

You should note the results on an inspection schedule, then draw up an inspection checklist to ensure all the requirements of the Regulations have been met

Appendix 6 of BS 7671 gives examples of model certificates and schedules. An example of an inspection schedule is given in page 477 and an inspection checklist is provided in the Appendix.

For domestic electrical installations, compliance with BS 7671, a non-statutory document, is the only requirement.

Initial testing

BS 7671 Regulation 610.1 states that 'Every installation shall, during erection and/or on completion before being put into service, be inspected and tested to verify, so far as is reasonably practicable, that the requirements of the Regulations have been met. Precautions shall be taken to avoid danger to persons, livestock, and to avoid damage to property and installed equipment during inspection and testing.'

BS 767 Regs 610.1 and 610.4

Regulation 610.4 goes on to state that we must verify that any alteration or addition to an installation does not impair the safety of the existing installation.

The sequence of tests

BS 7671 Regulation 612 lists the sequence in which tests should be carried out. Installation testing can be dangerous and the danger level can increase if tests are not carried out in the correct sequence. For safety, protective conductors should be in place before you inject current or carry out any live tests, and insulation resistance should be satisfactory before carrying out an earth loop impedance test. Some tests will be carried out using links between known conductors established from earlier tests in the sequence.

Initial tests should be carried out in the following sequence, where applicable, before the supply is connected or with the supply disconnected as appropriate:

- Continuity of protective conductors including main and supplementary bonding (612.2.1)
- Continuity of ring final circuit conductors (612.2.2)
- Insulation resistance (612.3)
- Protection by SELV, PELV or electrical separation (612.4)
- Protection by barriers/enclosures provided during erection (612.4.5)
- Insulation resistance/impedance of non-conducting floors and walls (612.5)
- Polarity (612.6)
- Earth electrode resistance (612.7)
- Protection by Automatic Disconnection of Supply (ADS) (612.8)
- Earth fault loop impedance (612.9)
- Additional protection (612.10)
- Prospective fault current (612.11)
- Phase sequence (612.12)
- Functional testing (612.13)
- Verification of voltage drop (612.14).

The test results should be recorded on an inspection schedule (see page 477) and compared to the design criteria. If any test gives a bad result, you need to rectify the fault, then redo that test and any preceding tests.

Key term

nulled out – to measure the total resistance value of a circuit it is necessary to measure the resistance of the test leads and deduct this value from the overall result – this is known as nulling out the leads.

Continuity of protective conductors including main and supplementary bonding (612.1)

Every protective conductor, including each bonding conductor, must be tested to verify that it is electrically sound and correctly connected.

This test is carried out with a low-resistance ohmmeter. As well as checking the continuity of the protective conductor, it also measures $R_1 + R_2$ which, when corrected for temperature and added to the value of Z_e, will enable the designer to verify the calculated earth fault loop impedance Z_s.

Test method 1

Before carrying out this test, the leads should be **nulled out**. If the test instrument does not have this facility, you should measure the resistance of the leads and deduct it from the readings. The line conductor and the protective conductor are then linked together at the consumer unit or distribution board.

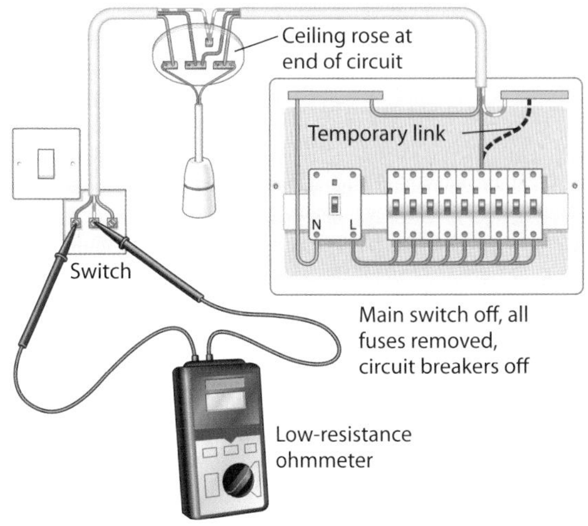

Figure 7.60: Test method 1 for a lighting circuit

Use the ohmmeter to test between the line and earth terminals at each outlet in the circuit. Record the measurement at the circuit's extremity: this is the value of $R_1 + R_2$ for the circuit under test.

This method should be carried out before any supplementary bonds are made.

Test method 2

Connect one lead of the continuity tester to the consumer's main earth terminals and then the other lead to a trailing lead, which you use to make contact with the protective conductor at light fittings, switches, spur outlets, and so on.

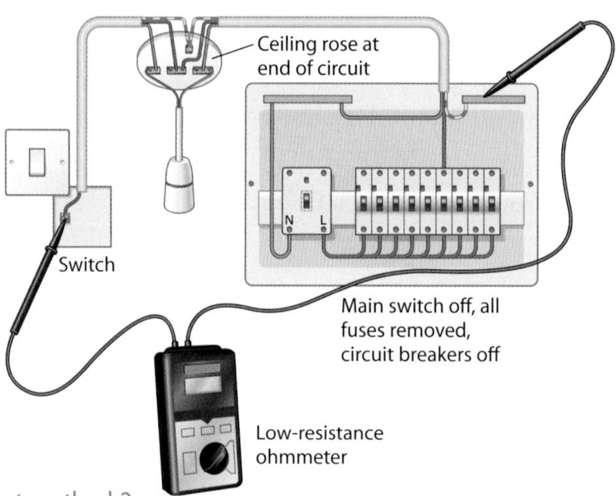

Figure 7.61: Test method 2

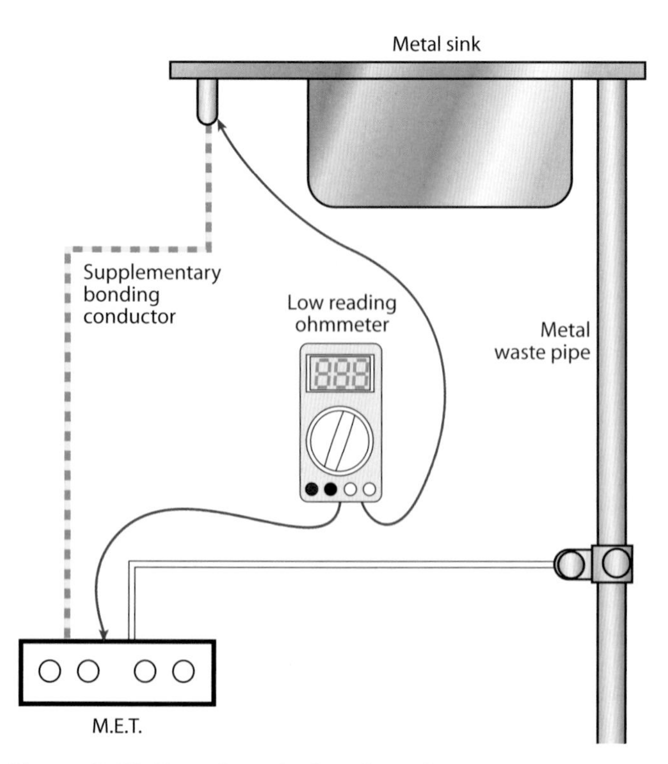

Figure 7.62: Test of continuity of supplementary bonding conductors

As you can see from Figure 7.61, the resistance of the test leads and wandering lead will be included in the result; therefore, if the instrument does not have a nulling facility, you must measure their resistance and subtract it from the reading obtained.

As you are only testing the protective conductor, only record R_2 on the installation schedule.

Test of the continuity of supplementary bonding conductors

Use test method 2, where the ohmmeter leads are connected between the points being tested, between simultaneously accessible extraneous-conductive-parts (such as pipework and sinks) or between simultaneously accessible extraneous-conductive-parts and exposed-conductive-parts (the metal parts of the installation).

This test will verify that the conductor is sound. To check this, move the probe to the metalwork to be protected as shown in Figure 7.62. This method is also used to test the main equipotential bonding conductors.

Where ferrous enclosures (such as conduit, trunking or steel-wire armouring) have been used as the protective conductors, you should:

- inspect the enclosure along its length to verify its integrity
- perform the standard ohmmeter test using the appropriate test method described above.

If you have any doubt about the soundness of this conductor, perform a further test using a phase-earth loop impedance tester after the connection of the supply. If there is still doubt, a further test can be carried out using a high-current, low-impedance ohmmeter, which has a test voltage not exceeding 50 volts and can provide a current approaching 1.5 times the design current of the circuit, but the current need not exceed 25 A.

Continuity of ring final circuit conductors

A test is required to verify the continuity of each conductor including the circuit protective conductor (cpc) of every ring final circuit. The test results should establish that the ring is complete, has no inter-connections and is not broken.

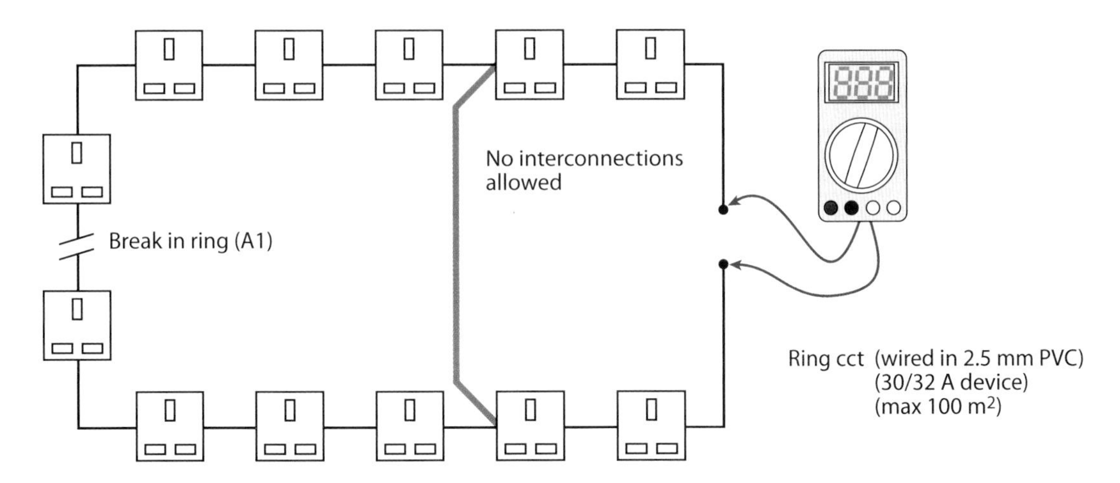

Figure 7.63: Test of continuity of ring final circuits

To establish that no interconnected multiple loops have been made in the ring circuit, you can visually check each conductor throughout its entire length. However, in most circumstances this will not be practicable, so you will need to follow these steps.

Step 1

The line, neutral and protective conductors are identified and the end-to-end resistance of each is measured separately (see Figure 7.64 on page 458). A finite reading confirms that there is no open circuit on the ring conductors under test.

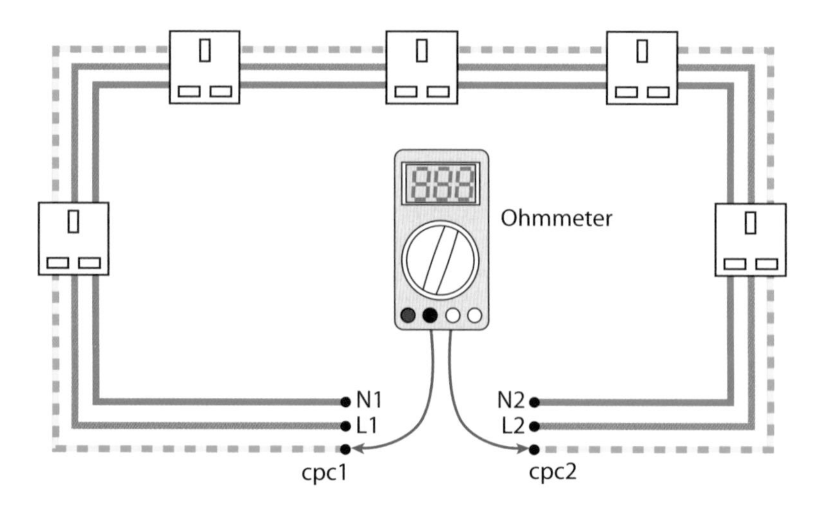

Figure 7.64: Step 1 Measurement of line, neutral and protective conductors

The resistance values obtained should be the same (within 0.05Ω) if the conductors are the same size. If the protective conductor has a reduced csa, its resistance will be proportionally higher than that of the line or neutral loop. If these relationships are not achieved, either the conductors are incorrectly identified or there is a problem at one of the accessories.

Step 2

The line and neutral conductors are then connected together so that the outgoing line conductor is connected to the returning neutral conductor and vice versa, as shown in Figure 7.65. The resistance between the line and neutral conductors is then measured at each socket outlet.

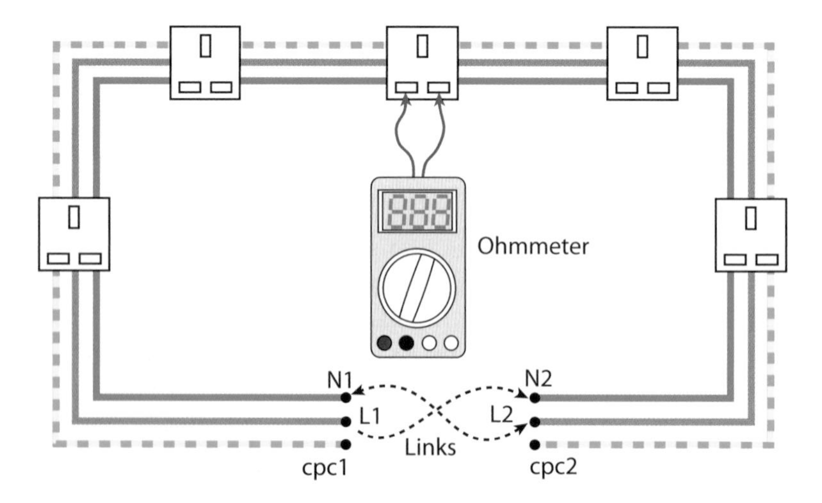

Figure 7.65: Step 2 Line and neutral conductors connected together

The readings obtained at each socket will be substantially the same provided they are connected to the ring (the distance around a circle is the same no matter where you measure it from); and the value will be approximately half the resistance of the line or the neutral loop resistance. Any sockets wired as spurs will have higher resistance value due to the extra length of the spur cable.

Step 3

Repeat Step 2, but this time with the line and cpc cross-connected as shown in Figure 7.66.

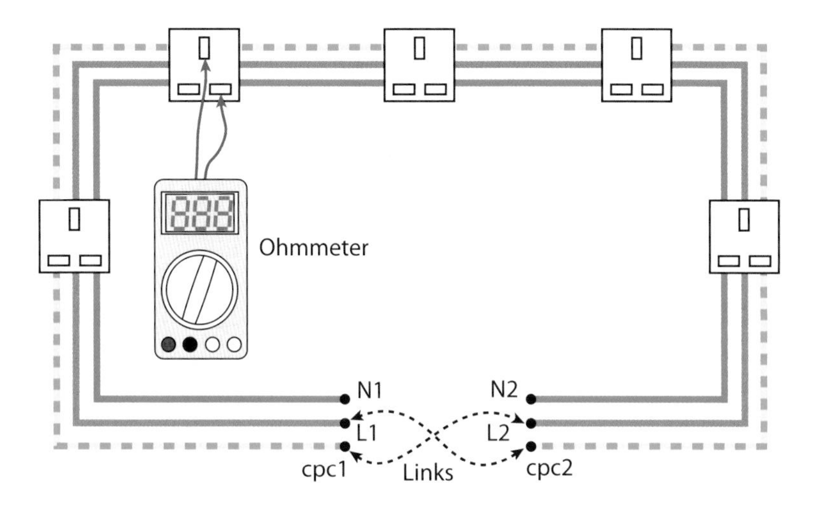

Figure 7.66: Step 3 Line and cpc cross-connected

Measure the resistance between the line and earth at each socket. Again, as they are connected on a ring, the readings should be basically the same and the value at each socket, with the value being about ¼ of the line plus cpc loop resistances, namely

$$\frac{R_1 + R_2}{4}.$$

The highest value recorded will represent the maximum $R_1 + R_2$ of the circuit; record this on the test schedule. This can also be used to determine the earth loop impedance (Z_s) of the circuit to verify compliance with the loop impedance requirements of the Regulations.

Insulation resistance

For compliance with BS 7671, insulation resistance tests verify that the insulation of conductors, electrical accessories and equipment is satisfactory and that electrical conductors and protective conductors are not short-circuited, or do not show a low insulation resistance (which would indicate faulty insulation). In other words, you are testing to see whether the insulation of a conductor is so poor as to allow any conductor to 'leak' to earth or to another conductor.

Before testing make sure that:

- pilot or indicator lamps and capacitors are disconnected from circuits to avoid an inaccurate test value
- voltage-sensitive electronic equipment (such as dimmer switches, delay timers and power controllers) is disconnected
- there is no electrical connection between any line and neutral conductor (for example, lamps left in).

Figure 7.67 shows why we remove lamps. The lamp filament effectively creates a short circuit between the line and neutral conductors – the same effect as if there were no insulation at all.

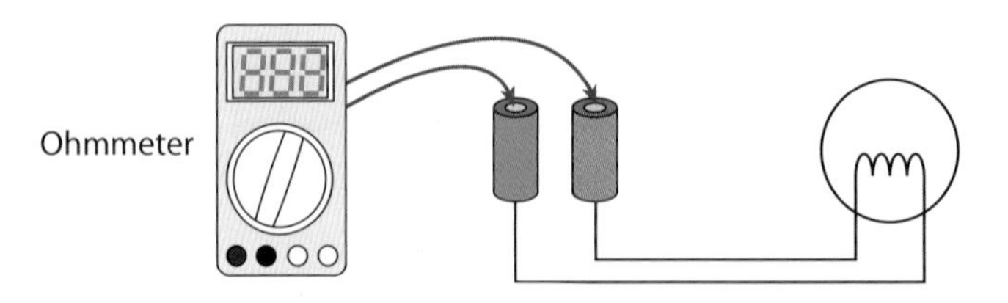

Figure 7.67: Why we remove lamps

BS 7671 Table 61

Here you use an insulation resistance tester meeting the criteria laid down in BS 7671, with insulation resistance tests carried out using the appropriate d.c. test voltage as specified in Table 61 of BS 7671.

The installation can be said to conform to the Regulations if the main switchboard and then each distribution circuit (tested separately, with all its final circuits connected but with current using equipment disconnected), have an insulation resistance not less than that specified in Table 61 of BS 7671 (see Table 7.26).

Circuit nominal voltage	Test voltage d.c. (V)	Minimum insulation resistance
SELV and PELV	250	0.5 MΩ
Up to and including 500 V with the exception of the above systems	500	1.0 MΩ
Above 500 V	1000	1.0 MΩ

Table 7.26: Circuit nominal voltage (from BS 7671 Table 61)

For a basic installation (one with only one distribution board), you would normally carry out the test on the whole installation, with the main switch off, all fuses in place, switches and circuit breakers closed, lamps removed, and fluorescent and discharge luminaires and other equipment disconnected.

Key term

strappers – the two conductors that interconnect between two 2-way or 2-way al intermediate switches.

Where you cannot remove lamps or disconnect any current-using equipment, the local switches controlling them should be open. But remember, on any two-way/intermediate circuits you will have to operate the two-way switch and retest the circuit to make sure that you have tested all of the **strappers**.

Although an insulation resistance value of not less than 1 MΩ complies with BS 7671, if you record a value of less than 2 MΩ, there could be a defect, so you should test each circuit separately.

You are now checking two things:

- conductors under test leaking to another conductor
- any conductor under test leaking to earth.

Most electricians prefer to test between individual conductors rather than to group them together (Figure 7.68 shows the ten readings that would need to be taken for a three-phase circuit).

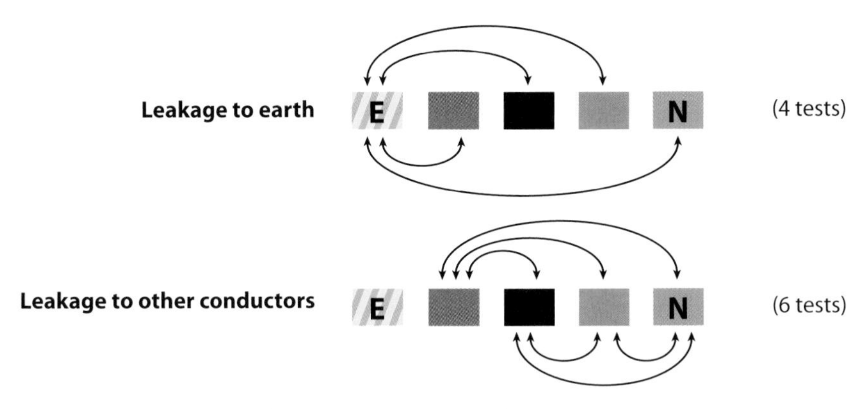

Figure 7.68: Readings for a three-phase circuit

PROGRESS CHECK

1. What are the first and last tests in the correct sequence?
2. Why will the readings obtained at each socket on a ring final circuit be substantially the same?
3. Why should you remove lamps before testing insulation resistance?

Protection by SELV, PELV or electrical separation (rarely required)

Where the protective measures of SELV, PELV or electrical separation have been used, you must carry out an insulation resistance test with the value being no less than that specified in Table 61 of BS 7671 for the circuit with the highest voltage present in the location. For example, if an area is being protected by SELV and that area also has low voltage circuits in it, the minimum insulation resistance value for the SELV circuit must be no lower than that of the low voltage circuit, namely no less than 1 MΩ.

For electrical separation, you should inspect the source of supply to prove that it does not exceed 500 V. The insulation between live parts of the separated circuit and any adjacent conductor (in the same enclosure or touching) and/or to earth must be tested at 500 V d.c. with insulation resistance being no less than 1 MΩ.

Polarity

You need to check the polarity of all circuits. This is a two-stage process, done before connection to the supply (using either an ohmmeter or the continuity range of an insulation and continuity tester) and then confirmed once the supply has been switched on (using an approved voltage indicator).

The polarity test verifies that:

- each fuse and single-pole control and protective device is connected in the line conductor only
- with the exception of ES14 and ES27 lamp holders to BS EN 60238, in circuits that have an earthed neutral, centre contact bayonet and ES lamp holders have the outer or screwed contacts connected to the neutral conductor
- wiring has been correctly connected to socket outlets and similar accessories.

As Figure 7.69 shows, having established the continuity of the cpc in an earlier test, you now use this as a long test lead, temporarily linking it out with the circuit line conductor at the distribution board and then testing across the line and earth terminals at each item in the circuit.

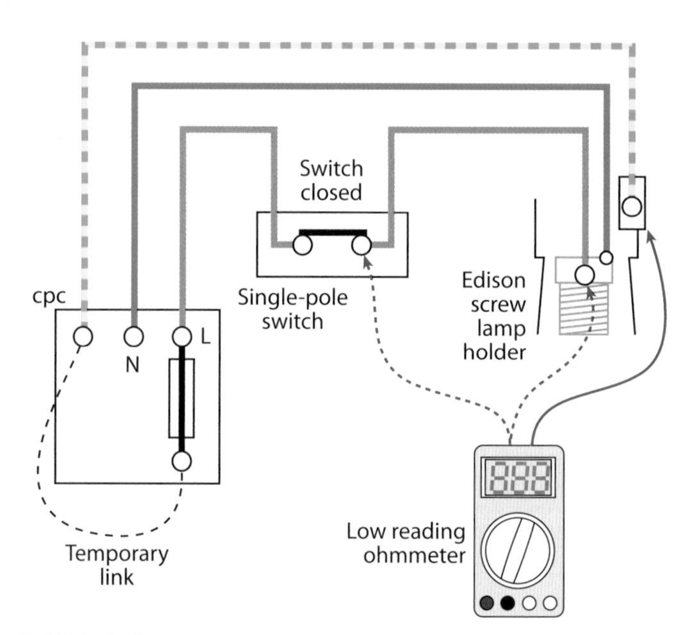

Figure 7.69: Polarity test

Remember

Remember to close lighting switches before carrying out the polarity test.

For ring circuits, if you have already carried out the tests required by Regulation 713-03 ring circuit continuity, the correct connections of line, neutral and cpc conductors will have already been verified so no further testing is required. However, for radial circuits you should measure the $R_1 + R_2$ at each point, using this method.

Earth electrode resistance

Where an earthing system incorporates an earth electrode as part of the system, as in rural areas, you need to measure the electrode resistance to earth.

Some of the types of accepted earth electrode are:

- earth rods or pipes
- earth tapes or wires
- earth plates

- underground structural metalwork embedded in foundations
- lead sheaths or other metallic coverings of cables
- metal pipes.

The resistance to earth will depend on the size and type of electrode used.

Remember

You want as good a connection to earth as possible. The connection to the electrode must be made above ground level.

Measurement by standard method

When measuring earth electrode resistances to earth where low values are required, as in the earthing of the neutral point of a transformer or generator, Test Method 1 below may be used, using an earth electrode resistance tester.

Test Method 1

Before starting the test, disconnect the earthing conductor to the earth electrode either at the electrode or at the main earthing terminal. This will make sure that all the test current passes through the earth electrode. However, as this will leave the installation unprotected against earth faults, switch off the supply before disconnecting the earth.

The test should be carried out when the ground conditions are at their least favourable, i.e. during a period of dry weather, as this will produce the highest resistance value. The test requires the use of two temporary test electrodes (spikes) and is carried out as shown in Figure 7.70.

Connect the earth electrode to terminals C1 and P1 of a four-terminal earth tester. To exclude the resistance of these test leads from the resistance reading, individual leads should be taken from these terminals and connected separately to the electrode. However, if the test lead resistance is insignificant, the two terminals may be short-circuited at the tester and connection made with a single test lead, the same being true if you are using a three-terminal tester. Connection to the temporary 'spikes' is now made as shown in Figure 7.70.

The distance between the test spikes is important, as if they are too close together their resistance areas will overlap. In general, you can gain a reliable result if the distance between the electrode under test and the current spike is at least ten times the maximum length of the electrode under test, in other words, 30 m away from a 3-m-long rod electrode.

Resistance area is important where livestock are concerned, as the front legs of an animal may be outside, and the back legs of the same animal inside the resistance area, thus creating

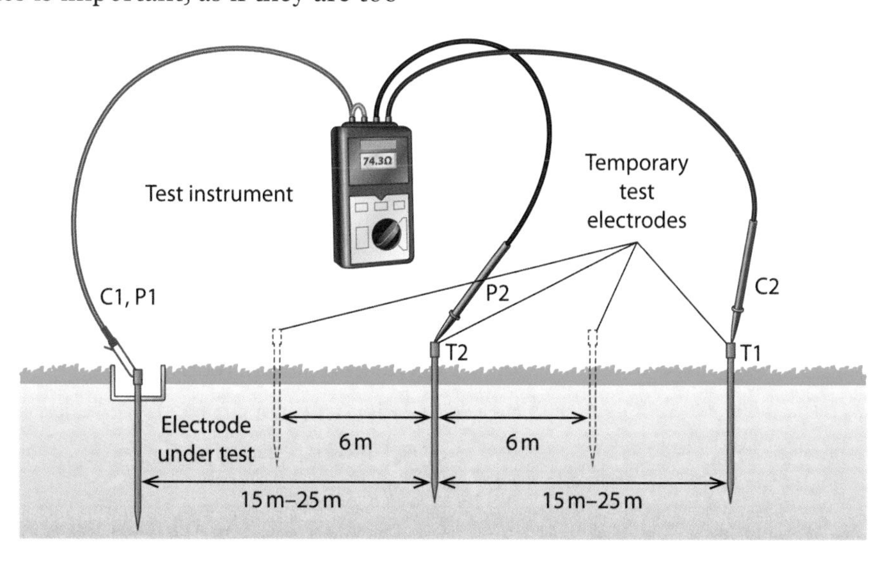

Figure 7.70: Using an earth electrode resistance tester

a potential difference. As little as 25 V can be lethal, so it is important to ensure that all of the electrode is well below ground level and that RCD protection is used.

Then take three readings:

- first, with the potential spike initially midway between the electrode and current spike
- second, at a position 10 per cent of the electrode-to-current spike distance back towards the electrode
- third, at a position 10 per cent of the distance towards the current spike.

By comparing the three readings, a percentage deviation can be determined. This calculation is done by taking the average of the three readings, finding the maximum deviation of the readings from this average in ohms, and expressing this as a percentage of the average.

The accuracy of the measurement using this technique is on average about 1.2 times the percentage deviation of the readings.

It is difficult to achieve an accuracy better than 2 per cent, and you should not accept readings that differ by more than 5 per cent. To improve the accuracy of the measurement to acceptable levels, the test must be repeated with larger separation between the electrode and the current spike.

Once the test is completed, make sure that the earthing conductor is re-connected.

Test method 2

Guidance Note 3 to BS 7671 lists a Test method 2 that uses an earth loop impedance tester. However, this is an alternative method for use on RCD protective TT installations only and if impractical, the measured value of external earth fault loop impedance may be used.

Protection by Automatic Disconnection of Supply (ADS)

This is an unusual test in the sequence as compliance will come from other required tests. However, you must check the worth of this protection as follows.

For TN systems

- Measure earth fault loop impedance.
- Confirm by visual inspection that overcurrent devices are the right type and set correctly.
- Check that for an RCD the disconnection times of BS 7671 can be met.

For TT systems

- Measure the resistance of the earth arrangement of the exposed-conductive-parts of the equipment for the related circuit.

- Confirm by visual inspection that overcurrent devices are the right type and set correctly.
- Check that for an RCD the disconnection times of BS 7671 can be met.

Earth fault loop impedance

When designing an installation, it is the designer's responsibility to ensure that, should a phase-to-earth fault develop, the protection device will operate safely and within the time specified by BS 7671. Although the designer can calculate this in theory, it is not until the installation is complete that the calculations can be checked.

You need to determine the earth fault loop impedance (Z_s) at the furthest point in each circuit and to compare the readings with either the designer's calculated values or the values tabulated in BS 7671 or the *Unite Guide to Good Electrical Practice*.

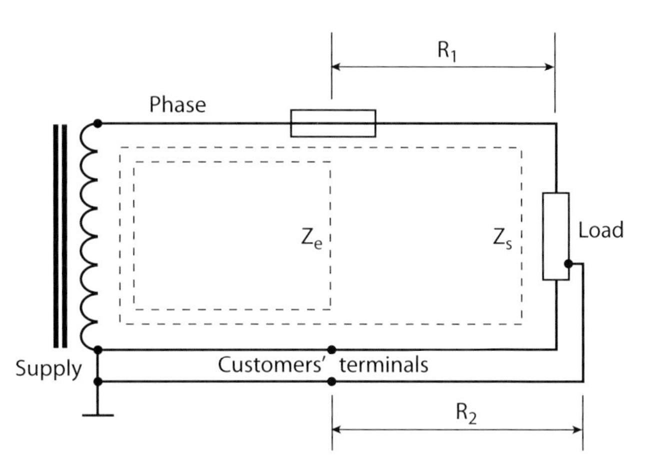

Figure 7.71
Earth fault loop

Figure 7.70 shows that, starting at the point of the fault, the earth fault loop is made up of the following elements:

- circuit protective conductor
- main earthing terminal and earthing conductor
- earth return path (depending on the nature of the supply, TN-S, TN-C-S, etc.)
- path through the earthed neutral of the supply transformer
- secondary winding (line) of the supply transformer
- phase conductor from the source of the supply to the point of the fault.

You can determine the value of earth fault loop impedance (Z_S) by one of three methods:

- direct measurement of Z_S
- direct measurement of Z_e at the origin of the circuit and adding to this the value of (R_1+R_2) measured during continuity tests:
  $$Z_S = Z_e + (R_1+R_2)$$
- obtaining the value of Z_e from the electricity supplier and adding to this the value of (R_1+R_2) as above. However, here you must carry out a test to ensure that the main earthing terminal is in fact connected to earth, using an earth loop impedance tester or an approved test lamp.

Direct measurement of Z_S

Direct measurement of earth fault loop impedance is done with an earth fault loop impedance tester, an instrument designed specifically for this purpose. It operates from the mains supply, so this measurement can only be taken on a live installation. The tester is usually fitted with a standard 13 A plug for connecting to the installation directly through a normal socket outlet, although test leads and probes are also provided for taking measurements at other points on the installation.

IET GN3 states that neither the connections with earth or bonding conductors are to be disconnected, but consequently readings may be less than $Z_e + (R_1+R_2)$ because of parallel paths. This should be taken into account when comparing with design readings.

For the very reason of eliminating any parallel earth return paths, a second school of opinion would argue that as the test is being conducted by a skilled, competent person the main equipotential bonding conductors should be disconnected for the duration of the test. This would then ensure that readings are not distorted by the presence of gas or water service pipes acting as part of the earth return path. Precautions must be taken, however, to ensure that the main equipotential bonding conductors are reconnected after the test has taken place.

Safety tip

Communication with other users of the building and the use of warning notices and barriers are essential.

Earth fault loop impedance testers are connected directly to the circuit being tested and care must be taken to prevent danger. Should a break have occurred anywhere in the protective conductor under test then the whole of the earthing system could become live. It is essential therefore that protective conductor continuity tests be carried out prior to the testing of earth fault loop impedance. Communication with other users of the building and the use of warning notices and barriers is essential.

Direct measurement of Z_e

You can measure the value of Z_e using an earth fault loop impedance tester at the origin of the installation. As this requires the removal of covers and the exposure of live parts, extreme care must be taken and the operation must be supervised at all times. The instrument is connected using approved leads and probes between the phase terminal of the supply and the means of earthing, with the main switch open or with all sub-circuits isolated.

Remember

Remember to reconnect all earthing connections on completion of the test.

To remove the possibility of parallel paths, disconnect the means of earthing from the equipotential bonding conductors for the duration of the test. Correctly connected and with the test button pressed, the instrument will give a direct reading of the value of Z_e.

Verification of test results

BS 7671 Chapter 4

The measured values of earth fault loop impedance obtained (Z_S) should be less than the values stated in BS 7671 Chapter 4, Tables 41.2, 41.3 and 41.4, reproduced below.

Please note that these values are for conductors at their normal operating temperature and if the conductors are at a different temperature at the point of test then a de-rating factor should be applied in line with the guidance given in Appendix 14 of BS 7671.

(a) General purpose (gG) fuses to BS 88-2.2 and BS 88.6						
Rating (amperes)	6	10	16	20	25	32
Z_s (ohms)	8.52	5.11	2.70	1.77	1.44	1.04
(b) Fuses to BS 1361						
Rating (amperes)	5	15	20	30		
Z_s (ohms)	10.45	3.28	1.70	1.15		
(c) Fuses to BS 3036						
Rating (amperes)	5	15	20	30		
Z_s (ohms)	9.58	2.55	1.77	1.09		
(d) Fuses to BS 1362						
Rating (amperes)	3	13				
Z_s (ohms)	16.4	2.42				

Table 7.27: BS 7671 Chapter 4 Table 41.2

(a) Type B circuit breakers to BS EN 60898 and the overcurrent characteristics of RCBOs to BS EN 61009-1														
Rating (amperes)	3	6	10	16	20	25	32	40	50	63	80	100	125	In
Z_s (ohms)	15.33	7.67	4.60	2.87	2.30	1.84	1.44	1.15	0.92	0.73	0.57	0.46	0.37	46/ In
(b) Type C circuit breakers to BS EN 60898 and the overcurrent characteristics of RCBOs to BS EN 61009-1														
Rating (amperes)		6	10	16	20	25	32	40	50	63	80	100	125	In
Z_s (ohms)		3.83	2.30	1.44	1.15	0.92	0.72	0.57	0.46	0.36	0.29	0.23	0.18	23/ In
(c) Type D circuit breakers to BS EN 60898 and the overcurrent characteristics of RCBOs to BS EN 61009-14														
Rating (amperes)		6	10	16	20	25	32	40	50	63	80	100	125	In
Z_s (ohms)		1.92	1.15	0.72	0.57	0.46	0.36	0.29	0.23	0.18	0.14	0.11	0.09	11.5/ In

Table 7.28: BS 7671 Chapter 4 Table 41.3

(a) General purpose (gG) fuses to BS-88.2 and BS 88-6								
Rating (amperes)	6	10	16	20	25	32	40	50
Z_s (ohms)	13.5	7.42	4.18	2.91	2.30	1.84	1.35	1.04
Rating (amperes)		63	80	100	125	160	200	
Z_s (ohms)		0.82	0.57	0.42	0.33	0.25	0.19	
(b) Fuses to BS 1361								
Rating (amperes)	5	15	20	30	45	60	80	100
Z_s (ohms)	16.4	5.00	2.80	1.84	0.96	0.70	0.50	0.36
(c) Fuses to BS 3036								
Rating (amperes)	5	15	20	30	45	60	100	
Z_s (ohms)	17.7	5.35	3.83	2.64	1.59	1.12	0.53	
(d) Fuses to BS 1362								
Rating (amperes)	3	13						
Z_s (ohms)	23.2	3.83						

Table 7.29: BS 7671 Chapter 4 Table 41.4

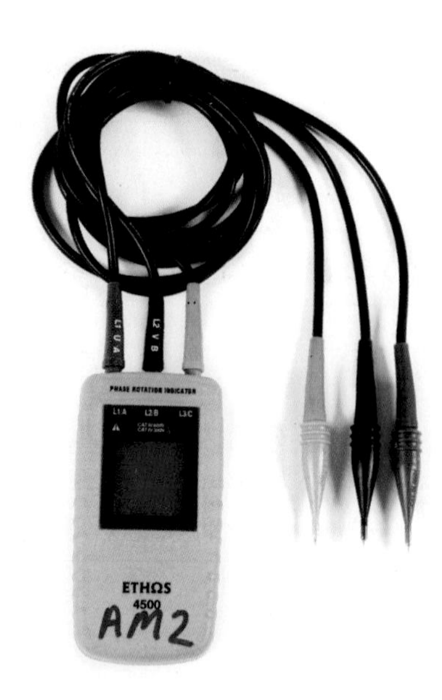

Digital display phase rotation tester sequence indicator

Additional protection

Where an RCD (≤30 mA) is being used to provide additional protection against contact with live parts, the operating time must not exceed 40 ms when tested at 5 × 30 mA.

The maximum test time must be no longer than 40 ms unless the potential on the protective conductor rises by less than 50 V.

Phase sequence

BS 7671 Regulation 612.12 requires that the phase sequence is maintained for multiphase circuits. This is important in terms of checking correct load balancing or motor rotation. To check phase sequence, the usual method is to use a phase rotation tester sequence indicator. There are typically two types using either a rotating disc system (basically a mini-induction motor) or a digital display.

Functional testing

BS 7671 Regulation 612.13.1 requires that where fault and/or additional protection are provided by a residual current device (RCD), the effectiveness of any test facility incorporated in the device shall be verified. Regulation 612.13.2 then requires that all assemblies are to be functionally tested to show that they operate as intended and are mounted and adjusted correctly. This includes: switches, motors, lights, heaters, PIRs and similar detectors, dimmers and other such controls.

RCD testing

Most RCDs have an integral test button, but even a successful test of this does not necessarily confirm that the RCD is working correctly. Basic testing of RCDs therefore involves determining the tripping time (in milliseconds) by inducing a fault current in the circuit. This test can be performed at distribution boards with test leads or at socket outlets.

The test is performed on a live circuit with all known loads disconnected and because some RCDs are more sensitive in one half cycle of the mains supply waveform than the other, the test must be carried out for both zero and 180 degree phase settings, and the longest time should be recorded. The measured tripping time is then compared with the maximum time permitted in the IEE GN.3.

The term 'residual current device' is defined in BS 7671 Part 2 as 'a mechanical switching device or association of devices intended to cause the opening of the contacts when the residual current attains a given value under specified conditions'.

Type of RCD		Description	Usage
RCCB	Residual current operated circuit breaker without integral overcurrent protection	Device that operates when the residual current attains a given value under specific conditions	Consumer units Distribution boards
RCBO	Residual current operated circuit breaker (RCCB) with integral overcurrent protection	Device that operates when the residual current attains a given value under specific conditions and incorporates overcurrent protection	Consumer units Distribution boards
CBR	Circuit breaker incorporating residual current protection	Overcurrent protective device incorporating residual current protection	Distribution boards in larger installations
SRCD	Socket outlet incorporating an RCD	A socket outlet or fused connection unit incorporating a built-in RCD	Often installed to provide supplementary protection against direct contact for portable equipment used out of doors
PRCD	Portable residual current device	A PRCD is a device that provides RCD protection for any item of equipment connected by a plug or socket. Often incorporates overcurrent protection	Plugged into an existing socket outlet PRCDs are not part of the final installation
SRCBO	Socket outlet incorporating an RCBO	Socket outlet or fused connection unit incorporating an RCBO	Often installed to provide supplementary protection against direct contact for portable equipment used out of doors

Table 7.30: Most common variations of an RCD

RCDs are now manufactured to harmonised standards and can be identified by their BS EN numbers. An RCD found in an older installation may not provide protection in accordance with current standards. The following list identifies the applicable current standards:

- BS 7071 (PRCD)
- BS 7288 (SRCD)
- BS EN 61008-1 (RCCB)
- BS EN 61009-1 (RCBO)

There are two RCD tests to carry out.

- With a test of half the rated tripping current (e.g. 15 mA for a 30 mA rated RCD) the device should not operate.
- With a test of the full rated tripping current (e.g. 30 mA for a 30 mA rated RCD) the device should trip in less than 300 ms unless it is Type 'S' (or selective) incorporating a time delay, in which case it must trip between 130 ms to 500 ms.

(Note: if the device was to BS 4293 then the full rated test should cause tripping within 200 ms.)

Additional protection

If an RCD is affording additional protection to either sockets up to 20 A or for sockets up to 32 A for mobile equipment in use outdoors, then a test of 5 × the full rated tripping current (e.g. 150 mA for a 30 mA RCD) means the device must trip within 40ms.

The integral test button

As we said earlier, most RCDs have an integral test button, but this basically only tests the operation of the mechanical parts and even a successful test of this does not necessarily confirm that the RCD is working correctly. Such a button only works if there is a supply present.

Unsatisfactory test results

Before you look at this topic, you need to understand the concept behind using a continuity meter or insulation resistance tester. A basic scale on an analogue insulation resistance tester can illustrate this.

When you touch the leads of the meter together, you are in effect creating a short circuit, which means negligible impedance between live conductors. In other words, there is basically no resistance between the two leads as they touch. Consequently the needle on the display shows zero (see Figure 7.72).

When you separate the leads, you have clearly created an open circuit – which is the same as with a broken conductor. Consequently the needle on the display shows infinity (see Figure 7.73).

Real tests require more detailed readings, but the logic holds good.

Continuity

When testing the continuity of circuit protective conductors or bonding conductors we should always expect a very low reading (near to zero), which is why we must always use a low-reading ohmmeter.

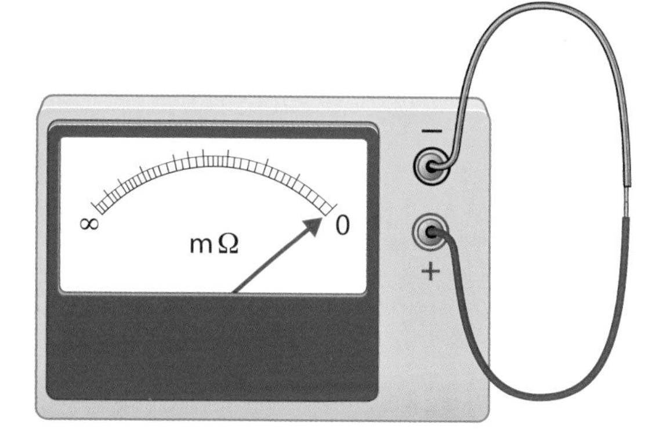

Figure 7.72: Creating a short circuit

Main and supplementary bonding conductors should have a reading of not more than 0.05 ohms while the maximum resistance of circuit protective conductors can be estimated from the value of (R_1 + R_2) given in the *Unite Guide to Good Electrical Practice*. These values will depend upon the cross-sectional area of the conductor, the conductor material and its length.

A very high (near to infinity end of scale) reading would indicate a break in the conductor itself or a disconnected termination that must be investigated. A mid-range reading may be caused by the poor termination of an earthing clamp to the service pipe, e.g. a service pipe which is not cleaned correctly before fitting the clamp or corrosion of the metal service pipe due to its age and damp conditions.

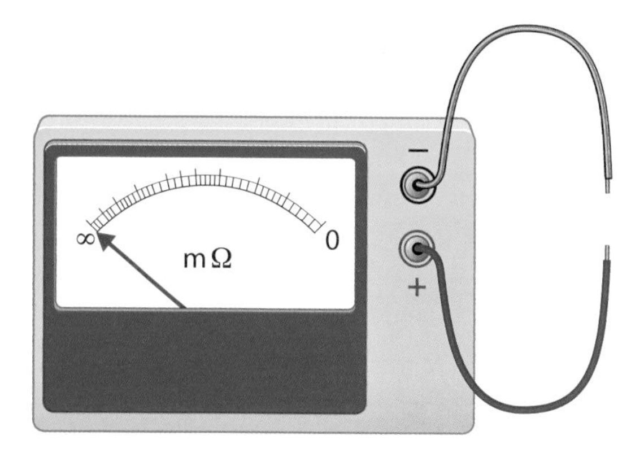

Figure 7.73: Creating an open circuit

When testing continuity of a ring final circuit, remember that the purpose of the test is to establish that a ring exists and that it has been correctly connected.

Remember that when measuring the end to end resistance of each of the conductors, a finite reading shows there is no break in the conductor, and as long as the conductors are the same size, then the readings should be the same. If the cpc is smaller, then its resistance will be higher.

In Step 2 of the test process we said that the readings obtained at each socket will be substantially the same provided they are connected to the ring (because the distance around a circle is the same no matter where you measure it from).

As you can see from Figure 7.74, should you test at a socket and get a higher reading, then this has probably been connected as a spur, and any sockets wired as spurs will have higher resistance value due to the extra length of the spur cable.

However, if while testing at each socket you found that your readings were increasing as you moved away from the distribution board, then it is likely that instead of having a ring, you have the ends incorrectly identified and are not cross connected between the outgoing leg live to the incoming leg neutral, but are instead 'linked out' across the live and neutral of

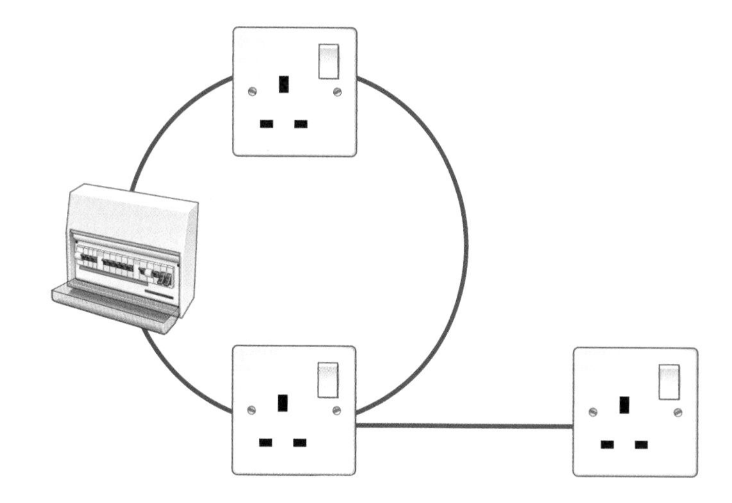
Figure 7.74: Testing at a socket

the same leg and are therefore measuring more cable at each socket. As can be seen from Figure 7.75, the reading taken at socket C will include more cable than that taken at socket A and therefore it will have a higher resistance reading.

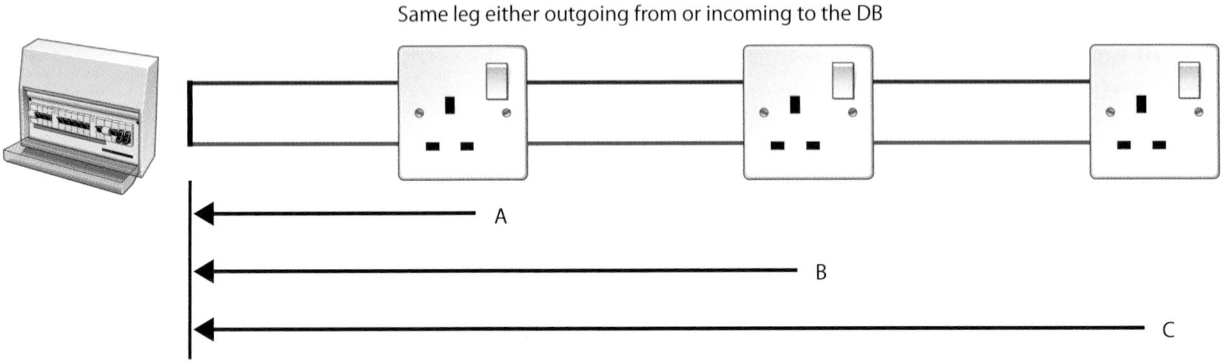

Same leg either outgoing from or incoming to the DB

A

B

C

Figure 7.75: Testing at each socket with readings increasing

Insulation resistance

The value of insulation resistance of an installation will depend upon the size and complexity of the installation and the number of circuits connected to it. When testing a small domestic installation you may expect an insulation resistance reading in excess of 200 MΩ while a large industrial or commercial installation with many sub-circuits, each providing a parallel path, will give a much smaller reading if tested as a whole. Bear in mind that if you double the length of a cable, you halve its insulation resistance. Length and insulation resistance are inversely proportional and longer cables will have more parallel paths and therefore a lower insulation resistance.

It is recommended that, where the insulation resistance reading is less than 2 MΩ, individual distribution boards or even individual sub-circuits be tested separately in order to identify any possible cause of poor insulation values.

An extremely low value of insulation resistance would indicate a possible short circuit between line conductors or a bare conductor in contact with earth at some point in the installation, either of which must be investigated. A reading below 1.0 MΩ would suggest a weakness in the insulation, possibly due to the ingress of dampness or dirt in such items as distribution boards, joint boxes or lighting fittings.

Although PVC insulated cables are not generally subject to a deterioration of insulation resistance due to dampness (unless the insulation or sheath is damaged), mineral insulated cables can be affected if dampness has entered the end of a cable before the seal has been applied properly. Other causes of low insulation resistance can be the infestation of equipment by rats, mice or insects.

Polarity

Correct polarity is achieved by the correct termination of conductors to the terminals of all equipment. This may be main intake equipment such as isolators, main switches and distribution boards or accessories such as socket outlets, switches or lighting fittings.

Polarity is either correct or incorrect; there is nothing in between. Incorrect polarity is caused by the termination of live conductors to the wrong terminals and is corrected by reconnecting all conductors correctly.

Earth fault loop impedance

As explained previously the earth fault loop path is made up of those parts of the supply system external to the premises being tested (Z_e) and the phase conductor and circuit protective conductor within the installation ($R_1 + R_2$), the total earth fault loop impedance being

$$Z_s = Z_e + (R_1 + R_2).$$

The path followed by a fault current as the result of a low impedance occurring between the phase conductor and earthed metal is called the earth fault loop and the current is 'driven' through the loop by the supply voltage. The loop in red is shown within the Figure 7.76 which shows a TT system.

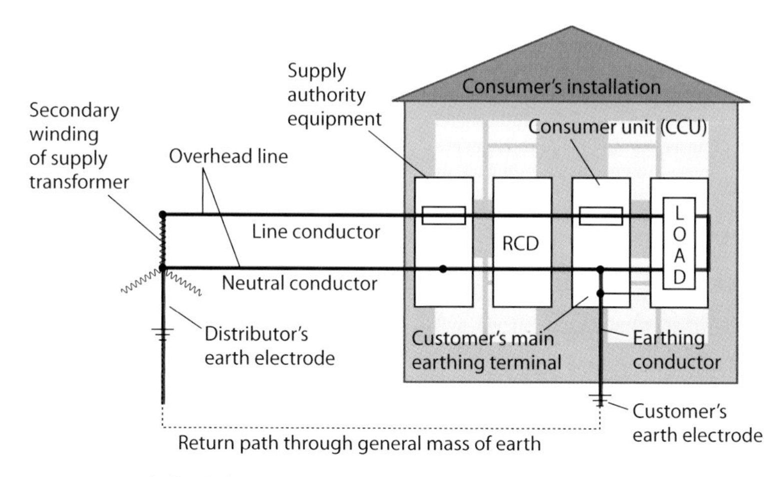

Figure 7.76: Earth fault loop in a TT system

Should the value of impedance measured be higher than that required by the design of the installation, then as we have no influence on the external value of impedance (Z_e) we can only reduce the value of Z_s by installing circuit protective conductors of a larger cross-sectional area or, if aluminium conductors have been used, by changing these to copper.

If the value were still too high to guarantee operation of the circuit protective device in the time required by BS 7671, then consideration would have to be given to changing the type of protective device (i.e. fuses to circuit breakers).

Residual current devices (RCDs)

Where a residual current device (RCD) fails to trip out when pressing the integral test button, this would indicate a fault within the device itself, which should therefore be replaced.

Should an RCD fail to trip out when being tested by an RCD tester, it would suggest a break in the earth return path, which must be investigated. If the RCD does trip out but not within the time specified then a check should be made that the test instrument is set correctly for the nominal tripping current of the device under test.

PROGRESS CHECK

1. The Electricity at Work Regulations require test instruments to be checked and records kept. Which method of checking instrument accuracy is not usually carried out by plumbing operatives?
2. When should a visual inspection of the electrical installation take place?
3. What is the most suitable instrument to check the resistance of the windings of a single-phase pump motor?
4. You are asked to test the insulation resistance of a 230 V single-phase feed to a pump pressurisation unit. What test voltage should the installation resistance tester apply?
5. Which instruments could be used to carry out a polarity test?
6. Where a BS EN 61008 30 mA RCD has been used as additional protection in the event of failure of basic protection, when should a test current of 150 mA trip the device?
7. What should be ensured before carrying out an insulation resistance test and why?

Certificates and reports

BS 7671 Chapter 631 requires the completion of certificates on completion of:

- a new installation
- changes to an existing installation
- a periodic inspection
- any minor electrical works.

BS 7671 Chapter 631
BS 7671 Appendix 6

BS 7671 Appendix 6 then shows examples of the forms to be used for certification and reporting.

The Electrical Installation Certificate

The Electrical Installation Certificate is designed for use when inspecting and testing a new installation, a major alteration or an addition to an existing installation. Where the design, construction, and inspection and testing of the installation are the responsibility of one person, that person can give a single signature instead of signing every box.

BS 7671 Appendix 6 gives the following guidance regarding the Electrical Installation Certificate. The first section is aimed at the person completing the certificate and the other, on page 476, at the person receiving the certificate.

ELECTRICAL INSTALLATION CERTIFICATE
(REQUIREMENTS FOR ELECTRICAL INSTALLATIONS - BS 7671 [IET WIRING REGULATIONS])

DETAILS OF THE CLIENT

INSTALLATION ADDRESS

DESCRIPTION AND EXTENT OF THE INSTALLATION Tick boxes as appropriate

Description of installation: | New installation ☐

Extent of installation covered by this Certificate: | Addition to an existing installation ☐

Alteration to an existing installation ☐

(Use continuation sheet if necessary) see continuation sheet No: ......

FOR DESIGN
I/We being the person(s) responsible for the design of the electrical installation (as indicated by my/our signatures below), particulars of which are described above, having exercised reasonable skill and care when carrying out the design hereby CERTIFY that the design work for which I/we have been responsible is to the best of my/our knowledge and belief in accordance with BS 7671:2008, amended to .... (date) except for the departures, if any, detailed as follows:

Details of departures from BS 7671 (Regulations 120.3 and 134.1.8):

The extent of liability of the signatory or the signatories is limited to the work described above as the subject of this Certificate.

For the DESIGN of the installation: **(Where there is mutual responsibility for the design)

Signature: .... Date: .... Name (IN BLOCK LETTERS): .... Designer No 1

Signature: .... Date: .... Name (IN BLOCK LETTERS): .... Designer No 2**

FOR CONSTRUCTION
I/We being the person(s) responsible for the construction of the electrical installation (as indicated by my/our signatures below), particulars of which are described above, having exercised reasonable skill and care when carrying out the construction hereby CERTIFY that the construction work for which I/we have been responsible is to the best of my/our knowledge and belief in accordance with BS 7671:2008, amended to .... (date) except for the departures, if any, detailed as follows:

Details of departures from BS 7671 (Regulations 120.3 and 134.1.8):

The extent of liability of the signatory is limited to the work described above as the subject of this Certificate.

For CONSTRUCTION of the installation:

Signature: .... Date: .... Name (IN BLOCK LETTERS): .... Constructor

FOR INSPECTION & TESTING
I/We being the person(s) responsible for the inspection & testing of the electrical installation (as indicated by my/our signatures below), particulars of which are described above, having exercised reasonable skill and care when carrying out the inspection & testing hereby CERTIFY that the work for which I/we have been responsible is to the best of my/our knowledge and belief in accordance with BS 7671:2008, amended to .... (date) except for the departures, if any, detailed as follows:

Details of departures from BS 7671 (Regulations 120.3 and 134.1.8):

The extent of liability of the signatory is limited to the work described above as the subject of this Certificate.

For INSPECTION AND TESTING of the installation:

Signature: .... Date: .... Name (IN BLOCK LETTERS): .... Inspector

NEXT INSPECTION
I/We the designer(s), recommend that this installation is further inspected and tested after an interval of not more than .... years/months

ORIES TO THE ELECTRICAL INSTALLATION CERTIFICATE

Company: .... Postcode: .... Tel No: ....

Company: .... Postcode: .... Tel No: ....

Company: .... Postcode: .... Tel No: ....

Company: .... Postcode: .... Tel No: ....

S AND EARTHING ARRANGEMENTS Tick boxes and enter details, as appropriate

Number and Type of Live Conductors		Nature of Supply Parameters	Supply Protective Device Characteristics
a.c. ☐	d.c. ☐	Nominal voltage, U/U_0 (1) V	Type: ....
1-phase, 2-wire ☐	2-pole ☐	Nominal frequency, f (1) .... Hz	
2-phase, 3-wire ☐	3-pole ☐	Prospective fault current, I_{pf} (2) .... kA	Rated current .... A
3-phase, 3-wire ☐	other ☐	External loop impedance, Z_e (2) .... Ω	
3-phase, 4-wire ☐		(Note: (1) by enquiry, (2) by enquiry or by measurement)	

LATION REFERRED TO IN THE CERTIFICATE Tick boxes and enter details, as appropriate

Maximum Demand

Maximum demand (load) .... kVA / Amps Delete as appropriate

Details of Installation Earth Electrode (*where applicable*)

Type (e.g. rod(s), tape etc) | Location | Electrode resistance to Earth .... Ω

Main Protective Conductors

material .... csa .... mm^2 Continuity and connection verified ☐

material .... csa .... mm^2 Continuity and connection verified ☐

s service ☐ To other elements:

Main Switch or Circuit-breaker

Current rating .... A Voltage rating V

Fuse rating or setting .... A

ent $I_{\Delta n}$ = ... mA, and operating time of ...... ms (at $I_{\Delta n}$) (applicable only where an RCD is suitable and is used as a main circuit-breaker)

INSTALLATION (in the case of an addition or alteration see Section 633):

SCHEDULES
The attached Schedules are part of this document and this Certificate is valid only when they are attached to it.
.... Schedules of Inspections and .... Schedules of Test Results are attached.
(Enter quantities of schedules attached)

Figure 7.77: The Electrical Installation Certificate

Notes on the Electrical Installation Certificate

1. The Electrical Installation Certificate is to be used only for the initial certification of a new installation or for an addition or alteration to an existing installation where new circuits have been introduced. It is not to be used for a periodic inspection, for which an Electrical Installation Condition Report form should be used. For an addition or alteration that does not extend to the introduction of new circuits, a Minor Electrical Installation Works Certificate may be used.

1. The original Certificate is to be given to the person ordering the work (Regulation 632.31). A duplicate should be retained by the contractor.
2. This Certificate is only valid if accompanied by the Schedule of Inspections and the Schedule(s) of Test Results.
3. The signatures appended are those of the persons authorised by the companies executing the work of design, construction, inspection and testing respectively. A signatory authorised to certify more than one category of work should sign in each of the appropriate places.

4. The time interval recommended before the first periodic inspection must be inserted.
5. The page numbers for each of the Schedules of Test Results should be indicated, together with the total number of sheets involved.
6. The maximum prospective fault current recorded should be the greater of either the short-circuit current or the earth fault current.
7. The proposed date for the next inspection should take into consideration the frequency and quality of maintenance that the installation can reasonably be expected to receive during its intended life, and the period should be agreed between the designer, installer and other relevant parties.

Guidance for recipients (to be appended to the Certificate)

This safety Certificate has been issued to confirm that the electrical installation work to which it relates has been designed, constructed, inspected and tested in accordance with British Standard 7671 (the IEE Wiring Regulations).

You should have received an original Certificate and the contractor should have retained a duplicate Certificate. If you were the person ordering the work, but not the owner of the installation, you should pass this Certificate, or a full copy of it including the schedules, immediately to the owner.

The original Certificate should be retained in a safe place and be shown to any person inspecting or undertaking further work on the electrical installation in the future. If you later vacate the property, this Certificate will demonstrate to the new owner that the electrical installation complied with the requirements of British Standard 7671 at the time the Certificate was issued.

The Construction (Design and Management) Regulations require that, for a project covered by those Regulations, a copy of this Certificate, together with schedules, is included in the project health and safety documentation.

For safety reasons, the electrical installation will need to be inspected at appropriate intervals by a competent person. The maximum time interval recommended before the next inspection is stated on Page 1 under 'Next Inspection'.

This Certificate is intended to be issued only for a new electrical installation or for new work associated with an addition or alteration to an existing installation. It should not have been issued for the inspection of an existing electrical installation. An 'Electrical Installation Condition Report' should be issued for such an inspection.

The Inspection Schedule

BS 7671 Chapter 631
BS 7671 Appendix 6

A requirement of Regulation 631, the Inspection Schedule provides confirmation that a visual inspection has been carried out as required by Part 6 of BS 7671. The completed Inspection Schedule is attached to and forms part of the Electrical Installation Certificate. Each item on the Inspection Schedule should be checked and either ticked as satisfactory or ruled out if not applicable. Once complete, the Inspection Schedule should be signed and dated by the person responsible for carrying out the inspection.

SCHEDULE OF INSPECTIONS **(for new installations only)**

Methods of protection against electric shock

Both basic and fault protection:

- ☐ (i) SELV
- ☐ (ii) PELV
- ☐ (iii) Double insulation
- ☐ (iv) Reinforced insulation

Basic protection:

- ☐ (i) Insulation of live parts
- ☐ (ii) Barriers or enclosures
- ☐ (iii) Obstacles
- ☐ (iv) Placing out of reach

Fault protection:

(i) Automatic disconnection of supply:

- ☐ Presence of earthing conductor
- ☐ Presence of circuit protective conductors
- ☐ Presence of protective bonding conductors
- ☐ Presence of supplementary bonding conductors
- ☐ Presence of earthing arrangements for combined protective and functional purposes
- ☐ Presence of adequate arrangements for alternative source(s), where applicable
- ☐ FELV
- ☐ Choice and setting of protective and monitoring devices (for fault and/or overcurrent protection)

(ii) Non-conducting location:

- ☐ Absence of protective conductors

(iii) Earth-free local equipotential bonding:

- ☐ Presence of earth-free local equipotential bonding

(iv) Electrical separation:

- ☐ Provided for **one item** of current-using equipment
- ☐ Provided for **more than one item** of current-using equipment

Additional protection:

- ☐ Presence of residual current devices(s)
- ☐ Presence of supplementary bonding conductors

Prevention of mutual detrimental influence

- ☐ (a) Proximity of non-electrical services and other influences
- ☐ (b) Segregation of Band I and Band II circuits or use of Band II insulation
- ☐ (c) Segregation of safety circuits

Identification

- ☐ (a) Presence of diagrams, instructions, circuit charts and similar information
- ☐ (b) Presence of danger notices and other warning notices
- ☐ (c) Labelling of protective devices, switches and terminals
- ☐ (d) Identification of conductors

Cables and conductors

- ☐ Selection of conductors for current-carrying capacity and voltage drop
- ☐ Erection methods
- ☐ Routing of cables in prescribed zones
- ☐ Cables incorporating earthed armour or sheath, or run within an earthed wiring system, or otherwise adequately protected against nails, screws and the like
- ☐ Additional protection provided by 30 mA RCD for cables concealed in walls (where required in premises not under the supervision of a skilled or instructed person)
- ☐ Connection of conductors
- ☐ Presence of fire barriers, suitable seals and protection against thermal effects

General

- ☐ Presence and correct location of appropriate devices for isolation and switching
- ☐ Adequacy of access to switchgear and other equipment
- ☐ Particular protective measures for special installations and locations
- ☐ Connection of single-pole devices for protection or switching in line conductors only
- ☐ Correct connection of accessories and equipment
- ☐ Presence of undervoltage protective devices
- ☐ Selection of equipment and protective measures appropriate to external influences
- ☐ Selection of appropriate functional switching devices

Inspected by .... Date ....

NOTES:

✓ to indicate an inspection has been carried out and the result is satisfactory

✗ to indicate an inspection has been carried out and the result is not satisfactory (applicable for a periodic inspection only)

NA to indicate that the inspection is not applicable to a particular item

LIM to indicate that, exceptionally, a limitation agreed with the person ordering the work prevented the inspection being carried out (applicable for a periodic inspection only).

Figure 7.78: The Inspection Schedule

The Schedule of Test Results

Also a requirement of Regulation 631, the Schedule of Test Results is a written record of the results obtained when carrying out the electrical tests required by Part 6 of BS 7671 and must be attached to the Electrical Installation Certificate.

DB reference no ....
Location ....
Zs at DB (Ω) ....
I_{pf} at DB (kA) ....
Correct polarity of supply confirmed yes / no
Phase rotation confirmed (where appropriate) ☐

Details of circuits and/or installed equipment vulnerable to damage when testing ....

Details of test instruments used (state serial and/or asset numbers)
Continuity ....
Insulation resistance ....
Earth fault loop impedance ....
Earth electrode resistance ....
RCD ....

Circuit details									**Test results**														
		Overcurrent device				Conductor details			Ring final Circuit continuity(Ω)					Continuity (Ω) $(R_1 + R_2)$ or R_2		Insulation Resistance (MΩ)		Correct polarity (yes / no)	Z_s (Ω)	RCD (ms)			Remarks (continue on a separate sheet if necessary)
Circuit number 1	Circuit Description 2	BS (EN) 3	type 4	Rating (A) 5	Pfc (kA) 6	Reference Method 7	Live (mm²) 8	cpc (mm²) 9	r_1 (line) 10	r_n (neutral) 11	r_2 (cpc) 12	Cross-connected line and neutral 13	Cross-connected line and cpc* 14	$(R_1 + R_2)$ 15	R_2 16	Live-Live 17	Live-E 18	19	20	@ $I_{\Delta n}$ 21	@ $5I_{\Delta n}$ 22	Test button operation 23	24

* Where there are no spurs connected to a ring circuit this value is also the $(R_1 + R_2)$ of the circuit.

Figure 7.79: Schedule of test results

The Minor Electrical Installation Works Certificate

The Minor Electrical Installation Works Certificate – or Minor Works Certificate – is intended to be used for additions and alterations to an installation that do not extend to the provision of a new circuit, such as adding a socket outlet or a lighting point to an existing circuit, or relocating a light switch. This Certificate may also be used for the replacement of equipment such as accessories or luminaires, but not for the replacement of distribution boards or similar items.

Appropriate inspection and testing should always be carried out, irrespective of the extent of the work undertaken.

The Certificate has two component parts.

Part 1 Description of minor works

1, 2 The minor works must be so described that the work that is the subject of the certification can be readily identified.

4 See Regulations 120.3 and 120.4. No departures are to be expected except in most unusual circumstances. See also Regulation 633.1.

Part 2 Installation details

2 The method of fault protection must be clearly identified: for example, Automatic Disconnection of Supply (ADS).

Part 3 Essential tests

The relevant provisions of Part 6 'Inspection and testing' of BS 7671 must be applied in full to all minor works. For example, where a socket outlet is added to an existing circuit it is necessary to:

1 establish that the earthing contact of the socket outlet is connected to the main earthing terminal
2 measure the insulation resistance of the circuit that has been added to, and establish that it complies with Table 61 of BS 7671
3 measure the earth fault loop impedance to establish that the maximum permitted disconnection time is not exceeded
4 check that the polarity of the socket outlet is correct
5 (if the work is protected by an RCD) verify the effectiveness of the RCD.

MINOR ELECTRICAL INSTALLATION WORKS CERTIFICATE
(REQUIREMENTS FOR ELECTRICAL INSTALLATIONS - BS 7671 [IET WIRING REGULATIONS])
To be used only for minor electrical work which does not include the provision of a new circuit

PART 1:Description of minor works

1. Description of the minor works
2. Location/Address
3. Date minor works completed
4. Details of departures, if any, from BS 7671:2008

PART 2:Installation details

1. System earthing arrangement TN-C-S☐ TN-S☐ TT☐
2. Method of fault protection
3. Protective device for the modified circuit Type .... Rating .... A
4. Omission of additional protection by 30 mA RCD
   for socket-outlets (see 411.3.3) ☐
   for cables concealed in walls (see 522.6.102) ☐

Comments on existing installation, including adequacy of earthing and bonding arrangements (see Regulation 134.1.9):

PART 3:Essential Tests
Earth continuity satisfactory ☐

Insulation resistance:
Line/neutral ....MΩ
Line/earth ....MΩ
Neutral/earth ....MΩ

Earth fault loop impedance ....Ω

Polarity satisfactory ☐

RCD operation (if applicable). Rated residual operating current $I_{\Delta n}$ ....mA and operating time of ....ms (at $I_{\Delta n}$)

PART 4:Declaration

I/We CERTIFY that the said works do not impair the safety of the existing installation, that the said works have been designed, constructed, inspected and tested in accordance with BS 7671:2008 (IET Wiring Regulations), amended to .... (date) and that the said works, to the best of my/our knowledge and belief, at the time of my/our inspection, complied with BS 7671 except as detailed in Part 1 above.

Name:	Signature: ....
For and on behalf of:	Position: ....
Address:	
....	Date: ....
....	

Figure 7.80: The Minor Electrical Installation Works Certificate

Part 4 Declaration

1, 3 The Certificate shall be made out and signed by a competent person in respect of the design, construction, inspection and testing of the work.

As with the Electrical Installation Certificate, the Minor Works Certificate goes on to give guidance to the person receiving the certificate.

Test instruments

Instrument standards

All instruments should be checked against BS EN 61010, which covers basic safety requirements for electrical test instruments. Older instruments may have been manufactured in accordance with BS 5458 but, provided these are in good condition and have been recently calibrated, they can still be used. Lead sets should conform to HSE GS38.

Key terms

Analogue – instruments fitted with a needle that gives a direct reading on a fixed scale.

Digital – instruments providing a numeric digital visual display of the actual measurement being taken.

Accuracy and calibration

Instruments may be **analogue** or **digital**. To ensure that the reading is reasonably accurate, all instruments should be accurate within 5 per cent; for analogue instruments accuracy of 2 per cent of full-scale deflection should be adequate.

All electrical test instruments should be calibrated regularly, at least every year, usually at a specialist test laboratory. A calibration label will then be attached, stating when the calibration took place and when the next calibration is due (see Figure 7.81). A calibration certificate will also be issued detailing the tests carried out and a reference to the equipment used. A further seal (Figure 7.82) is often placed over the joint in the casing, voiding calibration if the seal is broken.

Safety tip

Check to make sure an instrument is within calibration before using it.

Remember

Keep records of calibration and copies of certificates somewhere safe.

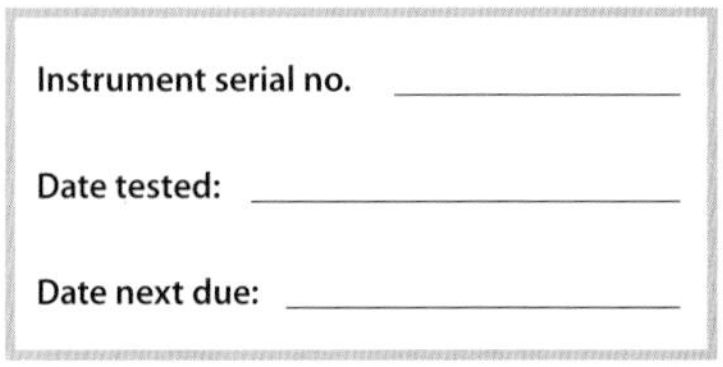

Instrument serial no. ______________

Date tested: ______________

Date next due: ______________

Figure 7.81: Calibration label

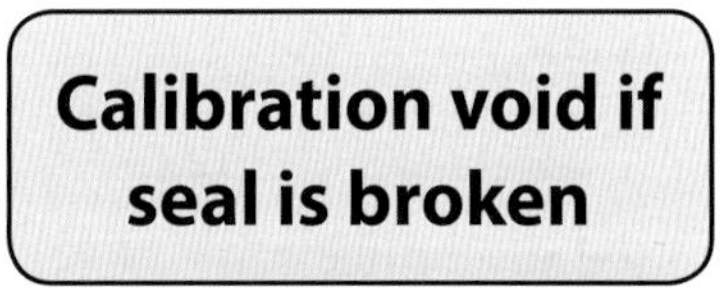

Calibration void if seal is broken

Figure 7.82: Adhesive label

Test equipment must be regularly checked to make sure it is in good and safe working order. Before starting work, follow this checklist.

Checklist

- Check the instrument and leads for damage or defects.
- Zero the instrument.
- Check the battery level.
- Select the correct scale for testing. If in doubt, ask or select the highest range available.
- Check the calibration date and record the serial number.
- Check leads for open and closed circuit before testing.
- Record the test results.
- After test, leave selector switches in off position. Some analogue instruments turn off automatically to save battery life.
- Always store instruments in their cases in secure, dry locations when not in use.

Types of instrument

Table 7.31 shows the different sorts of instruments you will need to be familiar with.

Instrument	Purpose and standard	Notes
Low-resistance ohmmeters	Used for low-resistance tests. May be either a specialised low-resistance ohmmeter, or the continuity range of an insulation and continuity tester. Test current may be d.c. or a.c. Recommended that current comes from source with no-load voltage between 4 V and 24 V, and short circuit current not less than 200 mA. [The low resistance ohmeter and insulation resistance tester are combined in the Megger MIT 320 used for illustration.]	Measuring range should cover 0.2 Ω to 2 Ω with resolution of at least 0.01 Ω for digital instruments. Instruments to BS EN 61557-4 will meet these requirements. **Accuracy** Errors can come from contact resistance or lead resistance. Contact resistance cannot be eliminated entirely; lead resistance can be eliminated by clipping leads together and zeroing the instrument before use or measuring resistance of the leads and subtracting from the reading. Thermocouple effects can be eliminated by reversing the test probes and averaging the resistance readings taken in each direction.
Insulation resistance ohmmeters (analogue or digital)	The instrument used should be capable of developing the test voltage required across the load. The d.c. test voltages required by Table 61 of BS 7671 are: • 250 V for SELV and PELV • 500 V for all circuits rated up to and including 500 V, but excluding extra-low voltage circuits mentioned above • 1000 V d.c. for circuits rated above 500 V up to 1000 V. Instruments conforming to BS EN 61557-2 will fulfil these requirements.	Can lock in 'on' position for hands-free operation. Automatic nulling device takes account of resistance of the test leads. **Accuracy** Errors from 50 Hz currents induced into cables under test and capacitance in test object cannot be eliminated. Instrument should have automatic discharge facility capable of safely discharging capacitance (up to 5 μF). After test, should be left connected until capacitance within installation has fully discharged.
Earth fault loop impedance testers (digital only)	Operate by circulating current from line conductor into the protective earth, raising the potential of the protective earth system. For circuits rated up to 50 A, use a line-earth loop tester with a resolution of 0.01 Ω. Instruments conforming to BS EN 61557-3 will fulfil these requirements.	To minimise electric shock hazard, keep test duration within safe limits. Often have facility for deriving prospective short circuit current, where current is calculated by dividing earth fault loop impedance value into the mains voltage. **Accuracy** Instrument accuracy is determined by the same factors as for loop testers, but here instrument accuracy decreases as scale reading reduces. Affected by transient variations of mains voltage, mains interference, test lead resistance and impedance measurement. Repeat test at least once.

Instrument	Purpose and standard	Notes
Earth electrode resistance testers	Can be four-terminal instrument, or three-terminal (where combined lead to the earth electrode would not have a significant resistance compared with the electrode resistance), to eliminate resistance of the test leads and temporary spike resistance from result. Some meters now offer 'stakeless' testing.	Should carry facility to check that resistance to earth of temporary potential and current spikes are within operating limits of instrument (instruments complying with BS EN 61557-5 incorporate this). Take care to position temporary spikes with reasonable accuracy. **Accuracy** Affected by effects of temporary spike resistance, interference currents and layout of test electrodes.
RCD testers (digital only)	Should be capable of applying full range of test current to an in-service accuracy as given in BS EN 61557-6 (including effects of voltage variations around nominal voltage of tester).	To check RCD operation and minimise danger, apply test current for no longer than two seconds.
Phase rotation instruments	BS EN 61557-7 gives the requirements for measuring equipment for testing the phase sequence in three-phase distribution systems whether indication is given by mechanical, visual and/or audible means.	Indication shall be unambiguous between 85 per cent and 110 per cent of the nominal system voltage or within the range of the nominal voltage and between 95 per cent and 105 per cent of the nominal system frequency. Should be suitable for continuous operation.
All-in-one test instruments	A recent innovation: 'all-in-one' instrument that can carry out all tests required by BS 7671.	Example shown can perform: continuity test, insulation resistance test, RCD test, loop impedance with no trip feature, prospective short circuit current (L-N, L-E), polarity test, phase rotation, earth electrode resistance and resistivity testing, power, energy, power factor, harmonic monitoring, light/lux measurements and cable/fuse location.

Table 7.31: Test instruments

12. Safely diagnosing and rectifying faults in electrically operated mechanical services components

Before you start

There is the potential for danger in any electrical work, but particularly in fault diagnosis and rectification work, as an electrical item (and its safety systems) may not be functioning properly. Follow this checklist to keep yourself and others safe.

Checklist

- **Think about others** Before stopping supplies, you need to take into account people's needs, such as using computers or washing machines.
- **Keep people informed** Liaise with everyone concerned to make sure that they fully understand the extent of work to be carried out, whether that be investigative or corrective.
- **Take safety measures** Do a full risk assessment. You may find that you need to isolate supplies, use barriers or temporary lighting on corridors or staircases, or place notices and signs.
- **Know your stuff** To investigate, diagnose and find faults on electrical installations and equipment is difficult, so you will need a thorough working knowledge of the installation or equipment involved.
- **Go step by step** Take a reasoned, logical approach to your investigation and subsequent remedy. Be aware of your limits and seek expert advice and support where necessary.
- **Forward plan if possible** An emergency or dangerous fault may allow little time for planning, but if faults are visible and straightforward, plan carefully and liaise with the client to limit disruption.
- **Inspect and test before energising a new installation** You may be able to rectify problems before circuits are energised: for example, pre-energising tests can be important when checking the function of circuits.

Stages of diagnosis and rectification

Some faults can be rectified very easily, especially when you have a working knowledge of the installation. But there are many occasions when you may be called out to a repair and there is no information available to you. It is on these occasions that your years of training and knowledge of systems and equipment have to be used. However, this knowledge is not always enough and a sequential and logical approach to rectifying a fault and gathering information is needed.

One popular approach is called the 'half-split technique', which simply means if you are told that there is a fault on a circuit, then split the circuit into two parts (halves) and see which half has the fault in it. You can then repeat the process inside the half that you know has the fault in it to narrow down the area where the fault is.

A sensible approach to finding any faults is outlined on the next page.

- **Identify the symptom:** this can be done by establishing the events that led up to the problem or fault on the installation or equipment.
- **Gather information:** this is achieved by talking to people and obtaining and looking at any available information. Such information could include manufacturer's data, circuit diagrams, drawings, design data, distribution board schedules and previous test results and certification.
- **Analyse the evidence:** carry out a visual inspection of the location of the fault and cross-reference with the available information. Interpret the collected information and decide what action or tests need to be carried out. Then determine the remedy.
- **Check supply:** confirm supply status at origin and locally. Confirm circuit or equipment when fault is isolated from supply.
- **Check protective devices:** check status of protective devices. If they have operated this would determine location of fault on circuit or equipment.
- **Isolation and test:** confirm isolation prior to carrying out sequence of tests.
- **Interpret information and test results:** by interpreting the test results, the status of protective devices and other information, the fault may be identified and remedied/rectified.
- **Rectify the fault:** this may be done quite simply or parts or replacement may be needed.
- **Carry out functional tests:** before restoring the supply, the circuit or equipment will need to be tested not only electrically but also for functionality.
- **Restore the supply:** care must be taken that the device has been reset or repaired to correct current rating. Make sure the circuit or equipment is switched off locally before restoring supply.
- **Carry out live and functional tests:** once the supply has been restored, it may be prudent to complete live tests to make sure the supply is stable, and also carry out any live functional tests such as operator switches, etc.

Remember

It may be the case that the original manufacturer is no longer in business and you have to look for an alternative. Most manufacturers will have their own design team who will help you find something suitable.

Interpreting test results

Some faults can be recognised at the installation stage when the testing and inspection process is carried out. You should be able to recognise typical test results for each non-live test and interpret the type of fault that may exist.

BS 7671 Part 6 (612)

BS 7671 Part 6 (612) lists non-live and live tests and as the non-live tests are carried out on the installation wiring circuits before they are ever energised, they will confirm the integrity of the circuit.

The non-live tests are:

- continuity of protective conductors including main and supplementary bonding

- continuity of ring final circuit conductors
- insulation resistance
- polarity.

Continuity of protective conductors including main and supplementary bonding

BS 7671 Regulation 411.3.1.1 requires that installations providing protection against electric shock using automatic disconnection of supply (ADS) must have a circuit protective conductor run to and terminated at each point in the wiring and at each accessory. An exception is made for a lamp holder having no exposed-conductive-parts and suspended from such a point.

BS 7671 Reg 411.3.1.1

Regulation 612.2.1 then requires that every protective conductor, including circuit protective conductors, the earthing conductor and main and supplementary equipotential bonding conductors, should be tested to verify that they are electrically sound and correctly connected.

BS 7671 Reg 612 2.1

Test methods 1 (*linking the circuit line to the protective conductor at the DB*) and 2 (*using a long wandering lead*) are alternative ways of testing the continuity of protective conductors and should be performed using a low-resistance ohmmeter. Remember that the resistance readings obtained include the resistance of the test leads and therefore the resistance of the test leads should be measured and deducted from all resistance readings obtained unless the instrument can auto-null.

Test method 1, as well as checking the continuity of the protective conductor, also measures ($R_1 + R_2$) which, when added to the external impedance (Z_e), enables the earth fault loop impedance (Z_s) to be checked against the design. However, as ($R_1 + R_2$) is the resistance of the line conductor R_1 and the circuit protective conductor R_2 the reading may be affected by parallel paths through exposed-conductive-parts and/or extraneous-conductive-parts.

Installations incorporating steel conduit, steel trunking, MI and SWA cables will introduce parallel paths to protective conductors and, similarly, luminaires fitted in grid ceilings and suspended from steel structures in buildings may create parallel paths.

In such situations, unless a plug and socket arrangement has been incorporated in the lighting system by the designer, the ($R_1 + R_2$) test will need to be carried out prior to fixing accessories and bonding straps to the metal enclosures and finally connecting protective conductors to luminaires.

Under these circumstances, some of the requirements may have to be visually inspected after the test has been completed. This consideration requires tests to be performed during the erection of an installation, in addition to tests at the completion stage.

When testing the effectiveness of main equipotential bonding conductors, the resistance between a service pipe/other extraneous-conductive-part and the main earthing terminal should be about 0.05 Ω. Similarly, supplementary bonding conductors should have a reading of 0.05 Ω or less.

Continuity of ring final circuit conductors

This test, carried out on the circuit wiring, is designed to highlight open circuits and interconnections within the wiring, and a low-resistance ohmmeter is used to carry out this test. Typical readings will be 0.01 to 0.1 ohms dependent on the conductor size and length of circuit wiring.

If the circuit is wired correctly, then the readings on the instrument will be the same at each point of test. If variable readings are found at each point of the test, this may indicate an open circuit or interconnections. If these types of fault exist, the consequence can be an overload on part of the circuit wiring.

Insulation resistance

This test is designed to confirm the integrity of the insulation resistance of all live conductors, between each other and earth. The type of instrument is known as an insulation resistance tester (not a megger) and typical test results would be around 50 to 100 megohms, i.e. 50 to 100 million ohms.

BS 7671 Table 61

Depending on the circuit, Table 61 of BS 7671 indicates minimum values between 0.5 MΩ and 1 MΩ. These are minimum values and, in practice if these values existed, the circuit may need further investigation. Generally, if a reading of less than 2 MΩ is recorded for an LV circuit then the circuit should be investigated.

Depending on where the test takes place, variable values of resistance may be recorded. In the case of a test on the supply side of a distribution board (Figure 7.83), a group value reading will indicate circuits connected in parallel (Figure 7.84), which invariably most are.

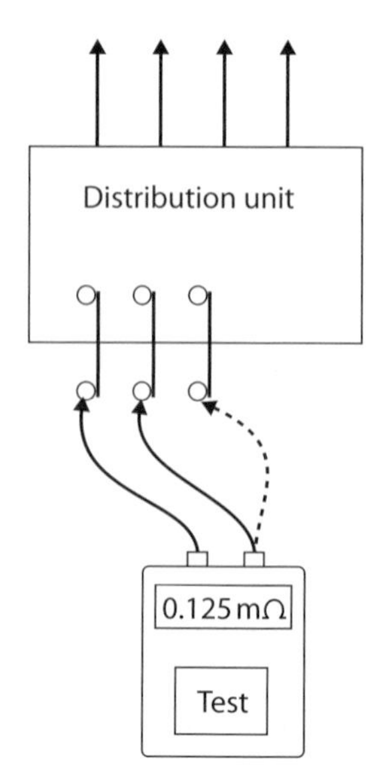

Figure 7.83: Test on supply side of distribution board

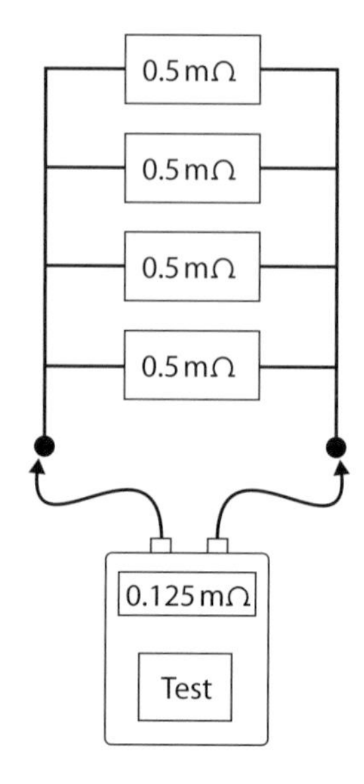

Figures 7.84: Test on circuits connected in parallel

From Figures 7.82 and 7.83 above, it can be seen that when testing grouped circuit conductors, a parallel reading would indicate a poor reading, but if each circuit were tested individually then the actual reading is likely to be acceptable.

Polarity

The polarity test can be carried out separately using a low-resistance ohmmeter as used for continuity testing. However, as polarity of a ring final circuit is confirmed in the ring circuit test mentioned earlier, polarity of switches and lighting can equally be confirmed during the test on the continuity of protective conductors including main and supplementary bonding.

Typical test results will depend on the resistance of the conductors under test but will generally be low, around 0.01 to 0.1 ohms. Values in excess of this may indicate an open circuit or incorrect polarity.

Live testing

Earth fault loop impedance testing

Overcurrent protective devices must, under earth fault conditions, disconnect fast enough to reduce the risk of electric shock. This can be achieved if the actual value of the earth loop impedance does not exceed the tabulated values given in BS 7671. The purpose of the test, therefore, is to determine the actual value of the loop impedance (Z_s) for comparison with those values.

The test procedure requires that all main equipotential bonding is in place. The test instrument is then connected by the use of 'flying' leads to the phase, neutral and earth terminals at the remote end of the circuit under test. Press to test and record the results.

Once Z_s and the voltage have been established at the remote point in the circuit, you can divide the voltage by Z_s to give you the fault current to earth. Apply this calculation to the time current characteristic graphs shown in BS 7671 and you will be able to determine the actual disconnection time of the circuit.

RCD test

Regulation 612.13.1 states that where fault protection and/or additional protection is to be provided by an RCD, the effectiveness of any test facility incorporated in the device shall be verified.

BS 7671 Reg 612.13.1

Such a test would use an appropriate RCD test instrument and the test should confirm the operation of the device independently of the device's integral test button. Residual current circuit breakers or devices are rated in milliamperes and the test should show that the device operates within the milliampere ratings of the device and within the time constraints. Typically if the rated current of the device was 30 mA then the instrument should prove operation at this value within 200 milliseconds.

Functional testing

Regulation 612.13.2 states that assemblies such as switchgear and control gear assemblies, drives, controls and interlocks shall be subjected to a functional test, to show that they are properly mounted, adjusted and installed in accordance with relevant requirements of BS 7671.

Typical functional tests should be applied to:

- lighting controllers, switches, etc.
- motors, fixings, drives, pulleys, etc.
- motor controllers
- controls and interlocks
- main switches
- isolators.

Full details of the various tests and related results, instruments and documentation are contained earlier in this unit, starting on page 449.

The limitation and range of instruments

Test	Instrument	Range
Continuity of protective conductors	Low-resistance ohmmeter	0.005 to 2 ohms or less
Continuity of ring final circuit conductors	Low-resistance ohmmeter	0.05 to 0.8 ohms
Insulation resistance	High-reading ohmmeter	1 MΩ to greater than 200 MΩ
Polarity	Ohmmeter Bell/buzzer	Low resistance None

Table 7.32: Range of test instruments

Many manufacturers make instruments with more than one facility. Inspectors and testers should fully understand the range of operation of their instruments, and know how to use them properly.

Before you test

- Check the instrument and leads for damage or defects.
- Zero the instrument.
- Check the battery level.
- Select the correct scale for testing; if in doubt, ask or select the highest range available.
- Check the calibration date and record the serial number.
- Check leads for open and closed circuit.

After you test

- Record test results.
- Leave selector switches in off position (some analogue instruments turn off automatically to save battery life).
- Always store instruments in their cases in secure, dry locations when not in use.

Knowledge and understanding of electrical systems and equipment

Faults can be highly dangerous. With a short circuit fault, high levels of fault current can develop, causing high temperatures and breakdown of insulation, and even starting fires. High temperature destroys the properties of insulation, and can lead to a short circuit. High currents damage equipment, and earth fault currents can cause electric shock.

Careful planning and thought can prevent most faults. Designers can build in fault protection and damage limitation, for example by installing fuses and protective devices, and making sure the system is easy to maintain and repair. Once installed, good maintenance will continue to limit potential faults.

However, faults will still occur. While some are common, it is important to remember that no two situations will be exactly the same. Also, as technology advances, faults can be harder to rectify. Understanding the electrical installation and the equipment you use is essential.

Some basic techniques can usually solve the most common faults that occur. However, there are occasions when it may be impractical to rectify a fault: for example, due to the cost of downtime, or the fact that replacement parts would be cheaper. You should monitor such situations and keep the client informed at every stage.

Careful planning can mean that most work is carried out efficiently with little or no disruption to the client.

Electrical system, installation and equipment

These can generally be categorised by:

- voltage – 230 V single-phase or 400 V three-phase
- installation type – in this case, domestic
- system type – power, heating, etc.

If a fault occurs on a system in the process of being installed, the data for the system should be readily available. When called out to a fault, you will need to know or have:

- type of supply – single-phase or three-phase
- nominal voltage – 230 volt or 400 volt
- type of earthing supply system – TT, TN-S, TN-C-S
- type of protective measure – ADS (Automatic Disconnection of Supply)
- types of protective device – HRC, MCB, RCD etc.
- ratings of devices
- location of incoming supply services
- location of electrical services
- distribution board schedules
- location drawings
- design and manufacturer's data
- the nature of the fault.

Using your own and others' experience

Your own knowledge of an installation can be an asset when fault finding process, and can show a client that you can be relied on to problem solve quickly. However, you need to make sure that you have the specialist knowledge of any equipment, and are insured to work on such systems.

You may have to work with other specialists, assisting each other in commissioning and testing such systems. Sometimes, you will be asked to repair faults on wiring and equipment caused by inexperienced personnel attempting to install or repair circuits or equipment.

When you first arrive at any installation, you should ask the personnel there to give you the background to the fault and relevant information, which could include:

- operating manuals
- wiring and connection diagrams
- manufacturer's product data/information
- maintenance records
- inspection and test results
- installation specifications
- drawings
- design data
- site diary.

You will need to understand and identify:

- position of faults (complete loss of supply at the origin of the installation and localised loss of supply)
- operation of overload and fault current devices
- transient voltages
- insulation failure
- arcing
- plant, equipment and component failure
- faults caused by abuse, misuse and negligence
- prevention of faults by regular maintenance.

Position of faults

The location of a fault can limit its severity in terms of disruption and inconvenience.

BS 7671 Reg 314.1

It is easier to find faults on installations where there are plenty of circuits. Regulation 314.1 of BS 7671 says that every installation shall be divided into circuits as necessary to:

- avoid hazards and minimise inconvenience in the event of a fault
- facilitate safe inspection, testing and maintenance (see also section 357)

- take account of hazards that may arise from the failure of a single circuit such as a lighting circuit.
- reduce the possibility of unwanted tripping of RCD's due to excessive protective conductor (PE) currents not due to a fault
- mitigate the effects of electromagnetic interferences (EMI)
- prevent the indirect energising of a circuit intended to be isolated.

Compliance with this Regulation will help you to locate faults more easily, by a process of elimination or simply looking at the device to see which fuse has operated.

Complete loss of supply at the origin of the installation

If there is no power at the main incoming terminals of the consumer unit, there is a fault with the equipment belonging to the supply company. These faults are usually the result of problems with the supply, such as an underground cable being severed by workers.

On rare occasions, faults on the installation wiring could result in the supply company's protective devices operating, as a consequence of poor design, misuse or overloading of the equipment.

Localised loss of supply

If a supply is present at the consumer unit, and a fault occurs on any one final circuit fed by that DB, only that circuit's protective device should have operated and only that circuit should be dead; the fault should not affect any other final circuit or any other cable.

Operation of overload and fault current devices

A fault is usually noticed because a circuit or piece of equipment has stopped working, because the protective device has done its job and operated.

The rating of a protective device should be greater than, or at least equal to, the rating of the circuit or equipment it is protecting. For example, 10 × 100 watt lamps equate to a total current use of 4.35 amperes, so a device rated at 5 or 6 amperes could protect this circuit; and a portable domestic appliance with a label rating of 2.7 kW equates to a total current of 11.74 amperes, so a fuse rated at 13 amperes should be fitted in the plug.

Protective devices are designed to operate when an excess of current (greater than the design current of the circuit) passes through it. The fault current's excess heat can cause a fuse element to rupture or the device mechanism to trip. These currents may not necessarily be circuit faults, but short-lived overloads specific to a piece of equipment or outlet. BS 7671 categorises these as overload current and overcurrent.

For conductors, the rated value is the current-carrying capacity. However, most excess currents are due to faults, either earth faults or short circuits, which cause excessive currents.

Here are some examples.

Overload and overcurrent faults:

- adaptors used in socket outlets exceeding the rated load of the circuit
- extra load being added to an existing circuit or installation
- not accounting for the starting current on a motor circuit.

Earth faults (see also page 397):

- insulation breakdown
- incorrect polarity
- poor termination of conductors.

Short circuit faults (see also page 397):

- insulation breakdown
- severing of live circuit conductors (for example, through penetration by a nail)
- inappropriate connection method
- wrong termination of conductors energised before being tested.

Insulation failure

Insulation is designed primarily to separate conductors and to ensure their integrity throughout their life. Insulation is also used to protect the consumer against electric shock by protecting against direct contact, and as a secondary protection against light mechanical damage, such as in the outer sheath of PVC/PVC twin and cpc cables. When insulation of conductors and cables fails, it is usually due to one or more of the following:

- poor installation methods
- poor maintenance
- excessive ambient temperatures
- high fault-current levels
- damage by a third party.

Arcing (see also page 402)

A faulty switching device that can be heard 'cracking' is probably arcing. An arc happens when electric current flows through air gaps between conductors or terminals, and can be extremely dangerous.

With a light switch, arcing occurs when loose or corroded connections make an intermittent contact, causing sparking between the connections.

Arc fault current and voltage concentrated in one place can generate large amounts of heat that can severely burn human skin, set clothing on fire and even produce a shock wave that can knock people off their feet. The high arc temperature can vaporise the conductors; in turn, the conductive vapour can help sustain the arc until overcurrent protective devices open the circuit. For fast acting fuses, this may take minutes; for other devices, it may take much longer.

Open circuit (see also page 471)

A circuit is 'open' when for some reason the circuit is incomplete and a break is preventing the flow of current.

Plant, equipment and component failure

Nothing lasts forever and this is certainly true of electrical equipment. Some faults are simply caused by wear and tear, although planned maintenance and regular testing and inspection can extend the life of equipment. Some common failures on installations and plant are:

- switches not operating – age-damaged mechanism
- motors not running – probably due to age of brushes
- socket outlets not working – worn contacts.

Faults caused by misuse, abuse and negligence

A common reason for faults on any electrical system or equipment is misuse, where the system or equipment is simply not being used as it was designed. Other faults are caused by carelessness during installation, such as poor termination or stripping of conductors. Following instructions for use and installation can stop these faults occurring.

Some common examples of such faults are:

- poor termination of conductors – overheating due to poor electrical contact
- wrong size conductors used – excess current could lead to overheating of conductors
- not protecting cables when drawing them into enclosures – damage to insulation
- overloading trunking capacities – overheating and insulation breakdown.

Prevention of faults by regular maintenance

A well-designed installation that is maintained, inspected and tested regularly should serve the consumer efficiently and safely for many years. Even a visual inspection can uncover simple faults which, if left, could lead to major problems in the future.

Specific types of fault

The manufacturer and the types of cables and accessories that you install must comply with BS and BS EN Standards and must then be installed in accordance with BS 7671. Part 5, in particular, covers:

- type of wiring system
- selection and erection of wiring systems in relation to external influences
- current-carrying capacity of cables
- cross-sectional area of conductors of cables
- voltage drop in consumers' installations
- electrical connections

- selection and erection of wiring systems to minimise the spread of fire
- proximity of wiring systems to other services
- selection and erection in relation to maintainability, including cleaning.

These items should have been accounted for in the initial design, but it is the responsibility of the installer to enforce them, through good installation practice and understanding the faults that could occur.

There are too many types of equipment and manufacturer to cover every type of fault here. However, some of the faults and solutions you may find with heating systems are listed below.

Heating system faults

Flame rectification devices (see also page 351)

Dirt, corrosion, or bad connections in the flame-sensing circuit can cause the controller to think that the flame is not lit, so the gas valve shuts down. A stable pilot flame engulfing the flame sensor and ground target is important. A fluctuating pilot flame or a sensing rod not making good flame contact can result in an intermittent flame signal. You can measure the current in a flame by putting an instrument (capable of measuring 1 to 10 micro amps d.c.) in series with the flame-sensing rod.

Central heating pumps (see also page 355)

A central heating pump should be reasonably quiet and the outer case should be cool to mildly warm. If the casing is hot, the overheating may be due to the pump having to work harder because of blockages inside it. Isolate the circuit, dismantle the pump and clean the assembly and motor spindle.

If the pump is rattling, check to see whether a fixing has come loose, then check internally for blockages or loose components.

Thermostats (see also page 357)

A thermostat is a switch that operates in response to temperature. If the thermostat is not working, clean it well and check the settings. If it is broken, you will need to replace the whole unit.

Thermocouple (see also page 353)

The thermocouple sits in the pilot light flame and delivers an electrical current when hot, which instructs the solenoid valve to open and allows the gas to flow. If the pilot goes out, the thermocouple cools, the electricity stops and the gas gets shut off.

The position of the tip of the thermocouple is essential, so look for movement. Look for dirt or corrosion that may be preventing the device from sensing correctly. Check that the joint of the two dissimilar metals has not broken down. Look also for broken or loose wiring connections.

Actuators and valves (see also page 358)

The actuator is the electric-mechanical means of opening or closing a valve. If a valve hasn't been used for a while, it could stick. You could use a hammer carefully, but may need to remove and refit the actuator.

Before fitting the motor unit to the valve, turn the valve spindle to the mid position, flush the system out and drain it to remove any foreign material in the water. Fill the system with water, adding corrosion inhibitor, if required. Next fit the actuator to the valve, making sure that the tongue on the valve spindle engages with the slot in the plastic coupling on the motor. Take account of the direction arrows on the side of the device.

Cable interconnections

Cable interconnections are usually seen as the first point to investigate in the event of a fault. This is because they are seen as the weak link in the wiring system and because they are usually readily accessible.

Cable interconnections are used in one or more of the following ways:

- power circuits, ring final socket outlet wiring and spurs, etc.
- general alterations and extensions to circuits when remedial work is being done
- rectification of faults or damage to wiring.

Where they must be used, they should be mechanically and electrically suitable, and accessible for inspection as laid down by Regulation 526.3.

BS 7671 Reg 526.3

There is an exception to Regulation 526.3 when any of the following is used:

- a joint designed to be buried underground
- a compound-filled or encapsulated joint
- a connection between a cold tail and heating element, as in ceiling heating, floor heating or a trace heating system
- a joint made by welding, soldering, brazing or appropriate compression tool
- a joint forming part of the equipment complying with the appropriate product standard.

The joints in non-flexible cables must be made by soldering, brazing, welding, mechanical clamps or of a compression type. The devices used should relate to the size of the cable and be insulated to the voltage of the system used.

Any connectors used must be to the appropriate British Standard and the temperature of the environment must be considered when choosing the connector.

Where cables with insulation of dissimilar characteristics are to be jointed (for example, XLPE thermosetting to general purpose PVC), the insulation of the joint must meet the highest temperature of the two cable insulating materials.

The most common types of terminating devices you will come across are:

- plastic connectors and porcelain connectors
- soldered joints
- screwits (older installations only, as they have been banned for a number of years)
- uninsulated connectors
- compression joints
- junction boxes.

The mechanical and electrical connection when joining two conductors together relies on good practice by the installer to ensure that the connection is made soundly. The terminating device used should:

- be the correct size for the cross-section of the cable conductor
- be at least the same current rating as the circuit conductor
- have the same temperature rating as the circuit conductor
- be suitable for the environment.

The most common fault at this point would probably be due to a poor/loose connection, which would produce a resistive joint and excessive heat that could lead to insulation breakdown or eventually fire.

When a fault occurs at a cable interconnection, it may not necessarily be due to the production of heat, but instead to a lack of support causing strain on the conductors, which may lead to the same outcome as above.

Cable terminations, seals and glands

Terminations

The rules for cable interconnections also apply to the termination of cables and the same consequences will occur if they are not adhered to.

BS 7671 Reg 522.8.5
BS 7671 Reg 526

- Take care not to damage the wires.
- BS 7671 522.8.5 requires that a cable should be supported to avoid any appreciable mechanical stress on the terminations of the conductor.
- BS 7671 526 gives detailed information regarding electrical connections.
- A termination under mechanical stress is liable to disconnect, overheat or spark.
- When current is flowing in a conductor a certain amount of heat is developed, the expansion and contraction may be sufficient to allow a conductor under stress, particularly tension, to be pulled out of the terminal or socket.
- A fault caused at a poorly connected terminal is known as a high-resistance fault.
- One or more strands or wires left out of the terminal or socket will reduce the effective cross-sectional area of the conductor at that point, which may result in increased resistance and likely overheating.
- Poorly terminated conductors in circuits that continue to operate correctly are known as latent defects.

Types of terminal

There is a wide variety of conductor terminations. Typical methods of securing conductors in accessories including pillar terminals, screw heads, nuts and washers, and connectors.

Seals and entries

Where a cable or conductor enters an accessory or piece of equipment, the integrity of the conductor's insulation and sheath, earth protection and of the enclosure or accessory must be maintained. Some cables and wiring systems have integrated mechanical protection, of varying effectiveness. However, their design capabilities should not be degraded and special glands and seals produced by the makers should be used where required.

Where PVC/PVC cables enter accessories, the accessory itself should have no loss of integrity and there should be no damage to any part of the cable.

When carrying out a visual inspection of an installation, either at the completion of works or at a periodic inspection, the checking of terminations of cables and conductors is an integral part of the inspection.

BS 7671 Regulation 611.3 states the minimum items for inspection, including:

BS 7671 Reg 611.2
BS 7671 Reg 611.3

- connection of conductors
- identification of conductors
- routing of cables in safe zones, or protection against mechanical damage, in compliance with section 522
- selection of conductors for current-carrying capacity and voltage drop, in accordance with the design
- correct connection of accessories and equipment
- presence of fire barriers, suitable to seals and protection against thermal effects
- methods of protection against electric shock
- selection of equipment and protective measures appropriate to external influences.

These checks are made to comply with BS 7671 Regulation 611.2, which states: 'The inspection shall be made to verify that the installed electrical equipment is in compliance with section 511, correctly selected and erected in accordance with the regulations and not visibly damaged or defective so as to impair safety.'

Accessories including switches, control equipment, contactors, electronic and solid state devices

Faults can appear on most items of equipment that you install. The most common fault is due to wear of the terminal contacts as they constantly make and break during normal operation. All items of equipment should be to BS or BS EN, and must be type-tested by the manufacturer.

Electronic and solid state devices

Solid state devices and electronic equipment work within sensitive voltage and current ranges in millivolts and milliamperes, and are consequently sensitive to mains voltage and heat.

With resistors on circuit boards, excessive heat produces more resistance and eventually an open circuit, when the resistor breaks down. You can use a low-resistance ohmmeter to check resistor values.

With capacitors, voltages in excess of their working voltage will cause them to break down, resulting in a short circuit. Such equipment requires specialist knowledge, but some basic actions can prevent damage to this type of equipment. When testing, disconnect the equipment before testing the circuits: some test voltages can damage these sensitive components, and their inclusion in the circuit would give an inaccurate reading.

Transient voltages

Transient voltages are a major cause of faults on components and equipment of this type. These voltages can arise from supply company variations or lightning strikes, so most large companies protect their equipment and install specialist lightning protection and filtering equipment.

Protective devices (see also page 390)

Protective devices operate in the event of a fault occurring on the circuit or equipment that the device is protecting. The most common reason for a protective device not operating is that the wrong type or rating of device has been used. When replacing or re-setting a device after a fault has occurred, use one with same type and rating. Switching on a device with an unrepaired fault will cause damage to the circuit wiring, the equipment and possibly the device itself.

Flexible cables and cords (see also page 350)

This type of conductor is used to connect many items within an electrical installation, such as:

- ceiling rose and pendant lamp-holders
- flex outlets and fused connection units
- fused plugs to portable appliances
- immersion heaters
- flexible connections to fans, motors and heating system equipment.

BS 7671 requires that all flexible cables and cords shall comply with the appropriate British and Harmonised Standard. The most common faults that occur with this type of conductor are arise from poor choice and suitability for the equipment and the environment.

Common faults relating to flexible cables and cords include:

- poor terminations into accessory, such as conductors showing
- wrong type installed, such as pvc instead of increased temperature type
- incorrect size of conductor – usually too small for load
- incorrectly installed when load bearing in luminaires.

Such problems should be identified during the visual inspection stage.

Damage to electronic devices due to over voltage

When testing or inspecting faults, you must take account of electronic equipment and their rated voltage. You should usually isolate or disconnect such equipment to avoid damage due to test voltages exceeding the equipment's rated voltage.

The voltage of an insulation resistance tester can reach levels between 250 and 1000 volts. Such voltage levels will cause components, control equipment, or data and telecommunication equipment to be faulty.

Note: Never isolate a circuit where computer equipment is connected. The loss of data is the most common occurrence when systems fail due to faults or power supply problems.

Factors that affect the repair process

When you have found a fault, you will need to consider the consequences to your company or your client before putting the fault right. In liaison with your engineer and the client, you can find out about any service agreements, cost concerns or other factors, and weigh up the options. The time and cost involved in carrying out a repair may outweigh those for a full replacement, so the best options could be full replacement, part replacement or full repair.

Here are some of the factors you may need to take into account.

Availability of replacement

Where a complete replacement of equipment is needed, you will need to find a replacement that closely matches the original specification of the faulty item. Here the manufacturer who has the item in stock and offers fast delivery will be given the order, so that you can mend the fault quickly.

Downtime under fault conditions

A fault on a circuit or item of equipment often costs the client money due to lost production, lost business or lost data. Sometimes the fault may be on the supply to the client's premises, but even here it may make good business sense to provide a temporary supply.

Legal responsibility

You will usually enter a contract with your client to avoid or help settle any disputes, which would cover costs, timings and guarantees. It is important that you stick to the terms of any such contract, and take note of it in the decisions you make.

Make an accurate cost analysis when giving an estimate or fixed price for repair work, which is usually charged on a daily rate plus costs for materials. If unforeseen work arises – such as finding dangerous wiring underneath an item of heavy plant to be repaired – you should clear costs and responsibilities with the client before going ahead.

Give the client a guarantee for the work done, covering quality and the materials used. If the client insists on reusing existing items, you may need to avoid writing a guarantee into the contract, as old items can be prone to failure.

Company procedures

Your company will have its own procedures when dealing with fault finding at any customer's premises. These may include:

- signing in
- wearing identification badges
- locating supervisory personnel
- locating data drawings
- liaising with the client and office before starting any work
- following safe isolation procedures.

Access to the system

There is nothing more costly or embarrassing than arriving at a customer's premises only to find the premises locked or unmanned. Make sure you make arrangements before you get there, including:

- access to the fault repair location
- access to the supply intake and isolation point
- security clearance, if needed
- notification of any personnel who may have their activities affected
- access outside normal working hours, if necessary.

Sequence for repairing a fault

- Identify the symptoms.
- Gather information.
- Analyse the evidence.
- Check supply.
- Check protective devices.
- Isolate and test.
- Interpret information and test results.
- Rectify the fault.
- Carry out functional tests.
- Restore the supply.
- Carry out live and functional tests.

Building fabric restoration

Depending on the fault and the repair needed, you may need to disturb the fabric of the building. It is vital that you discuss this with the client, so that they understand just what will be involved and approve any restoration work in advance. Always leave the site in a clean and safe condition, disposing of items in a safe and approved manner.

Check your knowledge

1. Reduced voltage is normally used on site. Which of the following is the normal voltage used?
   - **a** 25V
   - **b** 110V
   - **c** 230V
   - **d** 400V

2. When checking to ensure a circuit is isolated what should you do?
   - **a** Test with a lampholder and flex
   - **b** Approved voltage indicator
   - **c** Ask a colleague
   - **d** Switch off only

3. The supply earthing system which has a combined neutral and earth in part of the system is designated as what?
   - **a** TT
   - **b** IT
   - **c** TN-S
   - **d** TN-C-S

4. Why are three-phase supplies usually provided?
   - **a** Client requests it
   - **b** A large capacity is required
   - **c** Site engineer thinks it is a good idea
   - **d** Wiring is easier on three-phase systems

5. Which of the following would **not** be regarded as an extraneous-conductive part?
   - **a** Gas service pipe
   - **b** Water service pipe
   - **c** Metallic trunking
   - **d** Metal radiator

6. What is the minimum value of insulation resistance on a lighting circuit?
   - **a** 0.5 Ω
   - **b** 0.5 MΩ
   - **c** 1 Ω
   - **d** 1 MΩ

7. Where a 30 mA RCD is being used to provide additional protection against contact with live parts, what must the maximum operating time not exceed?
   - **a** 40 ms when tested at 1 x 30mA
   - **b** 40 ms when tested at 5x 30mA
   - **c** 100 ms when tested at 1 x 30mA
   - **d** 100 ms when tested at 5 x 30 mA

8. Which document lays down the safety requirements of test equipment leads?
   - **a** HSG 85
   - **b** HSE GS38
   - **c** BS7671
   - **d** BS5266

9. Which instrument is used for carrying out continuity tests
   - **a** RCD tester
   - **b** High resistance ohmmeter
   - **c** Low resistance ohmmeter
   - **d** Earth-fault loop impedance tester

10. Which of these should a Minor Electrical Installation Works Form **not** be used for?
    - **a** New circuit
    - **b** Moving a light position
    - **c** Adding a new lighting point to an existing circuit
    - **d** Adding an extra socket outlet to an existing circuit

11. What is the correct standard method for checking an approved test lamp or voltage indicator before safe isolation is attempted?
    - **a** Test on a ring socket
    - **b** Test on a light pendant
    - **c** Test on a known supply or proving unit
    - **d** Test on main incoming supply

Preparation for assessment

This unit introduces you to many of the craft and practical skills that you will require when carrying out electrical work on domestic plumbing and heating systems and components.

For this unit you will need to be familiar with:

- Electrical standards that apply to the mechanical services industry
- Principles of electricity supply to buildings
- Layout features of electrical circuits in buildings
- Electrical safe isolation procedure and carrying out electrical safe isolation procedure
- Site preparation techniques for the electrical connection of mechanical services components and applying site preparation techniques
- Installation and connection requirements of electrically operated mechanical services components
- Installing and connecting electrically operated mechanical services and components
- Inspection and testing requirements of electrically operated mechanical services components
- Carrying out inspection and testing on electrically operated mechanical services components
- Procedures for safely diagnosing and rectifying faults in electrically operated mechanical services components
- Carrying out diagnosis and rectify faults in electrically operated services components.

For each learning outcome, there are several skills you will need to acquire so you must make sure you are familiar with the assessment criteria for each outcome. For example, Learning outcome 4 you will need to be able to identify the correct electrical test equipment required to carry out safe isolation. You will also need to be able to state the industry agreed procedure for safe isolation. You will need to be able to describe the methods used to ensure circuits cannot be re-activated while work is taking place on them.

You must read each question carefully and take your time. To check your knowledge try the progress checks and multiple choice questions, without assistance initially, to see how much you have understood. Remember the answers are in the book and you can always clarify terms using the internet and the standard electrical installation guides and regulations. A useful resource is Part 2 of BS7671 that gives definitions of terms used in the electrical industry. There are some simple tips to follow when completing multiple choice exams and assignments:

Read all the questions – For written work and multiple choice exams it is a good idea to look through all the questions before you start – you don't want to miss any clues!

Describe with the aid of a diagram – a reasonably detailed explanation to cover the subject in the question should accompany any diagram requested.

Identify – Refer to reference material used in any assignment work.

This unit also has a range of practical skills in it, and you will need to be sure that you have carried out sufficient practice before you attempt a practical assessment. To ensure safety and timely completion of tasks it is wise to carefully plan all practical activities beforehand. Method statements for tasks are also a wise investment of time as they can ensure you are working correctly and avoid problems later on.

Remember – Relax, enjoy and take your time.

Appendix

Drawing symbols and their meanings			
Symbol name	**Symbol**	**Symbol name**	**Symbol**
Stop Valve	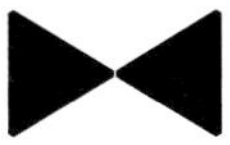	Service valve	
Drain valve		Capped end	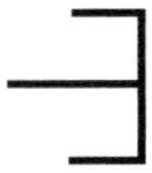
Hose Connection	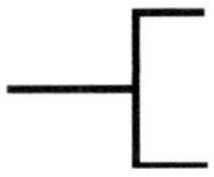	Single check valve	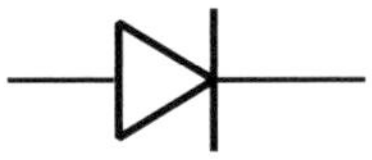
Pressure reducing valve		Double check valve	
Two port motorised valve	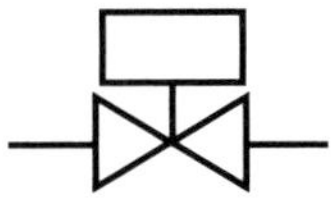	Three port motorised valve	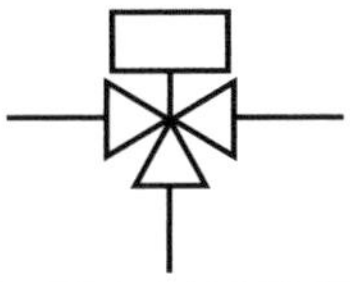
Pressure relief valve	 PR	Three port valve	
Temperature and pressure relief valve	 TPR	Strainer	
Circulating pump		Centrifugal pump	
Expansion vessel		Point of measurement (Pressure or temperature gauge)	
Wheelhead valve		Lockshield valve	

Water meter	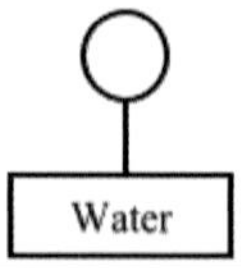 Water	Gas Meter	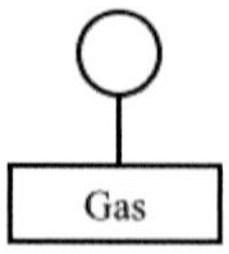 Gas
Float Valve	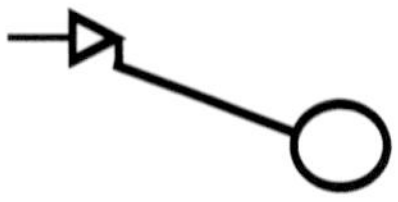	Tundish	
Visible pipe		Pipe behind duct (Hidden)	
Pipe at high level		Pipe above ceiling	
Direction of flow		Draw off point	
Close coupled toilet	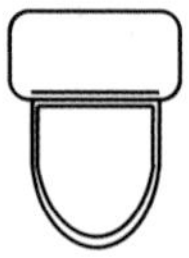	Bidet	
Sink	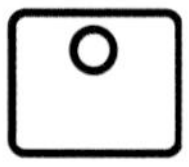	Bowl urinal	
Stall urinals	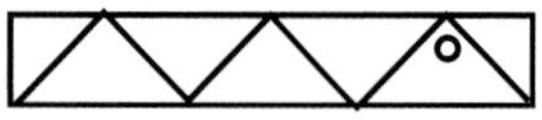	Bath	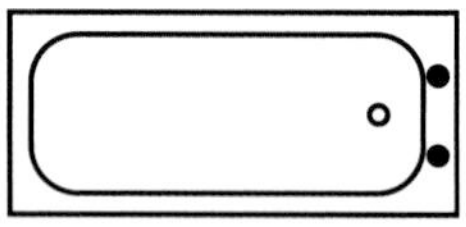
Wash Basin	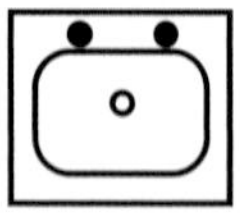	Left hand drainer sink	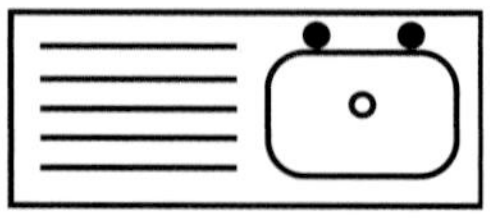
Shower tray (Rectangular)	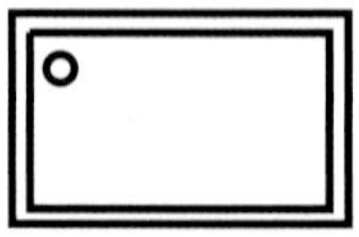	Pentagon shower tray	
Reclaimed water		Hot Water	
Cold water		Sizes of bands for colour coding.	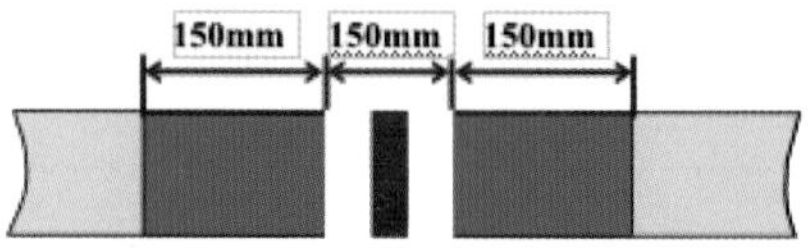 150mm 150mm 150mm

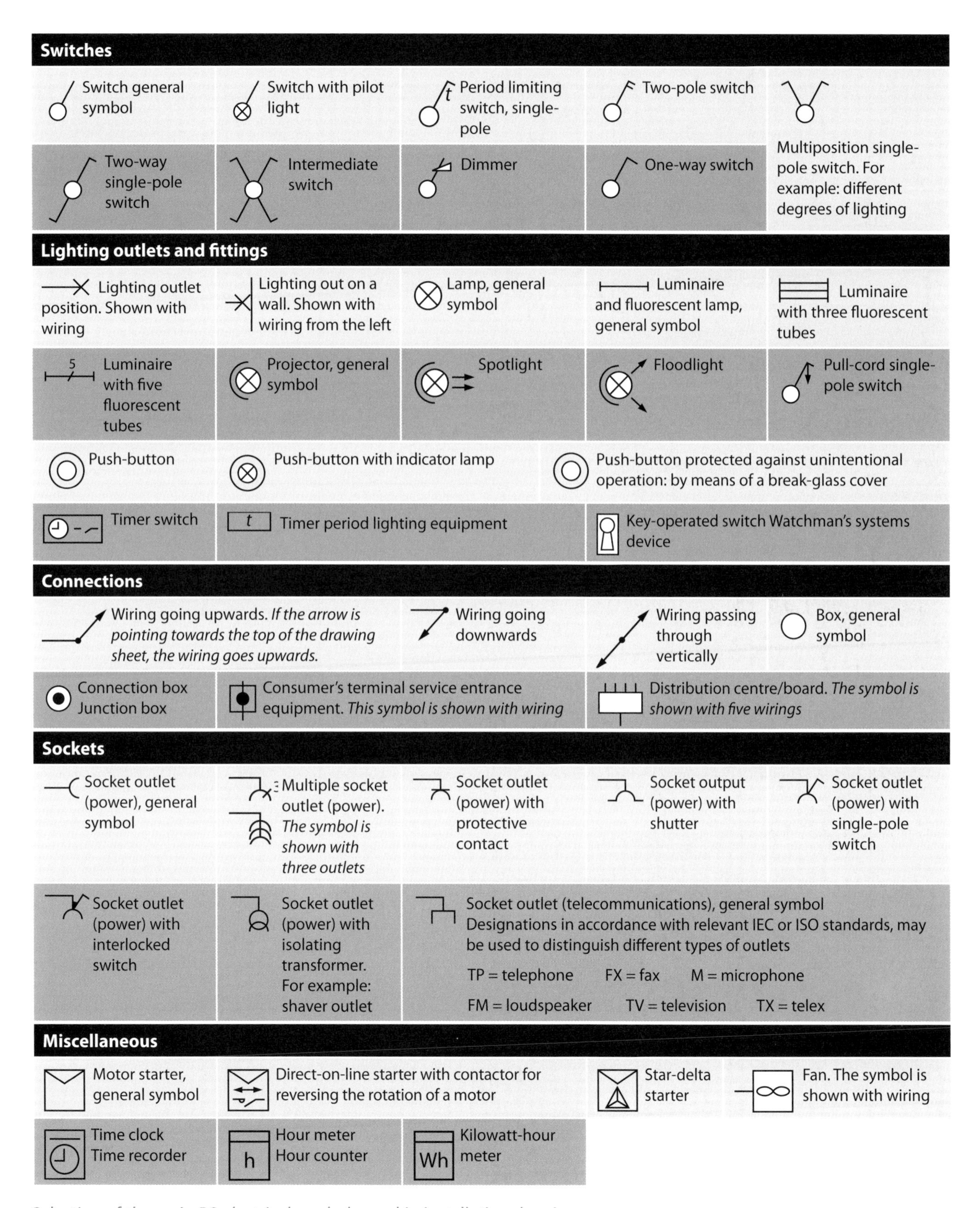

Selection of the main BS electrical symbols used in installation drawings

Appendix: Summary of backflow-prevention devices

Type	Description of backflow-prevention arrangements and devices	Fluid category for which suited	
		Back-pressure	Back-siphonage
AA	Air gap with unrestricted discharge above spillover level	5	5
AB	Air gap with weir overflow	5	5
AC	Air gap with vented submerged inlet	3	3
AD	Air gap with injector	5	5
AF	Air gap with circular overflow	4	4
AG	Air gap with minimum size circular overflow determined by measure or vacuum test	3	3
AUK1	Air gap with interposed cistern (e.g. a WC suite)	3	5
AUK2	Air gaps for taps and combination fittings (tap gaps) discharging over domestic sanitary appliances such as a washbasin, bidet, bath or shower tray shall be not less than the following – Size of tap or combination fitting — Distance of tap outlet above appliance spillover level Not exceeding G½ Exceeding G½ but not — 20 mm exceeding G¾ — 25 mm Exceeding G¾ — 70 mm	x	3
AUK3	Air gaps for taps or combination fittings (tap gaps) discharging over any higher risk domestic sanitary appliances where a fluid risk category 4 or 5 is present, such as (a) Any domestic or non-domestic sink (b) Any appliance in premises where a higher level of protection is required such as some appliances in hospitals or other health care premises The air gap shall be not less than 20 mm or twice the diameter of the inlet pipe to the fitting, whichever is the greater	x	5
DC	Pipe interrupter with permanent atmospheric vent		5

Notes:

1 X indicates that the backflow-prevention arrangement or device is not applicable or not acceptable for protection against back pressure for any fluid category within water installations in the UK.

2 Arrangements incorporating type DC devices shall have no control valves on the outlet side of the device, they shall be fitted not less than 300 mm above the spillover level of a WC pan, or 150 mm above the sparge pipe outlet of a urinal, and discharge vertically downward.

3 Overflows and warning pipes shall discharge through, or terminate with, an air gap, the dimension of which should satisfy a Type AA air gap.

Non-mechanical backflow-prevention devices acceptable under the Regulations

Type	Description of backflow-prevention arrangements and devices	Fluid category for which suited	
		Back-pressure	Back-siphonage
BA	Verifiable backflow preventer with reduced pressure zone (RPZ valve)	4	4
CA	Non-verifiable disconnector with difference between pressure zones not greater than 10%	3	3
DA	Anti-vacuum valve (or vacuum breaker)	x	3
DB	Pipe interrupter with atmospheric vent and moving element	x	4
DUK1	Anti-vacuum valve combined with verifiable check valve	2	3
EA	Verifiable single check valve	2	2
EB	Non-verifiable single check valve	2	2
EC	Verifiable double check valve	3	3
ED	Non-verifiable double check valve	3	3
HA	Hose union backflow preventor. Only permitted on existing hose union taps in house installations	2	3
HC	Diverter with automatic return (normally integral with domestic appliance applications only)	x	3
HUK1	Hose union tap which incorporates a verifiable double check valve. Only permitted for replacement of existing hose union taps in house installations	3	3
LA	Pressurised air inlet valve	x	2
LB	Pressurised air inlet valve with check valve downstream	2	3

Notes:

1 X indicates that backflow-prevention device is not acceptable for protection against back-pressure for any fluid category within water installations in the UK.

2 Arrangements incorporating a type BD device shall be fitted not less than 300 mm above the spill-over level of the appliance and discharge vertically downwards.

3 Types DA and DUK1 shall have no control valves on the outlet of the device and be fitted on a minimum 300mm type A upstand.

4 Relief outlet ports from types BA and CA backflow-prevention devices shall terminate with a type AA air gap.

Mechanical backflow devices acceptable under the Regulations

INSPECTION CHECKLIST	
General	
1 Complies with requirements 1-3 in Section 2.1 (133.1, 134.1)	
2 Accessible for operation, inspection and maintenance (513.1)	
3 Suitable for local atmosphere and ambient temperature. (Installations in potentially explosive atmospheres are outside the scope of BS 7671)	
4 Circuits to be separate (no borrowed neutrals) (314.4)	
5 Circuits to be identified (neutral and protective conductors in same sequence as line conductors) (514.1.2, 514.8.1)	
6 Protective devices adequate for intended purpose (BS.7671 - Ch. 53)	
7 Disconnection times likely to be met by installed protective devices (Ch. 41)	
8 Sufficient numbers of conveniently accessible socket-outlets are provided in accordance with the design (553.1.7). Note that in Scotland section 4.6.4 (socket outlets) in the domestic technical handbook published by the Scottish Building Standards Agency (SBSA) gives specific recommendations for the number ofsocket-outlets for various locations within an installation.	
9 All circuits suitably identified (514.1, 514.8, 514.9)	
10 Suitable main switch provided (Ch. 53)	
11 Supplies to any safety services suitably installed, e.g. Fire Alarms to BS 5839 and emergency lighting to BS 5266	
12 Environmental IP requirements accounted for (B5 EN 60529)	
13 Means of isolation suitably labelled (514.1, 537.2.2.6)	
14 Provision for disconnecting the neutral (537.2.1.7)	
15 Main switches to single-phase installations, intended for use by an ordinary person, e.g. domestic, shop, office premises, to be double-pole (537.1.4)	
16 RCDs provided where required (411.1, 411.3, 411.4, 411.5, 522.6.7, 522.6.8, 532.1, 701.411.3.3, 701.415.2, 702.55.4, 705.411.1, 705.422.7, 708.553.1.13, 709.531.2, 711.410.3.4, 711.411.3.3, 740.410.3, 753.415.1)	
17 Discrimination between RCDs considered (314, 531.2.9)	
18 Main earthing terminal provided (542.4.1) readily accessible and identified (514.13.1)	
19 Provision for disconnecting earthing conductor (542.4.2)	
20 Correct cable glands and gland plates used (BS 6121)	
21 Cables used comply with British or Harmonized Standards (Appendix 4 of the Regulations, 521.1)	
22 Conductors correctly identified (Section 514)	
23 Earth tail pots installed where required on mineral insulated cables (134.1.4)	
24 Non conductive finishes on enclosures removed to ensure good electrical connection and if necessary made good after connecting (526.1)	

25 Adequately rated distribution boards (BS EN 60439 may require de-rating)	
26 Correct fuses or circuit-breakers installed (Sections 531 and 533)	
27 All connections secure (134.1.1)	
28 Consideration paid to electromagnetic effects and electromechanical stresses (Ch. 52)	
29 Overcurrent protection provided where applicable (Ch. 43)	
30 Suitable segregation of circuits (Section 528)	
31 Retest notice provided (514.12.1)	
32 Sealing of the wiring system including fire barriers (527.2).	

Switchgear	
1 Suitable for the purpose intended (Ch. 53)	
2 Meets requirements of BS EN 61008, BS EN 61009, BS EN 60947-2, BS EN 60898 or BS EN 60439 where applicable, or equivalent standards (511)	
3 Securely fixed (134.1.1) and suitably labelled (514.1)	
4 Non conductive finishes on switchgear removed at protective conductor connections and if necessary made good after connecting (526.1)	
5 Suitable cable glands and gland plates used (526.1)	
6 Correctly earthed (Ch. 54)	
7 Conditions likely to be encountered taken account of, i.e. suitable for the foreseen environment (522)	
8 Where relevant correct IP rating applied (BS EN 60529)	
9 Suitable as means of isolation, where applicable (537.2.2)	
10 Complies with the requirements for locations containing a bath or shower (Section 701)	
11 Need for isolation, mechanical maintenance, emergency and functional switching met (Section 537)	
12 Fire-fighter's switch provided where required (537.6.1)	
13 Switchgear suitably coloured where necessary (537.6.4)	
14 All connections secure (Section 526)	
15 Cables correctly terminated and identified (Sections 514 and 526)	
16 No sharp edges on cable entries, screw heads, etc. which could cause damage to cables (522.8)	
17 All covers and equipment in place and secure (Section 522.6.3)	
18 Adequate access and working space (132.12 and Section 513).	

General (applicable to each type of accessory)	
1 Complies with BS 5733, BS 6220 or other appropriate standard (Section 511)	
2 Box or other enclosure securely fixed (134. 1.1)	
3 Metal box or other enclosure earthed (Ch. 54)	
4 Edge of flush boxes not projecting above wall surface (134.1.1)	
5 No sharp edges on cable entries, screw heads, etc. which could cause damage to cables (522.8)	
6 Non sheathed cables, and cores from which sheath removed, not exposed outside the enclosure (526.9)	
7 Conductors correctly identified (514.6)	
8 Bare protective conductors having a cross-sectional area of $6mm^2$ or less to be sleeved green and yellow (514.4.2, 543.3.2)	
9 Terminals tight and containing all strands of the conductors (Section 526)	
10 Cord grip correctly used or clips fitted to cables to prevent strain on the terminals (522.8.5, 526.6)	
11 Adequate current rating (133.2.2)	
12 Suitable for the conditions likely to be encountered (Section 522).	
Joint box	
1 Joints accessible for inspection (526.3)	
2 Joints protected against mechanical damage (526.7)	
3 All conductors correctly connected (526.1).	
Fused connection unit	
1 Correct rating and fuse (533.1)	
2 Complies with BS 1363-4 (559.6.1.1 vii).	

Trunking	
General	
1 Complies with BS 4678 or BS EN 50085-1 (521.6)	
2 Securely fixed and adequately protected against mechanical damage (522.8)	
3 Selected, erected and routed so that no damage is caused by ingress of water (522.3)	
4 Proximity to non-electrical services (528.2)	
5 internal sealing provided where necessary (527.2.4)	
6 Holes surrounding trunking made good (527.2.1)	
7 Band 1 circuits partitioned from Band 2 circuits or insulated for the highest voltage present (528.1)	
8 Circuits partitioned from Band 1 circuits or wired in mineral-insulated metal sheathed cables (528.1)	
9 Common outlets for Band 1 and Band 2 provided with screens, barriers or partitions	
10 Cables supported for vertical runs (522.8).	

Insulated cables	
Non-flexible cables	
1 Correct type (521)	
2 Correct current rating (523)	
3 Protected against mechanical damage and abrasion (522.8)	
4 Cables suitable for high or low ambient temperature as necessary (522.1)	
5 Non sheathed cables protected by enclosure in conduit, duct or trunking (521.10)	
6 Sheathed cables:	
• routed in allowed zones or mechanical protection provided (522.6.6)	
• in the case of domestic or similar installations not under the supervision of skilled or instructed persons, additional protection is provided by RCD having $I_{\Delta n}$ not exceeding 30mA (522.6.7)	
7 Cables in partitions containing metallic structural parts in domestic or similar installations not under the supervision of skilled or instructed persons should be:	
• provided with adequate mechanical protection to suit both the installation of the cable and its normal use	
• provided with additional protection by RCD having $I_{\Delta n}$ not exceeding 30mA (522.6.8)	
8 Where exposed to direct sunlight, of a suitable type (522.11)	
9 Not run in lift shaft unless part of the lift installation and of the permitted type (BS 5655 and BS EN 81-1) (528.3.5)	
10 Buried cable correctly selected and installed for use (522.6.4)	
11 Correctly selected and installed for use overhead (521)	
12 internal radii of bends not sufficiently tight as to cause damage to cables or to place undue stress on terminations to which they are connected (relevant BS, BS EN and 522.8.3)	
13 Correctly supported (522.8.4 and 522.8.5)	
14 Not exposed to water, etc. unless suitable for such exposure (522.3)	
15 Metal sheaths and armour earthed (411.3.1.1)	
16 identified at terminations (514.3)	
17 Joints and connections electrically and mechanically sound and adequately insulated (526.1 and 526.2)	
18 All wires securely contained in terminals, etc. without strain (522.8.5 and Section 526)	
19 Enclosure of terminals (Section 526)	
20 Glands correctly selected and fitted with shrouds and supplementary earth tags as necessary (526.1)	
21 Joints and connections mechanically sound and accessible for inspection, except as permitted otherwise (526.1 and 526.3).	

Flexible cables and cords (521.9)	
1 Correct type (521)	
2 Correct current rating (Section 523)	
3 Protected where exposed to mechanical damage (522.6 and 522.8)	
4 Suitably sheathed where exposed to contact with water (522.3) and corrosive substances (522.5)	
5 Protected where used for final connections to fixed apparatus, etc. (526.9)	
6 Selected for resistance to damage by heat (522.1)	
7 Segregation of Band 1 and Band 2 circuits (BS 6701 and Section 528)	
8 Fire alarm and emergency lighting circuits segregated (BS 5839, BS 5266 and Section 528)	
9 Cores correctly identified (514.3.2)	
10 Joints to be made using appropriate means (526.2)	
11 Where used as fixed wiring, relevant requirements met (521.9.3)	
12 Final connections to portable equipment, a convenient length and connected as stated (553.1.7)	
13 Final connections to other current-using equipment properly secured or arranged to prevent strain on connections (Section 526)	
14 Mass supported by cable to not exceed values stated (559.11.6).	
Protective conductors	
1 Cables incorporating protective conductors comply with the relevant BS (Section 511)	
2 Joints in metal. conduit, duct or trunking comply with Regulations (543.3)	
3 Flexible or pliable conduit to be supplemented by a protective conductor (543.2.1)	
4 Minimum cross sectional area of copper conductors (543.1)	
5 Copper conductors, other than strip, of 6mm2 or less protected by insulation (543.3.2)	
6 Circuit protective conductor at termination of sheathed cables insulated with sleeving (543.3.2)	
7 Bare circuit protective conductor protected against mechanical damage and corrosion (542.3 and 543.3.1)	
8 Insulation, sleeving and terminations identified by colour combination green and yellow (514.3.1, 514.4.2)	
9 Joints electrically and mechanically sound (526.1)	
10 Separate circuit protective conductors not less than 4mm^2 if not protected against mechanical damage (543.1.1)	
11 Main and supplementary bonding conductors of correct size (Section 544).	

Enclosures	
General	
1 Suitable degree of protection (IP Code in BS EN 60529) appropriate to external influences (416.2, Section 522 and Part 7).	

Index